P9-DYZ-993

DISCARDED
FROM
UNH LIBRARY

Microdosimetry
An Interdisciplinary Approach

Microdosimetry

An Interdisciplinary Approach

Edited by

Dudley T. Goodhead and Peter O'Neill
Medical Research Council, Radiation & Genome Stability Unit, Harwell, UK

Hans G. Menzel
European Commission, Directorate General for Science, Research & Development, Radiation Protection Research, Brussels, Belgium

The Proceedings of the Twelfth Symposium on Microdosimetry: An Interdisciplinary Meeting on Radiation Quality, Molecular Mechanisms, Cellular Effects and Health Consequences of Low Level Ionising Radiation, held on 29 September – 4 October 1996 at Keble College, Oxford, UK.

Organised by:

Medical Research Council, Radiation & Genome Stability Unit

European Commission, Directorate General for Science, Research & Development, Radiation Protection Research

United States Department of Energy, Office of Health & Environmental Research

GSF Centre for Environmental Health Research, Institute for Radiation

Special Publication No. 204

ISBN 0-85404-737-9

A catalogue record for this book is available from the British Library.

Published by The Royal Society of Chemistry,
Thomas Graham House, Science Park, Milton Road,
Cambridge CB4 4WF

Printed by Hartnolls Ltd, Bodmin, Cornwall, UK

Preface

The Twelfth Symposium on Microdosimetry was held at Keble College, Oxford from 29th September to 4th October 1996. The meeting was jointly organised by the Medical Research Council (UK), European Commission, GSF Centre for Environmental Health Research (Germany) and the Department of Energy (USA). We would also like to acknowledge the generous sponsorship provided by the following organisations who contributed to the overwhelming success of the meeting. Financial support was provided by Association for International Cancer Research, Cancer Research Campaign, International Association for Radiation Research, International Science Foundation, Nuclear Electric Ltd and Scottish Nuclear Ltd, Sir Scott of Yews Trust, Taylor and Francis Ltd and the United Kingdom Coordinating Committee on Cancer Research.

Microdosimetry plays an important role in multidisciplinary research of the physical, chemical and biological mechanisms involved in radiation effects in humans, in particular the induction of cancer as well as its treatment. The broad field of microdosimetry is concerned with the spatial and temporal correlation of radiation interactions over microscopic dimensions covering the scales of atoms, biomolecules and human cells. Of particular relevance is the role that these correlations play in directing the subsequent processes leading to the final molecular, cellular and tissue changes. Microdosimetric methods of analysis and instrumentation provide insights into the underlying biological processes and also have applications in the fields of practical radiation protection, radiotherapy and nuclear medicine.

The aim of the meeting was to provide a forum to bring together experts in physics, chemistry, molecular and cell biology and oncogenesis, with a common interest in understanding molecular mechanisms that can be gained from studies of the dependence of radiation effects on radiation quality, to advance the application of microdosimetry to other fields such as medicine, to promote and advance education and training in the field of radiation sciences.

This continues the eminent tradition of the eleven previous symposia on Microdosimetry, starting in 1967 and all co-organised by the European Commission, for rigorous analysis of the physical features of radiation quality and propagating these through interdisciplinary studies to a diversity of fundamental and practical endeavours.

These proceedings contain a selection of review articles presented by keynote speakers, complemented by original contributions from participants reflecting the interdisciplinary nature of the field. The specific topics have been regrouped into sections to give the reader a more balanced view of that area.

The contributions in the proceedings represent various research areas that contribute to the field of microdosimetry. They clearly emphasise the interdisciplinary nature of the field. The microscopic structures of radiation tracks set the initial conditions from which the subsequent biological effects follow. Therefore track structure simulations were presented as a starting point for many forms of analysis and biophysical modelling of effects. Significant information was included on the chemical processes leading to DNA damage, essential to improve the simulations. Expectations of clustered DNA damage, from close associations of ionizations within a track, were explored in presentations of theoretical and experimental data. DNA and chromosomal architecture impose their own sets of spatial and temporal constraints on the types and yields of chromosome aberrations and mutations that are induced by radiation. Therefore, the diverse components of the spatial features of track structure, the temporal aspects associated with the chemical stage of action, the subcellular architecture of DNA and chromosomes, DNA processing mechanisms, all contribute in concert to determining the final biological effectiveness of radiations of different quality. Understanding and modelling the key aspects of these mechanisms is necessary to describe radiation carcinogenesis and extrapolate risks to practical low doses and to other radiation qualities. Microdosimetry is highly relevant also to radiotherapy, at the subcellular level because of the dependence of cell killing on radiation quality and also over cellular or tissue dimensions for non-uniform radiation fields. Tissue non-uniformities are important also in assessment of radionuclides for protection or therapy. Under most practical situations of low-level radiation exposure individual cells experience essentially only single isolated particles. Microbeams have been developed to study the effects of single particles through cells, as well as to explore the spatial sensitivity of cellular components. A widely-used experimental tool in microdosimetry has been the low pressure gas proportional counter, especially to characterize radiation features on the scale of micrometers. These continue to have a variety of practical applications in protection and therapy. So called 'nanodosimetric' devices have also been under investigation because this is the scale of most relevance to assessment of DNA and chromatin damage. The responses of the variety of materials are potentially capable of giving microdosimetric information in the gas, liquid or solid phases.

It is hoped that these proceedings will stimulate the disciplines that contribute to microdosimetry, individually and jointly, to make even more advances in the field and be able to contribute strongly by guiding other developing fields that use radiation as probes, such as in DNA damage processing, or are involved in its practical consequences, such as in new therapy modalities or radiation protection.

Dudley Goodhead
Peter O'Neill
Hans Menzel

Contents

Track structure and damage simulation

Track structure calculations in radiobiology: How can we improve them and what can they do? 3
H. Nikjoo, S. Uehara and D.J. Brenner

Dynamics and recombination of positive ions and thermalized electrons in high-energy electron tracks calculated by computer simulations 11
L.D.A. Siebbeles, W.M. Bartczak, M. Terrissol and A. Hummel

A new approach to radiation transport in the complex DNA environment 15
M. Terrissol, M. Demonchy and E. Pomplun

Auger electron action inside hydrated DNA and nucleosome models 19
E. Pomplun, M. Demonchy and M. Terrissol

Interaction cross sections for electron inelastic scattering in liquid water 23
M. Dingfelder, D. Hantke and M. Inokuti

Total electron scattering cross sections of dimethyl ether 27
W.Y. Baek and B. Grosswendt

A track structure model based on measurements of radial dose distribution around an energetic heavy ion 31
S. Ohno, K. Furukawa, H. Namba, M. Taguchi and R. Watanabe

Radial dose model of SSB, DSB, deletions and comparisons to Monte-Carlo track structure simulations 35
F.A. Cucinotta, H. Nikjoo, J.W. Wilson, R. Katz and D.T. Goodhead

New Monte Carlo methods for evaluation of characteristics of nonadditive radiation action on microtargets 39
A.V. Lappa

Simulation of strand breaks and short DNA fragments in the biophysical model PARTRAC 43
W. Friedland, P. Jacob, H.G. Paretzke, M Perzl and T. Stork

Microdosimetric distributions for target volumes of complex topology 47
I.K. Khvostunov and S.G. Andreev

Monte Carlo simulation of the track interaction model applied to the TLD-100 response to 5.3 MeV α- particles 51
M. Rodríguez-Villafuerte, I. Gamboa-deBuen and M.E. Brandan

Chemical processes from radiation to DNA

Factors controlling the radiosensitivity of cellular DNA 57
J.F. Ward, J.R. Milligan and R.C. Fahey

Production yield of adenine from ATP irradiated with monochromatic X-rays in aqueous solution of different concentrations 65
K. Kobayashi, N. Usami, R. Watanabe and K. Takakura

The effect of electron energy on radiation damage 69
S.M. Pimblott and J.A. LaVerne

Oxygen decides: Double-strand breaks or DNA-protein crosslinks 73
L.V.R. Distel, B. Distel and H. Schüssler

The role of packaging in the radioprotection of DNA by highly charged ligands 77
C. Savoye, S. Ruiz, S. Hugot, D. Sy, C. Swenberg, R. Sabattier, M. Charlier and M. Spotheim-Maurizot

Strand break induction in DNA by aluminium k ultrasoft X-rays: Comparison of experimental data and track structure analysis 81
P. O'Neill, S.M.T. Cunniffe, D.L. Stevens, S.W. Botchway and H. Nikjoo

An X-ray photoelectron investigation of the effects of low-energy electrons on DNA bases 85
D. Klyachko, T. Gantchev, M.A. Huels and L. Sanche

A novel apparatus for low-energy electron (0-5000 eV) irradiation of lyophilized DNA in an ultra-clean UHV environment 89
M.A. Huels, J. Khoury, B. Gueraud, B. Boudaiffa, P.C. Dugal, D. Hunting, L. Sanche and A.J. Waker

The effect of dimethyl sulfoxide on inactivation, dsb induction and repair of V79 mammalian cells exposed to 252-Cf neutrons 93
T.J. Jenner, C de Lara and P. O'Neill

Calculation of G-value of Fricke dosimeter irradiated by photons of 100 eV - 10 MeV 97
H. Yamaguchi

Clustered DNA damage

Theoretical and experimental bases for mechanistic models of radiation-induced DNA damage 103
A. Ottolenghi and M. Merzagora

Ionising radiation induced clustered damage to DNA - A review of experimental evidence 111
K.M. Prise

Higher-order chromatin structures as potential targets for radiation-induced cell death 117
K.G. Hofer, X. Lin and M.H. Schneiderman

Dual spatially correlated nucleosomal double strand breaks in cell inactivation 125
A. Brahme, B. Rydberg and P. Blomquist

Rejoining kinetics of DNA double-strand breaks and variations in radiation quality 129
B. Stenerlöw, E. Höglund, E. Blomquist and J. Carlsson

Clustering of DNA breaks in chromatin fibre: Dependence on radiation quality 133
S.G. Andreev, I.K. Khvostunov, D.M. Spitkovsky and V. Yu Chepel

Survival of V79 cells to light ions: An analysis of the model system 137
G.F. Grossi, M. Durante, G. Gialanella, E. Mancini, M. Merzagora, F. Monforti, M. Pugliese and A. Ottolenghi

Chromosome architecture and aberrations

Nuclear architecture and its role in radiation-induced aberrations 143
C. Cremer, Ch. Münkel, S. Dietzel, R. Eils, M. Granzow, H. Bornfleth, A. Jauch, D. Zink, J. Langowski, T. Cremer

Modelling of chromosome exchanges in human lymphocytes exposed to radiations of different quality 152
V.V. Moiseenko, A.A. Edwards, H. Nikjoo and W.V. Prestwich

Modeling low and high LET FISH data on simple and complex chromosome aberrations 156
A.M. Chen, P.J. Simpson, C.S. Griffin, J.R.K. Savage, D.J. Brenner, J.N. Lucas and R.K. Sachs

Inter-chromosomal heterogeneity in the formation of radiation induced chromosomal aberrations 160
A.T. Natarajan, S. Vermeulen, J.J.W.A. Boei, M. Grigorova and I. Dominguez

Effectiveness of ultrasoft X-rays at inducing complex exchanges in human fibroblasts 164
C.S. Griffin, M.A. Hill, D.L. Stevens and J.R.K. Savage

On the nature of observed chromatid breaks 168
J.R.K. Savage and A.N. Harvey

Modelling the induction of chromosomal aberrations by ionising radiation 172
A.A. Edwards

Further evidence for the association of chromosome aberration yield coefficient α with "fast, short- range" and of coefficient β with "slow, long-range" pairwise interaction between DNA lesions 176
R. Greinert, E. Detzler, E. Bartels, K. Schulte, O. Boguhn, C. Thieke, R.P. Virsik-Peuckert and D. Harder

Implications of microdosimetry for a state vector model of chromosomal aberrations and cellular transformation: The case of multiple pathways to effects 180
D. Crawford-Brown, W. Hofmann , M. Nösterer and P. Eckl

Radiation quality and biological effectiveness

Undercounting of particle irradiation-induced DNA double-strand breaks by conventional assays 187
M. Löbrich

Cell inactivation, mutation and DNA damage induced by light ions: Dependence on radiation quality 191
F. Cera, R. Cherubini, M Dalla Vecchia, S. Favaretto, G. Moschini, P. Tiveron, M. Belli, F. Ianzini, L. Levati, O. Sapora, M.A. Tabocchini and G. Simone

The separation of spatial and temporal effects of high LET radiation 195
D.L. Stevens, S.J. Marsden, M.A. Hill, I.C.E. Turcu, R. Allott and D.T. Goodhead

The RBE of accelerated nitrogen-ions for apoptosis in human peripheral lymphocytes exposed *in vitro* 199
A.E. Meijer, U-S. E. Kronqvist, R Lewensohn and M Harms-Ringdahl

A versatile mammalian cell irradiation rig for low dose rate ultrasoft X-ray studies 203
M.A. Hill, D.L. Stevens, D.A. Bance and D.T. Goodhead

Microdosimetry of cells *in vitro* irradiated by alpha-particles 207
W.B. Li, W.Z. Zheng, X. Zhang, Y.F. Gong, J. Li and D.C. Wu

Prediction of cellular effects of high- and low-LET irradiation based on the energy deposition pattern at the nanometer level 211
R.W.M. Schulte

Risk extrapolation and cancer

Threshold dose for carcinogenesis: What is the evidence? 217
C. Streffer

Carcinogenic response at low doses and dose rates: Fundamental issues and judgements 225
R. Cox

An HSEF for murine myeloid leukemia 228
V.P. Bond, E.P. Cronkite, J.E. Bullis, C.W. Wuu, S.A. Marino and M. Zaider

The estimation of neutron quality factors: Future prospects based on further revisions of neutron doses in Hiroshima 232
P.R. Grimwood and M.W. Charles

A new paradigm for radiation risk assessment 236
M.M. Elkind and R.L. Wells

Cooperative behavior of irradiated cells in a three-dimensional model of radiation carcinogenesis 240
R. Bergmann, W. Hofmann, D. Crawford-Brown and H. Oberhummer

Modelling acute lymphocytic leukaemia using generalisations of the MVK two-mutation model of carcinogenesis: implied mutation rates and the likely role of ionising radiation 244
M.P. Little, C.R. Muirhead and C.A. Stiller

Analysis of lung cancer after exposure to radon using a two-mutation carcinogenesis model 248
H.P. Leenhouts, P.A.M. Uijt de Haag and K.H. Chadwick

Microdosimetry applied to radiotherapy

Microdosimetric considerations in the targeted radiotherapy of cancer 255
T.E. Wheldon and J. A. O'Donoghue

Relative biological effectiveness of Re- 188 beta particles: Implications for intravascular brachytherapy 262
H.I. Amols, R. Miller, J. Weinberger and E.J. Hall

Biophysical measurements at the COSY proton beam 266
R. Becker, P. Bilski, M. Budzanowski, W. Eyrich, D. Filges, M. Fritsch, J. Hauffe, H. Kobus, M. Moosburger, P. Olko, H. Paganetti, H.P. Peterson, Th. Schmitz, F. Stinzing

Microdosimetric characterization of clinical proton beams 270
F. Verhaegen and H. Palmans

Measurements of the LET in a proton beam of 62 MeV using the HTR-method 274
M. Noll, W. Schöner, N. Vana, M. Fugger and E. Egger

Microdosimetric investigations in the fast neutron therapy beam at Fermi National Accelerator Laboratory - work in progress 278
K. Langen, A.J. Lennox, T.K. Kroc, and P.M. DeLuca, Jr

Simulation of a microdosimetry problem: Behaviour of a pseudorandom series at a low probability 282
P. Meyer, J.E. Groetz, R. Katz, M. Fromm and A. Chambaudet

Microdistribution in tissues

The distribution of hits when lung cells are irradiated by alpha particles 289
J.A. Simmons and S.R. Richards

Microdosimetric modelling of damage to haemopoietic stem cells from radon decay in fat 293
T.D. Utteridge, D.E. Charlton, M.S. Turner, A.H. Beddoe, A.S.-Y. Leong, J. Milios, N.L. Fazzalari and L.B. To

Microdosimetry of tritiated particulates in alveolar sacs 297
R.B. Richardson and A. Hong

Elevated levels of Po-210 in human fetal tissues from mothers living near the Severn Estuary 301
D.L.Henshaw, J.E. Allen, P.A. Keitch and J.J. Close

Mean skeletal dose factors for ^{226}Ra and ^{239}Pu 305
S.L. Brooke and A.H. Beddoe

Comparison of experimental and calculated dose enhancement factors in tissue adjacent to high-Z implants for diagnostic X-rays 309
H. Průchová, D. Regulla, L.A.R. da Rosa and R. Seidlitz

Single particle effects

Microdosimetry and microbeam irradiation 315
L.A. Braby

Targeting cells individually using a charged-particle microbeam: The biological effects of single or multiple traversals of protons and $^3He^{2+}$ Ions 323
M. Folkard, K.M. Prise, B. Vojnovic, A.G. Bowey, C. Pullar, G. Schettino and B.D. Michael

Microbeam mediated cellular effects: single α particle induced chromosomal damage, cell cycle delay, mutation and oncogenic transformation 327
C.R. Geard, G. Randers-Pehrson, T.K. Hei, G.J. Jenkins, R.C. Miller, L.J. Wu, D.J. Brenner and E.J. Hall

Visualization of charged particle traversals in cells 331
N.F. Metting and L.A. Braby

Visualization of damage generated along alpha-particle tracks in irradiated rat tracheal epithelial cells 335
J.R. Ford, N.F. Metting, S.J. Marsden, D.L. Stevens, K.M.S. Townsend and D.T. Goodhead

Measured particle track irradiation of individual cells 339
E. Heimgartner, H.W. Reist, A. Kelemen, M. Kohler, J. Stepanek and L. Hofmann

Microbeam system for local irradiation of biological systems and effects of collimated beams on insect egg 343
Y. Kobayashi, H. Watanabe, M. Taguchi, S. Yamasaki and K. Kiguchi

The soft X-ray microprobe: A fine sub-cellular probe for investigating the spatial aspects of the interaction of ionizing radiations with tissue 347
G. Schettino, M. Folkard, K.M. Prise, B. Vojnovic, T. English, A.G. Michette, J.S. Pfauntsch, M. Forsberg and B.D. Michael

Proportional counter microdosimetry

A portable device for microdosimetric measurements 353
I. Almasi, E. Anachkova, T. Bartha, K. Erdelyi, A.M. Kellerer and H. Roos

Use of microdosimetry for mixed-field radiation measurements at the AFRRI TRIGA reactor 357
H.M. Gerstenberg, R.C. Bhatt, B.A. Torres and K.D. Bolds

Paired TEPCs for variance measurements 361
J.E. Kyllönen, L. Lindborg and G. Samuelsson

Comparative study of TE-gases and DME in a proportional counter 365
I.K. Bronić and B. Grosswendt

Monte Carlo simulations of electron motion in gas counters used in microdosimetry. Strong dependence of the results on the electron-molecule cross-sections 369
H. Průchová and B. Franěk

Nanodosimetric devices and other detectors

A nanodosimeter based on single ion counting 375
S. Shchemelinin, A. Breskin, R. Chechik, A. Pansky and P. Colautti

Theoretical basis for solid-state microdosimetry using photochromic alterations 379
D. Emfietzoglou and M. Moscovitch

Experimental microdosimetry with microstrip gas counters 383
J. Dubeau, M.S. Dixit, E.W. Somerville, A.J. Waker, R.A. Surette, F.G. Oakham and D. Karlen

A comparison of measured and calculated $\bar{y}_D$-values in the nanometre region for photon beams 387
J.-E. Grindborg and P. Olko

A SLDD based nanodosimeter for electrons and photons 391
T.M. Evans and C.K. Wang

Ionisation measurements in nanometre size sites with Jet Counter - Recent experimental results 395
S. Pszona

Microstrip Gas Chamber with TEG as a detector for microdosimetry 399
B. Bednarek, K. Jeleń, T.Z. Kowalski and E. Rulikowska-Zarębska

A radiation quality dosemeter based on thin organic films 403
C.E. Tucker, F.A. Smith and J. Oriel

Microdosimetry of LiF:Mg,Cu,P solid state detectors - what is the target? 407
P. Olko and P. Bilski

The peak-height ratio (HTR)-method for LET-determination with TLDs and an attempt for a microdosimetric interpretation 411
W. Schöner, N. Vana, M. Fugger and E. Pohn

Microdosimetry Research: A historic overview 415
H.H. Rossi

Use of new collection systems associated with a multicellular tissue equivalent proportional counter for individual neutron dosimetry 420
C. Hoflack, J.M. Bordy, Y. Charbonnier, T. Lahaye, M. Lemonnier, M.S. Dixit, J. Dubeau and P. Segur

Author Index 425

Subject Index 429

Track structure and damage simulation

TRACK STRUCTURE CALCULATIONS IN RADIOBIOLOGY: HOW CAN WE IMPROVE THEM AND WHAT CAN THEY DO?

H. Nikjoo[1], S. Uehara[2] and D. J. Brenner[3]

[1]MRC Radiation & Genome Stability Unit, Harwell, OX11 0RD, UK.
[2]School of Health Sciences, Kyushu University, Maidashi 3-1-1, Higashi-ku, Fukuoka 812, Japan.
[3]Center for Radiological Research, Columbia University, 630 W. 168th St., New York, NY 10032, USA.

1 INTRODUCTION

We examine the relative merit of various experimental data and theories adopted in Monte Carlo track structure codes and present a critical review of recent progress in the application of these codes in modelling of initial observable biological effects. The status of total ionization cross sections for electrons and the secondary electron spectrum in water has been tested in terms of experimental data and theoretical calculations. Four independent tests, namely radial distribution of interactions, dose distributions, frequency of energy deposition in DNA size targets and relative efficiency of strand break productions were used to assess the input data cross sections.

To date, a large body of scientific literature has been generated which employ Monte Carlo track structure calculations for predicting the measurable parameters from biological experiments for understanding mechanism(s) of damage in molecular radiation biology[1-4]. Our knowledge of spatial distributions of energy deposition in biological structures is mainly based on track structure studies in water and such data have been used in biophysical modelling of cellular effects of ionizing radiations. There are two aspects of track structure calculations and modelling which need to be considered. First, many of the essential input data for simulations of electron and ion transport in biological media are severely limited. Therefore, one question as how accurate and precise are the experimental input data and theoretical models describing the physics and chemistry of particle transport, and to what extent the differences between Monte Carlo track structure codes influence modelling of initial observed biological effects. Secondly, to what extent the use of Monte Carlo track structure codes facilitate and extent our understanding of the mechanism(s) of radiation damage in biomolecules? In the former, in the absence of rigorous experimental cross sections, various theoretical and analytical methods have been adopted to evaluate the required data. In the latter, biophysical models have made use of track structure descriptions of particle histories to predict initial observable effects, such as strand breakage, yield of chromosome aberrations and mutations.

Table 1 presents a list of Monte Carlo track structure codes most widely used in radiation biophysics. At the last microdosimetry symposium we presented a comparison between four of the Monte Carlo codes (two vapour and two liquid codes) as a first step towards a bench marking of cross sections used in track structure codes[17]. We concluded that all codes describing the transport of electrons in liquid water and water vapour showed strong similarities for larger size biomolecules. However, there were substantial differences between the liquid and vapour codes for DNA size targets. Such differences were attributed to the theoretical processes adopted in the codes. In this paper we examine the relative merits of ionization cross sections and the secondary electron spectra, using various experimental and theoretical treatment for generating electron tracks.

Table 1. Electron and Ion Track Structure Codes in vapour(v) or liquid(l)

Code	Particle	Medium	Energy Range	Author	Date	Ref
MOCA8B	e^-	H_2O (v)	10eV - 100keV	Paretzke	1970	5
OREC	e^- & ions	H_2O (l)	10eV - 1 MeV	Turner *et al*	1976	6
CPA100	e^-	H_2O (l)	10eV - 100keV	Terrissol *et al*	1976	7
MOCA14	ions	H_2O (v)	.30 - 10 MeV/u	Wilson, Paretzke	1980	8
BERKELEY	e^- & ions	H_2O (l)	10eV - MeV	Chatterjee, Magee	1980	9
DELTA	e^- & ions	H_2O (v,l)	10eV - 10 kev	Zaider,Brenner	1983	10
ETRACK	e^-	H_2O (v)	10eV - 1 MeV	Ito	1985	11
KIPC	e^-	H_2O (l)	10eV - 10 keV	Kaplan *et al*	1990	12
TRION	e^- & ions	H_2O (v)	10eV - 1 MeV	Lappa *et al*	1992	13
KURBUC	e^-	H_2O (v)	10eV - 10 MeV	Uehara, Nikjoo	1993	14
ETS	e^-	H_2O (v,l)	10eV - 10 keV	Hill, Smith	1994	15
PITS	ions	H_2O (v,l) DNA,C,ice	.30 - 10 MeV/u	Wilson *et al*	1994	16

2 BASIC CONCEPTS

The results obtained in Monte Carlo track structure simulations depend on details of the experimental cross section data and the theoretical models used. A complete model of track structure consists of descriptions for elastic and inelastic scattering, including stopping power, total cross sections, total elastic, total inelastic, total ionization, total excitation, singly and doubly differential ionization cross sections, partial excitation cross sections and secondary electron spectra. Not all cross sections are available for materials of interest in radiation biology. The majority of experimental data have been measured in water vapour. Data for other materials such as liquid water, DNA and proteins are scarce, as measurements are either very difficult or not practicable. In the absence of experimental data cross sections have been evaluated using various theoretical approach. A comprehensive treatment of all the above parameters has been given in recent review publications.[1-4,18]

2.1 Method of Comparison

Table 1 of Ref(17) and tables 9.4, 9.5 and 9.6 of Ref(3) show sources, method and assumptions used for the construction of cross sections in various codes. For example secondary electron spectra can be constructed using theoretical descriptions by Thomas-Fermi[19], Born-Bethe[20], Seltzer[21], Kim-Rudd[22], Jain-Khare[23], Gryzinski[24], and others[25]. Similar observations can be made for the construction of other cross sections. In this work we test the differences between electron tracks generated by the code KURBUC[14] using different sources of input data for cross sections. In this paper we limit our analysis to the test of total ionization cross sections and the secondary electron spectrum. In scheme A, the theoretical method of generating the secondary electron spectrum was changed, keeping all other cross sections unchanged. In scheme B, ionization cross sections were constructed by a least-square fit to all available experimental published data. Subsequently the secondary electron spectrum was changed as in scheme A. Experimental ionization cross sections included in the analysis were those of Hyashi[26], Djuric[27], Olivero[28], Bolorizadeh and Rudd[29], Schutten et al[30], and Gomet and Kastler[31]. Although there exist agreement between some of the measurements of ionization cross sections over part of the energy spectrum, there is considerable variation between some of experimental data in the range 50 eV to 1keV.

2.1.1 Test 1 - Radial Distribution of Interactions

The first comparison was made in terms of the radial distribution of interactions at 1nm intervals for 1 keV electrons. All electrons started at the origin were followed until the primary cut off energy at 10 eV and the secondary electrons at 1 eV. All distributions for 1000 starting electrons were summed up and normalised to the total number of interactions. Figure 1 shows two panels referring to scheme A and scheme B. Each panel shows the differences in radial distributions of interactions between the theoretical descriptions for secondary electron spectra. Comparison between the panels show a shift of the spectrum to the left when using experimental total ionization cross sections (TICS), indicating a lower frequency distribution.

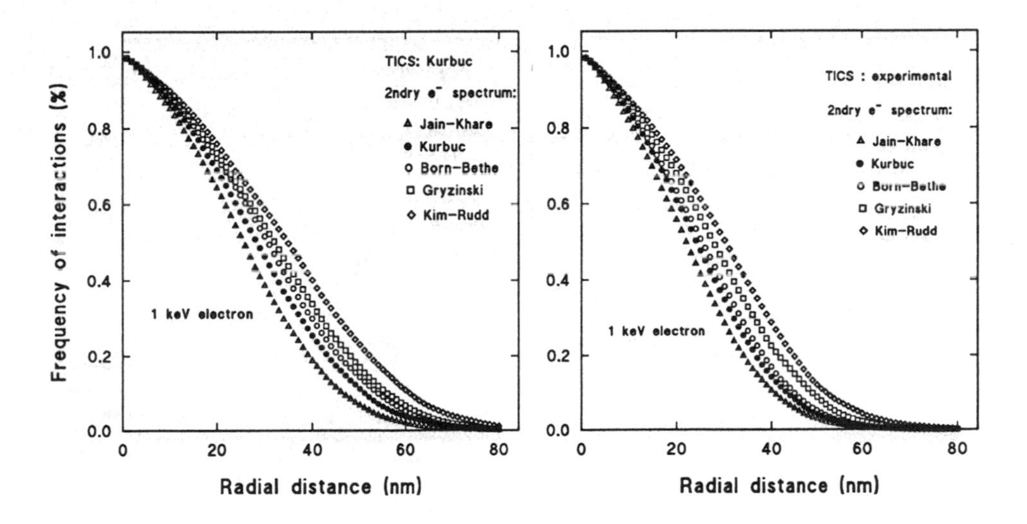

Figure 1 *Radial distribution of interactions presented as complement cumulative percentage of interactions, normalised to total number on interactions.*

2.1.2 Test 2 - Radial Dose Distribution

In a similar manner to test 1, radial dose distributions were scored in spherical shells of 1nm intervals from the origin, and normalized to the total energy of the tracks. The distributions (Figure 2) show the same trend as in test 1 in relation to the source of total ionization cross sections and the secondary electron spectrum description.

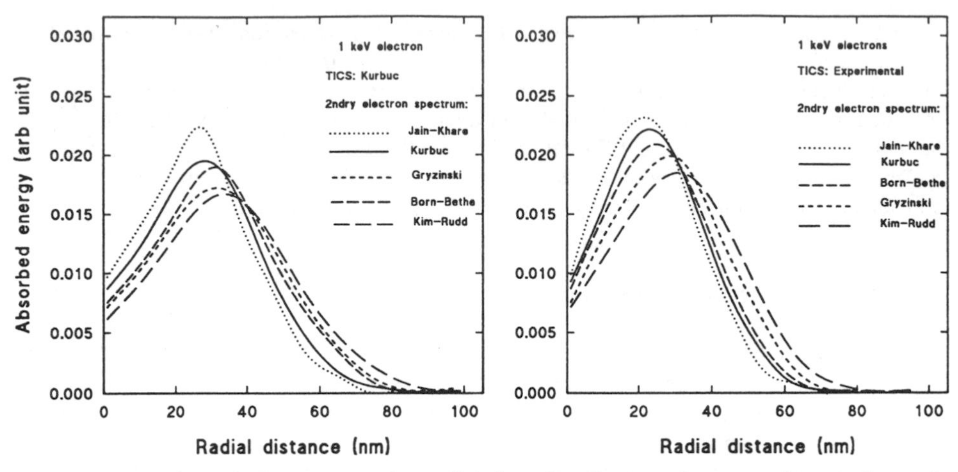

Figure 2 *Point kernel representing the radial dose distribution of energy deposited round the origin of 1 keV electrons generated by various cross sections.*

2.1.3 Test 3 - Frequency of Energy Depositions in DNA

The radial distributions of interactions and absorbed dose provide a macroscopic comparison of data. In this test we calculate the energy deposited by electrons in cylindrical volumes. Figure 3 shows the absolute frequency of energy depositions in DNA size volumes (2.3 nm height and diameter) positioned randomly in water irradiated uniformly with 1 Gy of monoenergetic electrons of 1 keV. Distributions are presented in 2 panels referring to case A and B (Figure 3). Both cases show similarity in energy depositions of 50 eV or less. Frequencies of energy depositions greater than 100 eV show appreciable variations between the sources of secondary electron spectra. On the other hand comparing the panels A and B shows a shift to lower frequencies for higher energy deposition events.

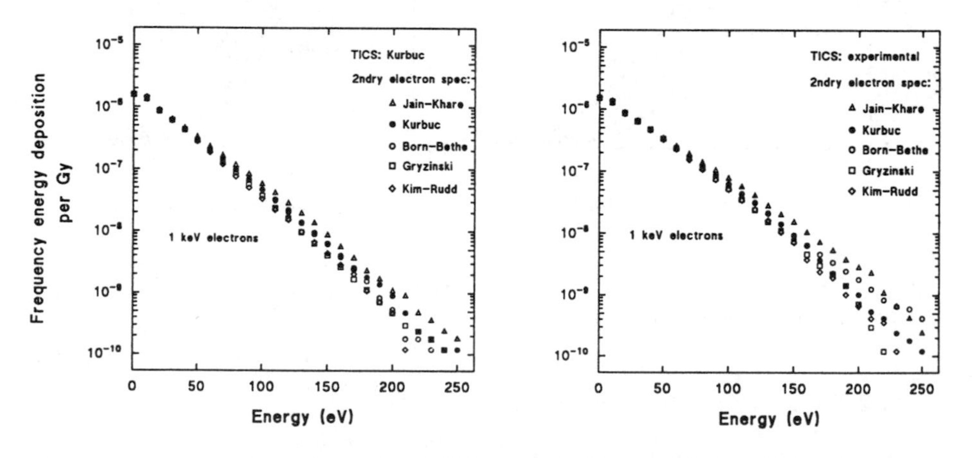

Figure 3 *Absolute frequency of energy deposition f(>E) greater than a given amount of energy E per Gy in a single cylindrical volume of diameter and length 2.3 nm placed at random in water homogeneously irradiated with tracks of 1 keV electrons.*

2.1.4 Test 4 - Strand Break Efficiency

A major application of track structure calculations is in modelling and calculations of the yield of DNA strand breaks, their complexes, and extension to yields of chromosome aberrations. In this section we have tested the efficiencies of strand breaks produced by electron tracks generated by the schemes A and B. The method of calculation has been described in detail elsewhere[32]. In this calculation it was assumed that at least a 17.5 eV energy deposition by a single event resulted in induction of a single strand break. A double strand break constituted the combination of two single strand break 10 base pairs or less apart. Table 2 shows relative frequencies of single (SSB) and double strand breaks (DSB) and the ratio of SSB/DSB. Again the shift in frequencies reflects the source of secondary electrons and the total ionization cross sections. The higher frequencies of SSB and DSB in scheme B indicates the smaller mean free path and harder collisions implied by the cross sections.

Table 2 Relative yield of strand breaks for various cross sections for 1 keV electrons

Source of Total ionization cross section	Method for 2ndry e^- spec	Y_{ssb} %	Y_{dsb} %	Y_{ssb}/Y_{dsb}
Kurbuc	Jain-Khare	40.8	4.4	9
Kurbuc	Kurbuc	39.2	3.8	10
Kurbuc	Born-Bethe	38.3	3.5	11
Kurbuc	Gryzinski	37.3	3.2	11
Kurbuc	Kim-Rudd	36.6	2.7	13
Experimental	Jain-Khare	43.5	5.2	8
Experimental	Kurbuc	43.6	5.0	8
Experimental	Born-Bethe	42.8	5.0	8
Experimental	Gryzinski	42.5	4.3	9
Experimental	Kim-Rudd	42.4	4.0	10

3 APPLICATION OF TRACK-STRUCTURE CALCULATIONS IN UNDERSTANDING RADIOBIOLOGICAL MECHANISMS

Since the pioneering work of Lea[33], there has been considerable interest in relating the results of track-structure calculations for different radiations and doses to corresponding measured biological effects. Many studies have investigated whether a single "critical" energy deposition event - such as a single ionization[33] or multiple ionizations within a small target[34] - can be correlated with measured biological effectiveness.

As an example, Brenner and Ward[35] concluded (see Figure 4) that yields of clusters of multiple ionizations within 2-3 nm sites correlate well with observed yields of DNA double-strand breaks (DSB). Similar approaches have been taken by various other authors, often identifying a DSB as two adjacent SSB, and often taking into account the early chemical development of the track, as well as the geometric structure of the target molecule (e.g., Refs. 32,36 and references therein).

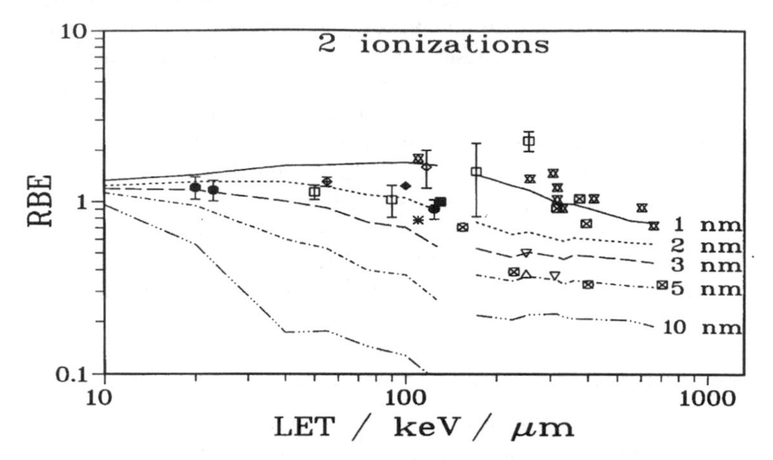

Figure 4 *(Updated from Ref 35) Points are experimental data for RBE for induction of DSB by protons (10-30 keV/μm), α particles (50-130 keV/μm), and heavy ions (Z=6-10,160-700 keV/μm). Curves correspond to calculated yields for clusters containing ≥2 ionizations, produced by α particles (10-130 keV/μm) and neon ions (170-700 keV/μm). The assumed cluster diameters are indicated on the right side of each curve.*

Such approaches have proved to be quite successful, and estimates of the nature of critical energy deposition events have been significantly refined since the advent of the current generation of track-structure codes in the early 1980's. However, the precision and accuracy of the track-structure codes are currently far out-stripping those of the corresponding measurements of, say, yields of DSB as a function of radiation quality (see Figure 4). Consequently it is not clear that further advances in the sophistication of track-structure calculations will necessarily yield more insight into this approach until the advent of more accurate data on DSB damage, both in terms of yields[37], and also more detailed DNA damage endpoints[38]

In this natural "critical-lesion" approach to interpreting track structure calculations, it is assumed that a single radiation-induced initial lesion is sufficient to initiate events leading to the endpoint of interest - in contrast to pairwise interaction models discussed below. As such, the critical-lesion approach inherently predicts dose-response relations that are both linear with dose and also dose-rate independent - because, at relevant doses, the critical energy depositions can only be produced within a single track of radiation, and not through the random cooperation of independent tracks. While the critical lesion model is appropriate for endpoints such as DSB induction, attempts have been made (e.g. Refs 39-40) to extend this model to situations where dose-response relations are non linear, such as survival data for mammalian cells; such approaches do, however, require further explanations for non-linear dose-response relations - such as saturation of repair mechanisms at low doses - for which there is, as yet, little direct evidence.

Track structure calculations have also been used to quantify pairwise "interaction" type approaches[41], in which survival or aberration formation is related to illegitimate reunions between DSBs or DSB free ends, the relative proximity of which is predicted by track-structure codes (e.g., Refs. 42-44, and references therein). Such models can inherently predict dose-rate effects, and non-linear dose-response relations, for low-LET induced endpoints such as aberration induction and clonogenic survival.

4 CONCLUSIONS

The accuracy and predictive power of current Monte Carlo track structure codes were examined in terms of their sensitivity to various experimental and theoretical input data cross sections. Similarities and differences were observed depending on the physical or biological parameters of interest. The objective tests adopted in this study provide a method for establishing a complete and accurate set of cross sections for Monte Carlo track structure codes.

For endpoints exhibiting linear dose-response relations (i.e. that unequivocally result from the effects of single tracks), track-structure calculations await only more accurate and detailed biological measurements to better quantify the appropriate critical energy deposition. However, for more complex endpoints, such as chromosome aberration formation and resulting endpoints, track-structure calculations have not, as yet, allowed either the "critical lesion" or "damage interaction" approaches to be rejected, and it is not clear that further sophistication in the computer codes will, of itself, accomplish this goal. Nevertheless, track-structure calculations are important tools for quantifying either approach and, at such time as the basic mechanistic questions discussed here are resolved, track-structure calculations would be a necessary first step in a complete theory of radiobiological action.

ACKNOWLEDGEMENTS

Authors thank Adrian Ford for help with the preparation of the manuscript. This work was partially supported by the Commission of the European Communities Contract F14P-CT95-0011 and by NIH grants ES-07361 and CA-63897. Part of calculations were carried out on Cray J32 at the Rutherford-Appleton Laboratories.

REFERENCES

1. D. T. Goodhead, in 'The Dosimetry of Ionizing Radiation' vol II, K.R Kase *et al*, (eds)., Academic Press, 1987.
2. M.N. Varma and A. Chatterjee, 'Computational Approaches in Molecular Radiation Biology, Monte Carlo Methods', Plenum Press, New York, 1994.
3. IAEA, 'Atomic and molecular data for radiotherapy and radiation research', IAEA-TECDOC-799, 1995.
4. J. W. Wilson et al, 'Transport Methods and Interactions for Space Radiations', NASA Reference Publication 1257, 1991.
5. H. G. Paretzke, in ' Kinetics of Nonhomogeneous Processes', J. Wiley & Sons, New York, Chapter 3, 1987.
6. R.N. Hamm, J.E. Turner, R.H. Ritchie, H.A. Wright, *Radiat. Res.* 1985, **104**, S-20.
7. M. Terrissol, J.P. Patau, T. Eudaldo. In 'Sixth Symposium on Microdosimetry' J. Booz and H.G. Ebert (eds). Harwood Academic Publishers Ltd, 1988, 169.

8. Wilson, W.E. and Paretzke, H.G., *Radiat. Res*., 1971, **47,** 359.
9. J. Magee, A. Chatterjee, Radiation Protection Dosimetry, 1986, **13**, 137-140.
10. M. Zaider, D.J. Brenner and W.E. Wilson, *Radiat. Res*., 1983, **95**, 231.
11. A. Ito. In 'Nuclear and Atomic Data for Radiotherapy and Related Radiobiology', International Atomic Energy Agency, Vienna, 1987.
12. I.G. Kaplan, A.M. Miterev, V.Y. Sukhonosov, *Radiat. Phys. Chem,* 1990, **36**, 493.
13. A.V. Lappa, E.A. Bigildeev, D.S. Burmistrov and O.N. Vasilyev, *Radiat. Environ. Biophys*, 1993, **32**, 1.
14. S. Uehara, H. Nikjoo, D.T. Goodhead, *Phys. Med. Biol,* 1993, **38**, 1841.
15. M.A. Hill and F.A. Smith, *Radiat. Phys. Chem,* 1994, **43**, 265.
16. W. E. Wilson, J. H. Miller, H. Nikjoo, In 'Computational approaches in molecular radiation biology'. Varma, M.N., Chatterjee, A. Eds. New York: Plenum, 1994, 137.
17. H. Nikjoo, M. Terrissol, R. N. Hamm, J. E. Turner, S.Uehara, H. G. Paretzke, D. T. Goodhead, *Radiation Protection Dosimetry*, 1994, **52**, 165.
18. M. Inokuti, 'Advances in Atomic, Molecular, and Optical Physics. Cross Section Data', Volume 33, Academic Press, Boston, 1995.
19. J. Lindhard, Reprint No 39 1964. Publication 1133. N.A. of Sciences; N.R.L.
20. M. Inukuti, Reviews of Modern Physics, 1971, **43**, 297.
21. S. M. Seltzer, in 'Monte Carlo transport of electrons and photons', eds T.M. Jenkins, W.R. Nelson and A. Rindi, Plenum Press, New York, pp.81-114, 1988.
22. Y. K. Kim, and M. E. Rudd, *Phys. Rev. A,* 1994, **7,** 1257.
23. D. K. Jain and S. P. Khare, *J. Phys. B,* 1976, **9,** 1429.
24. M. Gryzinski, *Phys. Rev.* 1965, **138,** A305.
25. ICRU Report 55, 1996.
26. M. Hayashi, 'Atomic and Molecular Data for Radiotherapy', IAEA-TECDOC-506, 1988, 192.
27. N. L. Djuric, I.M. Cadez, M.V. Kurepa, *Int.J.Mass Spec. Ion Proc.,* 1988, **83,** R7.
28. J. J. Olivero, R.W. Stagat and A.E.S. Green. *J. Geophys. Res*., 1972, **77,** 25.
29. M. A. Bolorizadeh and M. E. Rudd, *Phys. Rev. A*, 1986, **33,** 882.
30. J. Schutten, F. J. De Heer, H. R. Moustafa, A. J. H. Boer-Boom, J. Kistemakter, *Int. J. Radiat. Biol.,* 1966, **66,** 453.
31. J. C. Gomet and M. A. Kastler, C.A Acad. Sc. Paris, 1975, B-627.
32. H. Nikjoo, P. O'Neill, M. Terrissol, D.T. Goodhead, *Int.J.Radiat.Biol.* 1994, **66,** 453-457.
33. D.E. Lea, 'Action of Radiations on Living Cells', Cambridge University Press, London, 1955.
34. P.D. Holt, in 4th Symposium on Microdosimetry, edited by J. Booz, H.G. Ebert, R. Eickel and A Waker (Euratom, Luxembourg), 1974, 353.
35. D.J. Brenner and J.F. Ward, *Int. J. Radiat. Biol.,* 1992, **61,** 737.
36. W.R. Holley and A. Chatterjee, *Radiat. Res.,* 1996, **145,** 188.
37. J. Heilmann, G. Taucher-Scholz and G. Kraft, *Int. J. Radiat. Biol.,* 1995, **68,** 153-162.
38. B. Rydberg, *Radiat. Res.,* 1996, **145,** 200.
39. D.T. Goodhead and D.J. Brenner, *Phys. Med. Biol.,* 1983, **28,** 485.
40. D.T. Goodhead and H. Nikjoo, *Int. J. Radiat. Biol.,* 1989, **55,** 513.
41. R.K. Sachs, A.M. Chen and D.J. Brenner, *Int. J. Radiat. Biol.,* 1996.
42. D.J. Brenner and M. Zaider, *Radiat. Res.,* 1984, **99,** 492.
43. D.J. Brenner, *Radiat. Res.,* 1990, **124,** S29.
44. A.A. Edwards, V.V. Moiseenko and H. Nikjoo, *Radiat. Prot. Dosim.,*1996, **35,** 25.

DYNAMICS AND RECOMBINATION OF POSITIVE IONS AND THERMALIZED ELECTRONS IN HIGH-ENERGY ELECTRON TRACKS CALCULATED BY COMPUTER SIMULATIONS

L.D.A. Siebbeles,[a] W.M. Bartczak,[a,b] M. Terrissol[c] and A. Hummel[a]

a. Interfaculty Reactor Institute, Mekelweg 15, 2629 JB Delft, The Netherlands
b. Institute of Applied Radiation Chemistry, Technical University, Lodz, Poland
c. CPAT, Université Paul Sabatier, Toulouse, France

1 INTRODUCTION

When a high-energy charged particle loses energy along its path through a medium it produces a track of positive ions, thermalized electrons and electronically excited molecules. The reactions of these transients eventually give rise to the chemical and biological effects of high-energy radiation.

Several experimental studies have been devoted to the measurement of the number of ions escaping from recombination in high-energy electron tracks in saturated hydrocarbon liquids and to the production of excited states on charge recombination.[1-9] The present study involves computer simulations of the number of ions that escape from recombination and of the fraction of singlet excited states that are produced by the charges that recombine. Comparison of the calculated results with experimental results from the literature provides information on the initial spatial distribution of the charged species in the track.

2 COMPUTATIONAL METHOD

The initial spatial distribution of the ionizations in tracks of electrons with kinetic energies between 40 eV and 30 keV were calculated by Monte-Carlo simulations.[10, 11] The paths of the primary and secondary electrons were followed until they had reached an energy below 20 eV. Both inelastic scattering due to excitation of electrons in the medium and elastic scattering were brought into account. The cross section for excitation of the valence electrons was calculated by using the optical oscillator strength of polyethylene from Ref. 12, which was considered representative for saturated hydrocarbon liquids.

The secondary electrons in the track were assumed to thermalize in an exponential, f_{exp}, or Gaussian distribution, f_{Gauss}, around the positive ions from which they originate. The distributions are given by $f_{exp}(r)dr=1/(r_{av}-r_{reac})\exp[-(r-r_{reac})/(r_{av}-r_{reac})]dr$ for $r>r_{reac}$ and $f_{exp}(r)dr=0$ for $r\leq r_{reac}$ and $f_{Gauss}(r)dr=32r^2/(\pi r_{av}^3)\exp[-4r^2/(\pi r_{av}^2)]dr$, with r the distance between the electron and the positive ion, r_{av} the average thermalization distance and r_{reac} the reaction radius at which recombination of opposite charges occurs (r_{reac}=15Å). The paths of the positive ions and electrons were obtained by calculating random diffusive displacements and the displacements due to the drift in each other's

Coulomb field.[12] In this way the number of free ions is obtained as a function of time. On the time scale of the recombination process the orientation of the spins of the charged particles was assumed not to change. Recombination of an electron with the positive ion from which it originates (geminate recombination) then yields singlet excited states only, while recombination of an electron with another positive ion (non-geminate recombination) will produce singlets and triplets with a probability ratio of 0.25:0.75.

The track of an electron with sufficiently high energy consists of independent tracks of secondary electrons with lower energies. Therefore the number of escaped ions or the singlet fraction for a track of a primary electron with an energy above 30 keV was calculated by averaging the calculated results for the tracks of electrons with lower energies with weights given by the energy loss distribution of the primary electron.[11, 13]

3 RESULTS AND DISCUSSION

The calculated ion escape yields for exponential electron thermalization distributions are presented in Figure 1 for different r_{av} values, together with experimental results from Refs. 1-4. The ion escape yield, G_{esc}, is defined as the number of escaped ion pairs per 100 eV energy absorbed by the medium. The calculated ion escape yield is seen to vary remarkably with the energy of the primary electron. As the primary electron energy increases from a few tens of eV to a few keV the ion escape yield decreases, due to the fact that there are more ionizations close together. As the primary electron energy increases further, the ion escape yield increases due to the fact that the energy losses by the primary electron occur more separated from each other and the track consists of small groups of ions that have larger escape yields than the tracks in the keV region. The dashed parts of the curves for energies above 30 keV were obtained by averaging the yields at lower energies. It is seen that even in the MeV range the escape yield remains considerably lower than that for a single ion pair.

Figure 1 shows that the general behavior of the calculated curves agrees with the experimental results. Comparison of the calculated escape yields with the experimental results below 30 keV obtained from Refs. 3 and 4 allows the determination of the values of r_{av} for the saturated hydrocarbon liquids considered. The results are given in Table 1 for both an exponential and a Gaussian electron thermalization distribution. The average thermalization distances are seen to increase on going from n-hexane to 2,2,4 TMP and 2,2,4,4 TMP.

For the values of r_{av} in Table 1 the calculated escape yields in the MeV range are found to be larger than the experimental values. The calculated escape yields in the keV range are not very sensitive to the assumptions made in the calculations. However, the calculated track structures below 100 eV, and hence the escape yields, are uncertain, since the available electron scattering cross sections are inappropriate for low energies. Since an MeV electron loses about half of its energy in losses smaller than 100 eV the discrepancy in the MeV range could be due to a too large calculated escape yield for primary electron energies below 100 eV.

The calculated fractions of singlet excited states produced on recombination of positive ions and electrons are presented in Figure 2, together with experimental results from the literature.[5-8] The singlet fraction decreases as the primary electron energy increases from a few tens of eV to somewhat below 1 keV, reflecting that non-geminate recombination becomes more likely when there are more ionizations close together. As the energy of the

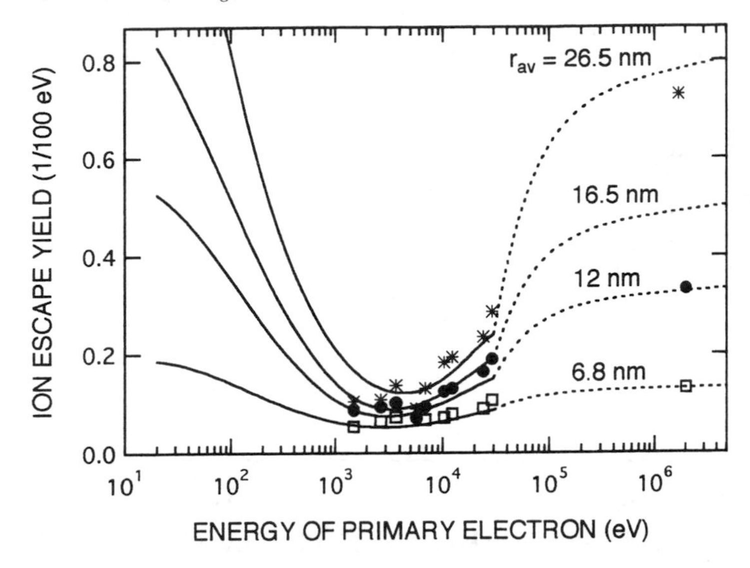

Figure 1 *Calculated ion escape yields for exponential electron thermalization distributions with average thermalization distances as indicated and experimental results for n-hexane (□), 2,2,4 trimethylpentane (2,2,4 TMP) (●) and 2,2,4,4 tetramethylpentane (2,2,4,4 TMP) (✳) from Refs. 1-4.*

primary electron increases further the singlet fraction goes up, since the energy losses by the primary electron occur at larger distances from each other and the track contains more independent groups of a few ion pairs which yield a larger singlet fraction. At low energies the calculated singlet fraction is insensitive to the average thermalization distance. For higher energies the singlet fraction is seen to decrease as the average electron thermalization distance increases, reflecting that non-geminate recombination is more likely when the electron thermalizes at larger distances. The singlet fractions calculated with r_{av}=6.8 nm show very good agreeement with the experimental results for lower energies. The experimental singlet fractions in the MeV range, however, exhibit considerable scatter, which makes a detailed comparison with the calculated results difficult.

Table 1. *Average thermalization distances, r_{av}, obtained by comparing calculated and experimental ion escape yields for primary electron energies below 30 keV.*

	exponential r_{av} (nm)	Gaussian r_{av} (nm)
n-hexane	8 ± 1	10 ± 1
2,2,4 TMP	15 ± 2	20 ± 3
2,2,4,4 TMP	> 22	> 26

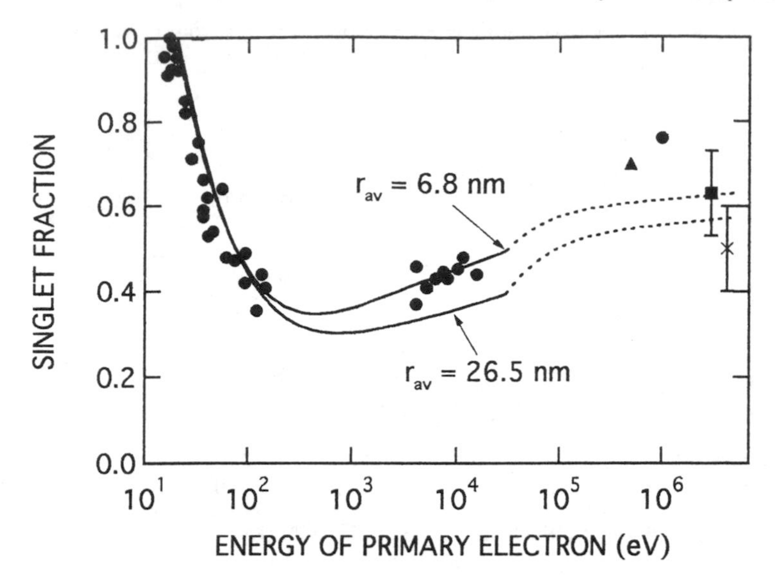

Figure 2 *Calculated singlet fractions for exponential electron thermalization distributions with average thermalization distances as indicated and experimental results from Refs. 6 (●), 7 (▲), 8 (×) and 9 (■).*

4 CONCLUSIONS

The calculated ion escape yield and the fraction of singlets produced on recombination of positive ions and electrons in high-energy electron tracks in saturated hydrocarbon liquids vary dramatically with the energy of the primary electron and agree with experimental results from the literature. Information on the thermalization distribution of the secondary electrons in the track was obtained by comparing the calculated and experimental results.

References

1. W. F. Schmidt and A. O. Allen, *J. Chem. Phys.*, 1970, **52**, 2345.
2. T. G. Ryan and G. R. Freeman, *J. Chem. Phys.*, 1978, **50**, 5144.
3. R. A. Holroyd and T. K. Sham, *J. Phys. Chem*, 1985, **89**, 2909.
4. R. A. Holroyd, T. K. Sham, B. X. Yang and X. H. Feng, *J. Phys. Chem.*, 1992, **96**, 7438.
5. L. H. Luthjens, H. C. deLeng, W. R. S. Appleton and A. Hummel, *Radiat. Phys. Chem.*, 1990, **36**, 213.
6. B. Brocklehurst, *Chem. Phys. Lett.*, 1993, **211**, 31.
7. J. A. LaVerne and B. Brocklehurst, *J. Phys. Chem.*, 1996, **100**, 1682.
8. M. C. Sauer and C. D. Jonah, *Radiat. Phys. Chem.*, 1994, **44**, 281.
9. P. Dorenbos , L. H. Luthjens and A. Hummel, preliminary result.
10. M. Terrissol, PhD Thesis, Université de Paul Sabatier de Toulouse, 1978.
11. L. D. A. Siebbeles, W. M. Bartczak, M. Terrissol and A. Hummel, *J. Phys. Chem.*, accepted.
12. W. M. Bartczak and A. Hummel, *J. Chem. Phys.*, 1987, **87**, 5222.
13. W. M. Bartczak and A. Hummel, *J. Phys. Chem.*, 1993, **97**, 1253.

A NEW APPROACH TO RADIATION TRANSPORT IN THE COMPLEX DNA ENVIRONMENT

M. TERRISSOL[*], M. DEMONCHY[*] and E. POMPLUN[**]

[*]C.P.A.T., Université Paul Sabatier, 118 route de Narbonne
31062 Toulouse Cedex, France
[**]Abteilung Sicherheit und Strahlenschutz, Forschungszentrum Jülich GmbH,
D-52425 Jülich, Germany

1 INTRODUCTION

DNA as a radiation target in cellular environment cannot be considered homogeneous. The physico-chemistry and the chemical evolution of species are function of the initial distribution of track. At DNA scale, each particle path can cover several regions. Energy migration and exchanges in the DNA, hydration shell and bulk water are important cause of damage and their relative positions are essential for repair. The interposition of hydration shell between DNA and its cellular environment is an important aspect which is of direct relevance to mechanisms of radiation damage. Moreover, we cannot use a simple sampling procedure since a free path can cover several media : bulk water, hydration shell, DNA, and histones. In this paper we show how the hydration shell can influence the energy deposits, chemical reactions, diffusions and finally DNA damages.

2 HYDRATION OF DNA

The degree of hydration 'Γ' is defined as the ratio of molecules of water to nucleotide. It plays an important part in DNA conformation [1] and damage induced by ionising radiation [2-5].

2.1 Crystallographic Data

We used the distribution of water molecules around a canonical B-DNA decamer to generate hydration around a specific DNA sequence. The crystallographic solvent analysis data[6] gives the water position around each sugar-phosphate groups and bases in terms of spherical co-ordinates. The reverse transformation of these data was applied to generate the hydration sites around the reference groups to the co-ordinate system of our linear DNA model formed with 41 base pairs.

2.2 Modelling of the Water Distribution

To overcome some limitations using the crystallographic data and to extend to hydration of a nucleosome[7] we used the molecular simulation. We generated distribution of water molecules around the B-DNA by molecular simulation up to 20 waters per nucleotide. We minimised the energy to obtain stable conformations. The molecular modelling of the hydration shell involves 7530 atoms for the linear B-DNA sequence and 26 594 atoms for the nucleosome model.

Table 1 : *Data for 4 different hydration shells. The degree of hydration "Γ" is associated with the total number of water molecules and the thickness of the hydration layer. "cryst." means that the layers were generated from the crystallographic data.*

	Linear Model		Nucleosome
Γ	no. of Water molecules	Thickness (nm)	no. of Water molecules
4	328	cryst.	1171
8	656	cryst.	2339
14	1148	0.21	4094
20	1641	0.30	5846

Each nucleotide pair is surrounded by a layer of solvent of a specified thickness (Table 1). Figure 1 shows models of DNA.

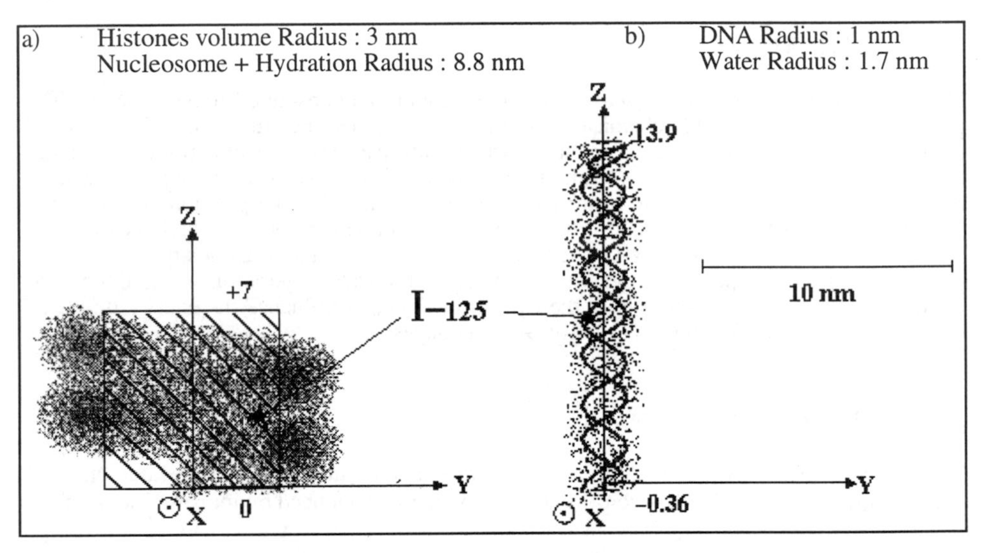

Figure 1 : *a) The nucleosome model formed with 146 base pairs (9056 atoms) with $\Gamma = 20$ (17 538 atoms). The hatched volume represents the histones. b) The linear model formed with 41 base pairs (2607 atoms) with a hydration $\Gamma = 20$ (4923 atoms).*

3 ELECTRON PATH THROUGH A COMPOSITE MEDIUM

Since low energy mean free paths are of the order of DNA dimension, we used a method to sample the path lengths in the composite medium : a) bulk water - b) hydration shell - c) the DNA structure itself (see Figure 2).

To sample an event, we first determined each distance to a boundary between two media. S_j is the partial distance travelled in the *j*th medium and S the total distance travelled. In the case of a homogeneous medium composed of n_j targets of type *j* per unit volume, σ_{Tj} being their total cross section, then :

$$\Sigma_j = n_j \sigma_{Tj} \tag{1}$$

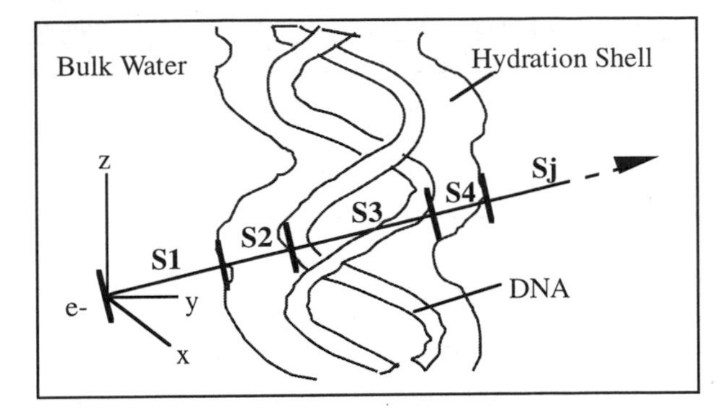

Figure 2 : *Modelling of the electron transport through several media.*

For a heterogeneous medium, we considered an arbitrary volume around the path containing n_j atoms or molecules with total cross section σ_{kj}, then

$$\Sigma_j = \sum_{k=1}^{n_j} \sigma_{kj} \tag{2}$$

If ξ is a uniform random number between 0 and 1, then the *m*th medium was determined such as :

$$\sum_{j=1}^{m-1} \Sigma_j S_j \leq -\log\xi < \sum_{j=1}^{m} \Sigma_j S_j \tag{3}$$

it gives the path

$$S = \sum_{j=1}^{m-1} S_j - \left(\sum_{j=1}^{m-1} \Sigma_j S_j + \log\xi \right) / \Sigma_m \tag{4}$$

with $\sum S_j = 0$ for $j = 1$

The electron interacts with the closest atom or molecule of its new sampled position.

When the medium is a set of n_j independent targets with cross section σ_{Tj}, the sampled path S is the path from the origin to the first target with distance d_j between this target and the direction of travel such that :

$$d_j^2 \leq \frac{\sigma_{Tj}}{\pi} \tag{5}$$

4 APPLICATION

To test the new approach we used Auger electrons emitted from ^{125}I incorporated in DNA. To sample the electron path, we used theoretical elastic cross-section and inelastic mean free path for liquid water and DNA[8]. The other components of the simulation model can be found in literature[9,10].

Table 2 : *Deposited energy (eV) per* ^{125}I *decay in nucleosome as a function of* Γ.

Γ	DNA	Hydration	DNA+Hydration	Histones
0	201	0	201	125
4	155	96	251	125
8	167	154	321	126
14	151	197	348	124
20	127	245	372	125

We present results on energy deposition at the physical stage (up to 10^{-15} s). Others results on strand breaks and base release will be found in this volume[11]. Figure 3 shows the distribution of the deposited energy for Γ=0 (dry DNA) and Γ=14.

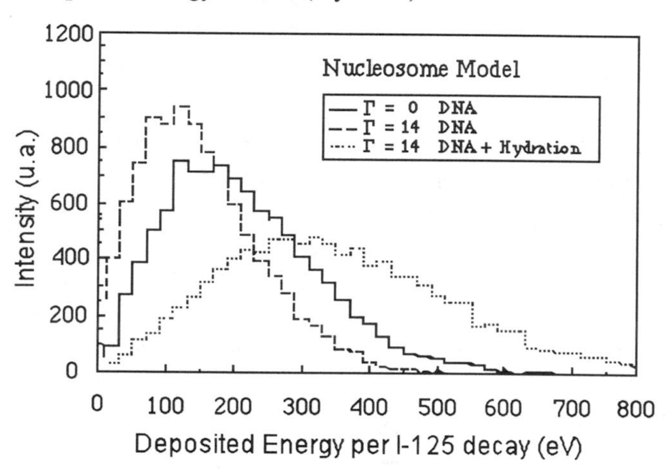

Figure 3: *Variation of the deposited energy in dry DNA, in hydrated DNA and in the 'DNA including hydration shell' (*Γ*=14).*

We see that hydrated DNA (doted curve) receives a larger amount of energy due to the increased volume, but the DNA itself (dashed curve) receive less energy. This is partly due to the energy spectrum of the ^{125}I where some very low energy electrons are stopped by the water in the grooves and do not reach the other strand. Table 2 gives the energy deposited as a function of hydration for a nucleosome. We conclude that the histones, taken as a cylinder of water (Figure 1), receive a constant amount of energy.

References

1. E.Westhof, *Ann.Rev.Biophys.Biophys.Chem.*, 1988, **17**, 125
2. N.Mroczka, W.A.Bernard. *Radiat.Res.*, 1993, **135**, 155
3. J.Hüttermann, M.Röhrig, W.Köhnlein, *Int.J.Radiat.Biol.*, 1992, **61**, 299
4. S.G.Swarts, M.D.Sevilla, D.Becker, C.J.Tokar, *Radiat.Res.*, 1992, **129**, 333
5. S.G.Swarts, D.Becker, M.D.Sevilla, K.T.Wheeler. *Radiat.Res.*, 1996, **145**, 304
6. Y.Umrania, H.Nikjoo, J.Goodfellow, *Int.J.Radiat.Biol.*, 1995, **67**, 145
7. E.Pomplun, 'Radiation Research', Chapman, Toronto, 1991, Vol.1, p.121.
8. J.A.La Verne and S.M.Pimblott, *Radiat. Res.*, 1995, **141**, 208
9. M.Terrissol, E.Pomplun, *Radiat.Prot.Dosim.*, 1994, **52**, 177
10. M.Terrissol, *Int.J.Radiat.Biol.*, 1994, **66**, 447
11. E.Pomplun, M.Demonchy, M.Terrissol, *this volume*.

AUGER ELECTRON ACTION INSIDE HYDRATED DNA AND NUCLEOSOME MODELS

E. Pomplun(*), M. Demonchy(**) and M. Terrissol(**)

(*) Abteilung Sicherheit und Strahlenschutz, Forschungszentrum Jülich GmbH, D-52425 Jülich, Germany
(**) C.P.A.T., Université Paul Sabatier, 118 route de Narbonne, F-31062 Toulouse Cedex, France

1 INTRODUCTION

Auger electrons emitted by DNA-incorporated radionuclides seem to be a most suitable probe to investigate individual aspects of the radiation mechanism. In particular, due to their low kinetic energies (few eV - some ten keV)[1] these electrons have a range comparable with the geometric dimensions of many sub-cellular biological structures so that significant fractions of their energies are deposited in DNA-neighbouring compartments, e.g. in the hydration layer, a sheath of water molecules closely bound to the DNA and responsible for the DNA structural integrity. On the basis of experimental studies, it has been suggested that an ionisation of water molecules from the hydration layer will cause the release of unaltered bases[2]. To simulate these experiments a complex computer model has been applied together with a linear DNA helix model[3,4] to have most realistic geometrical conditions.

Furthermore, to study the influence of different DNA structures on the efficiency of damage induction also a nucleosome model[4,5] was used and in addition to base release, single (ssb) and double (dsb) strand breaks have been investigated. The results to be obtained for both DNA models are expected to show significant differences. E.g. in the linear DNA, electrons once escaped from the helix have a small probability to attack it again whereas in the nucleosome there is a much greater chance to interact with the neighbouring double helix. Moreover, the hydration shell in the nucleosome occupies a larger volume and will be hit many more times than in the linear DNA.

2 METHODS

The main components of the simulation model are described elsewhere[5,6]. The new feature of considering the hydration shell is presented in this symposium[4]. Two Auger electron sources have been used here: (i) I-125 that emits about 12 electrons per decay and (ii) Carbon that releases a 274 eV Auger electron after x-ray ionisation of the K-shell. The targets (Linear DNA Model and Nucleosome Model with spherical volumes of r = 8 nm and r = 10 nm, respectively) received a dose of 12 kGy (Linear DNA) and 7 kGy (Nucleosome) per 274 eV Carbon Auger electron as well as 57 kGy and 35 kGy per I-125 decay.

The assumptions for the two different biological endpoints (see INTRODUCTION) have been the following: (a) for the release of an unaltered base such water molecules of the hydration shell have to be ionised which are associated with the phosphate or sugar groups of the DNA; (b) a ssb will be induced by an ionisation of a phosphate or sugar group atom or after a reaction of a deoxyribose-monophosphate with a radical during the physico-chemical phase. Two ssb within a distance of ten basepairs result in a dsb.

3 RESULTS

3.1 Release of Unaltered Bases

Although quite simple assumptions have been applied for the simulation of the release of bases, the obtained results correlate remarkable well with data derived from experiments[2] with Cs-137 irradiation. According to the experimental conditions the linear DNA model must be taken here for a comparison. The values for this model lie nearly completely within the range stretched by the two experimentally based curves for 2 and 90 kGy

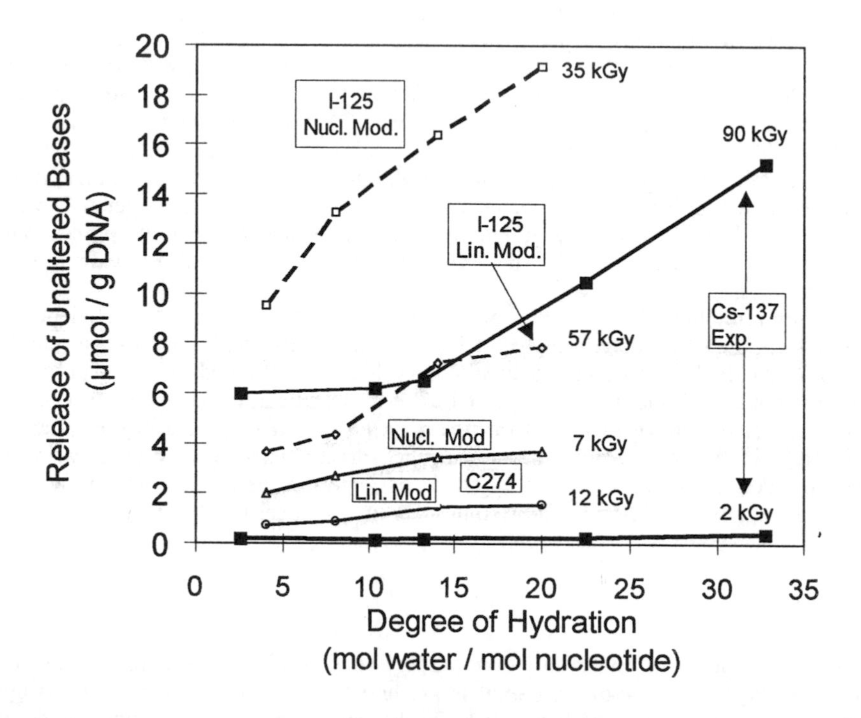

Figure 1: *Number of released unaltered bases after I-125 decay and K-shell ionisation of Carbon as a function of hydration's degree compared with experimental results from Swarts et al.*[2] *for Cs-137 irradiation; for calculation of dose values see METHODS*

low-LET radiation (Figure 1). The efficiency of I-125 decays in the nucleosome model, however, is about two times larger than the 90 kGy low-LET radiation, although the slope is similar. This is what can be expected since the volume occupied by the hydration shell is much larger than in the linear model.

Plotting the number of released bases as a function of absorbed dose (Figure 2) shows that all values for the linear DNA model fit very well into the experimental results. Being in a very high dose region, it is not unlikely that the secondaries created by incorporated I-125 and by external Cs-137 give the same distributions at the DNA level. This would explain the same yields per unit dose for the two different radiation sources. The values obtained with the nucleosome model are in significant excess due to the reason mentioned above.

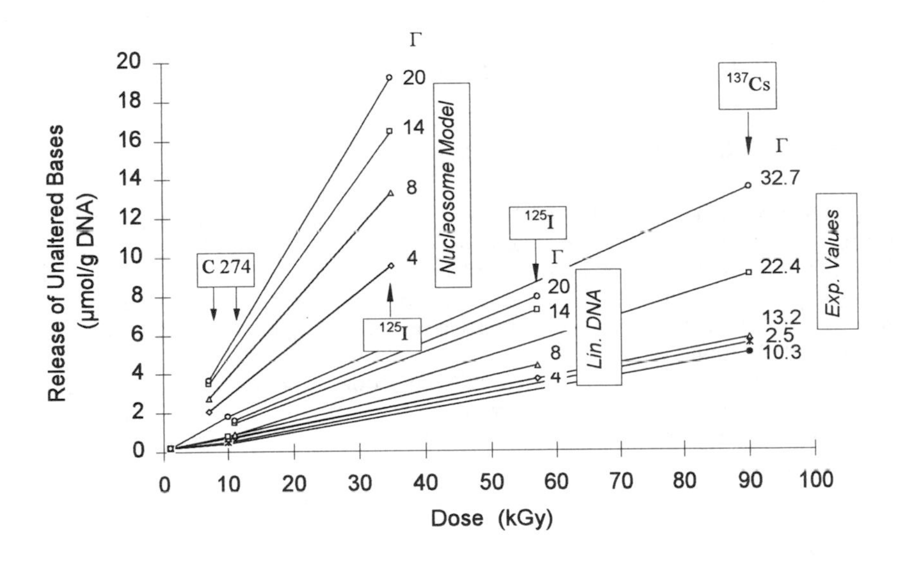

Figure 2: *Number of released unaltered bases after I-125 decay and K-shell ionisation of Carbon as a function of dose; experimental results for Cs-137 irradiation have been taken from Swarts et al.*[2] *; for dose calculations of this work see METHODS*

3.2 Single and Double Strand Breaks

For both DNA structure models the yields for ssb and dsb calculated in this work (Figures 3 and 4) are within the experimentally found range of 3 - 6 ssb per decay and about 1 dsb (for literature see reference 5) for low temperature conditions prohibiting repair processes which are not included in the simulation model. The break efficiency in the linear DNA does not seem to be sensitive to degree of hydration, whereas there is a decrease in the nucleosome model indicating a protective character of the hydration shell against direct radiation damage.

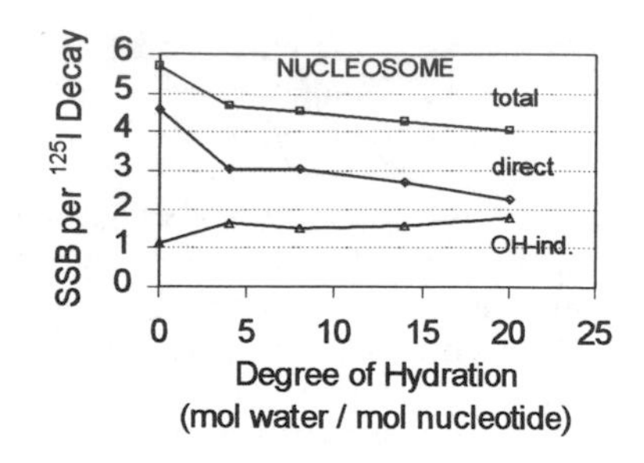

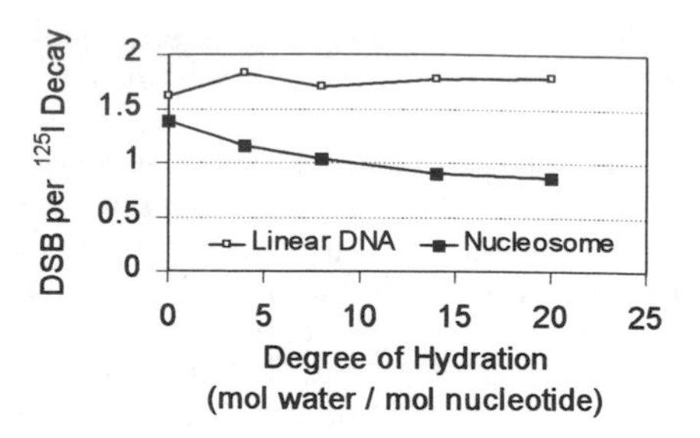

Figure 3: *Number of directly and indirectly induced ssb as a function of degree of hydration (this work)*
Figure 4: *Number of dsb as a function of degree of hydration (this work)*

4 CONCLUSIONS

A new approach of modeling the DNA hydration shell presented in this symposium[4] has been incorporated into a complex simulation model to trace the fate of low energy Auger electrons emitted from DNA-bound I-125 and from Carbon after K-shell photoionisation. Applying simple assumptions for the simulation of base releases as well as for the induction of strand breaks has resulted in reasonable damage efficiency data if compared with experimentally found values.

The introduction of the hydration shell in connection with the induction of DNA damages will touch the classical distinction between direct and indirect effects because the structural water molecules have to be considered as an integral part of the DNA molecule (see also Baverstock and Will[7]).

References

1. E. Pomplun, J. Booz, D.E. Charlton, Radiat. Res., 1987, **111**, 533
2. S.G. Swarts, M.D. Sevilla, D. Becker, C.J. Tokar, K.T. Wheeler, Radiat. Res., 1992, **129**, 333
3. E. Pomplun, Int. J. Radiat. Biol., 1991, **59**, 625
4. M. Terrissol, M. Demonchy, E. Pomplun, this volume
5. E. Pomplun, M. Terrissol, Radiat. Environ. Biophys., 1994, **33**, 279
6. M. Terrissol, E. Pomplun, Radiat. Prot. Dosim., 1994, **52**, 177
7. K.F. Baverstock, S. Will, Int. J. Radiat. Biol., 1989, **55**, 563

INTERACTION CROSS SECTIONS FOR ELECTRON INELASTIC SCATTERING IN LIQUID WATER

Michael Dingfelder, Detlef Hantke and Mitio Inokuti

GSF – National Research Center for Environment and Health
Institute of Radiation Protection
Ingolstädter Landstraße 1
85764 Neuherberg, Germany

1 INTRODUCTION

Track stucture simulations are often used to study the biological response of matter to radiation. These transport calculations require reasonable interaction cross sections for charged particles in biological target molecules and an adequate description of the geometrical structure of the key targets. Most of the currently used cross sections for electron inelastic scattering in liquid water are based on the dielectric response function, derived from measured optical data.

The potential for a wide range of applications of the cross-section data makes it desirable to document them precisely and in detail, so that a user can readily adopt them in applications. Full documentation of the data is also an initial step of updating them from time to time as new pertinent information becomes available. With this idea in mind, we have begun a new survey of cross sections and of the dielectric response function for water.

2 DIELECTRIC RESPONSE FUNCTION

Provided that an external charged particle is sufficiently fast, the dielectric response of matter upon sudden transfer of energy E and momentum $\hbar K$ is given in terms of the function $\epsilon(E,K)$, characteristic of matter. This function is in general a complex funcion consisting of a real part $\epsilon_1(E,K)$ and an imaginary part $\epsilon_2(E,K)$ which are real functions. $\epsilon(E,K)$ can be derived from various measurements. Basically we follow the general idea of Ritchie and co-workers[1] but correct its difficulty getting contributions below ionisation thresholds. For modelling $\epsilon_2(E,0)$ we use a linear superposition of Drude–like functions. In case of discrete excitations we use a derivative Drude function $D^*(E,E_k)$, which is sharply peaked around $E \approx E_k$, and thus is more nearly discrete than the Drude function. For ionisation we use a Drude function $D(E,E_j)$ multiplied with a step function $\Theta(E-E_j)$, which avoids contributions below the threshold, and a Gauss function $G(E,E_j)$ to represent in effect the width of the valence band including the broadening of electronic spectra due to coupling with phonons. Our model then reads

$$\epsilon_2(E,0) = E_p^2 \left\{ \sum_{k=1}^{4} D^*(E,E_k) + \sum_{j=1}^{5} \int_{E_j-\Delta}^{E_j+\Delta} D(E,\Omega)\,\Theta(\Omega-E)\,G(\Omega,E_j)\,d\Omega \right\} \quad (1)$$

Table 1 *Model fit parameters compared with experimental values from [2] and [3].*

excitations				ionisations				
k	E_k [eV]	γ_k [eV]	Zf_k	j	E_j [eV]	γ_j [eV]	Zf_j	E_{exp} [eV]
1	8.3	1.55	0.11	1	10.9	19.5	1.45	10.9[2]
2	10.1	2.35	0.20	2	13.5	14.0	5.05	13.5[2]
3	11.3	1.41	0.06	3	16.2	16.0	1.77	17.0[2]
4	12.8	4.50	0.72	4	32.0	160.0	2.65	32.4[3]
				5	540.0	200.0	2.12	539.7[3]

with

$$D(E,E_j) = \frac{f_j\gamma_j E}{(E_j^2 - E^2)^2 + \gamma_j^2 E^2} \quad , D^*(E,E_k) = \frac{2f_k\gamma_k^3 E^3}{\left[(E_k^2 - E^2)^2 + \gamma_k^2 E^2\right]^2} \quad , \tag{2}$$

$$\Theta(E) = \begin{cases} 0 & \text{for} \quad E < 0 \\ 1 & \text{for} \quad E \geq 0 \end{cases} \quad , G(E,E_j) = \frac{1}{\Delta}\exp\left(-\frac{(E-E_j)^2}{2\Delta^2}\right) \tag{3}$$

and $\Delta = 1\,\mathrm{eV}$. $E_p = 21.46\,(\mathrm{eV})$ is the plasma energy of liquid water.

For excitations we consider four discrete levels named $\tilde{A}^1B_1$, $\tilde{B}^1A_1$, Rydberg A+B and Rydberg C+D, while "diffuse bands" and other excitations[1] are not used because they overlap with the ionisation levels and a clear separation could not be done. The parameters $f_{k/j}$, $\gamma_{k/j}$ and $E_{k/j}$ are obtained in a fit of $\epsilon_2(E,0)$ to experimental data[4] also taking into account sum rules and relative shell contributions. They are displayed in table 1. The ionisation energies agree well with known experimental data from photoelectron spectroscopy at liquid water surfaces[2] and with calculation in water vapour for the inner shells[3].

After setting up $\epsilon_2(E,0)$ we calculate $\epsilon_1(E,0)$ using the Kramers–Kronig relation

$$\epsilon_1(E,0) = 1 + \frac{1}{\pi} P\int_{-\infty}^{+\infty} \frac{\epsilon_2(E',0)}{E'-E} dE' \quad . \tag{4}$$

Both, the real and imaginary part of the dielectric response function are shown in figure 1 and compared to experiment[4]. Finally, the probability of energy transfer E from a fast charged particle at the dipole limit $K = 0$ is given by

$$\eta_2(E,0) = \mathrm{Im}\left(\frac{-1}{\epsilon(E,0)}\right) = \frac{\epsilon_2(E,0)}{\epsilon_1^2(E,0) + \epsilon_2^2(E,0)} \quad . \tag{5}$$

Following Ritchie and co–workers[5] we intoduce the momentum–transfer dependence by using the generalized oscillator strength $f_{k/j}(K)$ instead of $f_{k/j}$ in Eq. 1 and by shifting the ionisation energies E_j to $E_j + (\hbar K)^2/2m$.

3 DIFFERENTIAL CROSS SECTIONS

We calculate the macroscopic cross section differential in energy transfer E in the First Born Approximation (FBA). It is based on the assumption that the kinetic energy

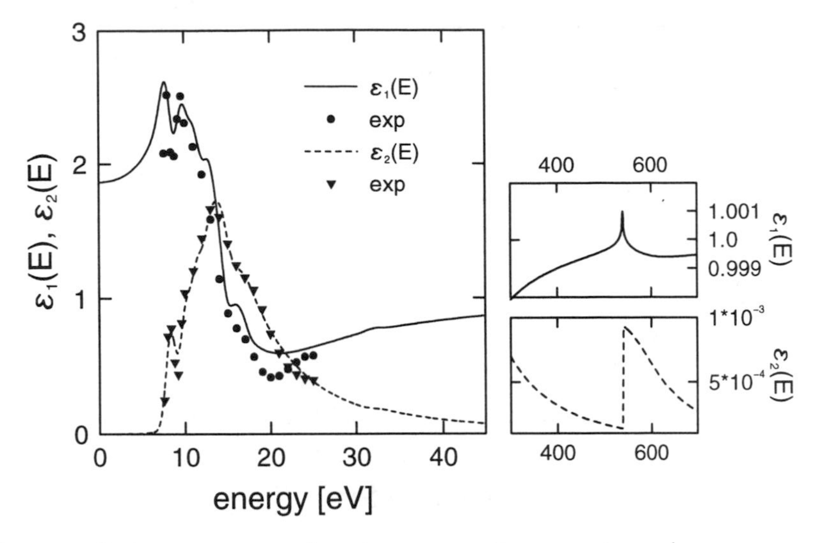

Figure 1 *Dielectric response function compared to experiment* [4]

of the incoming electron T is large compared to the kinetic energy of the "bound" electron, which is about 1 Rydberg, and still nonrelativistic, i. e.,

$$T \gg \mathrm{Ry}, \quad 1\,\mathrm{Ry} = 13.6\,\mathrm{eV}, \quad T \ll mc^2 = 511\,\mathrm{keV} \quad . \tag{6}$$

The differential cross section for a single excitation n is

$$\frac{d\Sigma^n}{dE}(E,T) = \frac{1}{\pi a_0 T} \int_{K_-(E,T)}^{K_+(E,T)} \eta_2^n(E,T)\, \frac{dK}{K} \tag{7}$$

where η_2^n is the contribution of excitation n to η_2. The momentum transfer is limited to $\hbar K_\pm(E,T) = \sqrt{2m}(\sqrt{T} \pm \sqrt{T-E})$. In case of ionisations we include "exchange" effects in the sense of Mott. This limits the allowed energy transfer to $E_{\min} = E_n$ to $E_{\max} = 1/2(T + E_n)$. Finally we get

$$\frac{d\Sigma}{dE}(E,T) = \sum_{\substack{k=1\\exc}}^{4} \frac{d\Sigma^k}{dE}(E,T) + \sum_{\substack{j=1\\ion}}^{5} \left(\frac{d\Sigma^j}{dE}(E,T) + \text{"exchange terms"} \right) \quad . \tag{8}$$

The differential cross section (dashed line) as well as the ionisation part (dashed line) are shown in figure 2 for an incident electron energy of $T = 2\,\mathrm{keV}$ over a wide range of energy transfer E.

The asymptotic behaviour for high incident energies T and $T \gg E$ is

$$\frac{d\Sigma}{dE} = \frac{1}{\pi a_0 T} \left[A(E) \ln\left(\frac{T}{\mathrm{Ry}}\right) + B(E) + O\left(\frac{E}{T}\right) \right] \tag{9}$$

where $A(E)$ and $B(E)$ are functions of E derivable from $\eta_2(E,K)$. This form is also known as the Bethe cross section.

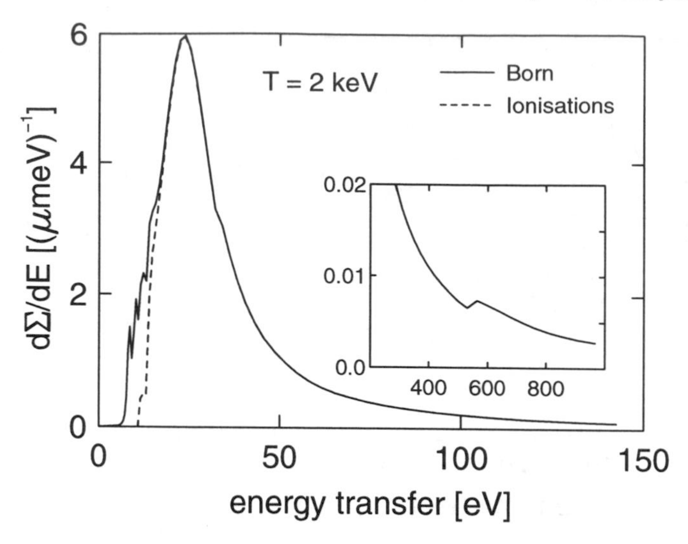

Figure 2 *Differential cross section in the First Born Approximation for an incident energy* $T = 2\,\text{keV}$.

Outside the valitidy of the First Born Approximation we need corrections, i. e., for relativistic and low energies. At relativistic energies we use the relativistic form of the Bethe cross section. A more detailed description of the whole procedure is in preparation

Acknowledgement

The work is supported by the European Community under Contract No. FI4P–CT95–0011 "Biophysical Models for the Induction of Cancer by Radiation"

References

1. R. H. Ritchie, R. N. Hamm, J. E. Turner, H. A. Wright, and W. E. Bloch, in 'Physical and Chemical Mechanisms in Molecular Radiation Biology', edited by W. A. Glass and M. N. Varma, Plenum Press, New York, 1991, p. 99 .

2. M. Faubel and B. Steiner, in 'Linking the Gaseous and Condensed Phases of Matter', ed. by L. G. Christophorou et al, Nato ASI Series B **326**, p. 517–523, Plenum Press, New York, 1994.

3. S. Uehara, H. Nikjoo and D. T. Goodhead, *Phys. Med. Biol.*, 1993, **38**, 1841.

4. J. M. Heller Jr., R. N. Hamm, R. D. Birkhoff and L. R. Painter, *J. Phys. Chem.*, 1974, **60**, 3483.

5. R. N. Hamm, H. A. Wright, R. H. Ritchie, J. E. Turner and T. P. Turner, in 'Fifth Symposium on Microdosimetry', Verbania, Pallanza, September 1975, EUR–5452 d–e–f, 1037.

TOTAL ELECTRON SCATTERING CROSS SECTIONS OF DIMETHYL ETHER

W. Y. Baek and B. Grosswendt

Physikalisch-Technische Bundesanstalt
Bundesallee 100, D-38116 Braunschweig, Germany

1 INTRODUCTION

The characteristic of a proportional counter is basically determined by charge transport processes taking place in its gas-filled volume. Important parameters that influence these processes are the electric field distribution inside the sensitive volume of the counter, the gas pressure and the atomic or molecular properties of the counting gas. Preliminary studies at several laboratories revealed that dimethyl ether (CH_3-O-CH_3) has properties which are useful for its application as counting gas in TEPCs; it is almost tissue-equivalent and shows an advantageous behaviour with respect to diffusion and avalanche formation.

The numerical modelling of TEPCs and the understanding of its properties require that the elastic and inelastic electron-molecule scattering cross sections of the counting gas are exactly known. Moreover, total scattering cross sections are often applied as reference values to put differential scattering cross sections on an absolute scale. Unfortunately, experimental data for electron scattering cross sections of dimethyl ether are not yet available.

In view of this fact, the total scattering cross sections of dimethyl ether for electrons are measured at energies from the first ionization potential[1] of 10 eV to 5.0 keV. This energy range is of particular interest because the calculation with the semiempirical model of Rudd et al.[2] shows that the major part of the secondary electrons produced, for instance, during the interaction of protons with matter has energies lower than 5.0 keV.

2 EXPERIMENTAL SETUP

Since the apparatus is the same as that described in our previous report,[3] it will be outlined only briefly. It consists of an electron gun, a gas chamber and an electron energy analyzer with a channel electron multiplier as detector. The electron gun is capable of producing electron beams with energies from 10 eV to 5.0 keV. The beam is bundled by a unipotential lens with variable focal length. The electron gun contains two pairs of deflector plates perpendicular to each other, enabling the direction of the beam to be modified.

The electron beam enters the gas chamber through an entrance aperture 0.5 mm in diameter. The gas chamber is 13.2 cm in length and filled via a regulating valve with the dimethyl ether gas at pressures up to 1 Pa. The pressures are measured by a capacitance manometer. The electrons undergo scattering processes with the molecules in the chamber und leave it in the forward direction through an aperture 0.5 mm in diameter. Since the output of the electron source may change during the measurement, the electron currents on the entrance aperture are recorded and later used for correcting the change in the initial electron current. In order to avoid the deflection of electrons by external magnetic fields, almost the whole path of the electrons from the source to the detector is shielded against the fields by mu-metal cylinders.

3 MEASUREMENT

The electrons leaving the gas chamber are analyzed with respect to their energy by an electrostatic 150° spherical deflector with a mean radius of 10 cm and detected by a channel electron multiplier. The total scattering cross sections are measured by setting the pass energy of the analyzer equal to the primary energy of the electrons and by recording the change of the count rates of the channel electron multiplier as a function of the pressure in the gas chamber. The measurements are repeated 10 times for each energy. In order to avoid multiple scattering effects, the range of the gas pressure is chosen such that the attenuation of the beam intensity does not exceed 60%. The overall energy resolution of the apparatus was better than 0.7 eV so that electrons inelastically scattered by dimethyl ether could be well discriminated from the non-scattered ones.

The total scattering cross sections σ_t are obtained by least square fits of the transmitted electron current as a function of the number density using the equation

$$\ln\left(\frac{I_0}{I_1}\right) = \sigma_t\,(n_1 - n_0)\,l + const \cdot r(n_1 - n_0)\;, \qquad \textbf{(1)}$$

which is based on Beer's law. I_0 and I_1 are the transmitted electron currents at the number densities n_0 and n_1 of the scattering gas, respectively, and l is the scattering length. The number density n_i is homogeneous in the gas chamber and calculated from the measured pressures p_i using the ideal gas law. The scattering length l is in the first order equal to the length of the gas chamber. The second term on the right side of Eq. 1 is introduced to take into account the additional beam attenuation due to the outstreaming gas in the vicinity of the apertures of the gas chamber and in the electron energy analyzer. The former is estimated using the results of Murphy[4] and the latter from the measured pressure increase in the analyzer as a function of the pressure in the gas chamber.

It is well known that the reading of a capacitance manometer has to be corrected for the so-called thermal transpiration effect[5] which results from a temperature gradient between gas and gauge head. The correction factor depends on the pressure, but not significantly on the specific target gas.[5] Hence, we used the correction factor for methane, which was determined before for our manometer to be 1.027 at the pressures occurring in this experiment.

4 RESULTS AND DISCUSSION

4.1 **Total scattering cross sections**

The results of the measurements are listed in Table 1 and shown in Fig. 1 in comparison with the total scattering cross sections calculated by using the additivity rule $\sigma_t(C_2H_6O)=\sigma_t(C_2H_6)+\frac{1}{2}\sigma_t(O_2)$. The cross sections for ethane[6] and oxygen[7] were taken from literature. While the additivity rule generally fails at low energies, it is a good approximation at high electron energies where the Born approximation is applicable. This can be clearly seen from Fig. 1. While our experimental values are about 10% higher than those calculated using the additivity rule below 20 eV, the discrepancies decrease with increasing energy to disappear at 400 eV. It can further be seen from Fig. 1 that the total scattering cross sections monotonically decrease within the uncertainties with the increase in the energy above 10 eV

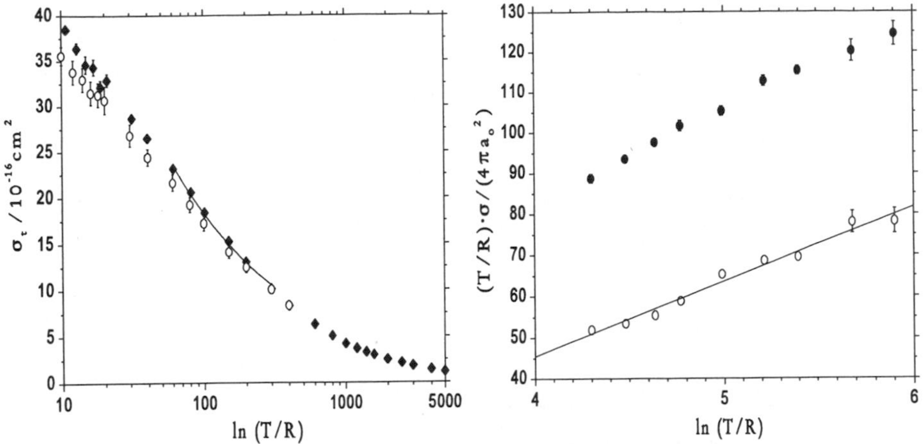

Figure 1 *Total electron scattering cross sections of dimethyl ether as a function of the energy. (♦): our experimental results, (○): values calculated from the total cross sections of ethane*[6] *and oxygen*[7] *using the additivity rule. The solid line is the result of the fit of Eq. 2 to our data.*

Figure 2 *Fano plot of the inelastic scattering cross sections (○) obtained by subtracting elastic scattering cross sections*[10] *from our data listed in Table 1. For comparison, the Fano plot of the total scattering cross sections (●) is shown. R is 13.61 eV, a_0 Bohr's radius.*

as is typical for organic molecules with ionization potentials of close to 10 eV.

For the intermediate energy range, it is found[8] that the total scattering cross sections of organic molecules can be fitted by

$$\sigma_t = b \,/\, \sqrt{T}\ , \tag{2}$$

where b is a specific constant for the molecule.[8] The least square fit of Eq. 2 to our data in the energy range from 60 eV to 300 eV yields $183.4\cdot10^{-16}\mathrm{cm^2eV^{1/2}}$ for b and $7.1\cdot10^{16}\mathrm{cm^2eV^{1/2}}$ for b/N_e, where N_e is the total number of target electrons. Our value of b/N_e is about 21% lower than that found by Nishimura and Tawara[8] for alkane and alkene molecules.

At high electron energies where the Born approximation is valid, the inelastic scattering cross sections generally yield a straight line in the so-called Fano plot,[9] which is shown in Fig. 2. The Fano plot is used here to check the internal consistency of our results. The inelastic scattering cross sections are obtained by subtracting theoretical elastic scattering cross sections from our experimental values. Since no elastic scattering cross sections of dimethyl ether are available for electrons, they are calculated from the elastic scattering cross sections[10] of its atomic constituents using the additivity rule. As can be seen from Fig. 2, our results are well represented within the uncertainties by a line in the Fano plot. It should be noted that in Fig. 2, only the uncertainties of our experimental data are considered and those of the elastic scattering cross sections (about 10%) are not taken into account.

4.2 Discussion of uncertainties

The statistical uncertainties of significance arise from the reading of the capacitance manometer and of the output of the channel electron multiplier. The statistical uncertainties and the uncertainties due to the drift in the electronics are given in Table 1. The sources of systematic uncertainties are:

(1) inaccuracy in determining the effective scattering length, which amounts to 0.8%,

Table 1 *Total electron scattering cross sections of dimethyl ether. The uncertainties given are due to statistical uncertainties and drift in the electronics. Other systematic uncertainties are discussed below.*

T / eV	σ_t / 10^{-16}cm^2	T / eV	σ_t / 10^{-16}cm^2	T / eV	σ_t / 10^{-16}cm^2
10.77	38.50±0.41	81.57	20.58±0.29	1204.0	3.72±0.03
12.75	36.40±0.59	101.5	18.36±0.29	1406.0	3.32±0.03
14.75	34.63±0.94	150.0	15.27±0.28	1607.0	3.03±0.04
16.76	34.34±0.87	200.0	13.03±0.26	2000.0	2.52±0.03
18.75	32.17±0.69	301.0	10.13±0.20	2500.0	2.16±0.03
20.79	32.90±0.70	402.0	8.32±0.06	3000.0	1.84±0.02
30.95	28.67±0.42	602.0	6.33±0.12	4000.0	1.44±0.03
39.83	26.49±0.14	804.0	5.10±0.03	4985.0	1.20±0.03
60.61	23.17±0.26	1004.0	4.23±0.05		

(2) the calibration of the capacitance manometer bears an uncertainty of 0.5%,
(3) the determination of the electron energy with a field retarding method was carried out with an uncertainty of 0.2 eV. This uncertainty is insignificant for energies higher than 100 eV, while at energies below 20 eV it causes an uncertainty of 0.5% in the total scattering cross sections. The square root of the quadratic sum of the single uncertainties gives 1.1% as the overall systematic uncertainty.

References

1. S. A. C. Clark, A. O. Bawagan and C. E. Brion, Chem. Phys., 1989, **137**, 407.
2. M. E. Rudd, Y.-K. Kim, D. H. Madison and T. J. Gay, Rev. Mod. Phys., 1992, **64**, 441.
3. W. Y. Baek and B. Grosswendt, Radiat. Prot. Dosim., 1995, **61**, 237.
4. D. M. Murphy, J. Vac. Sci. Technol. A, 1989, **7**, 3075.
5. T. Edmonds and J. P. Hobson, J. Vac. Sci. Technol., 1965, **2**, 192.
6. O. Sueoka and S. Mori, J. Phys. B: At. Mol. Phys., 1986, **19**, 4035.
7. M. S. Dababneh, Y.-F. Hsieh, W. E. Kauppila, C. K. Kwan, S. J. Smith, T. S. Stein and M. N. Uddin, Phys. Rev. A, 1988, **38**, 1207.
8. H. Nishimura and H. Tawara, J. Phys. B: At. Mol. Opt. Phys., 1991, **24**, L363.
9. M. Inokuti, Rev. Mod. Phys., 1971, **43**, 297.
10. M. E. Riley, C. J. MacCallum and F. Biggs, At. Data and Nucl. Data Tabl., 1975, **15**, 443.

A TRACK STRUCTURE MODEL BASED ON MEASUREMENTS OF RADIAL DOSE DISTRIBUTION AROUND AN ENERGETIC HEAVY ION

S. Ohno, K. Furukawa*, H. Namba**, M. Taguchi**, and R. Watanabe**

Inst. R & D, Tokai University, 2-28 Tomigaya, Shibuya-ku, Tokyo 151, Jpn
*Dept. Chem. & Fuel Res., JAERI, Tokai-mura, Ibaraki 319-11, **TRCRE, JAERI, Takasaki 370-12, Jpn

1 INTRODUCTION

In order to know basic radiation actions due to passage of particles of different mass, charges and energies, it is very important to have a good understanding of the spatial distribution of energy deposition in matter around the particle path. The track structure model developed by Katz and co-workers,[1-3] which is based on a number of simplifying assumptions, has been most successful in correlating the yield-dose curve and the LET-value. The model was constructed by considering the average radial dose around the particle path solely due to delta-rays. Monte Carlo[4,5] and semi-empirical[6,7] calculations have been made, but it should be remembered that the pattern of energy deposition is necessarily complex and the reliable cross section data, *e.g.* doubly differential cross section in energy and angle for the electron ejection, needed for the calculation are very limited.

Experimentally, the radial doses have been measured in gases using an ionization chamber method. This method uses a collimated beam of energetic particles, thereby ionization currents produced in a small mesh-wall ionization chamber located at varying distances from the beam path being measured. Measurements have been made for 1 and 3 MeV H and He in tissue-equivalent gas (TE)[10], 38.4[11] and 41[14] MeV O, 33 and 61.9 MeV I[11], 42 MeV Br[12] in TE as well as for fast particles such as 930 MeV He[15] and 377 MeV/n Ne[9] in air. But data are rather fragmentary. We think it worth while to add new measurements to get a more systematic view of the radial dose distribution.

The purpose of this paper is to present our results for 200 MeV Ni in Ar at variable pressures. We aim at obtaining data over as wide radial distribution as possible and discuss them together with those in the literature to get, if possible, a unified view of heavy ion track structure.

2 EXPERIMENTAL

Principle of the measurement is the same as reported by Wingate and Baum.[10] A collimated beam of 200 MeV Ni^{12+} ions from the JAERI-Tokai Tandem accelerator is introduced in a cylindrical vessel through a set of apertures (diameters: 0.1 and 0.5 mm) which serves also for differential pumping system. The cylindrical vessel, (diameter: 300mm; length: 1000mm) was filled with Ar gas at the pressure which ranged from 0.007

to 40 Torr and was automatically controlled by a leak valve (MKS 248K, Baratron). The wall-less ionization chamber consists of two Ni-mesh cylinders and a central wire each supported by a Teflon insulator. Dimension of the inner cylinder is 15mm in diameter and 85mm in length, and the outer cylinder serves as a shield and support. The radial distance from the beam track was adjusted by changing either the gas pressure or the radial position (35-70mm; by remote control) of the ionization chamber. Other details have been published.[8]

Dose in eV/g at the simulated radial distance t (to d_1: the density 1 g/cm^3) is given by

$$D(t) = (I_p / f I_i)\,(Z\,W_e / (v\,d_{gas}))\,(d_1/d_{gas})^2 \quad (1)$$

where Z is the charge of the incident ion, W_e is the W-value of Ar for electron (=26.2eV), v and f are the volume and transparency of the mesh-wall chamber(15.8cm^3, 52%), respectively, and I_p and I_i are the ionization current in the mesh-chamber and the incident ion beam current, respectively. I_p was obtained from the saturation curve (current *vs.* applied voltage) and I_i was obtained using a Faraday gauge before and after each ionization measurement.

3 RESULTS AND DISCUSSION

3.1 Results for 200 MeV Ni into Ar

The present results are shown in **Figure 1** as a function of simulated radial distance t which is calculated as the measured distance $\times\, d_{gas}/d_1$. The results indicate that the radial dose $D(t)$, as t increases, first decreases less steeply than the usual t^{-2} dependence and finally it shows t^{-3} dependence at larger distances. It is noted that the energy deposition is observed beyond the maximum range of the electron as calculated by Katz's method,[3] *i.e*, T_{max}=1740nm (See **Table 1**).

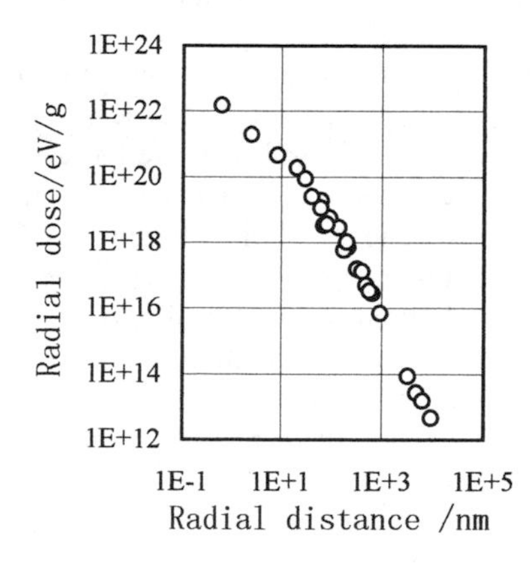

Figure 1 *Radial dose distribution in Ar traversed by 200 MeV Ni^{12+} ion*

3.2 A Scaled Radial Dose Curve Including the Available Data in the Literature

Table 1 includes the available data of radial dose measurements. Z^*, the effective charge of the ion, is calculated using Barkas equation:

Table 1 *Available Data of Radial Dose Measurements*

Ion	*Energy*/ MeV	β	Z^*	*Gas*	*Measured range*/ nm	T_{max}/ nm	*Reference*
H	1	0.046	1	TE	0.8 - 100	217	10
H	1	0.0	1	TE	0.8 - 80	1380	10
He	1	0.023	1.67	TE	0.6 - 30	30.9	10
He	3	0.04	1.91	TE	0.7 -50	135	10
He	930	0.6	2	Air	10^5 - 4 10^5	2.4 10^6	15
O	38.4	0.071	7.2	TE	0.2 - 140	940	11
O	41.1	0.074	7.21	N_2	0.6 - 280	1080	14
Ne	7540	0.702	10	Air	0.6 -10^6	5.94 10^6	9
Ni	200	0.086	19.2	Ar	0.6 - 9.4 10^3	1740	This work
Br	42	0.034	11.35	TE	0.5 - 60	75.7	12
I	33.25	0.024	9.3	TE	0.3 - 140	32.3	11
I	61.9	0.032	13.2	TE	0.3 - 110	67.2	11
U	1404	0.112	45.7	CH_4	33 - 100	4320	13

$$Z^* = Z\,[1 - \exp(-125\ \beta\ Z^{-2/3})],\quad \beta = v/c \tag{2}$$

T_{max} is the expected range in the liquid water of the electron of the kinematically limited (for free electrons only) maximum energy W and calculated using equations given by Katz *et al.*[3]

$$T_{max} = 6 \times 10^{-6}\ \mathrm{W}^{\ \alpha}\ \ \mathrm{g\,cm}^{-2} \tag{3}$$

$$\mathrm{W\,(keV)} = 2\ \mathrm{m\,c}^2\ \beta^{\ 2} / (1 - \beta^{\ 2}) \tag{4}$$

$$(\alpha = 1.667 \text{ if } \beta > 0.03;\ \ \alpha = 1.079 \text{ if } \beta < 0.03)$$

In the table the range of distances where the measurements were done is also shown. It should be noticed that only in limited cases, i.e. I and Ni, measurements have been made beyond the range of T_{max}.

The radial dose is generally a function of t, β and Z^*. If we assume that all the energy lost by the energetic particle is transferred to the medium and measured through ionization chamber method, the LET (= L) is described as:

$$L = \int_0^{\infty} D(t, \beta, Z^*)\, 2\, \pi\, t\, \mathrm{d}t \tag{5}$$

On the other hand, Bethe's stopping power is proportional to $4\ \pi\ e^4\ Z^{*2} / m_e\ c^2\ \beta^{\ 2}$. From this we can expect that $(\beta / Z^*)^2\, D(t, \beta, Z^*)$ is a function of t only, f(t). In **Figure 3** are plotted $(\beta / Z^*)^2\, D(t, \beta, Z^*)$ *vs* t. We can see that all the data are described roughly (the factor of 2) as $f(t) \fallingdotseq (1\sim2)\ 10^2\ t^{-2}$ within sufficiently well below the range of T_{max}. The Ni, I, Br, and He(1 MeV) cases show t^{-3} dependence at near T_{max}.

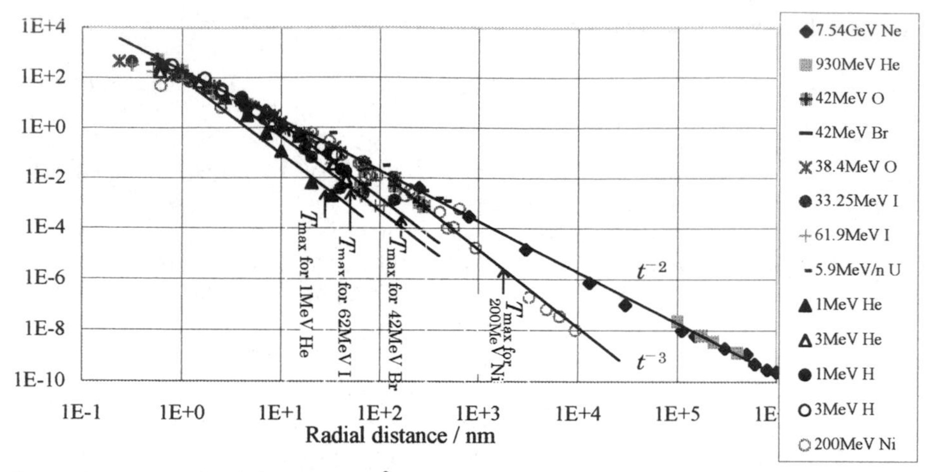

Figure 2 *Normalized dose* $(\beta / Z^*)^2 D(t, \beta, Z^*)$ */Gy of available data plotted as a function of simulated radial distance*

3.3 Conclusion

We have arrived at unified radial dose distribution using the unit $(\beta /Z^*)^2$ for all energetic particles. Thus, we may expect to use an approximation that L be equated to $(Z^*/\beta)^2(1{\sim}2)10^2 t^{-2}\, 2\pi\, t\, \mathrm{d}t$ integrated with respect to t as in Eq.5 and to proceed theoretical estimation of the yields using spatial dose distribution due to passage of different particles. But before this can be realized, we must clarify two points: dose distribution at t less than 0.5nm which closely relates to the so-called core-penumbra problem and how to know the transition point at which t^{-2}- changes to t^{-3}-distribution.

References

1. J. J. Butts and R. Katz, *Radiat. Res.*, 1967, **30**, 855.
2. C. Zhang, D.E. Dunn, and R.Katz, *Radiat. Protec. Dosim.*,1985, **13**, 215.
3. M. P. R. Waligorski, R. N. Hamm, and R. Katz, *Nucl. Tracks Radiat. Meas.*, 1986, **11**, 309.
4. R. N. Hamm, J. E. Turner, R. H. Ritchie, H. A. Wright, *Radiat. Res.*, 1986, **104**, s20.
5. M. Krämer and G. Kraft, *Radiat. Environ. Biophys.*, 1994, **33**, 91.
6. A. Chatterjee and W. R. Holley, *Int. J. Quantum Chem.*, 1991, **39**, 709.
7. J. Fain, M. Monnin, and M. Montret, *Radiat. Res.*, 1974, **57**, 379.
8. K. Furukawa, S. Ohno, H. Namba, and M. Taguchi, *JAERI M*, 1992, **91-170**, 61.
9. M. N. Varma and J. W. Baum, *Radiat. Res.*, 1986, **81**, 355.
10. C. L. Wingate and J. W. Baum, *Radiat. Res.*, 1976, **65**, 1.
11. M. N. Varma, J. W. Baum, and A. V. Kuehner, *Radiat. Res.*, 1975, **62**, 1.
12. M. N. Varma, J. W. Baum, and A. V. Kuehner, *Phys. Med. Biol.*, 1980, **25**, 651.
13. L. H. Toburen, L. A. Braby, N. F. Metting, G. Kraft, M. Scholz, F. Kraske, H. Schmidt-Böcking, R. Dörner, and R. Seip, *Radiat. Protect. Dosim.*, 1990, **31**, 199.
14. M. N. Varma, J. W. Baum, and A. V. Kuehner, *Radiat. Res.*, 1977, **70**, 511.
15. M. N. Varma, H. G. Paretze, J. W. Baum, J. T. Lyman, and J. Howard, *5th Symp. Microdosim.*, ed. by J. Booz, CEC, 1976, p.75.

RADIAL DOSE MODEL OF SSB, DSB, DELETIONS AND COMPARISONS TO MONTE-CARLO TRACK STRUCTURE SIMULATIONS

F. A. Cucinotta[1,2], H. Nikjoo[2], J. W. Wilson[1], R. Katz[3], and D. T. Goodhead[2]

[1]NASA, Langley Research Center, Hampton VA 23681-0001, USA
[2]MRC Radiation and Genome Stability Unit, Harwell, Didcot, OX11, 0RD, UK
[3]University of Nebraska, Lincoln NE, 68558, USA

1 INTRODUCTION

The initial lesions formed in DNA by ionizing radiation include base damage, single strand breaks (SSB), double strand breaks (DSB), DNA cross links, and deletions[1]. Deletions occur through energy deposition and perhaps more importantly through recombination repair[2] of DSB's. Several mechanisms for the formation of DSB's and deletions related to energy deposition can be considered. Track simulation codes have indicated the importance of clusters of ionizations in small volumes similar to the size of a nucleosome. These clusters have been related to several types of damage to DNA, including DSB and deletions resulting from multiple DSB's formed by single electron tracks[3–5]. The deletion size expected from clusters can be estimated at 2–100 bp as constrained by the wrapping of DNA about histones in the nucleosome and expected cluster regions of <5 nm. A second mechanism for deletion results from the higher order structure of DNA. Single ion tracks passing through cells will intersect several segments of DNA and deletions of kbp size as related to chromatin structure are expected and have recently been measured[6]. In heavy ion irradiation, the high densities of ionizations leads to the overlap of electron tracks suggesting an alternative mechanism for the formation of DSB's or deletions. For electron or photon irradiation, the contribution of electron overlap in causing DNA damage has been estimated to be small[4] at doses below 10^6 Gy. The radial distribution of dose from secondary electrons exceeds 10^6 Gy near an ions path and the lateral region of such energy deposition may extend to distances >100 nm for large ion charge suggesting an electron overlap contribution for formation of DSB's or deletions.

The radial dose model of track structure[7] considers the acute dose response of a biological system for energetic photons or electrons and the radial dose profile of ions to evaluate action cross sections for the same endpoint. This approach has been quite successful in fitting experimental data for inactivation and mutation by protons and heavy ions[7,8]. In this paper we discuss calculations of strand break and deletion formation using the radial dose model of track structure. The radial dose model is limited to the prediction of average quantities based on measurements for energetic photons or electrons. Such measurements exist for yields of SSB and DSB, and for limited information on the size distribution of large DNA fragments. Good predictions of SSB and DSB cross sections for SV-40 virus in EO buffer were found by Katz and Wesley[9] using this approach. The measurement of deletions as caused by energy deposition has proven difficult, therefore excluding the mapping procedure used in the radial dose model to make prediction of deletions from ions. In order to make estimates of deletion cross sections, we consider the results of Monte-Carlo track simulations for energetic electrons[3–5] to estimate the probability of ionization clusters including 2 DSB's within a small volume of the size of a nucleosome. In contrast to the radial dose approach, track simulations make detailed considerations of energy depositions in

DNA such as the stochastics of ionization events and the dependence on the secondary electron spectrum. Comparisons are made to measurements of RBE's for SSB and DSB, and to track simulation results for ions.

2 CROSS SECTIONS FOR DNA DAMAGE

The induction of SSB's is observed to increase linearly with dose for all radiation types. The cross section for ions is then modelled as a one-hit process as given by

$$\sigma_{SSB} = 2\pi \int t\,dt\,(1 - \exp(-D(t)/D_{SSB})) \tag{1}$$

where t is the impact parameter of the ion, D(t) is the radial dose of the ion, and D_{SSB} the D_{37} dose for the induction of SSB by X-rays[10]. The induction of DSB is also observed to increase linearly with dose for all radiation types. Because of the large ionization density at small t, we consider 2 mechanisms for the production of DSB by ions. In the first, clusters of ionizations from single electron tracks in a volume similar to a nucleosome lead directly to DSB's. The second mechanism considers the role of overlapping electron tracks near to the path of an ion by folding the probability function for SSB's using an inter-separation of up to 10 bp for the two SSB's. The cross section for DSB production is then written as

$$\sigma_{DSB} = \sigma_{DSB\text{–}clusters} + \sigma_{DSB\text{–}overlap} \tag{2}$$

where

$$\sigma_{DSB\text{–}cluster} = 2\pi \int t\,dt\,(1 - \exp(-D(t)/D_{DSB})) \tag{2a}$$

$$\sigma_{DSB\text{–}overlap} = 2\pi \int t\,dt\,(1 - \exp(-D(t)/D_{DSB}))\,(1/V) \int dr\,(1 - \exp(-D(t\text{–}r)/D_{SSB}) \tag{2b}$$

The cross sections for production of several breaks in the sugar-phosphate backbone of DNA could be defined if the related D_{37} dose were known. In order to investigate these effects in the radial dose model we consider the results of track simulations for electrons.

In the track simulation approach[3–5], the yield of DNA breaks is evaluated by relating the total energy deposited in DNA segments to the number of breaks of various types. A volume model of DNA is used which considers the volume of sugar-phosphate moieties and their rotation about histones. SSB formation is assumed to occur if energy deposition in the sugar-phosphate volume above a threshold value (~17.5 eV) occurs. Higher-order damage as determined by the occurrence of one or more SSB's on the same or opposite strands in various combinations are also scored. More recent calculations consider the early chemistry of water radicals[5]. The track simulation approach by considering the stochastics of ionization and excitation events is also able to consider the frequency distribution of breaks along DNA. Details of the model are given in ref. 3–5.

We assume that cross sections for these various types of damage have two contributions from clusters of ionizations in small volumes and from electron overlap in the ion's track. Cross sections for 2 DSB's within 10bp (denoted DSB^{++}) are found in a similar fashion to that of eq. (2) with D_{37} doses based on yields of SSB, DSB, and DSB^{++} for electrons from the track simulation model, scaled to D_{37} measurements[10] for SSB as shown in Table 1.

Table 1. *D_{37} Values for Strand Break Induction by Energetic Electrons*

D_{SSB}, Gy	D_{DSB}, Gy	D_{DSB++}, Gy
5	75	2200

3 RESULTS AND DISCUSSION

Comparisons for yields of strand breaks are shown in Figure 1. The radial dose model predicts lower yields for induction of SSB's as compared to the track simulation results with the differences greater for LET values corresponding to energies below 1 MeV/u. For DSB, contributions from the one-track mechanism from clusters in the radial dose model contributes about 1/2 of the yield at high energies, but is overcome by the electron overlap terms at low energies where the density of the track increases. The RBE in the radial dose model is general less than 1 when a one-hit mechanism is assumed which results from wastage of energy or over-kill effect. Both models predict a drop in break induction for energies below about 0.5 MeV/u, however in the radial dose model this drop is more rapid. In this region the maximum range of the secondary electrons falls below 10 nm and the effects of scaling all electrons to X-rays may become less appropriate[4]. Also, in the radial dose approach contributions from excitations of DNA are not considered. In the track simulation, excitations and ionizations are considered on equal footing in evaluating energy deposition. In Figure 2 we show comparisons to experiments[10–11] for RBE's for SSB in mammalian cells. There is some under-estimation of the RBE for He, while good agreement is found for the higher charged ions. The large decrease in RBE at high LET's for Ne, Ar, and U is reproduced by the model and occurs due to the decreasing range of the secondary electrons. In Figure 3 RBE's for DSB and small deletions in the models and experiments[10–13] are compared. The contribution from electron overlap in the radial dose model leads to RBE's above unity for lower charges in agreement with experiments. For higher charge ions, the RBE approaches unity at low LET and below unity at higher values. For the predictions of RBE's for small deletions, the radial dose model predicts RBE's greater than unity for all charges as electron contributions becomes dominate over the one track mechanism. For higher-order damage such as several DSB's or large deletions higher RBE's would be expected than found here for small deletions (DSB^{++}).

The radial dose model predicts a dominant role for electron overlap for ions, especially as the severity of damage increases, corresponding to larger volumes of energy deposition. The differences between the track models considered for low energy H and He results from the use of a stochastic approach in the track simulation model, including its treatment of short-ranged electrons and the description of excitations by ions. It is expected that continued analysis of the models and new measurements will be required to understand the role of each of these factors.

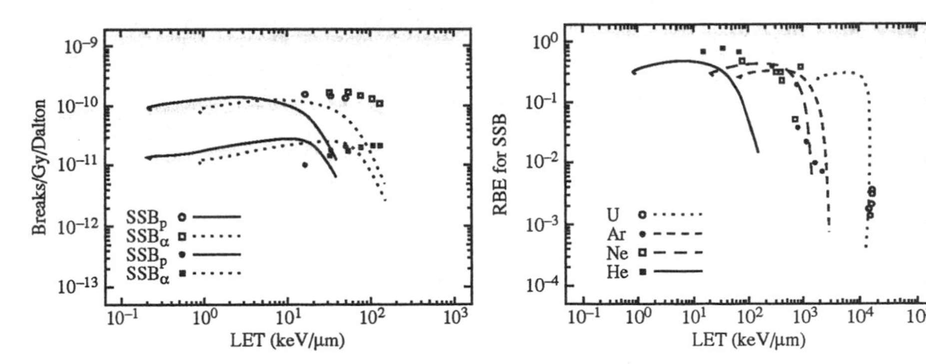

Figure 1. *Calculations of yields for strand breaks. Lines are radial dose and symbols track simulation model.*

Figure 2. *Comparisons of RBE's for SSB's in model radial dose model to experiments.*

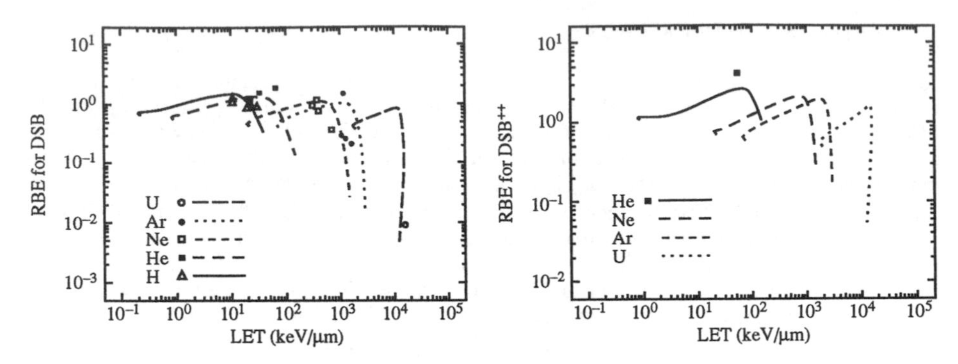

Figure 3. *Comparison of RBE's for DSB and DSB++ small deletions in radial dose model to experiments and result of the track simulation model for small deletions for He.*

4 ACKNOWLEDGEMENTS

This work was partially supported by NASA and EC Contract F14P-CT95-0011.

5 REFERENCES

1. D. T. Goodhead, *Int. J. Radiat. Biol.*, 1989, **56**, 623.
2. E. C. Friedberg, G.C. Walker, and W. Siede, DNA Repair and Mutagenesis, ASM Press, Washington D.C., 1995.
3. D. E. Charlton, H. Nikjoo, and J. L. Humm, *Int. J. Radiat. Biol.*, 1899, **56**, 1.
4. D. T. Goodhead and H. Nikjoo, *Int. J. Radiat. Biol.*, 1989, **55**, 513.
5. H. Nikjoo, et al., *Int. J. Radiat. Biol.*, 1994, **66**, 453.
6. B. Rydberg, *Radiat. Res.*, 1996, **145**, 200.
7. R. Katz, et al., *Radiat. Res.*, 1994, **140**, 402.
8. R. Katz and S. Wesely, *Radiat. Environ. Biophys.*, 1991, **30**, 81.
9. F. A. Cucinotta, et al., *Int. J. Radiat. Biol.*, 1996, **69**, 593.
10. J. Heilmann, H. Rink, G. Taucher-Scholz, and G. Kraft, *Radiat. Res.*, 1993, **135**, 46.
11. G. Kampf and K. Eichorn, *Studia Biophys.*, 1983, **1**, 17.
12. M. Belli, et al., *Int. J. Radiat. Biol.*, 1994, **65**, 529.
13. T. J. Jenner, et al., *Int. J. Radiat. Biol.*, 1992, **61**, 631.

NEW MONTE CARLO METHODS FOR EVALUATION OF CHARACTERISTICS OF NONADDITIVE RADIATION ACTION ON MICROTARGETS

A. V. Lappa

Chelyabinsk State University. Br.Kashirinykh, 129, Chelyabinsk, 454136, Russia
Chelyabinsk State Institute of Laser Surgery. Pr.Pobedy 287,Chelyabinsk,454026, Russia

1 INTRODUCTION

Detailed simulation of particle tracks is used in two approaches for evaluation of characteristics of radiation action on small sensitive targets. One is based on the idea of random superimposition of a target on previously simulated tracks; it includes the associated volume method and its (often far-reaching) modifications.[1,2] The other is based on exact mathematical models of statistical radiation transport theory, it includes two FD (Fluctuation Detector) methods implemented in the TRION code.[3,4] The second approach has a number of advantages: formulations of the methods for different types of radiation are identical, algorithms for track scoring are simpler and the processing is simultaneous with track simulation, the methods do not require storing a great amount of intermediate information. Due to these advantages TRION is a compact and fast code working successfully on personal computers.

The characteristics used for deterministic description of physical action of radiation on small targets are probabilistic characteristics (distributions, moments, and others) of certain stochastic responses of irradiated system to such action. A disadvantage of the FD methods is that the class of admissible responses is somewhat narrower than in the first approach. This class includes only responses of scalar additive type: deposited energy, number of ionisations or number of collisions of any other type. Joint distributions and mixed moments of several responses for one or several targets, and other characteristics associated with non-additive or vector responses can not be calculated by the FD methods. The disadvantage is to a large extent overcome in two generalizations of FD methods presented below.

2 METHODS

2.1 Purpose, Assumptions, Parameters

2.1.1 Purpose. Both methods may be used to calculate arbitrary probabilistic characteristics of an arbitrary response, scalar or vectorial, additive or non-additive. The target may be simple or composite (an aggregate of several subtargets, for example). Deposited energy Q in a subtarget, number of ionisations N in the same or another subtarget, the vector (Q,N) are examples of admissible responses; distribution and moments of Q or N, joint distribution and mixed moments of Q and N are examples of calculable characteristics. All characteristics are calculated per particle emitted by an external source or per event of interest (energy absorption, ionisation, and other events).

In general, the methods allow us to calculate the mean (expectation) of a response of a certain class to a track of a particle emitted by a source. The class of admissible responses includes those that are (i) responsive only to particle collisions in a small space region (target), (ii) independent of times of these collisions, and (iii) equal to zero if there is no collisions. We shall call them *collisional local stationary* responses. Any such response is simply a symmetric function of phase coordinates (excluding time) of all collisions of a track in a given target. Any sum of independent contributions from collisions is additive. Therefore, the responses Q and N are additive, and

$$Q^n,\ N^n\,(n>1),\ \mathbf{I}_A(Q),\ \mathbf{I}_B(N),\ Q^nN^m,\ \mathbf{I}_A(Q)\mathbf{I}_B(N)\ \left(\mathbf{I}_C(\omega)\equiv\begin{cases}1, \omega\in C\\ 0, \omega\notin C\end{cases}\right) \qquad (1)$$

are non-additive, admissible if $A, B \in (0,\infty)$. The last two responses, which define mixed moments and joint distributions, can not be considered in the FD methods.

2.1.2 Assumptions. It is necessary that:

(*A*) the target can be covered by a convex domain whose volume is not very large in comparison with the volume of the target, such domain is called the site;

(*B*) the site is so small that the spatial variation of particle fluence over it is negligible.

It is desirable that

(*C*) one or both of the following hold: 1) the site can be chosen highly symmetric, 2) angle distribution of charged particle fluence at the target is highly symmetric.

2.1.3 Parameters. No fitted parameters are used in the methods. The calculation accuracy weakly depends on the minimal distance δ between the surface of the site and the target. The value $\delta=0$ may lead to an error usually not exceeding 5% (most often 1%). The error rapidly tends to zero as δ increases, it becomes practically negligible when δ is equal to the average target chord.

2.2 Mathematical Models

All non-analog Monte Carlo methods are based on mathematical models of the quantities to be calculated. Let φ be a stochastic response to a particle track and $\mathbf{M}\varphi$ be its mean per particle emitted by a source. The classical pair of linear forms, which in the stationary case are written as

$$\sum_{\alpha}\int \Phi_\alpha(x)R_\alpha(x)dx = \sum_{\alpha}\int \overline{\varphi}_\alpha(x)S_\alpha(x)dx\,, \qquad (2)$$

are adequate models for $\mathbf{M}\varphi$ only if φ is additive. Here and below α is a particle type; sums with the index α are taken over all particle types; $x\equiv(\mathbf{r},\mathbf{\Omega},E)$ is the particle position, direction of motion, and energy; $\int dx \equiv \iiint d\mathbf{r}d\mathbf{\Omega}dE$; $\Phi_\alpha(x)$, $S_\alpha(x)$ are the fluence and the source density of type α particles at a phase point x per emitted particle; $R_\alpha(x)$ is a 're-

sponse function' determined by the response φ; $\overline{\varphi}_\alpha(x) \equiv \mathbf{M}_{x,\alpha}\,\varphi$ is the 'importance function', i.e. the mean of φ for a track of a type α particle emitted at a phase point x.

Models of the characteristic $\mathbf{M}\varphi$ for non-additive φ are much more complex. In previous our works a set of general models were constructed including the two models on which the FD methods are based. In the most general case the models are constructed for a response φ that is a non-random function of an additive response.[5] The following two representations generalize these models to arbitrary collisional stationary response (locality is not required here).

2.2.1 Model 1. It holds that

$$\mathbf{M}\varphi = \sum_\alpha \int \Phi_\alpha(x)F_\alpha(x)dx\,,\quad F_\alpha(x)=\sigma_\alpha(x)\mathbf{M}^c_{x\alpha}\{\varphi - \sum_i \varphi_i\}, \tag{3}$$

where $\sigma_\alpha(x) = \sigma_\alpha(\mathbf{r}, E)$ is the total macroscopic cross section; $\mathbf{M}^c_{x\alpha}$ is the mean per track initiated by a collision of a particle with fixed pre-collision parameters x,α. The sum of φ_i is taken over all particles leaving this initial collision, φ_i is the value of the response for the track of the i-th particle (φ is the value for the whole track including the initial collision). Formulas (3) give the first form of (2) for an additive φ.

2.2.2 Model 2. Divide the irradiated object into two nonintersecting domains V and W with piece-wise smooth boundary V' between them, and divide the set of all particle types into two groups B and Γ. Let $\mathbf{n}_{V'}$ be the normal to V' external with respect to V, $\int dp \equiv \iint d\,\Omega dE$. Then it holds that

$$M\varphi = \sum_{\alpha\in B}\{\int_V d\mathbf{r}\int dp S_\alpha\overline{\varphi}_\alpha - \int_{V'} dV'\int dp(\mathbf{\Omega n}_{V'})\Phi_\alpha\overline{\varphi}_\alpha\} + \sum_\alpha\{\int_V d\mathbf{r}\int dp\,\Phi_\alpha f_\alpha + \int_W d\mathbf{r}\int dp\,\Phi_\alpha F_\alpha\}, \tag{4}$$

where $f_\alpha(x) = \sigma_\alpha(x)\mathbf{M}^c_{x,\alpha}\xi_\alpha$, $\xi_\alpha \equiv \mathbf{I}_\Gamma(\alpha)\varphi - \sum_i \mathbf{I}_\Gamma(\alpha_i)\varphi_i$. This representation is new even for additive φ. In this case (4) gives the first form of (2) if $V=\varnothing$ or $B=\varnothing$, and the second form if $W=\varnothing$ and $\Gamma=\varnothing$.

2.3 Algorithms

A few two-step calculation algorithms follow immediately from each of the described models (the first step is calculation of either fluence Φ_α or functions $F_\alpha, \overline{\varphi}_\alpha, f_\alpha$). However, they can be practically used only if the fluence and/or the functions have some symmetry that reduces multiplicity of the integrals. Different symmetries and different choices of the domain V and the group B yield a number of two-step algorithms. Here we present two algorithms corresponding to the purpose and assumptions ***A,B,C*** given in 2.1. The symmetry is spherical for the site or isotropic for the fluence; V is the domain occupied by the site, B is the group of charged particles (Γ is the neutral group).

2.3.1 Algorithm 1. It follows from model 1 and generalizes the FD-1 method. At the first step actual spectra of particles of all types in the site, $\widetilde{\Phi}_\alpha(E) = \int \Phi_\alpha(\mathbf{r},\Omega,E)d\Omega$, $\mathbf{r}\in V$, are calculated and stored (that is a standard radiation transport problem, especially simple

in the case of electron equilibrium). At the second step the quantity of interest is calculated in accordance with the probabilistic representation

$$\mathbf{M}\varphi = |V| \sum_{\alpha} \|\tilde{\Phi}_{\alpha}\| \mathbf{M}^{c}_{\alpha V}\{\sigma_{\alpha}(\mathbf{r}_0, E_0)(\varphi - \sum_{i} \varphi_i)\}, \tag{5}$$

where $|V|$ is the volume of V, $\|\tilde{\Phi}_{\alpha}\| \equiv \int \tilde{\Phi}_{\alpha}(E)dE$. The means $\mathbf{M}^{c}_{\alpha V}$ are estimated on tracks initiated by a collision of type α particle with random direction Ω_0 and energy E_0 at a random point $\mathbf{r}_0$. These random variables are independent and their distributions are respectively isotropic, proportional to $\tilde{\Phi}_{\alpha}$, and uniform over V.

2.3.2 Algorithm 2. It follows from model 2 and generalizes the FD-2 method. The first step is the same as in algorithm 1. At the second step the quantity of interest is calculated in accordance with the representation

$$\mathbf{M}\varphi = |V| \sum_{\alpha} \|\tilde{\Phi}_{\alpha}\| \mathbf{M}^{c}_{\alpha V}\sigma_{\alpha}(\mathbf{r}_0, E_0)\xi_{\alpha} - \sum_{\alpha \in B} \{|V'| \|\tilde{\Phi}_{\alpha}\| \mathbf{M}_{\alpha,V'}(\Omega_0 \mathbf{n}_{V'}(\mathbf{r}_0))\varphi + |V| \|\tilde{S}_{\alpha}\| \mathbf{M}_{\alpha,V}\varphi\}, \tag{6}$$

where $|V'|$ is the area of V', $\|\tilde{S}_{\alpha}\| \equiv \int \tilde{S}_{\alpha}(E)dE$, $\tilde{S}_{\alpha}(E) = \int S_{\alpha}(\mathbf{r}, \Omega, E)d\Omega$, $\mathbf{r} \in V$. Symbols $\mathbf{M}_{\alpha,V}$, $\mathbf{M}_{\alpha,V'}$ denote averaging on tracks of two kinds, initiated respectively by emission of type α particles from the interior of the site V and from its boundary V'. Parameters $\mathbf{r}_0, \Omega_0, E_0$ of the particles at the moment of emission are random; Ω_0 is distributed isotropically, $\mathbf{r}_0$ uniformly (over V or V'), E_0 proportionally to $\tilde{S}_{\alpha}$ in the first case and to $\tilde{\Phi}_{\alpha}$ in the second.

2.3.3 Examples. The algorithms applied to responses (1) calculate, respectively, the higher moments of deposited energy Q and number of ionisations N in a given target, the probabilities $\mathbf{P}\{Q \in A\}$ and $\mathbf{P}\{N \in B\}$, the mixed moments of Q and N, joint probability $\mathbf{P}\{Q \in A, N \in B\}$. Dividing these characteristics by the probability of the energy absorption event $\mathbf{P}\{Q \in (0, \infty)\}$ or the ionisation event $\mathbf{P}\{N \in (0, \infty)\}$ we normalize calculated characteristics to one event. In this case the functions $\tilde{\Phi}_{\alpha}, \tilde{S}_{\alpha}$ in (5), (6) can be used in any normalization.

References

1. H.Nikjoo, D.T.Goodhead, D.E.Charlton, H.G.Paretzke, 'Energy Deposition by Monoenergetic Electrons in Cylindrical Volumes', MRC, Radiobiology Unit, Monograph 94/1, Chilton, U.K., 1994.
2. P. Olko, J. Booz, *Radiat. Environ. Biophys.*, 1990, **29**, 1.
3. A.V. Lappa, '5th All-Union Symp. on Microdosimetry', MIFI, Moscow, 1987, Vol.2, p.100 (In Russian).
4. A.V. Lappa, E.A. Bigildeev, D.S. Burmistrov, O.N. Vasilyev, *Radiat. Environ. Biophys.*, 1993, **32**, 1.
5. A.V. Lappa, 'The Monte Carlo Methods in Computational Mathematics and Mathematical Physics'. SOAN, Novosibirsk, Russia, 1991, p.73 (In Russian).

SIMULATION OF STRAND BREAKS AND SHORT DNA FRAGMENTS IN THE BIOPHYSICAL MODEL PARTRAC

W. Friedland, P. Jacob, H.G. Paretzke, M. Perzl and T. Stork

GSF - National Research Center for Environment and Health
Institute of Radiation Protection
85764 Neuherberg
Germany

1 INTRODUCTION

The formation of small single-stranded (SS-) and double-stranded (DS-) DNA-fragments resulting from pairs of single strand breaks (SSBs) and double strand breaks (DSBs) has recently attracted attention[1] in biophysical modelling, esp. since the experimental detection of such fragments has been reported[2]. On the other hand, the structure of the chromatin fibre is still a contentious issue[3]. The present paper is focused on simulation calculations of SSBs, DSBs and the formation of small SS- and DS-fragments due to low-LET radiation using different chromatin fibre structures. To this end, the biophysical model PARTRAC[4] has been improved to cope with higher-order DNA target structures. The simulation model results are discussed with the background of the experimental data.

2 METHODS

2.1 Modelling of Particle Tracks

In PARTRAC, the interaction of photons with biological matter is simulated considering all relevant processes using a mixture of cross section of the elemental composition of the materials considered. The Monte Carlo simulation of ionisations and excitations due to primary and secondary electrons is based on cross sections for water vapour[5] scaled to the density of the cell nucleus and the surrounding plasma inside a drop of blood. The resulting tracks inside the nucleus are overlaid with the DNA target model.

2.2 Modelling of the DNA Target

The DNA target model developed for the application in PARTRAC includes five levels of DNA organisation (nucleotide pair, double helix, nucleosome, chromatin fibre structure and chromatin fibre loop). The structure of the chromatin fibre is defined by a set of parameters, esp. fibre radius, linker DNA length, tilt angle, pitch and angle between succeeding nucleosomes. A stochastic arrangement of nucleosomes is established in the model by replacing the fixed values of some parameters by appropriate ranges of values[6]. In **Figure 1**, chromatin fibres with a solenoidal, a crossed-linker and a stochastic structure are presented. In order to describe chromatin fibre loops, two curved fibre elements are introduced in addition to the linear element which can be stacked together with smooth connections of all lower-order DNA structures. Thus, similar fibre loops with about 85 kbp are constructed for the three fibre structures. The total DNA inside a cell nucleus ($3.6 \cdot 10^{12}$ Dalton) is represented by about 70,000 identical fibre loops randomly distributed inside the cell nucleus.

Figure 1 *Solenoidal, crossed linker and stochastic structure of chromatin fibers*

2.3 Modelling of Strand Breakage and Fragments

The DNA strand breakage model is adapted to SSB and DSB yields for mammalian cells given in the literature[7,8]. Parameters in this adaptation are the DNA groups considered as target, the atomic radii and the minimum deposited energy needed for SSB induction, and the distance between strand breaks for DSB induction. Pairs of succeeding strand breaks on the same strand resulting from the same primary particle on the same fibre loop are assumed to produce SS-fragments, pairs of succeeding DSBs result in DS-fragments regardless of any SSBs in-between.

2.4 Simulation calculations

Electron radiation fields were simulated by tracks of electrons starting with set of energies between 0.1 and 100 keV in random directions at random positions in the cell yielding a homogeneous slowing down spectrum of electron energies in the target region. Additionally, secondary electrons from a parallel radiation field of 220 kVp x-rays incident on a layer of blood containing target cells were simulated. The simulations were performed in a geometry of three concentric homogeneous spheres describing a cell (10 μm ∅) with a nucleus (5 μm ∅) inside a drop of blood (0.4 mm ∅). In the simulations, a dose of 1 Gy corresponding to an energy of 552 keV deposited per cell nucleus was applied to 100 or 1000 cells. The three chromatin fibre structures displayed above were considered.

3 RESULTS

3.1 Single and Double Strand Breaks

Calculated numbers of SSBs and DSBs per cell nucleus and per Gy resulting from the 220 kVp x-ray simulation are listed in **Table 1** for different assumptions for strand break induction. The data using all strand atoms with twice the van-der-Waals radii as target with 10 eV threshold energy and 3 bp maximum distance for DSBs is in good agreement with experimental data[7,8] for both the SSB and the DSB yields, therefore, further simulation results are based on these assumptions for strand breakage. The energy and fibre structure dependence of SSB and DSB yields is given in **Figure 2**. Both yields do not depend on the fibre structure.

3.2 Yields of Single- and Double-Stranded DNA-Fragments

The yields of small SS- and DS-fragments for different energies and fibre structures are given in **Figure 3** together with experimental data[2] for 220 kVp x-rays. Simulation results agree for SS-fragments but are lower by a factor of 3 for DS-fragments. Below 1 keV the results for the solenoidal structure are constantly higher than for the other

Table 1 *Yields of SSBs and DSBs per nucleus in model calculations and measured data*
[1]) S: sugar group atoms; PS: phosphate group and sugar group atoms
[2]) 1: single van-der-Waals radius; 2: double van-der-Waals radius

Target Atoms[1]	Atomic Radius[2]	Threshold Energy [eV]	Break distance for DSBs [bp]	SSBs [Gy^{-1}]	DSBs [Gy^{-1}]	Ratio SSB/DSB
S	1	0	10	368	16	22
S	2	0	10	714	66	11
PS	1	0	10	542	31	17
PS	2	0	10	1140	130	9
PS	2	10	10	952	84	11
PS	2	0	3	1280	59	22
PS	2	10	3	1045	37	28
Experimental Data, Mean				1200	34	35
Experimental Data, Range				900 -1750	27 - 46	20 - 65

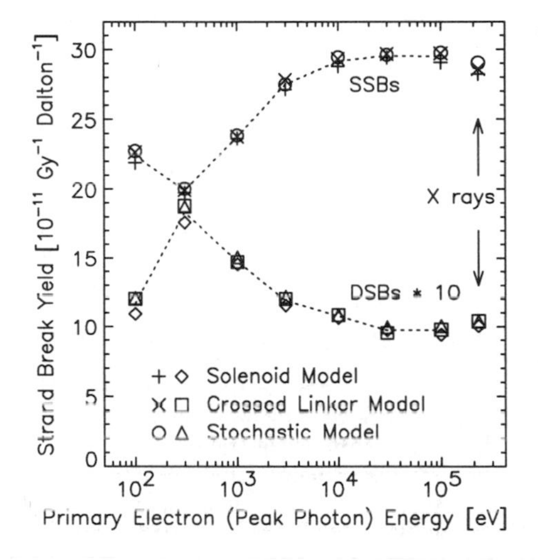

Figure 2 *Yields of SSBs and DSBs*

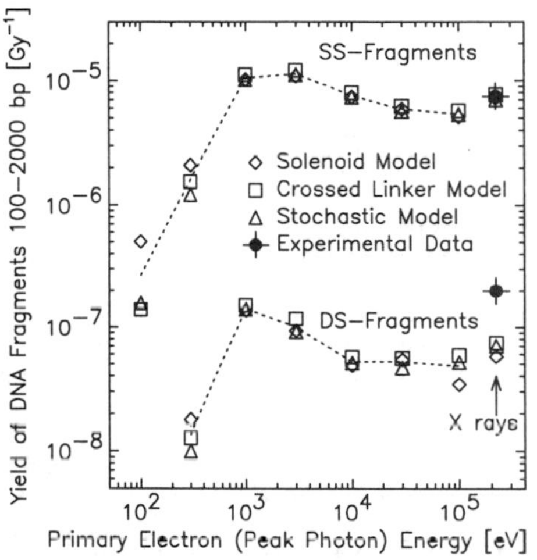

Figure 3 *Yields of small single- and double-stranded DNA fragments*

structures whereas for higher energies the differences between fibre structures reflect mainly statistical fluctuations.

3.3 Size Distributions of Single- and Double-Stranded DNA-Fragments

In **Figure 4**, size spectra of small DNA fragments simulated with crossed linker fibre structure are given as mass distributions. The spectra of SS-fragments are similar to DS-fragment spectra, both exhibit only a weak dependence on the energy beyond 1 keV. However, a strong dependence on the fibre structure is seen in **Figure 5**, in which smoothed fragment number distributions are displayed together with measured data[2]. The results for the solenoidal structure are similar to reported model results[1] being not consistent with the experimental data for high-LET radiation. The model results obtained using crossed linker structure show a different distribution of peaks, however, with no correspondence in the measured data. Better agreement is found for the stochastic model with respect to the lack of pronounced peaks, but the higher fragment yield around 0.3 kbp visible in the experimental spectrum is not reproduced in the simulation.

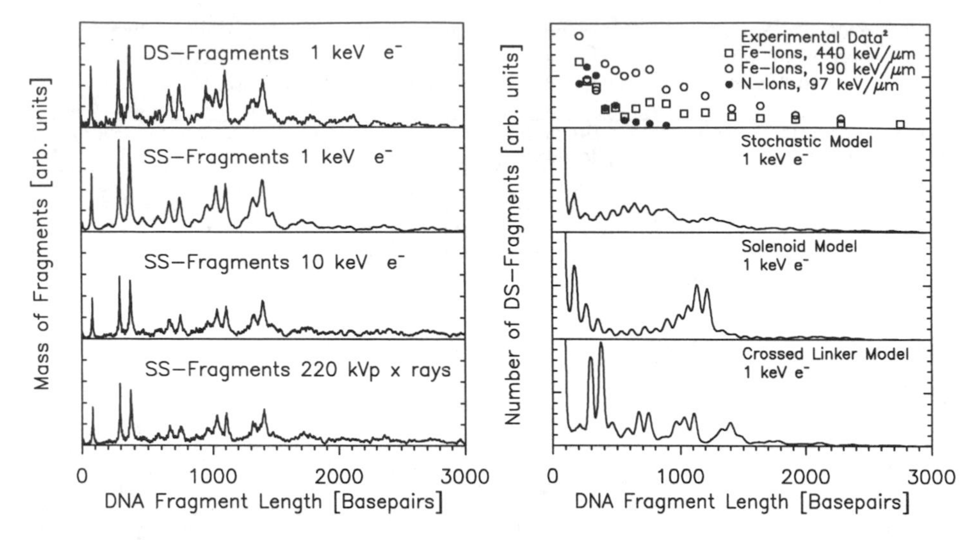

Figure 4 *Fragment mass spectra for the crossed linker model*

Figure 5 *Fragment number spectra for different chromatin fibre structures*

4 DISCUSSION

With a strand break model adapted to experimental data for strand break induction the yields of small fragments are found to be consistent for SS-fragments, but are too small for DS-fragments in comparison with measured yields. However, on the basis of the entire fragment distribution the DSB induction rate is reported[9] to be higher by a factor of 1.8 than the results of conventional FAR and hybridisation assays. Calibrating the strand break model with this higher DSB rate removes the discrepancy between simulated DS-fragment yields and experimental results. Measured size distributions of small fragments are closer to simulations using a stochastic structure than a regular solenoidal or crossed linker structure. A rather consistent simulation can be expected using a sophisticated choice of model parameters for a stochastic chromatin fibre structure.

ACKNOWLEDGEMENT

This work is supported by EC-Contract No. FI4P-CT95-0011.

REFERENCES

1. W.R. Holley and A. Chatterjee, *Radiat. Res.*, 1996, **145**, 188.
2. B. Rydberg, *Radiat. Res.*, 1996, **145**, 200.
3. K. van Holde and J. Zlatanova, *J. Biol. Chem.*, 1995, **270**, 8373.
4. S. Henß and H.G. Paretzke, in 'Biophysical Modelling of Radiation Effects', Adam Hilger, Bristol, 1992, p. 69.
5. H.G. Paretzke, in 'Kinetics of Nonhomogeneous Processes', Wiley, New York, 1987, p.89.
6 P. Jacob, W. Friedland, T. Stork, Proc. 27th Annual Meeting of the European Society for Radiation Biology, Montpellier, 1-4 Sept 1996.
7. G.L. Newton, J.A. Aguilera, J.F. Ward and R.C. Fahey, *Radiat. Res.*, 1996, **145**, 776.
8. M. Löbrich, S. Ikpeme and J. Kiefer, *Int. J. Rad. Biol,*. 1994, **65**, 623.
9. M. Löbrich, P.K. Cooper and B. Rydberg, *Int. J. Rad. Biol.*, 1996, in press.

MICRODOSIMETRIC DISTRIBUTIONS FOR TARGET VOLUMES OF COMPLEX TOPOLOGY

I.K. Khvostunov, S.G. Andreev*

Medical Radiological Research Centre, Korolev str.4, 249020 Obninsk, Kaluga Region
*Institute of Biochemical Physics, Kosygina str.4, 117977, Moscow, Russia

1 INTRODUCTION

Microdosimetric distributions can be sensitive to the shape and size of target volume. The sensitivity to the shape seems to be important for special type of targets like long flexible cylinders or polymers of given length (size). Biologically important example for that is a very long DNA polymer chain folded in a cell in various higher-order structures owing to DNA-protein interactions. It has been supposed that energy deposition distributions in DNA target may be sensitive to the DNA three-dimensional organization[1]. If so, this observation might have a radiobiological implication. Existing data suggest an important role of large-scale DNA structures in radiation damage[2] and cell death[3], but underlying mechanisms are still unclear. One of the mechanisms to explain different sensitivity of DNA in different topological states in the cell is the protective role of chromosomal proteins against OH radicals ("indirect" action)[2]. It has been shown by Monte Carlo simulation that under condition of inhibition of water radical diffusion ("direct" action) formation of DNA double strand break (dsb) clusters depends on overall DNA conformation[4]. To understand a role of target topology in radiation damage it is interesting to study the microdosimetric aspect of this problem. Several attempts were made in this direction. For spheres and thin cylinders of infinite length with the same mean chord length (2 nm) it was shown that probability distributions of the number of ionizations per energy deposition event are very similar[5]. Frequency of energy deposition in a short cylinders of different sizes was found to be uniquely different for high and low LET radiation[6]. Difference in microdosimetric quantities were observed for non-homogeneous model of nucleosome as compared with homogeneous one[7]. Models for more complex level of organization of biological target-DNA were developed recently[8-10]. This allows to study a sensitivity of microdosimetric distributions to changes of large-scale topology of target volume by means of variation of DNA packaging in higher-order chromatin structures.

2 METHODS

Different DNA target models - DNA double helix, decondensed and condensed chromatin fibre, looped and coiled domain were used. A volume model of DNA was applied for simulation of radiation damage in linear double helix segment[11]. Nucleosomes in chromatin fibre were arranged in condensed solenoid fibre of diameter 32 nm with 7.4 nucleosomes per turn[9] and in decondensed extended solenoid or zig-zag structure like "beads-on-a-string". The model of further folding of chromatin fibre in domain structure was described elsewhere[9]. The Monte Carlo track structure code DeTrack[12] was used to generate segments of heavy charge particle tracks in gas water approximation for unit density medium. The code follows straight line trajectory of heavy particle and individual histories of all secondary electrons until a lower electron cut-off energy of secondaries is reached (13.6 eV). Energy deposition events were scored by randomly superimposing target volumes on the individual tracks[6,11,13].

Sampling procedure included a calculation of coordinates of energy transfer points (ionisations and excitations) and selection of points inside both entire volume of DNA (see Figures 1-3) and sugar-phosphate subvolumes (see Figure 4). The coordinates and energy transfer of selected transfer points were recorded to estimate microdosimetric distributions. Calculation of distance density distribution between energy transfers in complex target was performed as follows. Cartesian coordinates of all energy transfer points within given sugar-phosphate subvolume of DNA (i.e. sugar-phosphate with nonzero energy imparted) were grouped. Then coordinates of this subvolume were recorded and they were referred further as a coordinates of single "point" of potential break $\boldsymbol{R}_i$. Next, distances x between any two "points" $\boldsymbol{R}_i$ and $\boldsymbol{R}_j$, $x = |\boldsymbol{R}_i - \boldsymbol{R}_j|$ were recorded for every event and frequency distribution $\psi(x)$ was calculated. $\psi(x)d^3x$ is the probability per single track of two "points" in target structure to be separated by distance x, x+dx in three dimensions. Distribution $\psi(x)$ calculated for complex target was compared with single track proximity function t(x)[14] in homogeneous medium. $t(x)d^3x$ is defined here as a probability of observing a distance x, x+dx in 3-D space between any two transfer points $\boldsymbol{r}_i$ and $\boldsymbol{r}_j$, $x = |\boldsymbol{r}_i - \boldsymbol{r}_j|$. t(x) was derived for track segment of 102 keV/μ m alpha particle. To compare both distributions they were normalized as follows,.$\int t(x)d^3x=1$ and.$\int \psi(x)d^3x=1$

3 RESULTS AND DISCUSSION

Comparison of microdosimetric single event spectra computed by DeTrack code with previous results[11] based on MOCA14 code is shown in Figure 1 for alpha particle with ≈ 100 keV/μm. Maximum difference between distributions generated by two track structure programs is not more than 25% in 0-40 eV interval.

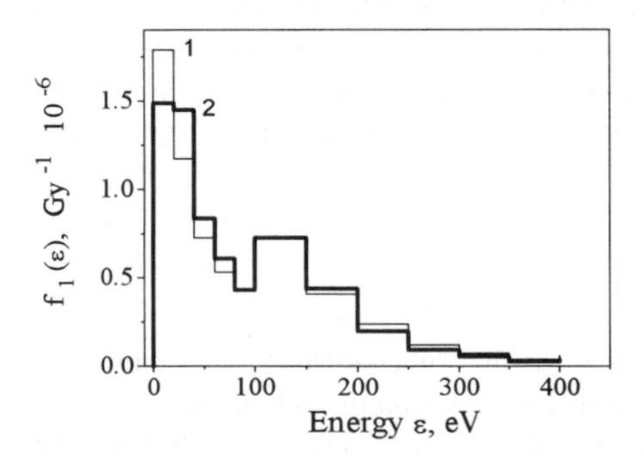

Figure 1 *Frequency distribution of event size per Gy per target for cylindrical volume similar to a linear DNA segment of 54 bp placed at random in water irradiated with alpha particle. (1) calculations from Charlton et al.[11], LET≈100 keV/μm, (2) calculation of this work, LET=102 keV/μm*

Integral single event frequency distributions F_1 (ε>E) of energy greater than a given amount E imparted in DNA within different higher-order structures are presented in Figure 2 for alpha particle track of 102 keV/μm. The variations of spectra are seen for 5.2 kbp DNA cylinder volume in linear, nucleosome chain and solenoid fibre conformations. Results for 50 kbp DNA folded in solenoid fibre, looped and coiled domain are presented in Figure 2b. Changes in distributions which are due to different DNA target sizes and are not the target shape are also shown in Figure 2a, curves 3,4.

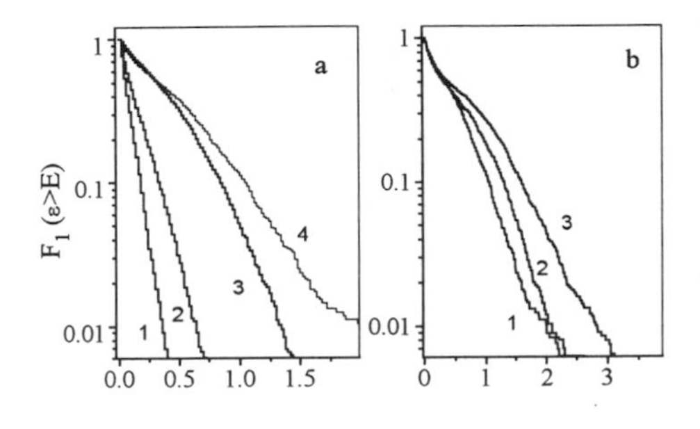

Figure 2 *Integral frequency distributions of energy imparted F_1 (ε>E) in DNA within a different higher-order structures irradiated with alpha particle track of 102 keV/μm.*
a *5.2 kbp (1) - linear segment, (2) - nucleosome chain, (3) - solenoid fibre. (4) - solenoid fibre of 50 kbp DNA.*
b *50 kbp (1) - solenoid fibre, (2) - looped domain, (3) - coiled domain.*

Probabilities of single hit in the target volume per unit dose were calculated for 5 kbp DNA as a linear cylinder and being folded in compact solenoidal fibre. Results presented in Figure 3 show that single hit probabilities differ in 1.5-50 times in the range of 5-10^4 keV/μm. Calculations were carried out for different types of ions - alpha particles, Ne, Ar, Xe. Corresponding points for different ion types are indicated in Figure 3.

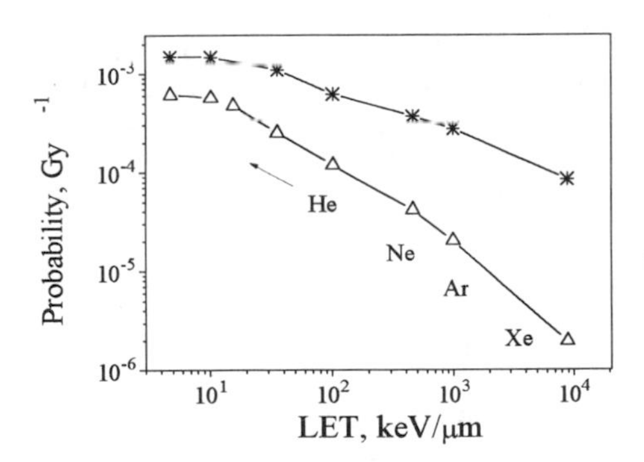

Figure 3 *Probability of single energy deposition event per Gy per target calculated for different heavy ions for 5.2 kbp DNA linear segment (stars) and DNA within solenoid fibre (up triangles).*

In Figure 4 distance density probability for target volume ψ(x) is compared with proximity function for homogeneous medium t(x). Influence of higher-order structure is manifested by increasing frequency of events with closer proximity than for track in homogeneous medium. On the other hand, role of complex target structure is characterised by less frequent events at larger distances.

Presented results support the notion[1] that microdosimetric quantities, both integral and differential, may be sensitive to the overall shape of complex structures formed by folding of flexible cylindrical and polymer target volumes. We have shown that microdosimetric distributions for biologically important targets like DNA molecule are different for various higher-order structures having the same DNA content or volume size. Transition of linear DNA molecule to a bead on a string structure and then to solenoid fibre results in marked change in the F_1 (ε>E), at the top of curves. Whereas the looping of solenoid fibre and subsequent coiling affect the tail of distribution. These results demonstrate that pathway of folding of polymer target in complex structures may be important determinant of the shape of microdosimetric distributions.

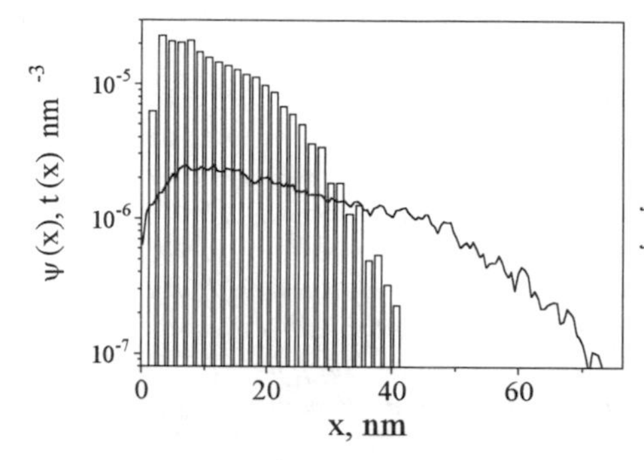

Figure 4 *Calculated probability density ψ(x) of distances between a pair of energy transfer "points" in sensitive target, chromatin 5.2 kbp fibre (columns) in comparison with proximity function t(x) in homogeneous medium for 102 keV/μm alpha particle track segment of 76.6 nm length (solid line).*

Comparison of functions for the same structure, solenoid, and different DNA sizes, 5 and 50 kbp, shows that difference between curves resembles surprisingly the behaviour of curves for 50 kbp DNA undergoing transition from looped fibre to coiled loop state. Precise interpretation of this observation is unknown. It seems likely that further interesting regularities could be observed under variation of topological state of target volumes.

References

1. S.G. Andreev, D.M. Spitkovsky, 'Proceedings of the Fifth All-Union Symposium on Microdosimetry', Ed. V.I. Ivanov, Moscow, MIPhI, 1987, Vol.2, p. 240.
2. N.O. Oleinick, U. Balasubramaniam, L. Xue, S. Chiu, *Int. J. Rad. Biol*., 1994, **66,** 523.
3. K.G. Hofer, N. VanLoon, M.H. Schneiderman, D.E. Charlton, *Rad. Res*., 1992, **130**, 121.
4. S.G. Andreev, I.K. Khvostunov, D.M. Spitkovsky, V.Yu. Chepel, *in this issue*.
5. A.V. Lappa, E.A. Bigeldeev, *Radiat. Prot. Dosimetry*, 1994, **52**, 85.
6. D.T. Goodhead, H. Nikjoo, *Int. J. Rad. Biol.*, 1989, **55**, 513.
7. V. Michalik, M. Begusova, *Int. J. Rad .Biol*., 1994, **66**, 267.
8. I.K. Khvostunov, V.Yu. Chepel, S.G. Andreev, 'The 24-th Annual Meeting of the European Society for Radiation Biology', Erfurt, Germany, 1992, Abstracts, p. 165.
9. V.Yu. Chepel, I.K. Khvostunov, L.A. Mirny, T.A. Talyzina, S.G. Andreev, *Radiat. Prot. Dosimetry*, 1994, **52**, 259.
10. H. Tomita, M. Kai, Y. Aoki, A. Ito, *Int. J. Rad. Biol*., 1994, **66**, 669.
11. D.E. Charlton, H. Nikjoo, J.L. Humm, *Int. J. Rad. Biol*., 1989, **56**, 1.
12. V.A. Pitkevich, V.G. Vidensky, V.V. Duba, *Atomic Energy*, 1982, **52**, 190.
13. A.M. Kellerer, D. Chmelevsky, *Rad. Environm. Biophys*., 1975 , **12**, 205.
14. A.M. Kellerer, H.H. Rossi, *Rad. Res.,* 1978, **75**, 471.

Monte Carlo simulation of the Track Interaction Model applied to the TLD-100 response to 5.3 MeV α-particles

Rodríguez-Villafuerte M., Gamboa-deBuen I. and Brandan M.E.

Instituto de Física, UNAM
A. P. 20-364, México 01000, D. F.
MEXICO

1 Introduction

The TL dose response curve of thermoluminescent materials under heavy charged particle (HCP) irradiation displays a linear region up to approximately a few Gy, followed by a supralinear region, before entering a sublinear region arising from saturation and radiation damage effects. The Track Interaction Model (TIM) is a model introduced in the late sixties,[1,2] and subsequently modified by Horowitz and co-workers,[3-5] to explain the supralinear behaviour of thermoluminescent materials under HCP irradiation. The TIM interprets supralinearity as a phenomenon arising during the heating stage in which the charge carriers produced during irradiation and trapped within a small radius of the initial particle track, can migrate away from the track where they were produced and recombine with an activated luminescent centre within a neighbouring track. The analytical expression of the supralinearity factor in the context of TIM assumes two-dimensional charge carrier mobility, parallel irradiation geometry and contributions up to the third neighbouring track, and does not include saturation effects at high fluences. The predictions of TIM have described qualitatively experimental results obtained with TLD-100 under α-particle irradiation.[3-6] In order to extend the scope of this analytical model, a Monte Carlo (MC) simulation has been developed (called MCTIM, herein), based on the same assumptions as the original TIM. The main advantages of MCTIM are the non-restricted charge migration, a simulation of the solid angle factor (implicitly taken into account during the charge carrier transport), and inclusion of saturation effects.

2 Methods

2.1 Supralinearity

The simulation is based on the assumption that electrons travel radially outwards from the ion track, perpendicular to the incident HCP trajectory. The electrons can recombine in the intertrack region either with competitive non-luminescent or non-activated luminescent centres, or in a neighbouring track with activated luminescent centres. Recombination in the intertrack region is governed by the electron mean free path, λ, which depends on the temperature. In this work, as in the TIM, λ is treated as a free parameter. Recombination with an activated luminescent centre is considered purely geometric, that is, if an electron reaches a neighbouring track, it is assumed that recombination occurs. The ion track is assumed to be cylindrical with an effective radius

r_{eff} (defined in some works as the radial distance from the track axis in which approximately 98% of the dose is deposited[4]) and a length equal to the CSDA range of the HCP. MCTIM calculations have been carried out in the 1×10^8 to 9×10^{10} particles/cm^2 fluence interval. For each fluence, N ion tracks were randomly sampled in a circular area with a radius $\geq 3\lambda$. A total of 1×10^6 electrons histories were traced for each fluence. Two separate calculations have been performed to include saturation effects due to track overlap at high fluences. In the first calculation, the electrons were able to migrate from the parent track whether it was intersecting any other track or not. The second calculation considered that those electrons produced within both the parent track and any other overlapping track cannot migrate from the parent track and cannot contribute to the supralinearity effect.

2.2 Saturation

The linear region of the TL response curve is assumed to be proportional to the area of the ion tracks in the irradiated region.[7] Thus, the response in the linear region of this curve is proportional to N times the response of a single track, where N is the total number of ion tracks in the irradiated area. At very high fluences (> 1×10^{10} particles/cm^2), the effective area presented by the tracks is smaller than N times the response of a single track due to track overlap. Therefore, separate simulation was performed to calculate the area covered by the ion tracks as a function of fluence. This calculation required only one free parameter, the effective radius (r_{eff}) of the ion track.

2.3 Normalised TL Response

In this work, we propose to express the normalised TL response per unit dose in the following manner:

$$f(n) = 1 + f^{\text{sup}}(n) + f_{\text{sub}}(n) \tag{1}$$

where n is the fluence, the first term represents the TL arising from a single track (linear response), $f^{\text{sup}}(n)$ is the contribution of the supralinearity effect and $f_{\text{sub}}(n)$ is the contribution of the sublinearity (saturation) effect, which is expected to be negative.

3 Results and Discussion

The MCTIM supralinear components will be presented in the following sections. They have been normalised using a factor of 2.5, which arose from the comparison with experimental data (presented below). We interpret this number as the ratio between the number of electrons with a high probability of migrating to neighbouring tracks and the number of electrons recombining within the parent track (N_e and N_w respectively[5]). The ratio N_e/N_w can be estimated using the analytical expression of the radial dose distribution proposed by Butts and Katz[8] produced by 5.3 MeV α-particles in LiF; we have obtained preliminary values between 1 and 5, depending on the target size. Thus, the normalisation factor used in our calculations, and required by the data, might carry information about the size of the ion core radius.

3.1.1 Supralinearity - without Track Overlap

The linear and supralinear components of the TL response as calculated with MCTIM are shown in figure 1a for four sets of parameters (r_{eff},λ). These parameters were obtained from a fit to experimental data previously reported by our group[6,9]. It can be observed that: a) supralinearity increases and begins at lower fluences as the effective radius increases. This effect can be attributed to an increase in the solid angle presented by neighbouring tracks with respect to the parent track. b) supralinearity increases and it begins at lower fluences as the electron mean free path increases.

3.1.2 Supralinearity - with Track Overlap

Figure 1b shows the TL response (linear and supralinear components) as a function of fluence, when track overlap was taken into account. It can be observed that the curves follow initially the same tendency as the curves in figure 1a, but as the fluence increases, track overlap inhibits the electrons from migrating to neighbouring tracks. This effect produces a sudden decrease of the TL response as the fluence increases, which is much more important as the effective radius of the ion tracks increases.

3.2 Saturation

Figure 1c shows the effect of the sublinear component on the TL response. It can be observed, as expected, that the sublinearity factor is a decreasing function of fluence. The fall in TL response is certainly more important at higher r_{eff}.

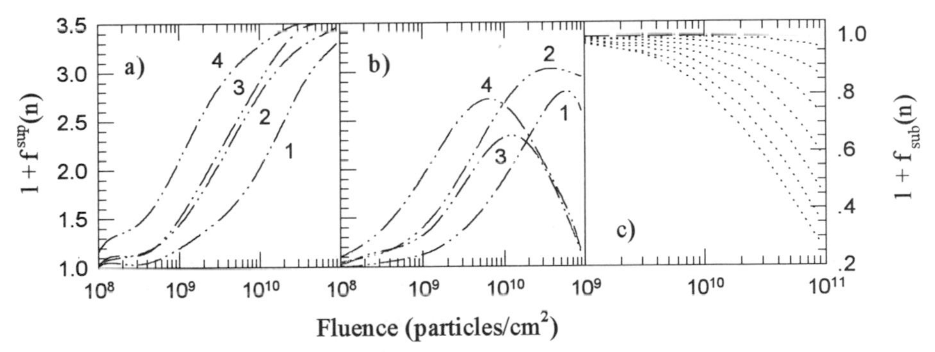

Figure 1. *a) and b) Linear and supralinear components without and with track overlap, respectively. (r_{eff}, λ)= (110,3e3), (110,1e4), (330,3e3), (330,1e4) Å in ascending order. c) Linear and sublinear components, r_{eff}=50 Å to 350 Å from top to bottom in 50 Å intervals.*

3.3 TL Response

The normalised TL response including the linear, supralinear and sublinear components is presented in figure 2a. It can be observed that the inclusion of the three components results in $f(n) \approx 1$ at low fluences, followed by an increasing and then decreasing behaviour. The maximum TL response depends very strongly on the effective radius of the ion track. Figure 2b shows MCTIM results compared with experimental data obtained with 5.3 MeV α-particles incident on TLD-100. It also shows the predictions obtained with the analytical expression of TIM. In this work, only peak 8 was compared with the MCTIM calculations.

4 Conclusions

The results obtained in this work reproduce qualitatively the TIM predictions. However, the MCTIM calculations have different shape, probably due to a more accurate geometric description during the electron transport simulation. The inclusion of track overlap changed drastically the behaviour of the TL response, and pointed out that large effective radii cannot reproduce the experimental data. The MCTIM showed a closer description of the experimental data than the TIM, in spite of considering different irradiation geometries.

Our results indicate that the Monte Carlo technique is a powerful tool to study the response of TL materials since several experimental conditions can be included, which cannot be considered otherwise. In order to further understand the TL response to HCP, extensive data under different irradiation conditions are needed. Experiments at our Institute using a Pelletron accelerator to measure the TL response to HCP as a function of linear energy transfer are in progress.

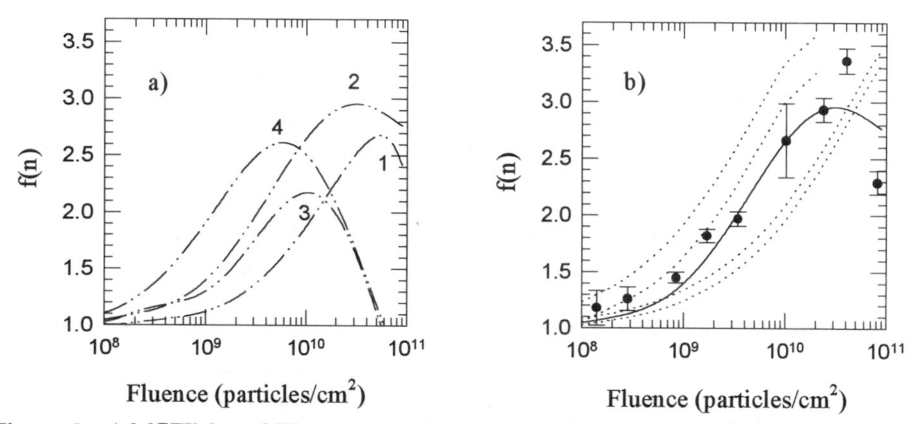

Figure 2. *a) MCTIM total TL response. Experimental data (symbols) for TLD-100, peak 8; MCTIM results (solid line, r_{eff}=110 Å, λ=1e4 Å). TIM calculations (dotted lines) with (r_{eff}, λ)=(330,1e4), (330,3e3), (110,1e4) and (110,3e3) Å from top to bottom.*

Acknowledgements

This work was supported by DGAPA-UNAM (IN100396) and CONACyT (E0077-E).

References

1. E.W. Claffy, C.C. Klick and F.H. Attix, Proc. 2nd. Int. Conf. Luminescence Dosim., AEC-CONF 680920, 1968, 302.
2. F.H. Attix, *J. Appl. Phys.*, 1975, **46**, 81.
3. M. Moscovitch and Y.S. Horowitz, *Radiat. Prot. Dosim.*, 1986, **17**, 487.
4. Y.S. Horowitz, *Radiat. Prot. Dosim.*, 1990, **33**, 75.
5. Y.S. Horowitz., M. Rosenkrantz, S. Mahajna and D. Yossian, *J. Phys. D. Appl. Phys.*, 1996, **29**, 205.
6. I. Gamboa-deBuen, A.E. Buenfil, M. Rodríguez-Villafuerte, C.G. Ruiz, A. Zárate-Morales and M.E. Brandan, *Radiat. Prot. Dosim.*, 1996, **65**, 13.
7. Y.S. Horowitz, M. Moscovitch and A. Dubi, *Phys. Med. Biol.*, 1982, **27**, 1325.
8. J.J. Butts and R. Katz, *Radiat. Res.*, 1967, **30**, 855.
9. I. Gamboa-deBuen, PhD Thesis, Universidad Nacional Autónoma de México, 1996.

Chemical processes from radiation to DNA

FACTORS CONTROLLING THE RADIOSENSITIVITY OF CELLULAR DNA.

J.F. Ward[*], J.R. Milligan[*] and R.C. Fahey[#].

[*] Department of Radiology, School of Medicine 0610
[#] Department of Chemistry and Biochemistry
University of California, San Diego
La Jolla, CA 92093
U.S.A.

1 INTRODUCTION

Modelling of ionizing radiation damage to cellular DNA is carried out using two approaches: Biophysical and Biochemical. Biophysical models begin with descriptions of radiation energy deposition patterns and simulate, using a computer code, the reactions of the various species produced from the energy deposition events with the DNA structures[1,2]. On the other hand Biochemical models [3-6] measure the chemical alterations produced in model DNA structures by the radiation in question. In the former case both the radiation energy deposition pattern and the DNA structure are modelled. In order to extrapolate findings of these models to a mammalian cell it is necessary to input into the model the structure and neighboring molecules involved with the cellular DNA. Thus both Biophysical models and Biochemical models require that assumptions be made about the cellular environment of DNA in terms of the nearest neighboring molecules.

The total number of radiation-induced free radicals reacting with DNA and their sites of reaction will be modulated by the structures assumed to be associated with the DNA. The accuracy of the Biophysical model depends on the accuracy of the database. The validity of any Biochemical models depends on how well the systems used mimic the intracellular environment of the DNA: A model will only be valid if it accurately simulates the relationships between DNA and its neighboring molecules.

Several factors are known to affect the level of radiation damage to DNA; chemical, such as the level of free radical scavenger present[6] or the presence of oxygen[7] ; physical, such as masking of sites from the approach of free radicals. This paper will discuss the latter topic. In a previous paper the effects of physical self-masking of the DNA from radical attacks were discussed[8]. It was shown that in the double helical polymer there are major effects on the accessibility of sites to attack by free radicals produced by irradiation of water. Recent evidence indicates that there is additional shielding caused by the binding of nucleohistone[6] and polyamines[9] to the DNA. Such physical protection may not be homogeneous throughout the genome and may vary in extent throughout the cell cycle. The possible variations are considered in the following in the context of extrapolations from models to cellular DNA.

2 SOURCES OF DNA DAMAGE IN CELLS

It is generally accepted that ionizing radiation damage to cellular DNA is produced as a consequence of the direct ionization of the DNA and from reactions of reactive species produced by ionization of neighboring molecules. Here we hypothesize that the damage which results from the deposition of energy directly in the DNA, or in the first layer of water molecules (pseudo-direct effect[10]) are unaffected by neighboring molecules. Early work[11] indicated that electron adducts of DNA are not biologically significant so that the important species produced by the direct plus pseudo-direct effects are the cation radicals produced on the DNA. The yield of these cation radicals would be expected to be independent of the presence of neighboring molecules - unless these proximate molecules transfer holes to DNA more efficiently than DNA's neighboring water molecules. There is the additional possibility that the neighboring molecules could modulate the subsequent reactions of the cation radicals thereby affecting the yield and/or the type of damage produced from this route. However, it is clear that the major effect of variations in packaging of DNA is the modulation of its accessibility to diffusing radicals.

It was shown several years ago that diffusing $^{\bullet}OH$ radicals produced by radiolysis of cellular water are the source of ~⅔ of the cell killing and ⅔ of the DNA single strand breaks (SSB) produced by low LET radiation[11-13]. Thus for the cell overall and for the DNA overall, the **average** contribution to the total damage is ⅔ $^{\bullet}OH$ radical and ⅓ non-scavengable. These relative contributions also apply to DNA double strand breaks[14,15] chromosome aberrations[16] and mutation at the HGPRT and 6-TG loci of V79 cells[15]. In these studies the mean distance that the $^{\bullet}OH$ radicals diffuse before causing the damage can be calculated from the concentration of scavenger necessary to reduce the scavengable damage by a factor of two. Most of the studies are not carried out over a sufficient range of scavenger concentrations to permit the mean distance travelled by $^{\bullet}OH$ radicals to be estimated with any precision. However the measurements are relatively consistent with a $^{\bullet}OH$ radical life time of ~1.3 ns. Using the expression developed by Roots and Okada[12] (Diffusion distance = $(16/\pi\ Dt)^{1/2}$, where D is the diffusion constant of $^{\bullet}OH$ radicals) it is calculated that during 1.3 ns an $^{\bullet}OH$ radical diffuses 4 nm. Of course this value is not absolute since the diffusion constant of an $^{\bullet}OH$ radical close to DNA in a cellular environment may not have the same diffusion constant as in aqueous solution.

Since the yield of double strand breaks in specific chromosomes is equal to that in the genome overall[17] it can be concluded that the distance travelled by $^{\bullet}OH$ radicals in causing biologically significant damage is probably fairly constant and uniform for all types of damage. This generality is for the <u>mean</u> distance that an $^{\bullet}OH$ radical travels and assumes that the contribution of non-scavengable damage is constant for the DNA regardless of the way it is packaged.

Recently we have detected novel routes to DNA damage production in a model system which could also occur in a cell. We have found for instance that a carbon centered radical $CH_3^{\bullet}$ can react with DNA to cause a SSB[18] and that an alkyl peroxyl radical $CH_3O_2^{\bullet}$ can react with DNA to cause base damage[19] but not to produce strand breaks. The latter finding brings into question the source of the damaged bases found endogenously in DNA[20]. These altered bases could be produced by oxidative processes involving peroxyl radicals rather than ˙OH radicals and, if so a corollary would be that these endogenous oxidation processes may not produce strand breaks.

3 REASONS FOR SHORT ˙OH LIFE TIME IN MAMMALIAN CELLS

It has been generally assumed that the major factor controlling the yield of ˙OH radical induced damage in cellular DNA is the presence of molecules which compete for these radicals. Steric factors such as those alluded to above have not been considered. However, there has always been a discrepancy between the general assumption and the data obtained when the molecules associated with DNA within a cell are dissociated in a stepwise fashion prior to irradiation[21-24]. The radiosensitivity of the DNA in cell nuclei increased by 1.7[21] for SSB and 1.2[22], 1.2[23], and 2.1[24] for DSB over that of DNA in the intact cell. Presumably the cell nuclei, isolated in buffer of low ˙OH scavenging capacity, have lost endogenous low molecular weight ˙OH scavengers. For a 2.1[24] increase in sensitivity it can be calculated that the ˙OH scavenging capacity of the material lost on preparation of nuclei has a scavenging capacity of $5.2 \times 10^8 s^{-1}$. For an increase of 1.2[22,23] the calculated scavenging capacity of the low molecular weight material is $3.4 \times 10^8 s^{-1}$. (It is of interest that several years ago working from *a priori* assumptions of cell contents Hunt *et al*[25] arrived at a cellular scavenging capacity of $8 \times 10^8 s^{-1}$). The distance a ˙OH radical diffuses at a scavenging capacity of $4 \times 10^8 s^{-1}$ is ~ 5nm.

The residual radiosensitivity of DNA in the cell nuclei (e.g. in the absence of low molecular weight ˙OH radical scavengers) is still low (compared to isolated DNA); hence the radioresistance of cellular DNA is not determined solely by general ˙OH scavenging capacity and other factors must be involved. Physical shielding of the DNA by closely associated molecules is the probable additional factor.

It can be concluded that cellular DNA is protected from ionizing radiation damage by two mechanisms: The presence of low molecular weight scavengers of ˙OH radicals, and condensation and aggregation (which could be considered to be self protection). The former limits the number of diffusing ˙OH radicals which migrate to and react with the DNA, while the latter limits the volume of water surrounding the DNA in which radicals are produced and react with the macromolecule.

4 PHYSICAL PROTECTION OF DNA

The accessibility of the sites in DNA which react with ˙OH radicals is markedly reduced in double stranded DNA compared to single stranded DNA[8]. The surface area of the reactive sites in deoxyribose is reduced by a factor of ~2.5, while that of the reactive double bonds of the bases is reduced by a factor of ~10. This difference goes a long way to explaining why the ratio of base damage to strand breaks is ~20 in ss

DNA but is ~4 in ds DNA. This physical protection occurs because the deoxynucleotide moieties shield each other.

Milligan et al.[6] compared the sensitivity of SV40 DNA to that of SV40 DNA within minichromosomes. They measured the yield of SSB produced by radiation induced ·OH radicals. They found that over a range of scavenging capacities the DNA in the minichromosome was protected by a factor of ~5 from ·OH radical damage, e.g. by the presence of nucleosome structures. This amount of protection is insufficient to explain the radioresistance of DNA in cell nuclei preparations referred to above.

Recently Newton et al. reported[9] that the presence of polyamines such as spermine can markedly protect DNA in dilute solution. A protection factor of 145 was achieved by increasing the spermine concentration from 10^{-3}M to 3 x 10^{-3}M. This protection can not be explained on the basis of scavenging of ·OH free radicals since spermine reacts relatively slowly and is not present in high enough concentration. Instead the authors hypothesize that the DNA is physically protected by condensation and aggregation (they coined the acronym PICA, polyamine induced condensation and aggregation, for this phenomenon). In other words the polyamine, by neutralizing the negative charge on the DNA permits the macromolecules to pack closer together, an effect presumably also occurring within cells facilitating the packaging of the DNA into the cell nucleus. Indeed the radiosensitivity of the DNA when condensed and aggregated was within a factor of two of that of cellular DNA. Of particular interest in the present context was the finding that the PICA effect in SV40 minichromosomes occurs at 0.6mM spermine compared to 1.6mM for DNA. The effectiveness at lower concentration in the case of the minichromsome was interpreted as due to the neutralization of ~50% of the DNA negative charge by the histones present. This differential between the concentration of spermine necessary to condense free DNA and minichromosomes may be significant when one considers the level of spermine present within the mammalian cell is 1mM[26].

5 RADIOSENSITIVE SITES IN CELLULAR DNA

Now that we have examined the various possible means of modulating radiosensitivity of DNA we can approach the data which indicate variable radiosensitivity of specific regions of DNA in a cell. Several publications have suggested that specific sites in cellular DNA may be more sensitive than DNA in general[27-30].

Warters and Childers[27] compared the yield of oxidised thymine molecules in bulk and newly replicated DNA as functions of radiation dose. They found the yield to be 3.3 fold greater in the latter and ascribed this to the greater accessibility of this DNA to ·OH radicals. As argued above, any increase in damage in uncondensed DNA is limited by the scavenging capacity of the low molecular weight cellular constituents. Thus such an increase in yield of damage is possible, however, there also exists the possibility that some of the base damage is caused by reactions of peroxyl radicals as we have shown recently[19] . We have previously hypothesized that the anomalously high yield of radiation induced base damage in mammalian cells is due to reactions of peroxyl radicals[31]. In the present situation, these radicals being more bulky than $^{\bullet}OH$ radicals would be expected to exhibit preferential reactivity with any more open DNA structures. However, we also showed that the peroxyl radicals do not give rise to strand breaks in DNA[19].

Other work has examined the radiosensitivity of actively transcribing genes. In one study[28] there is an increase in double strand breaks (DSB) at the lowest dose; however, the amount of DSB present at zero dose is not quoted in the table. Plotting yield versus dose of these DSB indicates that the increase in the yield of DSB in actively transcribing DNA is at most a factor of 1.4.

A recent publication[30] also indicates that the DNA in actively transcribing genes is up to three times more sensitive to SSB formation than DNA in general. These authors used a novel alkaline unwinding/Southern blotting method to compare the sensitivity of DNA in different genomic regions. The assay measures SSBs in regions of about 1 mega base pair (Mbp) which include the gene of interest. They found that the yield of SSB was 2-3 fold greater in the length of DNA containing the expressed c-*myc* gene than that in DNA overall or in satellite DNA. However, the yield of SSB was also ~2 fold greater in the β-globin gene; this gene is not expressed but has the "potential for active transcription". Thus the authors conclude that genes which are actively transcribed and those which have the potential to be transcribed are hypersensitive to radiation damage. (It should be noted that the assay used to measure damage in bulk DNA was not the same as that for damage in a specific region - a normal alkali unwinding assay was used.)

A mechanistic explanation of the greater sensitivity is that the increased yield is caused by an increase in the relative number of $^{\bullet}OH$ radicals reacting with the DNA i.e. the DNA is more open to free radical attack. It is unlikely that the yield of damage from direct ionization (+quasi-direct ionization) can vary significantly in different regions of the genome - see above. In order to achieve a 3-fold increase in the yield of SSB, the number of $^{\bullet}OH$ radicals reacting with the DNA must increase by a factor of 4. This would indicate that the sensitive DNA is exposed to a lower scavenging capacity. To determine the level of scavenging capacity necessary to explain the increased yield we can utilize the data of Milligan et al.[3] in which the yield of SSB was measured as a function of scavenging capacity. Thus the scavenging capacity in the cell, $8 \times 10^8\ s^{-1}$, corresponds to 0.125M DMSO, and to increase this yield of damage by a factor of 4 requires that the scavenging capacity be reduced to $6 \times 10^7 s^{-1}$. How this could be achieved is not clear since this scavenging capacity is lower by a factor of 8 than that calculated for the scavenging capacity of low molecular weight compounds present in the cell nucleus (see above Section 3).

Additional arguments against the generality of the hypothesis that actively transcribing DNA and potentially actively transcribing DNA are hypersensitive to radiation are as follows. There are 10,000-20,000 actively transcribing genes within a mammalian cell[32]. If these genes are considered to a first approximation to be uniformly distributed and if a region of 1Mbp around each of these genes is hypersensitive then the total genome would be hypersensitive. In addition Bunch et al.[29] probed the unwinding of the DNA within the region close to the gene of interest and within the neighboring regions. As their unwinding times increased the distance accessed becomes more remote from the gene of interest. If hypersensitivity occurs close to the gene it would be expected that as the lengths of DNA remote from the gene are accessed in the assay, the radiosensitivity would decrease. This was not observed. A source of concern in the data shown by these investigators is the difference in initial damage in unirradiated DNA for the gene of interest. The authors state that "the rate of baseline unwinding is greater within the c-*myc* region than in the bulk of the genome." (This had been noticed earlier by Warters et al.[28].) Thus the effect of pre-existing breaks (equivalent to a dose of about 2 Gray) within the region of interest within the genome could have unknown effects.

In the absence of condensation and aggregation the amount of damage to cellular DNA would be controlled by the scavenging capacity of molecules in its vicinity, be they enzymes involved in transcription, replication or low molecular weight compounds. The calculation of the general scavenging capacity of low molecular weight compounds can be used to obtain an upper limit to the number of $^{\bullet}OH$ radicals reacting with cellular DNA in the absence of condensation and aggregation: This scavenging capacity ($4.8 \times 10^8 s^{-1}$) corresponds to a yield of DNA strand breaks only 15% higher than that in the presence of the low molecular weight scavengers (calculated from Milligan et al[3].)

6 SPACINGS OF LESIONS WITHIN MULTIPLY DAMAGED SITES

We have previously indicated that there are three major variables of multiply damaged sites (MDS)[31]:

a. Multiple possibilities of combinations of lesions.
b. Multiplicity of lesions present in MDS.
c. Distance apart of lesions on opposite strands.

Now we consider what effect the accessibility of the site has on these complexities.

MDS are made up of combinations of singly damage sites which result from directly ionized DNA and attack by secondary radicals ($^{\bullet}OH$). The number and composition of MDS made up of two directly ionized individual sites would not be affected by the distance that the $^{\bullet}OH$ radicals migrate before reacting.

If a doubly damaged site contains one damage from direct ionization and one damage from $^{\bullet}OH$ radical attack, the distance from the DNA that the $^{\bullet}OH$ radical is produced is limited by the dimensions of the energy deposition event in which it is produced, since by definition the direct ionization must occur on the DNA. Thus this distance from the DNA is at maximum the diameter of the spur - i.e. 10nm. For bulk DNA which is protected by both the PICA effect and the presence of low molecular weight scavengers the distance that $^{\bullet}OH$ radicals migrate before reacting is of the same order i.e.10nm. The probability that these radicals would react with DNA would not be affected by the absence of the PICA effect e.g. in regions hypothesized to be depleted in tightly bound histones such as transcribing/replicating DNA.

In the absence of the PICA effect the number of MDS formed from two diffusing ${}^{\bullet}OH$ radicals could be increased in yield by increasing the volume of water from which ${}^{\bullet}OH$ radicals can be scavenged by the DNA. However, since the radicals from the larger volume would necessarily travel further before reacting with the DNA, they will be a lower probability that they react with the DNA at sites close together. Arguments can be made that doubly damaged sites with the lesions well separated are not so biologically significant as those where the damages are close together[33].

7 CONCLUSIONS

Cellular DNA is protected by two mechanisms of approximately equal effectiveness: Low molecular weight ${}^{\bullet}OH$ radical scavengers limit the number of these radicals which can react with the DNA; Compaction and aggregation of the macromolecule and chromatin limits the volume of water from which ${}^{\bullet}OH$ radicals can be scavenged by the DNA. Conditions to simulate cellular DNA radiosensitivity can be achieved in aqueous solution using relatively low concentrations of the polyamine spermine which compacts and aggregates DNA and chromatin. Effects on chromatin are achieved at subcellular spermine concentrations while higher concentrations are needed to achieve the PICA effect for free DNA. Thus it is possible that conditions can exist within the mammalian cell whereby chromatin is condensed but free DNA is not compacted. Such conditions have the potential to lead to an increase in the radiosensitivity of the uncompacted DNA by permitting access of a greater number of ${}^{\bullet}OH$ radicals to the DNA. Such an increase in radiosensitivity is limited by the ${}^{\bullet}OH$ scavenging capacity of the cell in terms of low molecular weight compounds. This scavenging capacity can be calculated from literature data to be of the order of $3\text{-}5 \times 10^{8}\ s^{-1}$. There are data in the literature which indicate that actively transcribing genes are more radiosensitive than bulk DNA although some of these data can be questioned. It can be argued that any increased damage of the MDS type in DNA resulting from a greater access to ${}^{\bullet}OH$ radicals would not be of the biologically significant type, because of the distance separating the individual lesions within the MDS on the DNA. We have previously argued that DNA DSB (and particularly those which have closely opposed SSB) are the important radiobiological lesion both for cell killing and for mutagenesis[34].

8 ACKNOWLEDGEMENTS

The authors would like to express their thanks to Ray Warters (University of Utah) for stimulating discussions. Work in the authors' laboratories is supported by grants from NIH; CA 46295 (JFW) and CA39582 (RCF).

9 REFERENCES

1. A. Chatterjee and W.R. Holley, Adv. Radiat. Biol. 1993, **17**, 181.
2. H. Nikjoo, P. O'Neill, M. Terrissol and D.T. Goodhead, D.T. Int. J. Radiat. Biol. 1994, **66**, 453.
3. J.R. Milligan, J.A. Aguilera and J.F. Ward, Radiat. Res. 1993, **133**, 151.

4. R.E. Krisch, M.B. Flick, and C.N. Trumbore, Radiat. Res., 1991, **126**, 251.
5. K.M. Prise, S. Davies and B.D. Michael, Radiat. Res., 1993, **134**, 102.
6. J.R. Milligan, J.A. Aguilera and J.F. Ward, Radiat. Res. 1993, **133**, 158.
7. I.S. Ayene, C.J. Koch and R.E. Krisch, R.E. Radiat. Res. 1995, **144**, 1.
8. J.F. Ward, Biochemistry of DNA Lesions. Radiat. Res. 1985, **104**, S103.
9. G.L. Newton, J.A. Aguilera, J.F. Ward and R.C. Fahey, Radiat. Res. 1996, **145**, 776.
10. D. Becker and M.D. Sevilla, Adv. Radiat. Biol. 1993, **17**, 121.
11. R. Roots and S. Okada, Int. J. Radiat. Biol. 1972, **21**, 329.
12. R. Roots, and S. Okada, Radiat. Res.1974, **64**, 306.
13. J.D. Chapman, A.P. Reuvers, J. Borsa, and C.L. Greenstock, Radiat. Res. 1973, **56**, 291.
14. J.W. Evans, J.F.Ward and C.L. Limoli, 1985, Abstr. Eh12, 33rd Radiat. Res. Soc. Mtg., Los Angeles, CA..
15. O. Sapora, F. Barone, M. Belli, A. Maggi, M. Quintilliani and M.A. Tabocchini, Int. J. Radiat. Biol. 1991, **60**, 467.
16. L.G. Littlefield, E.E. Joiner, S.P. Colyer, A.M. Sayer and E.L. Frome, Int. J. Radiat. Biol. 1988, **53**, 875.
17. J.W. Evans, X.F. Liu, C.U. Kirchgessner and J.M. Brown, Radiat. Res. 1996, **145** 39.
18. J.R. Milligan and J.F. Ward, Radiat. Res. 1994, **137**, 295.
19. J.R. Milligan, J.Y.-Y. Ng., C.L.L. Wu, J.A. Aguilera, J.F. Ward, Y.W. Kow, S.S. Wallace and R.P. Cunningham, Radiat. Res. 1996, **146**, 434.
20. Q. Chen, A. Fischer, J.D. Reagan, L.J. Yan and B.N. Ames, Proc. Nat. Acad. Sci (USA) 1995, **92**, 4337.
21. M. Ljungman, S. Nyberg, J. Hygren, M. Erikkson and G. Ahnström, Radiat. Res. 1991, **127**, 171.
22. L.-Y. Xue, L.R. Friedman, N.L. Oleinick, N.L. and S.-M. Chiu, Int. J. Radiat. Biol. 1994, **66**, 11.
23. J. Nygren, M. Ljungman and G. Ahnström, Int. J. Radiat. Biol. 1995, **68**, 11.
24. R.L. Warters and B.W. Lyon, Radiat. Res. 1992, **130**, 309.
25. H.B. Michaels. and J.W. Hunt, Radiat. Res. 1978, **74**, 23.
26. A.E. Pegg and P.P. McCann, Am. J. Physiol. 1982, **243**, C212.
27. R.L. Warters and T.J. Childers, T.J. Radiat. Res. 1982, **90**, 564.
28. R.L. Warters, B.W. Lyons, S.M. Chiu and N.L. Oleinick, Mutat. Res. 1987, **180,** 21.
29. S.M. Chiu, N.L. Oleinick, L.R. Friedman and P.J. Stambrook, P.J. Biochim . Biophys. Acta 1982, **699,** 15.
30. R.T. Bunch, D.A. Gewirtz and L.F. Povirk, Oncol. Res. 1992, **4**, 7.
31. J.F. Ward, Int. J. Radiat. Biol. 1994, **66**, 427.
32. B. Alberts, D. Bray, J. Lewis, M. Raff, K. Roberts, and J.D. Watson, In "Molecular Biology of The Cell" Garland Publishing N.Y. 1983,Chapter 8, p. 409.
33. J.F. Ward, Prog. Nuc. Acids and Molec. Biol. 1988, **35**, 95.
34. J.F. Ward, Radiat. Res. 1995, **142**, 362.

PRODUCTION YIELD OF ADENINE FROM ATP IRRADIATED WITH MONOCHROMATIC X-RAYS IN AQUEOUS SOLUTION OF DIFFERENT CONCENTRATIONS.

Katsumi Kobayashi, Noriko Usami, Ritsuko Watanabe[#] and Kaoru Takakura[$]

Photon Factory, National Laboratory for High Energy Physics, Tsukuba 305 Japan
[#]Takasaki Establishments, Japan Atomic Energy Research Institute, Takasaki 370-12 Japan
[$]International Christian University, Osawa, Mitaka 181 Japan

1 INTRODUCTION

From viewpoint of energy, essence of biological action of radiation lies in the non-uniformity in the spatial distribution of deposited energy due to the stochastic nature of the interaction between radiation and matter. Physical and chemical reactions which are considered to terminate in less than micro second produce molecular changes in biological system. We have been investigating the production mechanism of DNA damage by using monochromatic synchrotron light as a tool to control the energy deposition events. Among the processes included in the production of DNA damage, chemical process in aqueous system plays an important role in determining the nature and yield of the damage. Determination of the yields of reactive intermediates and products is very important in understanding cellular mechanism of radiation. However, experimental observation of the intermediates and their reaction is very difficult due to the short lifetime of them. Only way to study these processes has been considered to simulate energy deposition events and following prechemical and chemical reactions in aqueous solutions by computer codes.[1,2] They have evolved with accumulation of experimental data on the yield of intermediate species.[3,4] As one of the well known examples oxidation yield of ferrous solution (Fricke solution) have been simulated, which gives a measure of absorbed dose in aqueous solution through the yield of water radicals. The yield was calculated to decrease with the energy of X-rays or electrons down to 1 keV.[5,6] Our experimental data with monochromatic soft X rays decreased with the energy between 10 to 1.8 keV,[7] supporting the idea that higher LET radiation gives lower yield of reactive intermediates and, presumably, of chemical products.

On the contrary, we already reported that the Auger effect occurring at the phosphorus atom gives higher yield of production of adenine from ATP,[8] extent of which was higher in the more concentrated solution.[9] Considering that the Auger effect gives high density of ionization around the atom, these results suggest the idea of higher effect with higher LET or ionization density. This apparent discrepancy led us to a hypothesis that there exist two types of fields where radiation chemical reactions take place. The first one might be called as "cold field" where diffusive water radicals attack biologically important molecules. The other might be called as "hot field", where radicals exist densely and transiently, localized along the charged particle trajectory. In the “hot field”, some of them will disappear with the interaction between them, and survived radicals diffuse out of “hot field” to “cold field”. Due to the short lifetime of “hot field” and due to high reaction rates between radicals, chemical reactions between radicals and added solute molecules could be observable only in extremely high concentration. According to this hypothesis, the higher chemical yield in higher LET radiation field might be observable only in concentrated solutions. Reactions in hot field may become dominant in high LET radiation and the products might be different from those produced in the cold field, due to the multiple reactions with radicals.

In order to test the hypothesis, we irradiated adenosine triphosphate (ATP) solutions of different concentrations with monochromatic soft X-rays from 2 keV to 6 keV. 2.153

keV X-rays which correspond to the resonance absorption peak of K-shell of phosphorus[10] were also irradiated to produce more localized ionization around the molecule. Opposite energy dependence was expected between the solutions of high concentration and of low concentration.

2 MATERIALS AND METHODS

Adenosine triphosphate (Sigma, St. Louis) was commercially obtained and used without further purification. ATP solution was prepared with distilled, double-deionized water (≥ 18 MΩ). Concentrations of the solution were 150 mg/ml and 1.5 mg/ml, which correspond to 0.27 M and 2.7 mM, respectively. Irradiation was performed in the same manner as in our previous report.[8] Briefly, prepared solution was put in an acrylic sample cell which has a thin Kapton window in front and is 2 mm deep along the beam. Soft X-rays in this energy region are completely absorbed by 2 mm thick layer of water. Monochromatic synchrotron soft X-rays, introduced into air through thin Kapton windows, were irradiated on the sample cell in the atmospheric condition. X-ray flux was measured with a specially designed ionization chamber and absorbed dose was calculated assuming that incident X-ray energy was absorbed uniformly in the sample since the solution was stirred with a micro stirrer chip during irradiation. Irradiated samples were stored in a deep freezer until analyzed using an HPLC system equipped with an ion exchange column. Among the products, adenine was focused since it had been found to be produced most efficiently.[8] The yield was determined by integrating the peak area and calibrated with the area of the authentic molecule analyzed at the same condition.

3 RESULTS AND DISCUSSIONS

Purpose of this work is to see how X-ray energy dependence, or LET dependence, of chemical yield changes with the concentration of the solution. We already obtained that the enhancement of production yield of adenine associated with K-shell photoabsorption of phosphorus increases as shown in Table 1.[9] It was shown that the enhancement in the chemical yield by highly localized energy deposition becomes more remarkable in more concentrated solution.

Results on energy dependence of adenine yield in the solutions of different concentration are summarized in Figure 1. Three independent experiments were done in energy region from 2 to 6 keV. Relative yields, normalized at 2 keV or 2.147 keV within each experiment, were shown and mean values were shown above the bars in order to see the energy dependence more clearly. (2 keV X-rays and 2.147 keV X-rays give same level of localization of energy deposition since there lies no absorption edge of constituent atoms in the sample between two energies.) Two points can be seen from the figure. Firstly, when we compare the energy dependence around the phosphorus resonance absorption peak, enhancement in the yield was observed with the concentrated (150 mg/ml) solution, which confirmed our previous observation. In the dilute solution (1.5 mg/ml), however, a lower yield was observed with phosphorus photoabsorption. X-ray photoabsorption at K

Table 1 *Concentration Dependence of Enhanced Adenine Yield from ATP with Absorption of Soft X-ray by Phosphorus K-shell.[9] Yield is Expressed in Number of Molecules per Absorbed Energy of 100 eV (G value)*

ATP Concentration	*Irradiated X-ray Energy*	
	2.147 keV	*2.153 keV*
150 mg/ml	1.1	1.6
530 mg/ml	1.4	2.5

shell of phosphorus produces a hole in the K shell, and is followed by Auger cascade which emits three low energy electrons from the phosphorus atom. Considering that these electrons produce highly localized energy deposition around the molecule, we could conclude that the chemical yield in aqueous solution is enhanced with LET or ionization density, and that this enhancement increases with the concentration of solution as shown in Table 1. In the case of 0.27 M solution, average distance between the molecule can be estimated as 18 Å, which is almost the same with diffusion distance of solvated electron in 1 ns. In such case, the solute molecule is possible to participate into the fast reactions between radicals.

Secondly, it can be seen in Figure 1 that the yield increases with the X-ray energy except the resonance absorption peak of phosphorus (2.153 keV) in both concentrations. Similar tendency in the chemical yield was predicted by Bolch et al.[11] They simulated free ammonia production in glycylglycine solution of various concentrations (0.025 - 1.2 M) irradiated with electrons of different initial energy (1 - 140 keV). Yield of ammonia increased with the X-ray energy in all the concentration tested. They explained this tendency with an idea that a larger number of intratrack reactions occur between radicals in the field where reactive intermediates in higher initial density are produced by higher LET radiation. This is the same idea with ours concerning the dilute solution. It should also be pointed out from their results that calculated energy-dependent variation became less remarkable in high density solution. This seems to suggest that if they had simulated the case of higher concentration or of higher LET, opposite dependence might have been found which we found in concentrated ATP solution.

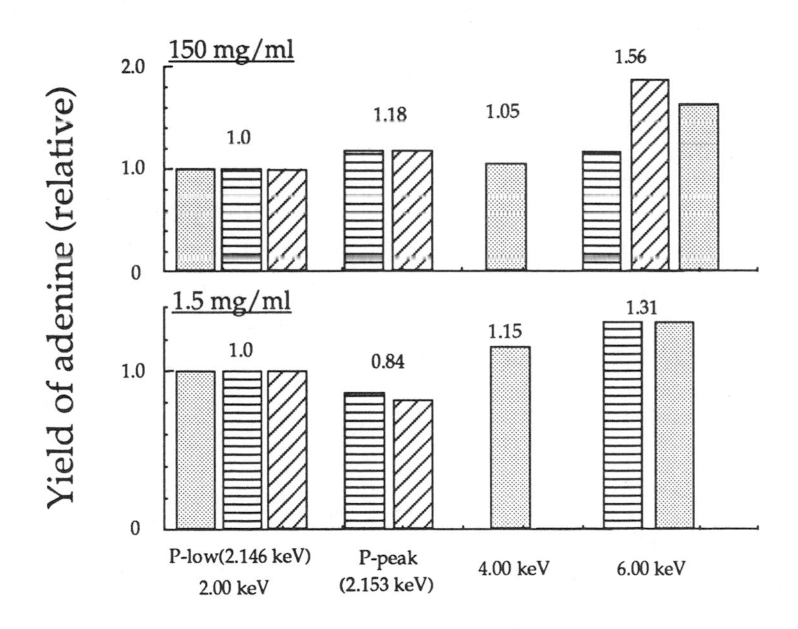

Figure 1 *Yield of adenine normalized to that of 2.00 keV or 2.147 keV X-rays in each experiment. Bars hatched with same pattern indicate that they were obtained in the same experiment. Variation of the values between each experimental set was less than 10 %.*

In conclusion, we demonstrated that the LET dependence of chemical yield shows opposite tendency depending upon the concentration of solution, which supports our hypothesis that there exist two reaction fields which contribute to the yield. Research using highly concentrated solution may reveal the chemical reactions in the transiently produced "hot field" containing reactive intermediates of high concentration. Since this transient field becomes more remarkable in high LET radiation, it may give us a clue to understand the mechanism of action of high LET radiation.

References

1. A. Mozumder and J. L. Magee, J. Chem. Phys., 1966, **45**, 3332.
2. J. E. Turner, R. N. Hamm, H. A. Wright, R. H. Ritchie, J. L. Magee, A. Chatterjee and W. E. Bolch, Radiat. Phys. Chem., 1988, **32**, 503.
3. C. D. Jonah, M. S. Matheson, J. R. Miller and E. J. Hart, J. Phys. Chem., 1976, **80**, 1267.
4. C. D. Jonah, and J. R. Miller, J. Phys. Chem., 1977, **81**, 1974.
5. H. Yamaguchi, Radiat. Phys. Chem., 1989, **34**, 801.
6. M. A. Hill and F. A. Smith, Radiat. Phys. Chem., 1994, **43**, 265.
7. R. Watanabe, N. Usami and K. Kobayashi, I. J. Radiat. Biol., 1995, **68**, 113.
8. R. Watanabe, M. Ishikawa, K. Kobayashi and K. Takakura, 'Biophysical Aspects of Auger Processes', American Institute of Physics, New York, 1992, p. 24.
9. R. Watanabe, PhD Thesis, Graduate University for Advanced Studies, 1993
10. K. Kobayashi, K. Hieda, H. Maezawa, Y. Furusawa, M. Suzuki and T. Ito, Int. J. Radiat. Biol.,1991, **59**, 643.
11. W. E. Bolch, J. E. Turner, H. Yoshida, K. B. Jacobson, H. A. Wright and R. N. Hamm, Radiat. Res., 1990, **121**, 248.

THE EFFECT OF ELECTRON ENERGY ON RADIATION DAMAGE

Simon M. Pimblott and Jay A. LaVerne

Radiation Laboratory
University of Notre Dame
Notre Dame, IN 46556, U.S.A.

1 Introduction

One of the principal goals in the theoretical study of radiolysis is the elaboration of the kinetics of the spatially non-homogeneous distribution of highly reactive radicals and ions created along the tracks of radiation particles.[1] The short-time chemistry of these radiation-induced reactants takes place on the sub micro-second time scale[2-5] and contains direct information about the fundamental physical, physico-chemical and chemical processes that are the basis of observable radiation damage. To extract this information requires precise and accurate models for all stages in the development of the system, that is for the energy transfer from radiation to medium, for the physical consequences of the transfer on the molecules of the medium, and for the diffusion and reaction of the radiation induced species.[6] This study focuses on the first of these problems.

2 Energy Loss Properties of Energetic Electrons

A recent series of papers has described the formulation for the energy loss properties of energetic electrons using the dipole oscillator strength distribution (DOSD) of the medium.[7-12] This approach has the advantage that all the effects of phase on the energy loss properties are included in the DOSD. The DOSD of water[13] and of hydrocarbons[12] all exhibit significant changes due to condensation. In consequence, the energy loss properties, such as stopping power (S), inelastic mean free path ($\Lambda_{inelastic}$), continuous slowing down approximation (*csda*) range and number density of energy loss events, also demonstrate effects due to the phase of the medium.[9, 12] The differences in the energy loss properties are most significant for electron energies smaller than ~ 1 keV. This energy regime is where S increases rapidly with decreasing electron energy and where electrons are believed to be the most effective at causing biological damage.[14, 15] Understanding this effect is qualitatively straightforward : increasing the rate of energy loss results on an increased density (or local concentration) of radiation-induced reactants. Unfortunately, while S is a convenient parameter for qualitatively describing energy loss, experiments using heavy ions have established that it is a poor descriptor for quantifying radiation damage.[16]

3 Tracks of Electrons in Water

The energy attenuation of electrons is a stochastic phenomenon, which is amenable to simulation using Monte Carlo methods. A variety of papers describing the modeling of the structure of electron tracks have been presented.[17-20] The simulation techniques are essentially the same with the only significant differences being in the cross-sections used to described the elastic and inelastic interactions between the energetic electron and the medium. The methodology followed here is described in ref.17. It makes extensive use of experimentally based cross-sections, in particular for the electronic contribution to inelastic collisions[9] and for determining the outcome (i.e. ionization or excitation) of this type of event.[17,21]

Averaged energy loss parameters such as the stopping power and the *csda* range reflect that the rate at which energy is lost as a function of distance traveled and do not take into account trajectory deviations due to elastic collisions. Such deviations are significant for low energy electrons as the cross-section for elastic collisions, $\sigma_{elastic}$, is very much greater than that for inelastic collisions, $\sigma_{inelastic}$. Even at high electron energies where the ratio of $\sigma_{elastic} : \sigma_{inelastic}$ is ~ 1:4,[17] the effects of elastic collisions are not negligible.

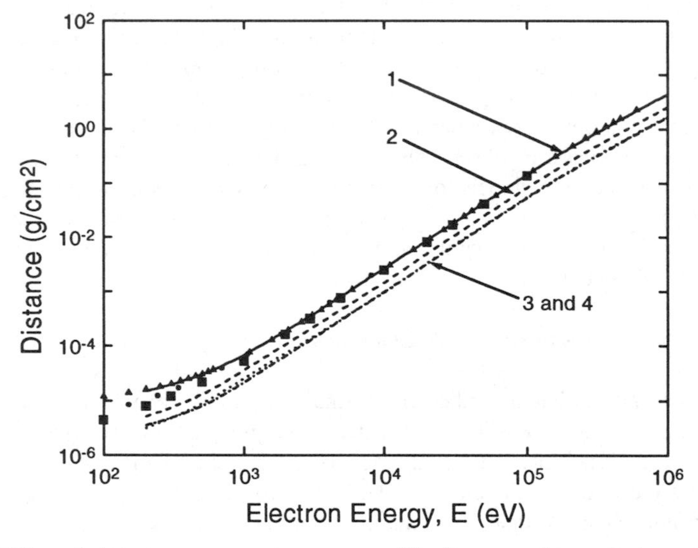

Figure 1. *Effect of electron energy on mean range. The lines are the predictions of Monte Carlo simulation for (1) the path length, (2) the separation between the initial and final positions, (3) the axial penetration, and (4) the radial penetration. The* (■) *are the recommended ranges of ICRU16, while the* (●) *and the* (▲) *are the csda ranges calculated using the energy dependent stopping powers of Paretzke et al.*[18] *and Pimblott et al.,*[9,11] *respectively.*

The effect of electron energy on the mean range of the electron in liquid water is shown in Figure 1. The total inelastic path length ($\Sigma\ \Lambda_{inelastic}$) is unaffected by elastic collisions and is the same as the *csda* range calculated from the energy dependence of the stopping power. For all energies, the mean separation between the initial and final positions ($|\boldsymbol{X}_{final}-\boldsymbol{X}_{initial}|$) of an electron is about a factor of two smaller than the total inelastic path length. This difference shows that the effects of elastic and large-energy-loss inelastic collisions are significant even for high-energy electrons. The mean axial and radial penetrations appear similar for electrons of energy greater than 1 keV, however, it should be noted that the probability distribution functions for the axial and the radial penetrations are not the same.[17] The former is strongly determined by inelastic collisions and the latter by elastic collisions.

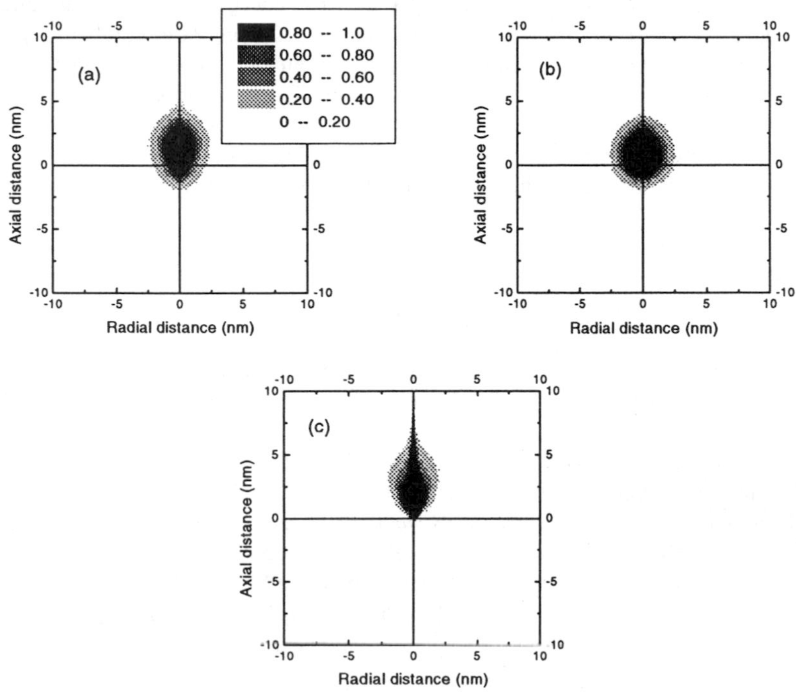

Figure 2. *Dose distribution for 100 eV electrons in liquid water. The dose is presented in units of eV nm^{-3}. Elastic cross-sections : a) experimental, b) screened Moliere, and c)* $\sigma_{elastic} = 0$.

The range of an electron reflects its complete degradation to thermal energy, but it does not provide information about the spatial distribution of energy loss events, which is the property of importance in understanding radiation damage. There is a marked dependence of the spatial distribution of the energy deposition, i.e. dose, on initial energy.[22] At low energies, the distribution is nearly spherical and is centered on the origin, but as the initial energy increases the distribution passes through an ellipsoidal to a teardrop shape.[17] This change in shape reflects the change in the ratio $\sigma_{elastic} : \sigma_{inelastic}$. and is reflective of the angular dependence of the cross-section for elastic collisions. Figure 2 shows the spatial distributions of dose for 100 eV electrons in liquid water calculated for three different $\sigma_{inelastic}$: (a) experimental $\sigma_{elastic}$,[17] (b) screened Moliere $\sigma_{elastic}$,[23] and (c) $\sigma_{elastic} = 0$. For cases (a) and (b), the total $\sigma_{inelastic}$ are the same, but the angular dependences are different. The experimental cross-sections result in a dose distribution that is more forward directed and has less of a radial spread than the screen Moliere cross-sections. For energetic electrons (E > ~10 keV), the angular dependences of the experimental and screened Moliere cross-sections are the same and the form of $\sigma_{elastic}$ does not affect the spatial distribution of dose. Excluding elastic collisions effectively prevents back scattering of the electron and the dose distribution is pointed in the initial direction of travel with little radial spread. These effects are evident in Figure 2c for an electron energy of 100 eV, but are found for all electron energies.

Track structure simulations of electron tracks (and stochastic diffusion-kinetic calculations based upon these tracks) reveal that electron energy has a significant effect on radiation damage. These studies hold the potential for providing considerable insight into the fundamental processes that determine observable chemical and biological outcome.

Acknowledgment

The research described herein was supported by the Office of Basic Energy Science of the U.S. Department of Energy. This is contribution NDRL-3964 of the Notre Dame Radiation Laboratory.

References

1. R. L. Platzman, in 'On the primary processes in radiation chemistry and biology', ed. J. J. Nickson, New York, 1950.
2. C. D. Jonah, E. J. Hart, and M. S. Matheson, *J. Phys. Chem.*, 1973, **77**, 1838.
3. C. D. Jonah, M. S. Matheson, J. R. Miller, and E. J. Hart, *J. Phys. Chem.*, 1976, **80**, 1267.
4. C. D. Jonah and J. R. Miller, *J. Phys. Chem.*, 1977, **81**, 1974.
5. A. C. Chernovitz and C. D. Jonah, *J. Phys. Chem.*, 1988, **92**, 5946.
6. S. M. Pimblott and N. J. B. Green, in 'Recent advances in the kinetics of radiolytic processes', ed. R. G. Compton and G. Hancock, Amsterdam, 1995.
7. J. C. Ashley, *J. Electron Spectrosc. Relat. Phenom.*, 1988, **46**, 199.
8. N. J. B. Green, J. A. LaVerne, and A. Mozumder, *Radiat. Phys. Chem.*, 1988, **32**, 99.
9. S. M. Pimblott, J. A. LaVerne, A. Mozumder, and N. J. B. Green, *J. Phys. Chem.*, 1990, **94**, 488.
10. S. M. Pimblott and J. A. LaVerne, *J. Phys. Chem.*, 1991, **95**, 3907.
11. J. A. LaVerne and S. M. Pimblott, *Radiat. Res.*, 1995, **141**, 208.
12. J. A. LaVerne and S. M. Pimblott, *J. Phys. Chem.*, 1995, **99**, 10540.
13. J. A. LaVerne and A. Mozumder, *J. Phys. Chem.*, 1986, **90**, 3242.
14. H. Nikjoo, D. T. Goodhead, D. E. Charlton, and H. G. Paretzke, *Physics in Medicine and Biology*, 1989, **34**, 691.
15. H. Nikjoo, D. T. Goodhead, and D. E. Charlton, *Int. J. Radiat. Biol.*, 1991, **60**, 739.
16. J. A. LaVerne and R. H. Schuler, *J. Phys. Chem.*, 1994, **98**, 4043.
17. S. M. Pimblott, J. A. LaVerne, and A. Mozumder, *J. Phys. Chem.*, 1996, **100**, 8595.
18. H. G. Paretzke, J. E. Turner, R. N. Hamm, H. A. Wright, and R. H. Ritchie, *J. Chem. Phys.*, 1986, **84**, 3182.
19. M. A. Hill and F. A. Smith, *Radiat. Phys. Chem.*, 1994, **43**, 265.
20. M. Zaider, D. J. Brenner, and W. E. Wilson, *Radiat. Res.*, 1983, **95**, 231.
21. S. M. Pimblott and A. Mozumder, *J. Phys. Chem.*, 1991, **95**, 7291.
22. A. Mozumder and J. L. Magee, *J. Chem. Phys.*, 1966, **45**, 3332.
23. B. Grosswendt and E. Waibel, *Nucl. Instrum. Meth.*, 1978, **155**, 145.

OXYGEN DECIDES: DOUBLE-STRAND BREAKS OR DNA-PROTEIN CROSSLINKS

L.V.R. Distel, B. Distel and H. Schüssler
Institut für Medizinische Physik
Klinik für Strahlentherapie
Universität Erlangen Nürnberg
Germany

1 INTRODUCTION

The oxygen effect is important in the radiation therapy of tumours. For understanding the mechanism of this effect, the analysing of basic experimental data and the modelling of DNA damage is an essential precondition. We provide data, which give evidence for an involvement of proteins in the oxygen effect.

It is assumed that the DNA is the main target of the cell for radiation induced damage. In cells the DNA is highly condensed in very close contact to proteins. The two main protein groups are histones and non histone proteins. The histones pack the DNA and they are regarded as highly protective against DNA damage.[1] However, the non histone proteins form the nuclear matrix and are involved in regulative actions and are suspected to protect DNA poorly. A close contact between non histone proteins and DNA is especially given in the highly sensitive regions of active DNA and at the bases of the DNA-loops.[2]

We applied a very simple cell model with isolated DNA and bovine serum albumin to mimic the active DNA. In the cell nucleus the weight ratio of non histone proteins to DNA varies between 0.5 to 1 and 1.5 to 1, while we used a ten fold excess of protein. We studied double-strand breaks and DNA-protein crosslinks, two DNA lesions which are supposed to be very important.

To study the influence of the protein on the oxygen effect, the DNA-protein solution was irradiated under aerobic and anaerobic conditions.

For double-strand break measurements constant field gel electrophoresis was applied, whilst for DNA-protein crosslink measurements the nitrocellulose filter assay was used. The nitrocellulose filter assay is interfered by the induction of double-strand breaks and can only detect the first protein crosslinked to one DNA molecule. A stochastic model was developed to correct the measured values of DNA-protein crosslinks.

2 MATERIALS AND METHODS:

Highly polymerized double stranded DNA from calf thymus with an average molecular weight of 16×10^6 Da from Calbiochem (Bad Soden/Ts, Germany) and lyophilized bovine serum albumin (BSA) from Boehringer GmbH (Mannheim) were used. All other chemicals were of analytical reagent grade and purchased from Merck (Darmstadt). Water was purified by deionization, reverse osmosis and a Milli-Q-system. Nitrocellulose filters BA 85 with a pore size of 0.45μm were purchased from Schleicher and Schüll (Dassel, Germany).

Solutions of DNA with and without proteins in 10^{-1}, 10^{-2}, 10^{-3} mol dm^{-3} phosphate buffer pH 7 were cooled down to 4°C, saturated with N_2 or air and irradiated while being slowly stirred with the appropriate gas passing over the surface. Irradiation was done with a Stabilipan X-ray machine (Siemens, Erlangen) at 200 kV, 20 mA and 2mm Al filter. The dose rate was about 24 Gy min^{-1}.

2.1 *Constant field electrophoresis.* Samples of 1 μg DNA were separated on agarose gels of different concentrations (0.5-1.2% (w/v)) in 0.5 TBE (Tris-borate-EDTA buffer) in a constant field with a voltage gradient of 85 Vm^{-1} at a temperature of 15° C for 16 hours. Gels were stained for 20 minutes with ethidium bromide (1μg ml^{-1}) and destained in 0.5 TBE for 1 h. The fluorescence of these gels was photographed. The negatives were scanned with a densitometer and the data were transferred to a personal computer and processed. Standard curves for the molecular weights of the nucleic acids were obtained by the separation of restriction fragments of λ-DNA and made possible the estimation of the number average molecular weight and the average number of double strand breaks per DNA-molecule.

2.2 *Nitrocellulose filter assay.* The DNA- protein (500μl) samples were diluted with 5500μl of 1mol dm^{-3} NaCl, 10mmol dm^{-3} Tris and 10mmol dm^{-3} EDTA at pH7.2. Then 2 ml of the diluted sample were pipetted onto the filter and drained slowly into test tubes. The optical density of the filtrates and references were measured at 260nm and thus the DNA contents of the filtrates were determined. The binding of the protein to the filter was proved by testing the filtrate with Coomassie Brilliant Blue G-250 according to the method of Sedmak and Grossberg.[3]

Under the conditions used, the protein is bound to the nitrocellulose filter, while the DNA pass unhindered the filter. If a DNA-protein crosslink is formed, the DNA is retained by the protein on the filter. The reduction of the DNA contents in the filtrate is measured. By this method only the first DNA-protein crosslink of one DNA molecule is measured and the induction of DNA double-strand breaks causes a decrease of retained DNA. Therefore the induced DNA double-strand breaks were measured and a mathematical model was developed to calculate the DNA-protein crosslinks induced.

3 RESULTS

3.1 Mathematical model for DNA-protein crosslinks

In order to quantify the measured DNA-protein crosslinks (see 2.2) a mathematical model was developed.

The number of crosslinks on one DNA-molecule X is distributed binomially:

The probability, that this number is i is therefore calculated according to

$$P(X=i) = \binom{\nu \cdot D}{i} \cdot \left(\frac{1}{m}\right)^{i} \cdot \left(1 - \frac{1}{m}\right)^{\nu D - i} \qquad (1)$$

where m is the number of DNA-molecules in solution and ν is the total number of protein molecules crosslinked to DNA per Gray.

The total number of crosslinks per dose $\nu \cdot D$ is very high and the probability for one crosslink per DNA-molecule $\frac{1}{m}$ is very small. Therefore, the binomial distribution can be approximated by a Poisson distribution:

$$P(X=i) = \frac{\left(\frac{\nu \cdot D}{m}\right)^{i}}{i\,!} \cdot \exp\left(-\frac{\nu \cdot D}{m}\right) \qquad (2)$$

The probability that at least one DNA-protein crosslink is induced $P\,(X \geq 1)$ is calculated by exactly the opposite $P\,(X=0)$, which is subtracted from 100%.

$$P(X \geq 1) = 1 - P(X=0) = 1 - \frac{\left(\frac{\nu \cdot D}{m}\right)^0}{0!} \cdot \exp\left(-\frac{\nu \cdot D}{m}\right) = 1 - \exp\left(-\frac{\nu \cdot D}{m}\right) \quad (3)$$

By the induction of double-strand breaks the number of DNA molecules m is not constant. So we have to take into account that $Z_{DSB_{tot}}$ double-strand breaks produce $Z_{DSB_{tot}}+1$ DNA molecules. According to $Z_{DSB_{tot}} = a \cdot D + b \cdot D^2$ and the initial m DNA molecules, we get $(a \cdot D + b \cdot D^2 + 1) \cdot m$ DNA molecules by dose D.

Therefore the probability that the number of DNA-protein crosslinks on one DNA molecule is one or more is:

$$P(X \geq 1) = 1 - \exp\left(-\frac{\nu \cdot D}{(a \cdot D + b \cdot D^2 + 1) \cdot m}\right) \quad (4)$$

where $Z_{DSB_{tot}} = a \cdot D + b \cdot D^2$ is measured by electrophoresis and $m = \frac{N_A \cdot [DNA]}{MW_{DNA}}$

N_A is the Avogadro number, $[DNA]$ is the concentration of DNA molecules per volume and MW_{DNA} the molecular weight of the DNA molecules.

If the number of double-strand breaks are known and DNA-protein crosslinks are estimated by the nitrocellulose filter assay, the DNA-protein crosslinks ν per dose can be calculated according to equation 4.

3.2 Experimental data of double-strand breaks and DNA-protein crosslinks

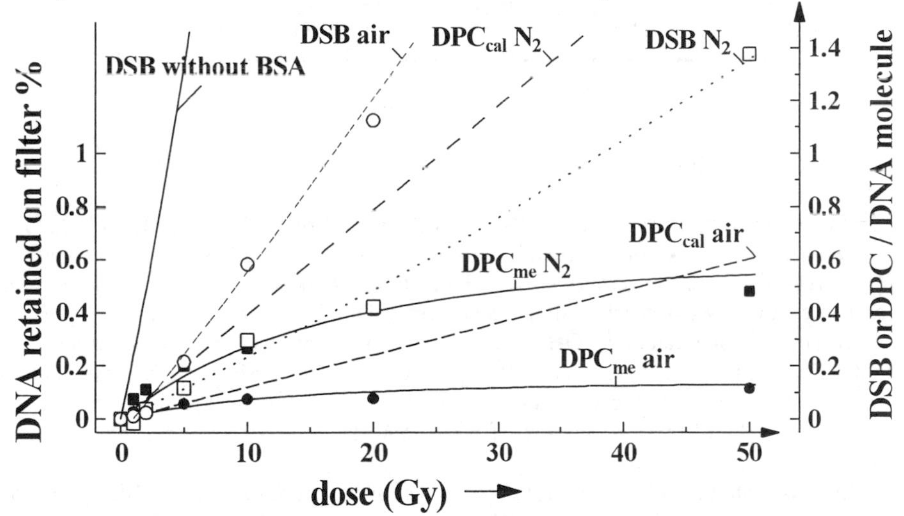

figure 1 Irradiation of DNA and BSA in 10^{-2} mol dm^{-3} phosphate buffer. Measured DPC_{me} (DPC-measured) values in dependence of dose, left ordinate and continuous line. DPC_{me} are fitted by equation 4. The resulted DPC_{cal} (DPC-calculated) are given as dashed lines, the $DSB = a \cdot D + b \cdot D^2$ as spotted lines and right ordinate.

To study the influence of proteins on the radiation induced DNA damage, different concentrations of phosphate buffer were used. At higher salt concentration (10^{-1} mol dm^{-3})

the accessibility of protein to DNA is less than at low salt concentration (10^{-3} mol dm^{-3}), because of shielding the DNA by cations. The proteins scavenge due to homogenous reaction kinetics more than 87% of the primary radicals.
The DNA-protein crosslinks were measured by the nitrocellulose filter assay and the obtained values were fitted by equation 4. The strong influence of the aerobic or anaerobic conditions for DSB and DPC can be seen in figure 1. The results of radiation induced DSB and DPC are condensed in table 1. If DNA is irradiated without a scavenger, the OER of DSB's is inverse. The addition of protein to DNA causes an OER of about 2 for DSB, while for DPC's the OER is inverse. The DSB - DPC ratio is smaller than 1 for nitrogen and higher than 1 for air.

table 1 The α-values for $DSB = a \cdot D + b \cdot D^2$ and the DPC's per DNA molecule and Gy are listed. To get DSB or DPC per $1x10^6$ nucleotides, the values have to be multiplied by 20.25

Gas	PO_4 (mmol dm^{-3})	α DSB without BSA	α DSB + BSA	DPC	ratio DSB:DPC
Nitrogen	1	0.227	0.033	0.070	0.47
	10	0.210	0.022	0.039	0.57
	100	0.116	0.010	0.030	0.35
Air	1	0.186	0.051	0.011	4.47
	10	0.167	0.046	0.012	3.80
	100	0.069	0.015	0.015	1.00
OER	1	0.82	1.56	0.16	
	10	0.80	2.05	0.31	
	100	0.59	1.44	0.50	

4 CONCLUSION

Since one protein crosslinked to DNA is sufficient to retain DNA on the filter, further crosslinked protein is not recorded by this assay. In addition DSBs reduce the amount of DNA retained by proteins on the filter. The equation 4 takes into consideration both facts and calculates the exact values of actually formed DPC's per dose.

DSBs are mainly caused by OH-radicals and radiolysis of DNA without scavengers does not show an OER. In the presence of BSA oxygen enhances DSB, though under anaerobic and aerobic conditions the same amounts of OH-radicals are produced and in both cases more than 87% of these radicals react with BSA. Therefore, this effect must be protein related.

In oxygenated solutions the proteinperoxyl-radicals produce mainly DSB, while under anaerobic conditions the proteinradicals form predominantly DNA-protein crosslinks.

1. Ljungman M., Radiat. Res. 1991, **126**, 58-64.
2. Chiu S.-M., L.R. Friedman, N.M. Sokany, L.Y.Xue and N.L. Oleinick; Radiat. Res., 1986, **107**, 24.
3. Sedmak, J.J. & Grossberg, S.E. 1977 Analytical Biochemistry 79, 544-552.

THE ROLE OF PACKAGING IN THE RADIOPROTECTION OF DNA BY HIGHLY CHARGED LIGANDS

C. Savoye, S. Ruiz, S. Hugot, D. Sy, C. Swenberg*, R. Sabattier$, M. Charlier and M. Spotheim-Maurizot

Centre de Biophysique Moléculaire, CNRS, Rue Charles Sadron 45071 Orléans Cedex 2
*Applied Cellular Radiobiology Department, Armed Forces Radiobiology Research Institute, Bethesda, MD 20889, USA
$ CHR d'Orléans, Service d'Oncologie et de Radiothérapie, 45067 Orléans cedex 2

1 INTRODUCTION

The majority of DNA in all living organisms is present in a compact form. The compaction of the DNA and the stabilization of the compact structures are ensured by metal ions, polyamines, proteins, macromolecular crowding or DNA supercoiling.

Polyamines are implicated in many biological processes (initiation of DNA and RNA synthesis, increase of the fidelity of translation) leading to cell growth and proliferation. They are more abundant in cells with fast proliferation than in quiescent ones[1] and are supposed to cause radioresistance of tumor cells, and therefore to diminish the efficiency of radiotherapy[2]. That is the reason of proposing drugs that decrease the intracellular concentration of polyamines as anti-tumoral strategies.

Another way of increasing the efficiency of radiotherapy is to increase doses. In this case, it is necessary to protect specifically normal cells located around the tumor. A drug presenting such an ability is the phosphorothioate Ethyol which acts *via* its two active forms : the aminothiol WR-1065 and the disulfide WR-33278.

Here, we compare the radioprotective effects of one metal ion[3], the Al^{3+}, of two polyamines[4] : the spermidine (NH_3^+-$(CH_2)_4$-NH_2^+-$(CH_2)_3$ -NH_3^+) (Z=+3) and the spermine (NH_3^+- $(CH_2)_3$- NH_2^+- $(CH_2)_4$- NH_2^+- $(CH_2)_3$- NH_3^+) (Z=+4) and one aminodisulfide[5], the WR-33278 (NH_3^+-$(CH_2)_3$-NH_2^+-$(CH_2)_2$-S-S-$(CH_2)_2$-NH_2^+-$(CH_2)_3$-NH_3^+) (Z=+4) on the radiolysis of a plasmid DNA and of a DNA fragment by fast neutrons. The similarities are discussed in terms of a general phenomenon induced by highly charged counterions : the packaging of long DNA molecules into radioresistant condensed structures. The differences are discussed in terms of the influence of the charge and the size of the counterion.

2 MATERIALS AND METHODS

The pOP203 plasmid (4566 bp) and the fragment of 120 base pairs were prepared as previously described[6]. DNA was irradiated in 1 mM phosphate buffer, pH 7.25 at a concentration of $1.4.10^{-4}$ M nucleotides (plasmid) or 5.10^{-6} M nucleotides (fragment).

Solutions of $AlCl_3$ (Merck), spermidine, spermine (Sigma) and WR-33278 (Walter Reed Army Institute, Washington, DC) were prepared in 1mM potassium phosphate buffer, pH 7.25, immediately prior to use and were added to each DNA sample under stirring .

Irradiations were performed with fast neutrons obtained by the nuclear reaction of 34 MeV protons on a semithick beryllium target (Centre d'Etude et de Recherche par Irradiation, CNRS, Orléans). For experiments with plasmids the mean dose rate of fast neutrons monitored with transmission chambers was 0.3 Gy/min and the dose mean lineal energy in water at the point of interest was 87.7 KeV/μm. For experiments with fragments the mean dose rates was 10 Gy / min.

For plasmid DNA, horizontal agarose (1.2%, Gibco BRL) minigels in Tris-acetate buffer (Tris 20 mM, sodium acetate 10 mM, EDTA 1 mM, pH 8.1) were run for 2 hours at 830 $V.m^{-1}$ at 20°C. The electrophoresis separates the supercoiled, the circular relaxed and the linear forms of the plasmid. The fractions of the three forms in each lane were assayed, after ethidium bromide staining, using a video-camera system (Bioprobe Systems) and the Image Quant software (Molecular Dynamics). The average number of strand breaks per plasmid was determined as previously described[7].

DNA fragment samples were loaded after heat-denaturation on polyacrylamide sequencing gel 8% (acrylamide/bisacrylamide : 19/1) supplemented with 7 M urea and 20% formamide. Electrophoresis was performed for 105 min at 40 W in TBE buffer (89 mM Tris, 89 mM borate, 2 mM EDTA, pH 8). To identify the bands, Maxam-Gilbert sequencing of purines and pyrimidines of the same fragment was performed on the same gel. Gels were then fixed, dried, exposed onto PhosphorImager plates and analysed using the Image Quant software.

The circular dichroism spectra were recorded at 4°C with a Jobin-Yvon autodichrograph Mark V.

3 RESULTS AND DISCUSSION

3.1 Radiolysis of plasmid DNA

Samples of supercoiled plasmid were irradiated with 30 Gy neutrons in the absence (control) and in the presence of increasing amounts of Al^{3+}, spermidine, spermine and WR-33278. Figure 1 presents the plots of the ratio of the number of strand breaks induced in presence and in absence of cations (1 / PF) as a function of the concentration of ligand. The order of protection efficiencies is : spermine and WR-33278 > spermidine and Al^{3+}.

To explain these protections, we have studied by circular dichroism the effects of the cations on the conformation of the plasmid. Large changes in circular dichroism spectra were observed in the ligand concentration range inducing protection (figure 2). These changes were attributed to packaging of DNA into condensed structures.

Trivalent and tetravalent cations are able to condense around DNA until they neutralize respectively 92% and 94% of charges of phosphate groups[8]. From 89 % of neutralization, condensation of DNA starts[9]. The process passes through metastable states (the Ψ-forms observed at certains wavelengths in our CD measurements for spermidine and WR-33278) to reach well-ordered condensates. They are formed by the intra and intermolecular condensation allowed by the decrease of electrostatic repulsive forces between the strands of the same or of different molecules. Intermolecular condensation by long cations can also occur by the cross-linking of DNA helices. For instance, the amine groups of polyamines may get in contact with the phosphates of different DNA molecules and the fully extended aliphatic chains may serve as bridges[10]. The condensed structures are essentially toroids in which DNA is circumferentially wrapped[11] so that helical segments are quasi- parallel and closely placed. Widom and Baldwin[12] showed that toroids formed from pBR322 (a plasmid only 5% shorter than pOP203), contain approximately 10 DNA molecules and have a diameter of 0.1 μm.

We conclude that the similar radioprotection by the four studied ligands is due to a general phenomenon: the packaging of DNA by highly charged counterions that reduces the accessibility of $OH^{\bullet}$ to the radiolytic attack sites. Because of their greater efficiency of phosphate neutralization, spermine and WR-33278 induce packaging at lower concentrations and are thus better radioprotectors than spermidine and Al^{3+}. Recently, Newton et al.[13] have reported a similar radiopotective effect of spermine in physiological conditions.

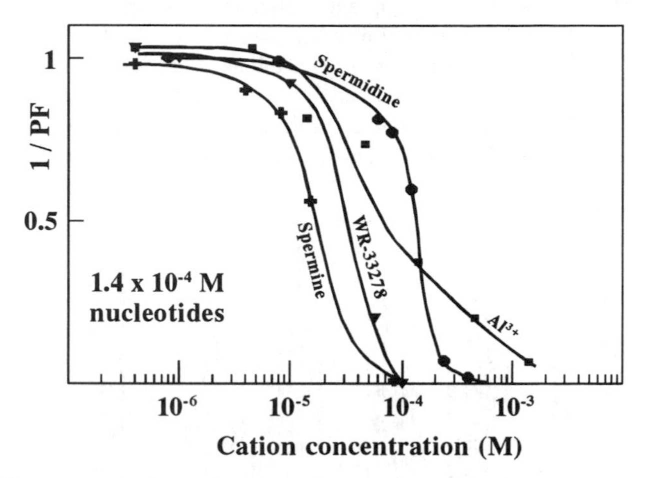

Figure 1 : Variation of the inverse of the protection factor (1/ PF) as a function of cation concentration.

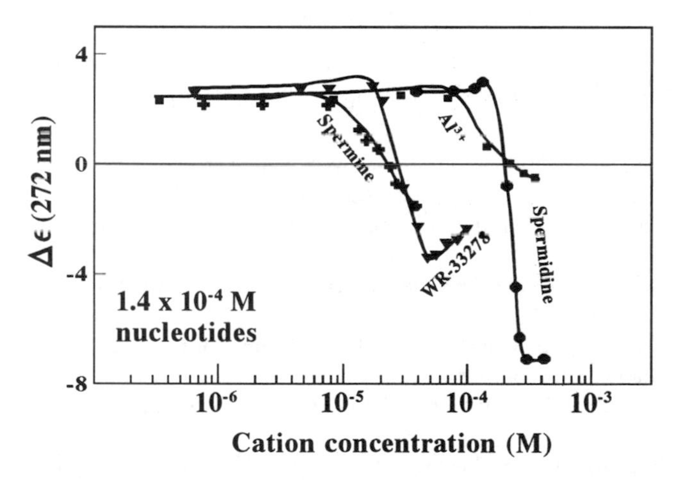

Figure 2 : Variation of the circular dichroism signal at 272 nm as a function of cation concentration.

3.2 Radiolysis of DNA fragments

To evaluate the effects of Al^{3+}, spermidine, spermine and WR-33278 on a short linear DNA fragment of 120 base pairs, we have determined the breakage probabilities at each nucleotide of the samples irradiated with and without a high concentration of cations. Figure 3 shows these values for DNA-spermine complex. The wave-like breakage patterns observed at high concentrations of spermidine, spermine and WR-33278 recall the periodic one observed for the 146 bp DNA of irradiated core nucleosome[14]. The fact that such a pattern is not observed for Al^{3+}, together with results of molecular modelling showing WR-33278 induced conformational changes of the helix (narrowing of the minor groove and contraction of DNA)[5], suggests two explanations. One is the possible bending of monomolecular DNA-bulky ligand complexes characterized by periodically variable structural parameters, as shown for the roll of a bent neutralized DNA[15]. If not only the roll but also the width of the minor groove varies periodically (period of around 10 bp), the wave-like pattern of DNA with the discussed ligands would be entirely explained. The second explanation involves the formation of multimolecular complexes involving bulky ligands mediated crosslinks in which the radiolytic attack sites would be only periodically accessible.

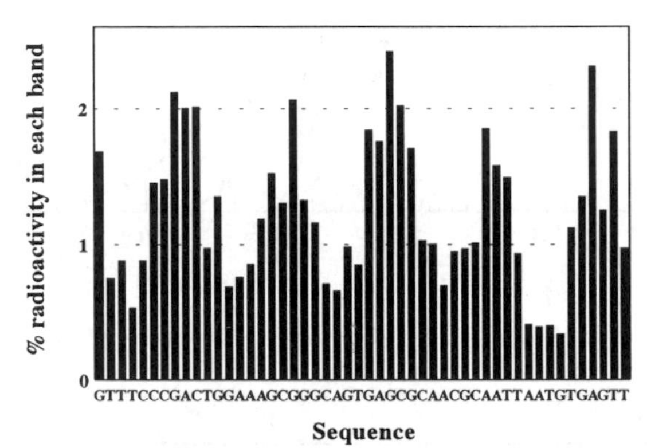

Figure 3 : Breakage probability at each nucleotide site of a part of a 120 base pairs fragment irradiated with 1 added spermine per nucleotide.

1 K.A. Abraham and A. Pihl, Trends Int. Biochem. Sci., 1981, **6**, 106

2 R.D. Snyder and K.K. Schroeder, Radiat. Res., 1994, **137**, 67

3 M. Spotheim-Maurizot, F. Garnier, R. Sabattier and M. Charlier, Int. J. Radiat. Biol., 1992, **62**, 659

4 M. Spotheim-Maurizot, S. Ruiz, R. Sabattier and M. Charlier, Int. J. Radiat. Biol., 1995, **68**, 571

5 C. Savoye, C. Swenberg, S. Hugot, D. Sy, R. Sabattier, M. Charlier and M. Spotheim-Maurizot, Int. J. Radiat. Biol., in press

6 C. Savoye, R. Sabattier, M. Charlier and M. Spotheim-Maurizot, Int. J. Radiat. Biol., 1996, **70**, 189

7 M. Spotheim-Maurizot, J. Franchet, R. Sabattier and M. Charlier, Int. J. Radiat. Biol., 1991, **59**, 1313

8 G.S. Manning, Quarterly Review of Biophysics, 1978, **11**, 179

9 R.W. Wilson and V.A. Bloomfield, Biochemistry, 1979, **18**, 2192

10 J.A. Schellman and N. Parthasarathy, J. Mol. Biol., 1984, **175**, 313

11 K.A. Marx and G.C. Ruben, J. Biomol. Struct. Dynam., 1984, **1**, 1103

12 J. Widom and R.L. Baldwin, J. Mol. Biol., 1980, **144**, 431

13 G.L. Newton, J.A. Aguilera, J.F. Ward and R.C. Fahey, Radiat. Res., 1996, **145**, 776

14 J. Franchet-Beuzit, M. Spotheim-Maurizot, R. Sabattier, B. Blazy-Baudras and M. Charlier, Biochemistry, 1993, **32**, 2104

15 S.R. Sanghani, K. Zakrzewska and R. Lavery, Biological Structure and Dynamics, Eds R.H. and M.H. Sarma, 1996, 267

STRAND BREAK INDUCTION IN DNA BY ALUMINIUM K ULTRASOFT X-RAYS: COMPARISON OF EXPERIMENTAL DATA AND TRACK STRUCTURE ANALYSIS

P. O'Neill, S.M.T. Cunniffe, D.L. Stevens, S.W. Botchway and H. Nikjoo

MRC Radiation & Genome Stability Unit, Harwell
Didcot
Oxfordshire OX11 0RD, UK

1 INTRODUCTION

An essential aspect of the development of quantitative, mechanistic models for DNA damage induction by ionising radiation is a comparison of the calculated yields of strand breaks with those determined experimentally. Since the lowest doses used to obtain experimental data on cellular DNA damage is generally ≥ 1 Gy, the development of these models is essential to make predictions at doses relevant to radiological protection.

Characteristic aluminium K ultrasoft X-rays (Al_K USX) with energy 1.5 keV have been used as probes[1] for low energy track ends for low LET radiation such as ^{60}Co γ-rays. Since the relative biological effectiveness of Al_K USX compared with ^{60}Co γ-rays for DNA double strand break (dsb) induction in V79-4 mammalian cells under aerobic conditions is 2.6 (see ref. 2), these two radiations were chosen to compare the yields of DNA single strand breaks (ssb) and dsb determined from track structure[3] and experimentally.

The experimental data were taken from studies on the induction of radiation induced DNA strand breaks in V79-4 cells[2] and in plasmid DNA[4] irradiated under conditions which mimic the mean diffusion distance of the hydroxyl (OH) radical in the cellular environment.[5] Further, a significantly higher proportion of cellular DNA damage produced by Al_K USX compared with damage induced by γ-radiation is suggested to be present as clustered damage.[2] Therefore, the ssb:dsb ratio represents a further benchmark for comparison of calculated and experimental data, since the ssb:dsb ratio decreases as the proportion of clustered DNA damage increases.[6]

From comparison of the yields of DNA strand breaks determined experimentally and theoretically, some factors which may influence the yields determined by these three approaches are discussed.

2 METHODS USED FOR DNA STRAND BREAK DETERMINATION

2.1 Experimentally determined DNA strand breaks

The induction of DNA dsb was determined as previously described[2] in irradiated V79-4 mammalian cells using pulsed field gel electrophoresis (at pH 8.3 and 278K, voltage 45V, pulse time 60 min, run time 48h, chromosomal grade agarose (Biorad)). The yields of dsb/Gy/Da were calculated from the percentage of DNA extracted from the well and comparison with molecular weight markers. The yield of ssb, determined by alkaline filter elution[7] (C. deLara, unpublished data), were normalised assuming a yield of 10^3 ssb/cell/Gy for γ-radiation.[8]

The yield of ssb and dsb induced in pUC18 DNA by Al_K USX and γ-radiation were determined [4,5] by constant field electrophoresis following irradiation of an aerated, aqueous solution of plasmid DNA in the presence of 0.1 mol dm^{-3} ethanol or 0.2mol dm^{-3} Tris HCl, corresponding to a mean diffusion distance of OH radicals of ~6nm. The irradiations and subsequent handling of the DNA was performed at 277K to minimise the formation of heat labile sites in the irradiated samples.

2.2 Track structure analysis of DNA strand breaks

Monte Carlo track structure calculations were used to determine the yield of dsb and ssb based on a 1.5 keV electron track (for Al_K USX) and 100 keV electron tracks (^{60}Co γ-ray) as previously described.[3] Briefly, the model of the DNA used was described by Charlton et al.[9] The track structure analysis of DNA damage incorporated both the direct effects of energy deposition in the DNA[3,9] and the damage arising from diffusible OH radicals on interaction with DNA.[10] The water radicals were allowed to diffuse for 10^{-9}s or, if they encountered DNA or other water radicals within 10^{-9}s, interact and the DNA damage recorded.

This time was chosen since it corresponds to a mean diffusion distance of the OH radical of ~4nm, its diffusion distance estimated in cells.[11] The probability of an OH radical producing a ssb was assumed to be 0.13.[12] Based radicals produced by either direct effects or OH radicals were assumed not to lead to strand breaks.

3 COMPARISON OF THE YIELDS OF STRAND BREAKS CALCULATED AND DETERMINED EXPERIMENTALLY

The yields of DNA ssb and dsb induced in V79-4 cells[2] and plasmid DNA[4] by Al_K USX and γ-radiation are shown in the Table together with the yields determined by track structure analysis.[3] The yields of dsb induced by both radiations are similar whereas the yields of ssb show more variability particularly for γ-radiation. With all three approaches, the yield of dsb induced by Al_K USX is significantly greater (1.6 - 2.7) than that induced by γ-radiation, consistent with the notion that the dsb induced by γ-radiation arise predominantly through low energy (track end) electrons.[2] From the experimental determined ssb:dsb ratio (Table 1), this increased yield of DNA dsb by Al_K USX is clearly at the expense of ssb, consistent with an increased probability of producing clustered damage by Al_K USX.[2] The ssb:dsb ratio from track structure analysis is considerably less than the experimentally determined ratio for γ-radiation whereas the ratios show less variability for Al_K USX. Although the yields of ssb and dsb determined by these three approaches are similar, there remains some variability between the yields for a given radiation. Some factors which may influence the yields of strand breaks are outlined below. The factors noted are by no means an exhaustive list but reflect experimental conditions or modelling parameters which probably influence the yields.

Table 1 *The yield of dsb and ssb induced by Al_K USX and ^{60}Co γ-radiation*[2-4]

	ssb/Gy/Da	dsb/Gy/Da	ssb:dsb
^{60}Co γ-radiation			
plasmid	4.0×10^{-10}	10×10^{-12} 8.0×10^{-12} * 14×10^{-12} **	40:1
V79-4 cells	1.6×10^{-10}	$4\text{-}6 \times 10^{-12}$	32:1
track structure	1×10^{-10}	6×10^{-12}	17:1
Al_K USXR			
plasmid	$< 3.6 \times 10^{-10}$ ***	1.6×10^{-11}	$<22:1$
V79-4 cells	1.5×10^{-10}	1.3×10^{-11}	12:1
track structure	2.4×10^{-10}	1.6×10^{-11}	15:1

* ref 12 , ** ref 13, *** an upper estimate since higher scavenging capacity required to mimic radical diffusion of cellular environment

In the determination of strand breaks by PFGE or elution, a heat treatment is generally used at a temperature sufficiently high to induce heat labile sites. In fact, the yield of dsb is known to be approximately doubled with a heat treatment[4,6] at 310K. Additionally, with alkaline elution, ssb may arise through alkali labile sites. Therefore, the yields of cellular ssb and dsb may be overestimated since they contain contributions from heat and alkali labile sites. In contrast, only frank strand breaks are determined with plasmid DNA or calculated by track structure analysis. Recent studies by Rydberg[14,15] have indicated that dsb may be produced non-randomly leading to the formation of small fragments, especially for high LET radiation. Therefore the yields of cellular dsb, as generally reported from PFGE analysis of DNA damage, may be an underestimate. Due to the small size of plasmid DNA (a few kilo- basepairs), the contribution of non-random effects should be less significant. Several factors as discussed above may lead to an underestimation (heat labile sites) or an overestimation (formation of small fragments) of the dsb yields.

Some factors which may lead to an overestimation of the yield of plasmid DNA strand breaks, when compared with the yield of cellular dsb, are limitations associated with the use of high scavenger concentrations to mimic the cellular environment with respect to OH radical diffusion distances. The data for plasmid DNA in the Table reflect a mean diffusion distance for OH radicals of ~6nm compared with ~4nm estimated in cells. Further, secondary radicals, produced on reaction of the scavenger with OH radicals, may induce strand breaks[16] but with a significantly reduced probability compared with that for OH-radicals.

In the track structure analysis, several parameters, which are variables, are used relating to the chemistry as outlined below.
i) The average energy required to produce a DNA ssb is taken to be 17.5eV.
ii) The probability of an OH radical to produce a ssb on interaction with DNA is taken to be 0.13. This parameter has been determined from plasmid DNA in aqueous solution.[12]
iii) To simulate a mean diffusion distance in cells[11] of 4nm for the OH radical, the water radicals are allowed to diffuse for 10^{-9}s or, if they encounter DNA or other water radicals within this time, interact. Alternative approaches, by including scavenger molecules in the calculation, may be used to mimic the cellular environment with respect to diffusion. The effect of variation of some of these parameters has been discussed in detail.[3]

In summary, the incorporation of DNA damage induced by water radicals into the models has significantly increased their predictive power as seen from the agreement with experimentally determined yields of DNA damage especially for Al_K USX. Furthermore, several parameters have been discussed for which more experimental data are required.

ACKNOWLEDGEMENTS

We would like to thank Professor Goodhead for helpful discussion. The study was partly funded by the Commission of the European Union (Contract No F14P-CT95-0011).

References

1. D.T. Goodhead, K.M. Prise, M. Folkard, S. Davies and B.D. Michael, *Radiat. Res.*, 1989, **117**, 489; D. Frankenberg, D.T. Goodhead, M. Frankenberg-Schwager, R. Harbich, D.A. Bance and R.E. Wilkinson, *Int. J. Radiat. Biol.*, 1986, **50**, 727.
2. S.W. Botchway, PhD thesis University of Leicester, 1996.
3 H. Nikjoo, P. O'Neill, M. Terrisol and D.T. Goodhead, *Int. J. Radiat. Biol.*, 1994, **66**, 453; *Int. J. Radiat. Biol.*, submitted.
4. S.M.T. Cunniffe, D.L. Stevens and P. O'Neill, in preparation.
5. P.S. Hodgkins, M.P. Fairman and P. O'Neill, *Radiat. Res.*, 1996, **145**, 24.
6. G.D.D. Jones, T.V. Boswell and J.F. Ward, *Radiat. Res.*, 1994, **138**, 291.
7. T.J. Jenner, C.M. deLara, P. O'Neill and D.L. Stevens, *Int. J. Radiat. Biol.*, 1993, **64**, 265.
8. J.F. Ward, *Prog. Nucl. Acid Res. Mol. Biol.*, 1988, **35**, 95.
9. D.E. Charlton, H. Nikjoo and J.L. Humm, *Int. J. Radiat. Biol.*, 1989, **56**, 1.
10. M. Terrissol and A. Beaudré, *Rad. Prot. Dos.*, 1990, **31**, 175.
11. R. Roots and S. Okada, *Radiat. Res.*, 1975, *64*, 306.
12. J.R. Milligan, J. A. Aguilera and J.F. Ward, *Radiat. Res.*, 1993, **113**, 151.
13. R.E. Krisch, M.B. Flick and C.N. Trumbore, *Radiat. Res.*, 1996, **126**, 251.
14. B. Rydberg, *Radiat. Res.* , 1996, **145**, 200.
15. M. Loebrich, P.K. Cooper and B. Rydberg, *Int. J. Radiat. Biol.*, 1996, **70**, 493.
16. J. R. Milligan and J. F. Ward, *Radiat. Res.*, 1994, **137**, 295.

AN X-RAY PHOTOELECTRON INVESTIGATION OF THE EFFECTS OF LOW-ENERGY ELECTRONS ON DNA BASES

D. Klyachko, T. Gantchev, M. A. Huels and L. Sanche

Groupe du CRM en Sciences des Radiations, Faculté de Médecine
Université de Sherbrooke
Sherbrooke, Québec, Canada J1H 5N4

1 INTRODUCTION

Radiation damage to DNA and its constituents induced by UV, X-ray and γ–radiations has been the topic of numerous studies in the past[1]. The radiation effects of high energy photons contain the contributions from the primary quanta as well as the low-energy secondary electrons generated by them. Nevertheless, the destructive effect of the latter on DNA and its constituents has not been studied to any extent. Due to their substantial scattering cross-sections, low-energy electrons penetrate into organic substances to a depth of 10 -100 nm, to which the radiation damage is confined. However, traditional methods of radiation chemistry such as Electron Paramagnetic Resonance, Nuclear Magnetic Resonance, etc. lack surface sensitivity. Thus, in this study, we apply methods of surface analysis, and, particularly, X-ray photoelectron spectroscopy (XPS)[2] to study the radiation damage in DNA bases induced by low-energy electrons.

2 EXPERIMENTAL METHODS AND RESULTS

The apparatus used for the present study is described in some detail in a concurrent contribution[3]. Nucleotide bases are evaporated onto the surface of a MoS_2 crystal using a miniature oven, held at T=105 C for thymine, T=135 C for cytosine, T=150 C for adenine and T= 275 C for guanine. Thin layer chromatography is used to verify that the composition of the deposited films is identical to the original compound. The thicknesses of the films are in the range of 100-200 Å which correspond to almost complete extinction of the characteristic spectra of the MoS_2 substrate.

The bases are irradiated with a $\sim 4 \times 10^{16}$ cm^{-2} dose density of 100 eV electrons at an angle of 25° from the surface and are characterized before and after irradiation using XPS.

Survey XPS spectra of the films of nucleotide bases are shown in Fig. 1. They reveal the C 1s, N 1s and O 1s photoelectron lines as well as the O KVV, N KVV and C KVV Auger lines. Each 1s peak is accompanied by a satellite located at a higher binding energy which originates from multielectron processes[2]. Using the tabulated coefficients of elemental sensitivity[4] we calculate the chemical composition of the films which are found to be in excellent agreement with the chemical formula of the compound.

High resolution C 1s spectra (in Fig. 2 we show those only of thymine and cytosine) exhibit a complicated structure due to the contributions of several C atoms occupying nonequivalent molecular sites. The spectra are fitted with an appropriate number of identical gaussians corresponding to the number of individual C atoms in a nucleotide base[5]. Subsequently, the spectral components are assigned to individual atomic sites using previous ab-initio calculations of the C 1s binding energies[6] in bases and XPS studies of different substituted nucleotide bases[7]. The labels in Fig. 2 represent the atom numbers in the chemical formula which are shown in the left upper corner of each figure. The N 1s spectra (not shown) also reveal multicomponent structure, but the separations between the components are lower than those in the C 1s spectra.

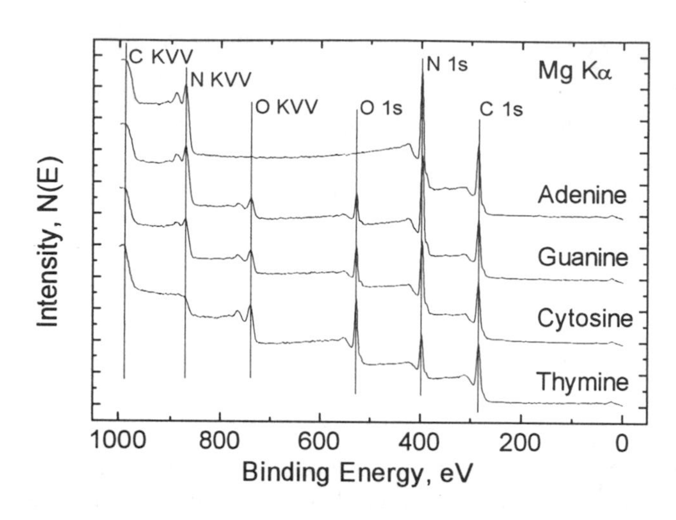

Fig. 1. Survey XPS spectra of the films of nucleotide bases. deposited in vacuum via thermal evaporation

The two components of the O 1s spectrum of thymine are unresolved. The variations of the C, O and N concentrations for each of the studied compounds are presented in Fig. 3. The amplitudes of the variations have been determined using three sets of data: (1) the area under the main 1s peaks in the high resolution spectra (left bar for each compound in Fig. 3), (2) the area under the main 1s peak in the low resolution spectra (middle bar for each compound in Fig. 3) and the area under the low

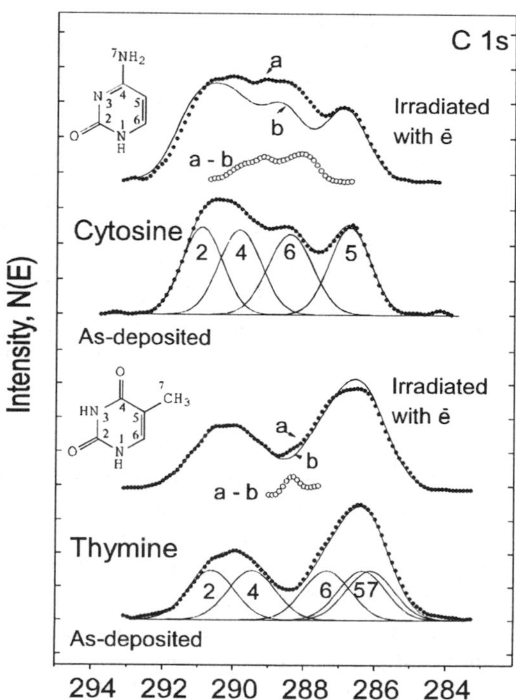

Fig. 2. High resolution C 1s spectra of cytosine and thymine. The spectra of as-deposited compounds are shown below those irradiated with electrons (E_e = 100 eV, $\Phi = 4\times10^{16}$ cm^{-2}). The peak labels correspond to the atom numbers in the chemical formula; a - the spectra after electron irradiation, b - broadened spectra of as-deposited compounds, a - b - their difference, corresponding to radiation products.

resolution 1s spectra including the satellite peaks (right bar for each compound in Fig. 3)[8]. According to Fig. 3 the O concentration drops in cytosine and guanine by ~12-15% but does not change in thymine. The small loss of N in guanine, of C and O in thymine, and C in adenine which can be traced from Fig. 3 is close to the experimental error[9] but is supported by measurements at higher electron doses.

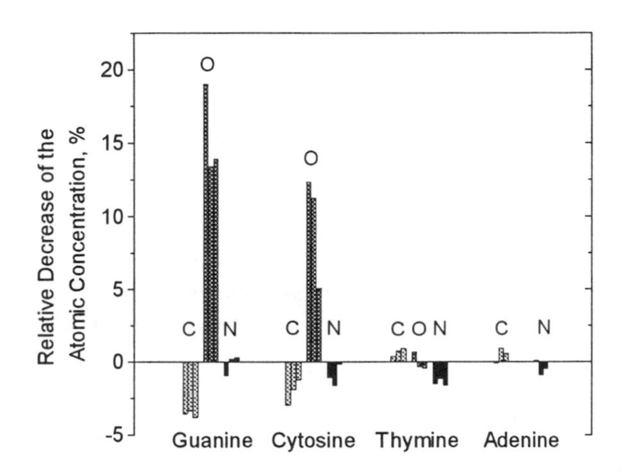

Fig. 3. The changes of the O, C and N atomic ratio in nucleotide bases irradiated with 100 eV electrons ($\Phi = 4 \times 10^{16}$ cm^{-2})

It should be noted that the C 1s spectra of irradiated nucleotides contain the spectral features of undamaged compounds as well as those of radiation products (Fig. 2) and are broadened by a few hundredth of an eV compared to those of the as-deposited films.

By adjusting the width of the spectra of as-deposited films and superimposing them onto the spectra of irradiated compounds the difference spectra, corresponding to the radiation products, can be extracted (open circles in Fig. 2).

In cytosine, the peaks corresponding to the radiation products, are superposed with the C4 and C6 spectral components. In thymine, the peak of the radiation product forms in the space between the C4 and C6 peaks, and its formation is accompanied by the relative reduction of the C5 and/or C7 spectral components. The C 1s spectrum of guanine (not shown) shows reduction of the C4, C5 and C8 components and their shift towards each other. In adenine, the only well resolved feature, corresponding to the C5 atom, reduces upon irradiation. Nevertheless, the spectra of the irradiated film can be fitted with a broadened spectrum of the as-deposited film, indicating that the reduction of the C5 peak is not related to the decay of the adenine molecule.

The identification of radiation products formed under the electron bombardment, and observed by XPS, is complicated by their large number as identified previously[1]. Most of these products have been identified in UV irradiated films of DNA bases, while the data on radiation products created by X- rays and electrons is scarce [1a, p.151]. In general, organic molecules loose hydrogen under electron and X-ray irradiation and develop intermolecular links [10].

A salient feature of our experiments is a rather high concentration of radiation products which comprise ~5% of the total number of molecules in the films (dictated by the detection limits of XPS), which is significantly larger than those, obtained in the experiments on UV- or γ-ray radiolysis of DNA bases. The interaction between defects might lead not only to the dimerization of molecules, but also to more complex multimer structures.

In thymine, the formation of the radiation products is accompanied by the reduction of the C5 and/or C7 components without any changes in the chemical composition. We can speculate that methyl groups and/or C5 atoms are involved into such links between molecules. In cytosine and guanine, the formation of the radiation products is accompanied by the loss of the oxygen signal. It is likely that detached oxygen atoms react with neighboring H atoms and form molecules of water, which desorb from the surface. The remaining dangling bonds recombine forming links between cytosine molecules.

3 CONCLUSIONS

We conclude that thin films of nucleotide bases can been formed without decomposition by thermal evaporation in vacuum and characterized with XPS. Irradiation of DNA bases with 100 eV electrons results in preferential desorption of oxygen from the films of cytosine and guanine. No desorption of oxygen is observed in thymine under the same conditions. The changes in the C 1s and N 1s XPS spectra in irradiated compounds show a selective effect of electrons on certain molecular sites.

The authors are grateful to Dr. S. Kudrevich for chemical analysis of the samples.

The work has been supported by Fonds de la Recherche en Santé du Québec and the Medical Research Council of Canada and the Atomic Energy Commission of Canada Ltd.

4 REFERENCES

1. (a) C. Von Sonntag, The Chemical Basis of Radiation Biology, Taylor and Francis, London, 1987 ; (b) A.O. Colson, M.D. Sevilla, Int. J. Radiat. Biol., 67, 627 (1995).
2. For a review of the method of X-ray photoelectron spectroscopy see K. Siegbahn, C. Nordling, A. Fahlman, *et al* Electron Spectroscopy for Chemical Analysis, Uppsala, 1967; for applications of XPS to organic molecules: D.T.Clark in Characterization of Metal and Polymer Surfaces, Ed. Lieng-Huang Lee, Academic Press, v. 2, p. 5;
3. M. A. Huels, J. Khoury, B. Gueraud, B. Boudaiffa, C. P. Dugal, D. Hunting, and L. Sanche, contribution to the 12th Symposium on Microdosimetry, Oxford, England, Sept. 29 - Oct. 4 (1996).
4. C.D.Wagner, W.M.Riggs, L.E.Davis, J.F. Moulder, Handbook of X-ray photoelectron Spectroscopy, Perkin-Elmer Corp., 1979.
5. In the case of cytosine, gaussians with the same area but with slightly different FWHM have been used to achieve better agreement with the experiment.
6. M. Barber, D.T. Clark, Chem. Commun., 23, 24 (1970); B. Mely, A. Pullman, Theor. Chim. Acta, 13, 278 (1969); A. van der Avoird, Chem. Commun., 727, (1970), K. Ishida, H. Kato, H. Nakatsuji, T. Yonezawa, Bull. of the Chem. Soc. Japan, 45, 1574 (1972).
7. J. Peeling, F.R. Hruska, N.S. McIntyre, Can. J. Chem. 56, 1555 (1978).
8. The presence of radicals in irradiated compounds might affect the probability of multielectron transitions in XPS. The latter produce a broad spectra which is superimposed with the background. By varying the resolution of the spectrometer and the width of the energy window in which the intensity of the core peaks is counted, it is possible to estimate the importance of the latter effect in the measured chemical composition.
9. Desorption of any chemical element from the surface results in a relative decrease of its concentration and in a relative increase of the concentration of the other chemical elements. Near zero changes of the N concentration in guanine and O in thymine can therefore be explained by small losses of these elements from the surface.
10. Radiation Chemistry of Hydrocarbons, Ed. G. Földiak, Elsevier, 1981.

A NOVEL APPARATUS FOR LOW-ENERGY ELECTRON (0-5000 EV) IRRADIATION OF LYOPHILIZED DNA IN AN ULTRA-CLEAN UHV ENVIRONMENT.

M. A. Huels, J. Khoury, B. Gueraud, B. Boudaiffa, P. C. Dugal, D. Hunting, L. Sanche, and A. J. Waker§

Canadian MRC Group in Radiation Sciences, Dept. of Nuclear Medicine, Faculty of Medicine, Univ. of Sherbrooke, Sherbrooke, Québec, Canada J1H 5N4
§ Atomic Energy Commission Limited, Chalk River Laboratories, Chalk River, Ontario, Canada K0J 1J0.

1 INTRODUCTION

It is generally believed that many catastrophic effects of ionizing radiation can be attributed to structural and chemical modifications of cellular DNA. This damage may take the form of, e.g., single and double strand breaks (ssb and dsb),[1] base and sugar modifications,[2] or DNA - protein cross links;[3] it is therefore crucial to understand the various mechanisms of this damage.

In the past, many research efforts have concentrated on DNA damage induced by initial radiation products, such as energetic primary and secondary electrons (MeV to keV range), or cations and radicals produced along their tracks. Only a few direct measurements[4] exist on DNA damage by *low*-energy electrons (LEE, << 1 keV); however, none used *in situ* analysis, or ultra-high vacuum (UHV) methods. The latter is required, since condensed phase electron - molecule reactions are easily perturbed by the presence of even small quantities of different molecular species[5, 6, 7]. That LEE, which comprise the majority along radiation tracks, may indeed be subject for concern, has already been demonstrated by the many electron impact studies of solids consisting of various small, biologically relevant, molecules[8]: it is generally found that (a) LEE, even at sub-ionization energies, induce formation of highly reactive atomic[9], or molecular[10], anion and neutral[11] fragments, within 10^{-13} s or less, and with large cross sections, and (b) molecular dissociation by LEE may be dramatically enhanced[7], reduced[6], or even quenched[5], depending on the composition of the molecular environment. Finally, recent LEE impact studies on isolated thymine (T) and cytosine (C) have evidenced molecular anion fragment, *and* stable T^- and C^-, formation at energies well below 5 eV.[12] The well-established surface science techniques utilized in many of the above studies can now be extended, and combined with standard biological methods, to probe the damaging effects of LEE on DNA, or its components.

2 EXPERIMENTAL METHOD AND PRELIMINARY RESULTS

2.1 Substrate / Sample Preparation and Post-Irradiation Analysis

The Au substrates are prepared in a custom made evaporation chamber by *in vacuo* deposition on freshly cleaved mica, or chemically clean glass. This recently constructed high vacuum chamber allows the production of 300 cm^2 of substrate surface per batch, under identical conditions, corresponding to 60 - 70 samples. Freshly cleaved mica (muscovite type) is placed in the evaporator, pumped to 10^{-8} Torr, and heated at ~350°C for 15-20 hours prior to Au deposition. During the Au evaporation, the mica is held at ~450°C, and then annealed at this temperature for another 1-2 hours. The quality, and surface structure of the prepared films has been verified by scanning tunneling microscopy;[13] the above method results in large surface planes (several 100 nm) of (111) orientation, with atomic steps clearly visible. Further chemical or structural analysis of these substrates may also be performed via SEM (Auger), X - ray fluorescence, ESCA and LEED. To eliminate residual organic contamination, the substrates are cleaned with sulfochromic acid, and copiously rinsed with distilled - deionized water.

The plasmid DNA is deposited onto the Au via lyophilization: $2x10^{-6}$g of either pGEM, or M13 (both freshly purified), are deposited from a solution of spectrograde H_2O onto the Au films, held at about - 80 C, which are subsequently dried at less than 5 mTorr by a *clean hydrocarbon free sorption pump*. Good adhesion, with minimal damage, is also obtained with small added quantities of DMF (< 5%), PEG or Glycerol (< 0.1%), or SDS (< 1%).The area of deposition is 6 mm in diameter. To recuperate the DNA after electron irradiation, the samples are rehydrated with a standard buffer (TE pH 7.5) for at least one hour. Subsequently, ssb's and dsb's, induced by electron impact, are detected via gel electrophoresis. Thus, structural damage to DNA can be quantified as functions of incident electron energy and total dosage by comparison with identical samples held in solution (unlyophilized and unirradiated); background signal of ssb's and dsb's, resulting from mechanical and chemical manipulations, are determined by comparison with samples that were lyophilized on Au, introduced into the experimental chamber but *not* irradiated, and recuperated from the Au substrate. The above protocol results only in a limited quantity of ssb's, comparable to that found already in solution, and no dsb's are observed due to the chemical or mechanical manipulations. Similarly, protocols using *radiolabeled DNA* will allow characterization of ssb's and dsb's as functions of *site specificity*.

2.2 DNA Irradiation Chamber

The device used for irradiation of DNA samples consists of two parts: a sealed dry N_2 glove box, and a UHV chamber, where the samples are irradiated by electrons from a high current LEE gun; they are connected via an o-ring seal to an extension of the UHV chamber's main flange. In the glove box about 32 (radiolabeled) samples can be prepared under a clean inert atmosphere and subsequently loaded onto a rotatable sample wheel, seen in Figure 1, which is mounted on a UHV flange. The latter is held on a rolling support rack, which allows the entire assembly to be easily attached to or removed from the UHV chamber. A Faraday detector slit (0.3 mm) allows measurements of the LEE beam density and spatial distribution; these measurements are crucial in order to determine the total cross sections of DNA damage. The samples are loaded onto the wheel, introduced into the UHV chamber, and irradiated by LEE fluxes of up to $2x10^{-4}$ A. The highly collimated beam, with a FWHM of 0.8 mm at 1 keV, is scanned via the X, Y deflectors over the sample area covered by the DNA. A shield (not shown in Fig. 1) protects the samples from stray charges.

Prior to sample introduction the UHV chamber, including the *empty* sample wheel, is pumped to 10^{-9}-10^{-10} Torr and degassed at about 100 C. Subsequently, the chamber is vented with dry N_2, and the sample wheel is retracted into the glove box where the DNA samples are ready to be loaded. *Thus, samples can be prepared in an inert atmosphere and introduced into a 'precleaned' UHV chamber*. This eliminates (a) the necessity of baking the experimental chamber with samples *in situ*, which is usually necessary to achieve the UHV pressures at which contamination of the samples by background gases is negligible, but which is not advisable when working with fragile DNA, and (b) long pumping times.

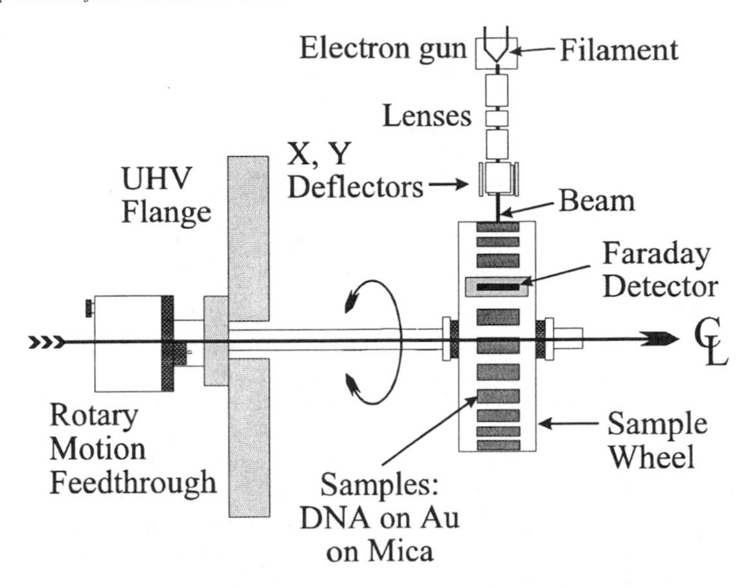

Figure 1 *Simple schematic of the device for irradiation of DNA by LEE as discussed in the text.*

Preliminary results, obtained at incident energies of 50, 100, 200, and 1000 eV, and LEE fluxes of 1 - 37x10^{-6} A, indicate almost complete conversion of supercoiled M13 to a relaxed (ssb) form for high doses at 1 keV, with a small signal of dsb's (linear form) visible at the highest dose of 1.1x10^{18} electrons cm^{-2}. At 50 - 200 eV we observe conversion of pGEM to relaxed (60-70%) and linear (about 30%) forms at doses between 2.1x10^{15} and 6.2x10^{15} electrons cm^{-2}.

2.3 System for *in Situ* Analysis

The apparatus for the *in situ* analysis of low energy electron damage to bio-molecules is a substantially modified version of a previous system used to study electron stimulated oxidation[14] and fluoridation[15] of Si substrates via X-ray photoelectron spectroscopy (XPS). It now consists of three sections: a *dry N_2 glove box* for the preparation of samples and their introduction into, and extraction from, the *load - lock* chamber. The latter is separated from the *main chamber* via a gate valve, and allows the introduction of samples, or an evaporation source for DNA bases, into the main chamber; the base pressures of the load - lock and main chamber are about 10^{-9} and 10^{-11} Torr, respectively. *This effectively minimizes the contamination of the DNA samples by background gases.*

Samples of plasmid DNA on Au are attached to a cryogenically cooled sample holder which is mounted on a rotary platform that permits positioning at various experimental stations in the main chamber. A gas/vapor dosing system allows condensation of other molecules (e.g. H_2O, O_2, hydrocarbons, or thiols) onto the DNA samples. Individual samples are introduced directly from the load - lock via a linear transfer rod. Sample carrousels with 12 samples may also be introduced into the main chamber in which case individual samples are moved onto the sample holder by means of a multi - motion wobble stick. Sample manipulation is monitored via several view ports, and a movable Faraday detector allows measurements of the electron beam profiles at different incident energies.

Two electron guns produce beam intensities between 1x10^{-5} and 2x10^{-4} A, for energies from

0 to 5 keV. Chemical changes and degradation of the samples, as well as chemical reactions of DNA fragments with a metal substrate (e.g. Pt or Si), can be monitored before and after electron irradiation via a high resolution XPS/ESCA system (PHI). This method has been recently used to measure LEE (100 eV) damage, i.e. Oxygen loss, to DNA bases condensed on MoS_2; the results are presented in a separate paper.[16] In this study it is also found that the (low dose) XPS method itself does not induce significant dissociation of the adsorbed molecules.

A high resolution quadrupole mass spectrometer, currently being added to the system, permits measurements of anion, cation, and neutral fragments (up to 500 amu) which may desorb from biological solids during LEE impact. This device is critical in identifying specific mechanisms of DNA damage, e.g. dissociative electron attachment, dipolar dissociation, or others, induced by the low energy electron impact or the reactive products thereof. The structural integrity of certain substrates, e.g. Si or Au, or the ordering of small biological molecules on other substrates such as MoS_2, may be probed with a reverse view LEED system.

Two quartz view ports allow bombardment of the DNA samples by laser photons (femto second pulses or CW mode - the laser is located in an adjacent laboratory). This permits measurements of (a) photo stimulated desorption of ionic or neutral fragments produced by photoelectrons at the substrate - molecule interface where contamination is minimized, and (b) energy spectra of photoelectrons produced at the substrate and scattered in the biological solid.

This research is funded in part by the Atomic Energy Commission of Canada Limited.

References

1. C. von Sonntag, 'The Chemical Basis of Radiation Biology', Taylor & Francis, London, 1987.
2. J. F. Ward, Ad. Radiat. Biol., 1975, **5**, 181.
3. O. Yamamoto, in 'Aging, Carcinogenesis and Radiation Biology', ed. K. Smith, Plenum, New York, 1976, p. 165.
4. M. Folkard, K. M. Prise, B. Vojnovic, S. Davies, M. J. Roper, and B. D. Michael, Int. J. Radiat. Biol., 1993, **64**, 651.
5. M. A. Huels, L. Parenteau, and L. Sanche, Chem. Phys. Lett., 1993, **210**, 340.
6. M. A. Huels, L. Parenteau, and L. Sanche, J. Chem. Phys., 1994, **100**, 3940.
7. P. Rowntree, H. Sambe, L. Parenteau, and L. Sanche, Phys. Rev. B, 1993, **47**, 4537; M. Huels, L. Parenteau, and L. Sanche, Nucl. Instrum. and Methods. B, 1995, **101**, 203.
8. e.g. H_2O, O_2, hydrocarbons, and various others. For reviews, see: L. Sanche, IEEE Trans. Elec. Insul., 1993, **28**, 789; Scan. Micros., 1995, **9**, 619; and references therein.
9. H. Sambe, D. E. Ramaker, M. Deschenes, A. D. Bass, and L. Sanche, Phys. Rev. Lett., 1990, **64**, 523; P. Rowntree, L. Parenteau, and L. Sanche, J. Chem. Phys., 1991, **94**, 8570.
10. L. Sanche and L. Parenteau, J. Chem. Phys., 1990, **93**, 7476; P. Rowntree, L. Parenteau, and L. Sanche, J. Phys. Chem., 1991, **95**, 4902.
11. P. Rowntree, P.-C. Dugal, D. Hunting, and L. Sanche, J. Phys. Chem., 1996, **100**, 4546.
12. M. A. Huels, I. Hahndorf, E. Illenberger, and L. Sanche, to be published.
13. STM work by Prof. P. Rowntree, Dept. of Chemistry, Univ. of Sherbrooke, QC, Canada.
14. D. Klyachko, P. Rowntree, and L. Sanche, Surf. Sci., 1996, **346**, L49.
15. W. Di, P. Rowntree, and L. Sanche, Phys. Rev., 1995, **B52**, 16618.
16. D. Klyachko, T. Gantchev, M. A. Huels, and L. Sanche, concurrently submitted paper to the 12th Symposium on Microdosimetry, Oxford U.K., Sept. 29 - October 4, 1996.

THE EFFECT OF DIMETHYL SULFOXIDE ON INACTIVATION, DSB INDUCTION AND REPAIR OF V79 MAMMALIAN CELLS EXPOSED TO 252-CF NEUTRONS

T J Jenner, C deLara and P O'Neill

MRC Radiation & Genome Stability Unit, Harwell, Didcot, Oxon OX11 0RD, UK

1. INTRODUCTION

Ionising radiation induces a variety of lesions in DNA such as single strand and double strand breaks (SSB and DSB), base modifications and DNA-protein crosslinks.[1] The biological effect of these lesions is thought to lead to cell inactivation, transformation and mutation. An increase in the linear energy transfer (LET) of the radiation leads to an increased efficiency in inactivation of cells, e.g., α-particles are more effective at inactivation of mammalian cells than low LET photons. Although the initial yields of DSB are similar for both types of radiation even taking account of the possible non-random distribution of DSB,[2] the extent of rejoining is significantly different resulting in more residual damage for high LET radiation. It was suggested that these differences reflect the complexity of DNA damage due to differences of the track structure. For high LET radiations the energy is deposited predominantly in discrete clusters whereas for low LET radiations the energy is more sparsely distributed.[3-7] If the DNA is considered to be an important target, then it is necessary to include the effect of radiation on the water molecules associated with the DNA.[8] When radiation interacts with soft tissue, the DNA can be damaged by two pathways, either directly or by the secondary particles interacting with a water molecule to produce a hydroxyl radical, which may diffuse to and cause damage to the DNA. It has been estimated that the hydroxyl radical must be produced within a few nanometers of the DNA.[9] This study was undertaken to extend our understanding of DNA complexity, through investigation of the effects of 252-Cf neutron exposure on cell inactivation, DSB induction and repair in V79-4 Chinese hamster cells. Dimethyl sulfoxide (DMSO), as an hydroxyl radical scavenger, was used to assess the involvement of water radicals.

2. EXPERIMENTAL METHODS

252-Californium was used as a source of neutrons. 252-Cf decays by α-emission and spontaneous fission, emitting neutrons of average energy of ~ 2MeV. When neutrons interact with a hydrogen nucleus in soft tissue, recoil protons are produced which have a range of LET values of between 10 and 90 $keV.\mu m^{-1}$ and a mean value of ~ 45 $keV.\ \mu m^{-1}$. The neutrons also interact with the heavier elements in tissue resulting in α particles of high LET, but these represent a small contribution to the total dose. There is an accompanying low LET photon component which accounts for one third of the total dose. Dosimetry was performed using Fricke dosimetry, assuming that the low LET component is 30%.

Chinese hamster cells, line V79-4, were used throughout this study as previously described.[4] For all experiments the cells were detached and irradiated in suspension. When used with DMSO, the cells were incubated for 15 min at 310K in the presence of 0.75M DMSO. For survival experiments the cell suspensions were exposed to neutrons or γ radiation at room temperature to a maximum dose of 16 Gy at a dose rate of 3 $Gy.h^{-1}$, or 10 $Gy.h^{-1}$ respectively. After irradiation, the cells were assayed for colony formation. For DSB determination they were labelled with 3-H thymidine 24h before use. The cell suspensions were irradiated as above but at 277K to minimise repair during the relatively long exposure times. DNA DSB were assayed using a CHEF-II pulsed field gel electrophoresis system.

3. RESULTS

Survival of V79-4 mammalian cells exposed to 252-Cf neutrons and low LET γ-radiation in the absence and presence of 0.75M DMSO is shown in Figure 1. The Relative Biological Effectiveness (RBE) at 10% survival level is 2.3 for neutrons compared with 5.3 for α-particles.[4] The protection factor (PF) of 0.75M DMSO at the 10% survival level is ~ 1.5 for both the neutron and γ-radiation which compares with a PF of 1.4 obtained for high LET α radiation at 0.5M DMSO.[8]

From the linear induction of DSB on dose by 252-Cf neutrons, the RBE for induction of DSB in V79-4 cells by neutrons is 0.93. Using DMSO as an OH radical scavenger, the extent of protection on DSB induction by 0.75M DMSO for neutron irradiation is ~ 44% and for low LET γ-radiation ~ 62% (Figure 2).

The effect of 0.75M DMSO on the yields of DNA DSB remaining after 60 min incubation of the cells at 310K following neutron and γ-irradiation at a dose of 20Gy are shown in Figure 2. In the radiation only case, there is significantly more rejoining for low LET compared with neutrons i.e. 80% compared with 55%. In the presence of 0.75M DMSO the yield at 1h with neutrons is 30% compared with 17% with γ-radiation. When the repair time was extended to 16h, most of the damage induced by both radiations is rejoined. Earlier results from high LET α-radiation gave a RBE for induction of dsb of 0.87, protection of 32% by 0.5M DMSO and a residual yield of ~ 65% with and without 0.5M DMSO.[8]

4. DISCUSSION

It is well documented that high LET particles are more effective at cellular inactivation than low LET photons. The RBE obtained at 10% survival (2.3) after irradiation with 252-Cf neutrons compares with that obtained previously with neutrons of similar energy [5-7, 10] This RBE for cellular inactivation by neutrons is significantly less than the value of 5.3 for α-particles.[4] An RBE of ~1.0 for DSB induction is in agreement with other studies using either neutral election or PFGE. [7,10] The 252-Cf neutrons are not as effective at cellular inactivation as high LET α-particles. This would be consistent with formation of less severe damage by neutrons than that by α-particles.

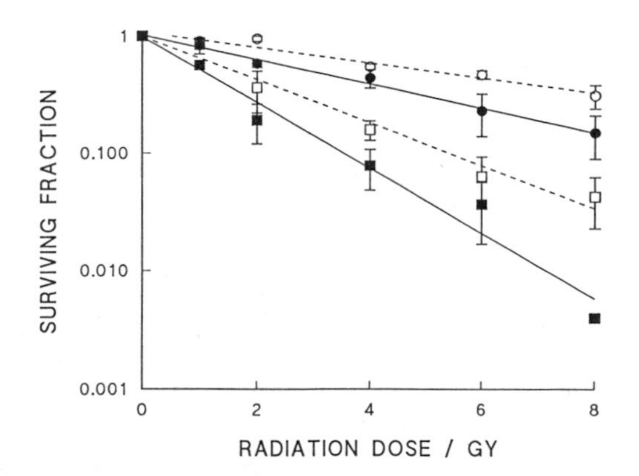

Figure 1 *The dose dependence for inactivation of V79-4 cells by (O ●) γ radiation and (□ ■) 252-Cf neutron radiation in the absence (● ■) and the presence (O □) of 0.75M DMSO.*

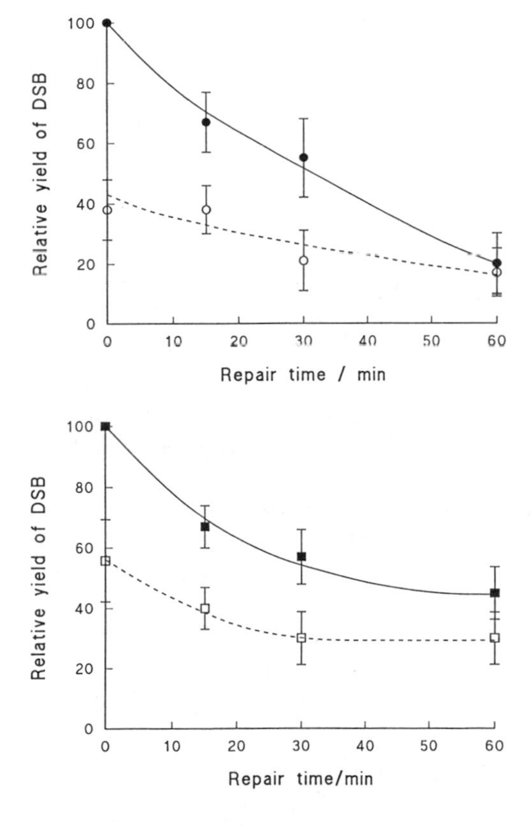

Figure 2 *Time course for rejoining DSB$_s$ at 310K induced in V79-4 cells (i) with 20 Gy of γ radiation in the absence (●) or the presence (O) of 0.75M DMSO (ii) with 20 Gy of 252-Cf neutron radiation in the absence (■) or the presence (□) of 0.75M DMSO.*

Even though the yield of DSB induced is not significantly dependent on the LET of the radiation, the extent of rejoining of DSB induced by neutrons is greater than that for α-partcles[4,7] but less than that for γ-radiation. If damage severity is reflected in the degree of DSB rejoining, then the damage caused by these neutrons is less severe than for α-particles. The inactivation of V79-4 mammalian cells is reduced in the presence of the radical scavenger DMSO for both neutron and γ-irradiation.[8] Earlier studies using high-Z ions of various LET have also shown that radical scavengers have a protective effect on cell inactivation but at a level significantly less than that for low LET radiation.[11]

From this protective effect with DMSO on DSB induction by neutrons it is inferred that diffusible hydroxyl radicals are involved. From the PF on DSB induction for neutrons, the direct effect is less than that with α-particles[7] but still greater than that with γ-radiation. Further evidence for the reduced severity of DNA damage induced by neutrons compared with α-particles is reflected in the extent of DSB rejoining within 1h. From the yields of DSB at 1h in Figure 2 there is an indication that some of the non-scavengeable DSB induced by neutrons are rejoined within 1h in contrast to the stability of non-scavengeable DSB produced by α-particles.[7]

In agreement with the conclusion for α-particles, neutrons produce clustered damage in DNA. but this damage is less severe than that observed with α-particles and hydroxyl radicals contribute to the complexity of DSB induced directly.

REFERENCES

1. J.T. Lett and W.K. Sinclair, Eds. *Adv. Radiat. Biol.*, 1993, **17**.
2. B. Rydberg, *Radiat. Res.*, 1996, **145**, 200.
3. D.T. Goodhead, *Int. J. Radiat. Biol.*, 1989, **56**, 623.
4. T.J. Jenner, C.M. deLara, P. O'Neill and D.L. Stevens, *Int. J. Radiat. Biol.*, 1993, **64**, 265.
5. Grdina, D.J., Sigdestad, C.P., Dale, P.J. and Perrin, J.M., Brit. J. Cancer, 1989, **59**, 17.
6. Peak, M.J., Wang, L., Hill, C.K. and Peak, J.G., *Int. J. Radiat. Biol.*, 1991, **60**, 891.
7. B.P. Kysela, J.E. Arrand and B. D. Michael, , *Int. J. Radiat. Biol.*, 1993, **64**, 531.
8. C.M. deLara, T.J. Jenner, K.M. Townsend, S.J. Marsden and P. O'Neill, Radiat. Res., 1995, **144**, 43.
9. R. Roots and S. Okadu, *Radiat. Res.*, 1975, **64**, 306.
10. Prise K.M., Davies. S and Michael B. D, *Int. J. Radiat. Biol*, 1987, **52**, 893.
11. R. Roots, A. Chatterjee, P. Chang, L. Lummel and E.A. Blakely, *Int. J. Radiat. Biol*, 1985, **47**, 157

CALCULATION OF G-VALUE OF FRICKE DOSIMETER IRRADIATED BY PHOTONS OF 100 EV - 10 MEV

H. Yamaguchi

Space and Particle Radiation Science Research Group
National Institute of Radiological Sciences
9-1, Anagawa 4, Inage-ku, Chiba 263 JAPAN

1. INTRODUCTION

Measurements of G-value of Fricke dosimeter have been elaborated for photons of low energies, 1.5 keV[1], 8.9 and 13.6 keV[2] and 1.8-10 keV[3]. Those values are basis for dosimetry at irradiation to biological samples in solution by low energy photons from synchrotron radiations. Those experimental G-values appear stronger energy dependence than those such as ^{137}Cs or ^{60}Co γ - rays. To study this "radiation quality" there are two approaches, one is Monte Carlo track simulations[4,5] and the other the prescribed diffusion model[6]. The Monte Carlo method is powerful to simulate track structure of electrons , but it is still difficult to simulate fully processes of subsequent chemical reactions[7]. The prescribed diffusion model, on the other hand, may deal with the chemical reaction processes, though it has some difficulty to take into account property of track structure in the model.

Magee and Chatterjee[8,9] estimated the Fricke G-value of electrons of energy from 0.1 keV to 10 MeV based on the model and the results for electrons are claimed to explain the experimental data.

Yamaguchi [10,11] has studied the model based on that of Schwarz[6], multi-radical system for neutral water. It assumed single size of spur, got differential yields of electron path segment and estimated integral yields by folding the differential yields with degradation spectra of electron path length. The calculated Fricke G-values agreed well with experiments by ^{3}T β -rays and photons of energies above 10 keV.

This paper revises those phenomological calculations[10,11] to estimate Fricke G-values of photons from 0.1 keV to 10 MeV for an aerobic Fricke solution: it includes explicitly O_2 and Fe^{2+} solutes in the calculation. The results explain well the current experimental data and are discussed in relation to the recommended Fricke G-values for photon of 5 keV - 150 keV by ICRU (1970)[12].

2. MATERIAL AND METHOD

Spurs are generated along a passage of an electron and contain dissociated water species. Size of a spur (number of species) is assumed to be identical and constant. Chemical reactions in the spur start around 10^{-12} sec reactions (1)-(11) among water species, and reactions (12)-(15) between water species and solutes O_2 and Fe^{2+}. The reaction (11)' is in an equilibrium and was included in this calculation. Reaction rate constants of these processes are given in the literature[13,14].

		rate constant $\times 10^{10}(M^{-1}S^{-1})$
(1)	$e^-_{aq} + e^-_{aq} + 2H_2O \rightarrow H_2 + 2OH^-$	0.55
(2)	$e^-_{aq} + H + H_2O \rightarrow H_2 + OH^-$	2.5
(3)	$e^-_{aq} + H^+ \rightarrow H$	1.7
(4)	$e^-_{aq} + OH \rightarrow OH^-$	2.5
(5)	$e^-_{aq} + H_2O_2 \rightarrow OH^- + OH$	1.3
(6)	$H + H \rightarrow H_2$	1.0
(7)	$H + OH \rightarrow H_2O$	2
(8)	$H + H_2O_2 \rightarrow H_2O + OH$	0.01
(9)	$H^+ + OH^- \rightarrow H_2O$	10
(10)	$OH + OH \rightarrow H_2O_2$	0.6
(11)	$e^-_{aq} + O_2 \rightarrow O_2^-$	1.9
(11)'	$H^+ + O_2^- \rightleftarrows HO_2$	-
(12)	$H + O_2 \rightarrow HO_2$	2.1
(13)	$OH + Fe^{2+} \rightarrow Fe^{3+} + OH^-$	0.05
(14)	$H_2O_2 + Fe^{2+} \rightarrow Fe^{3+} + OH + OH^-$	0.05
(15)	$HO_2 + Fe^{2+} \rightarrow Fe^{3+} + H_2O_2$	0.12

This reaction processes are described by diffusion controlled reaction differential equations,

$$(2.1)\quad dC_i/dt = D_i \nabla^2 C_i - \Sigma_j k_{ij} C_i C_j + \Sigma_{j,\, k \neq i} k_{jk} C_j C_k - k_s C_i C_s$$

where C_i is the concentration of the i-th species, D_i is its diffusion constant, k_{ij} is the rate constant between species i and j, C_s is the concentration of solute s, and $k_s C_i C_s$ is the reaction rate between species i and the solute s. It is assumed that spatial distribution of species is Gaussian around its center of the spur (prescribed), and that spurs are produced linearly along a electron path. Integration of eq.(2.1) over space coordinates reduces differential equations for number of species N_i in single spur,

$$(2.2)\quad dN_i/dt = -\Sigma_j k_{ij} N_i N_j f_{ij} + \Sigma_{j,\, k \neq i} k_{jk} N_j N_k f_{jk} - k_s N_i N_s$$

$$(2.3)\quad f_{ij} = [1 + \{\pi(b_i^2 + b_j^2)\}^{(1/2)} / Z_1] / \{\pi(b_i^2 + b_j^2)\}^{(3/2)}$$

where Z_1 is the average distance between successive spurs in an electron path, b_i^2 is the parameter of width of the spur in space: we assume here $b_i^2 = r_{i0}^2 + D_i t$, where r_{i0} is the initial radius of the species i.
We assumed Z_1 as,

$$(2.4)\quad Z_1 = Es / L_\Delta$$

where Es (in eV) is the energy to produce a spur and L_Δ is the restricted stopping power of energy cutoff Δ of the electron. We assumed that Es = Δ. The eq. (2.2) were solved for the initial yields $N_i = G_{0i} \times Es/100$ from the time of 10^{-12} sec to 10^{-6} sec. The initial values of G_{0i} were optimized in the previous work[10]. Concentration of dissolved oxygen in aerobic condition was assumed as 1.48×10^{-3}mol/l by Oswald dissolved constant. Concentration of Fe^{2+} was assumed 10^{-3}mol/l as does usual dosimetric solution and constant over the solution.

The G-values of species i, G(i,t,E) and yields of Fe^{3+}, G(Fe^{3+},t,E) of electron of energy E at the time $t = 10^{-6}$ sec in the unit molecules/100 eV become, $G(i,t,E) = (N_i /Es)x100$ for $i= Fe^{3+}$. Original dissociated water species e^-_{aq}, H^+ and H_2O_2 remain at the time $t = 10^{-6}$ sec: they may continue to react with Fe^{2+}. Thus we estimated Fricke G-value by yields of Fe^{3+} itself and remaining species at the time of 10^{-6} sec as.

$$(2.5) \qquad G(Fe^{3+},E) = G(Fe^{3+},t,E) + 3G(e^-_{aq},t) + 2G(H_2O_2,t) + 3G(H^+,t)$$

For photon of energy $h\nu$, we obtain the G-values $G(Fe^{3+},h\nu)$,

$$(2.6) \qquad G(Fe^{3+},h\nu) = \int f(E,h\nu)\, G(Fe^{3+},E)dE$$

The electron energy spectra $f(E,h\nu)$ were obtained by Monte Carlo method for water phantom of four sizes in cubic: lengths of a side were 1 cm, 2 cm, 3cm and 20 cm, respectively. In comparison with experiments all calculated results are normalized by the experimental yields of ^{60}Co γ -rays(the normalization constant $C= 16.07x10^{-7}$ mol/J)/ $G(Fe^{3+}, h\nu = 1250$ keV).

3. RESULT AND DISCUSSION

The comparisons are shown in Figure 1. The present calculations agree well with the experimental data and are consistent with the recommended values of ICRU[12]. The difference between the present calculations and the recommended values are within 5 %. Calculated Fricke G- values of photons from the results of electrons by Magee and Chatterjee[8,9] show some disagreement between their estimates and experiments. As shown in Figure 1, the previous calculations[11] does not explain the experiments for photon of energy below 5 keV. The previous method may estimate differential yields of only component of intra track, spurs on the same electron path, but not include component of inter track, spurs near by but not on the same electron path, which may become important for the energy below 5 keV. The present model, usage of the restricted stopping L_Δ, may be an alternative to take into account intra and inter track component as a whole for the low energy region..

The calculated yields for spurs of different sizes (Es) in the same water phantom (1 cm cubic) indicated that yields of Es=110 eV was the highest and Es=30 eV the lowest: results of the other sizes, 50.7,65,80 eV lay between these two. The case of Es=30 eV(C=0.603) was the best fit (Figure 1). And the estimated Fricke G-values in cubic of different size of container also indicated that there was no significant difference among cubic of 1 cm, 2 cm and 3 cm containers: difference in primary electron spectra affected less the yields within these sizes. The case of 1 cm is shown in Figure 1. The present results show a specific energy dependence, " a shoulder and a depression", in a region of 10 keV - 600 keV. It is attributed to the primary electron spectra.

The present calculations can reproduce the experimental values well, but it needs more studies on points, 1) the model itself: how to take into account spatial distribution of spurs, distribution of spur size[15] and its relation to the spatial distribution, and 2)to deal with as more realistic Fricke solution as possible such as reaction (11)' and reaction rate constants.

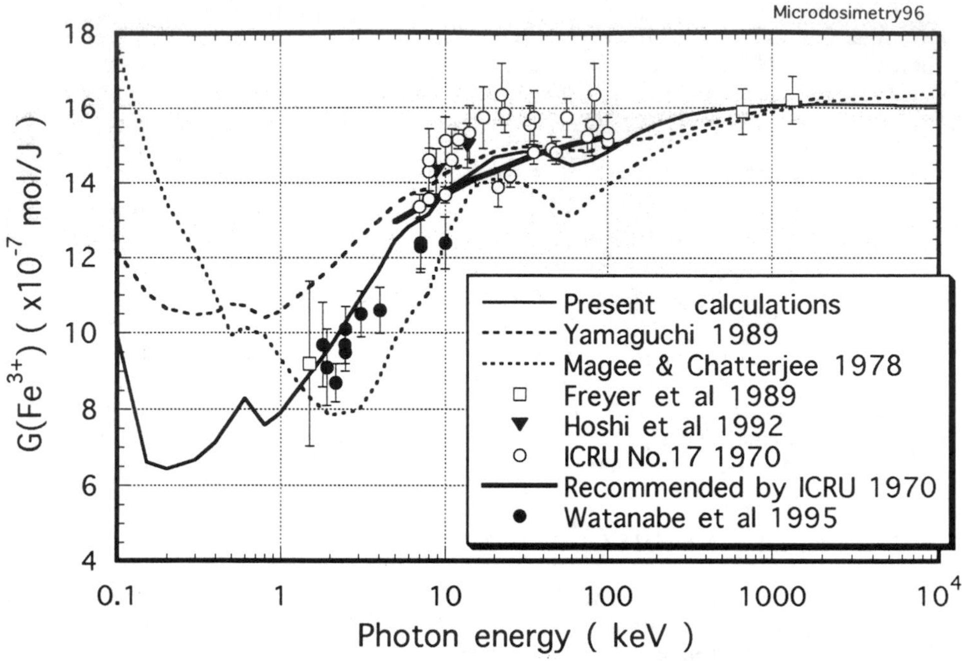

Figure 1 *Fricke G-value for photons*

References

1 J. P. Freyer, M. E. Schiliiaci and M. R. Raju, *Int.J.Radiat.Biol.*, 1989, 56, 885.

2 M. Hoshi, S. Uehara, O. Yamamoto, S. Sawada, T. Sato, K. Kobayashi, H. Maezawa, Y. Furusawa, K. Hieda, and T. Yamada, T., *Int.J.Radiation Biol.*, 1991, 61, 21.

3 R. Watanabe, N. Usami and K. Kobayashi, *Int.J.Radiat.Biol.*, 1995, 68, 113

4 J. E. Turner, J. L. Magee, H. A. Wright, A. Chatterjee, R.N.Hamm, and R. H. Ritche, *Radiat.Res.*, 1983, 96, 437.

5 H. Tomita, M. Kai, T. Kusama, Y. Aoki, Y and A. Ito, *Int.J.Radiat.Biol.*, 1994, 66, 669.

6 H: A. Schwarz, *J.Phys.Chem.*, 1969, 73, 1928.

7 H. Tomita, PhD thesis to University of Tokyo, 1995

8 J. L. Magee and A. Chatterjee, *J. Phys. Chem.*, 1978, 82, 2219.

9 H. J. L. Magee and A. Chatterjee, *Radiat.Phys.Chem.*, 1980, 15, 125.

10 H. Yamaguchi, *Radiat.Phys.Chem.*, 1988, 30, 279.

11 H. Yamaguchi, *Radiat.Phys.Chem.*, 1989, 34, 801.

12 ICRU Report 17, 1970, International Commission on Measurements Units and Measurements.

13 C. Von Sonntag, 'The chemical basis of radiation biology', Taylor & Francis, London - New York -Philadelphia, 1987.

14 Y. Tabata, Y. Ito, S. Tagawa, 'Handbook of Radiation Chemistry', CRC Press, Boca Raton Ann Arbor Boston, 1991.

15 S.M.Pimblott and A.Mozumder, *J.Phys.Chem.* 1991, 95, 7291.

Clustered DNA damage

THEORETICAL AND EXPERIMENTAL BASES FOR MECHANISTIC MODELS OF RADIATION-INDUCED DNA DAMAGE

A. Ottolenghi and M. Merzagora

Dipartimento di Fisica Università
di Milano and INFN sezione di Milano
Via Celoria 16, 20133 Milano Italy

1 INTRODUCTION

One of the general aims of radiobiology and radiation biophysics is a better understanding of the mechanisms that lead to biological endpoints such as inactivation, mutation, neoplastic transformation etc. The pathway which leads to cellular damages starts from initial energy depositions in biological matter, and goes through the production and diffusion of free radicals, direct and indirect DNA damage, repair processes, down to chromosome aberrations, mutations, etc. Many of the mechanisms governing each different step are still only roughly understood; it is also possible that some important steps are neglected in the descriptions currently adopted and it cannot be excluded that other pathways contribute to the final effect. Nevertheless, great improvements have been achieved in the last decades.

The objective is to develop mechanistic models strongly supported by both experimental results and theoretical biophysics. On these bases, the hope is also to single out physical quantities which are more related to the mechanisms leading to biological endpoints. Dose, LET, RBE, etc. are indeed of scarce help for understanding radiation damage, especially when dealing with low doses of particle radiation or mixed fields.[1,2]

Models of radiation action have also to justify and be consistent with experimental evidence such as shoulders observed in low LET survival curves, complex behaviours of mutation and inactivation cross sections vs LET (e.g. hooks at high LET), differences in the effectiveness of different particles of equal LET or the kinetics of chromosome aberration formation and its dependence on radiation quality.

We will concentrate here on DNA damage, particularly on double strand breaks (dsb), widely recognized as an important initial radiation effect, although not sufficient to explain most important features.[3,4,5] Indeed a scarce dependence of dsb induction on LET has been observed experimentally, whereas a significant dependence was found for mutations, inactivation and neoplastic transformation. The importance of different levels of DNA damage complexity[6] will be considered and the problems in comparing experimental data with results of simulations will be analyzed.

2 DNA DAMAGE

The first problem when trying to relate experimental results with theoretical predictions is to clarify the meanings of terms used by different people involved in this very interdisciplinary matter. In this case the first question is: what is a double strand break? The problem is not to agree on a formal definition such as "two strand breaks on the opposite helixes, within ten base pairs", but to clarify what is actually measured and what is simulated, considering the different approaches and techniques used by experimentalists and theoreticians, or by biologists, physicists, chemists, physicians etc.

2.1 Measurements of dsb

The first important aspect of experimental data on dsb is that they generally come from indirect measurements. Moreover, the rationales of the techniques conventionally used are very different: sedimentation of DNA molecules according to their molecular weight (Low Speed Sedimentation), measurements of the rate of DNA elution from a filter, assuming that it is controlled fundamentally by the number of dsb (Filter Elution in Non Denaturing Conditions), quantification of dsb through the fraction of DNA entering the gel under the influence of an electric field (Constant and Pulsed Fields Gel Electrophoresis). In all cases small fragments are generally not detected, doses are very high, and the smaller the fragments measured, the higher the dose required. With all these methods dsb are calculated on the basis of the hypothesis of a random distribution of double strand breaks. This assumption has recently been invalidated by Lobrich *et al*[7] that showed a non random distribution of dsb induced by N and Fe-ions, by analyzing the total fragment size distribution. Peaks in segment size distributions and their dependence on the quality of radiation are also predicted theoretically, on the basis of simulations that take into account various hypotheses on higher order structures of DNA.[8,9] As an effect of non-random distribution of dsb, the yields at high LET might be underestimated. On the contrary, if the lysis is too mild, only more complex dsb might be observed and dsb induction might be underestimated at low LET. In general all these very ticklish aspects of dsb measurements make it necessary to be particularly careful when using dsb data for modelling, and the different experimental procedures have to be taken into account, together with the methods used to calculate the number of dsb from the quantities measured.

2.2 Modelling of dsb

Since the early work of Lea,[10] which focused on energy depositions in specific cellular targets even before the characteristics and the role of the DNA molecule were known, there have been many improvements in modelling the pattern of energy deposition of different radiation fields and the relevant targets for radiation. It is in fact the connection between these two aspects that enabled the development of models to investigate the mechanisms governing radiation effects. It was soon realized that the spatial distribution of initial energy depositions plays an important role in determining radiation effectiveness, and improvements in track structure simulations (from the amorphous track structure of the Katz model[11] to full three dimensional stochastic description of tracks in different media[12,13]) allowed the analysis of the relevant features characterizing different radiations fields.

Two main approaches are being used for drawing the connection between track structures and DNA damage. One consists in an appropriate mathematical treatment of track structure data (such as proximity functions in the theory of dual radiation action[14,15] or cluster analysis techniques[16,17]) aimed at defining relevant parameters that can correlate with observed yields of DNA lesions such as dsb. Another approach explicitly takes into account biological targets, which can be described with different levels of detail.[18,19,20,21,22]

Particular attention is now focused on the development of mechanistic description of the processes occurring in the DNA after radiation insult. This means improving track structure simulations in the cellular environment, which in turn requires reliable cross section for the liquid phase and for the DNA molecule itself,[23] and the treatment of indirect effects by describing the processes of production, diffusion and recombination of water radiolysis products.[24,25,26] On the other hand, atom-by-atom models of the DNA structure, taking into account its organisation in higher order structures[8,9] and the effect of hydration shells[27] are under development and have already produced interesting results. However, despite the high level of detail achieved in the models, there still exist many factors which are only roughly taken into account. These include DNA higher order structure itself,[28] radiation induced chemical and structural changes in DNA[29], charge migration (mainly towards guanine),[30] the role of hydration shell,[31] repair mechanisms, etc.

The choice of the appropriate level of detail to be included in the simulations is indeed a very delicate point. Although truly mechanistic models remain an important goal, it should be considered that if too many microscopic quantities included in the model rely only on indirect experimental checks, the resulting description could be misleading, even if it gives correct results. For this reason, also less sophisticated models relying on some adjustable parameters can still provide valuable information, and refining of the models should again be paralleled by a clear understanding of the non-trivial relationship between what is modelled and what is measured.

2.3 Dsb and complex lesions

We have used the code MOCA15 by Paretzke and Wilson and the DNA geometrical model proposed by Charlton *et al*[20] to simulate DNA damage induced by protons and alpha particles of energies ranging from 0.3 to 4 MeV/u.[21] To quantify the level of complexity of double strand breaks we have studied the distributions of the energy deposited in DNA segments of 30 base pairs where a dsb occurred (see Figure 1). Areas under the curves represent the total yields of DSB/Gy/dalton. It is clear that there is no significant dependence of the total yields of dsb on LET or particle type. However, if one considers distributions the picture changes dramatically. Analyses of such distributions can be performed by means of various statistical parameters (mean values, medians etc.). If it is assumed that only damage of a certain complexity is indeed relevant, a better description is obtained by studying the integrals of the distributions starting from a threshold value of energy deposited in the segment. By increasing this threshold value, LET and particle dependences become more pronounced. As an example, protons and alpha particles of equal LET are compared in Figure 2. The ratio of the areas under the two curves increases significantly if only the tails of the distributions are considered.

Approaches based on mean values or thresholds are in any case often misleading: they introduce a cut off with little biological interpretation, and may mask the fact that even "simple" dsb could give rise to important biological endpoints, even if with low probability. We tested a different approach and operatively defined as a "complex" lesion (cl) damage to a DNA molecule characterized by at least two dsb within 30 bp (like the dsb++ of Charlton *et al*[20]). Results on total yields of dsb and complex lesions are reported in Figure 3. The corresponding distributions of energy deposited when a cl occurs are reported in Figure 1. As LET increases (and moving from α-particles to protons of equal LET) the yield of cl increases and the corresponding distributions are shifted to the right. This indicates that not only the total yield, but also the complexity of cl depends on LET and particle type. It

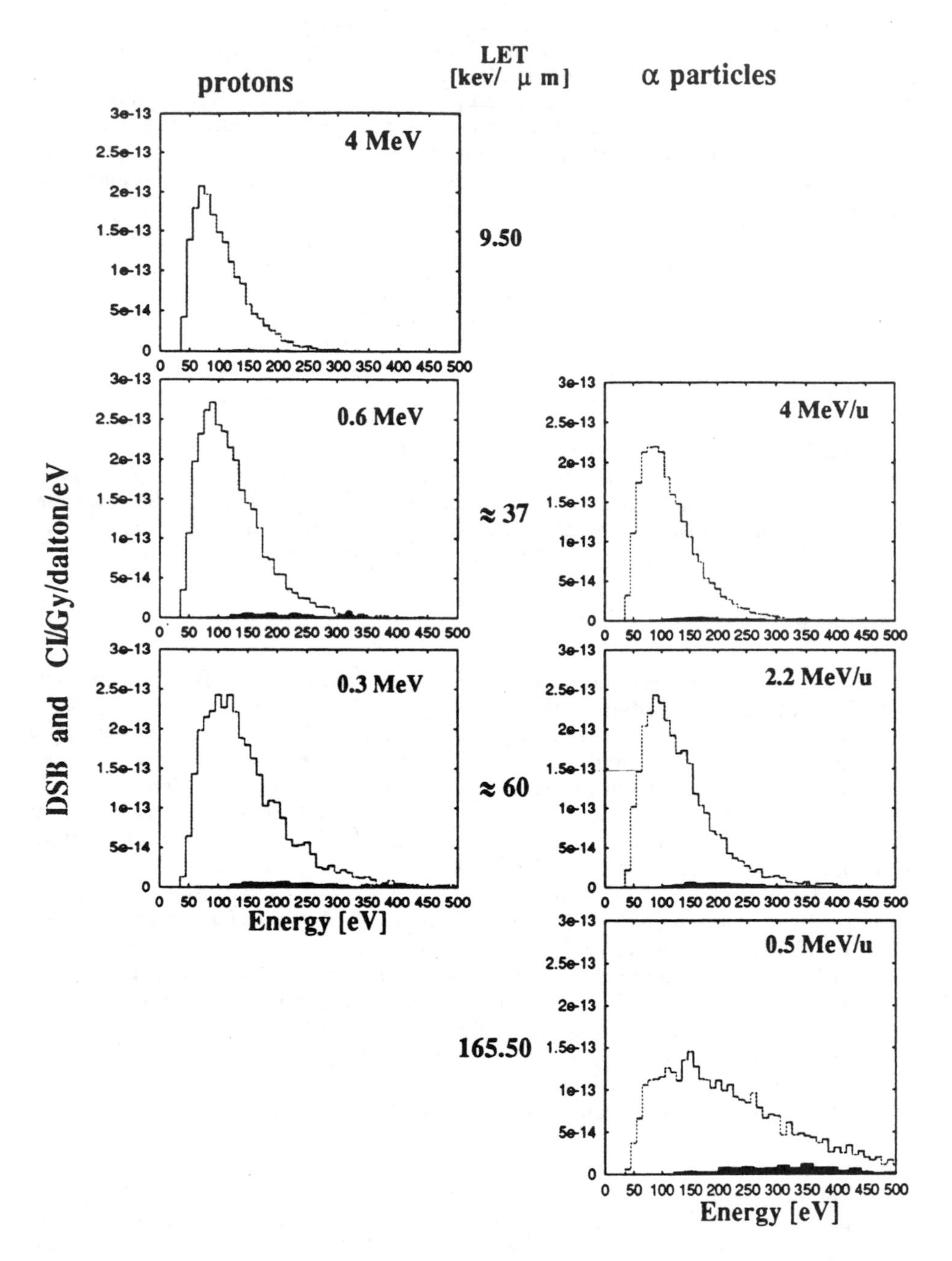

Figure 1 *Calculated distributions of energy deposited in DNA segments (30 bp) where a dsb (empty histograms) or a cl (filled histograms) has occurred. Areas represent the total yields of dsb and cl per Gy per dalton, as reported in fig 3.*

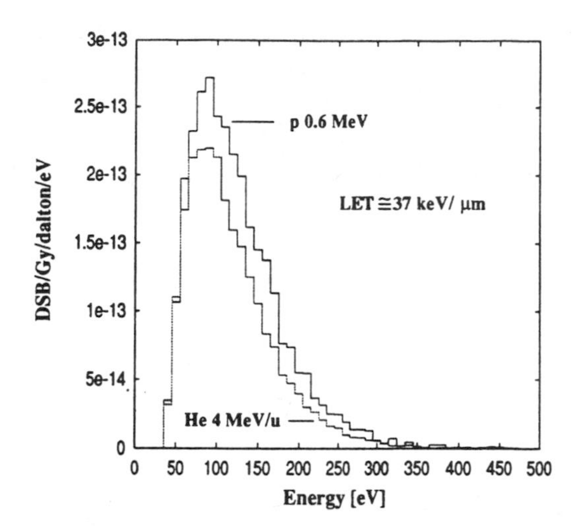

Figure 2 ***Calculated distributions for energy deposited in DNA segments (30 bp) where a dsb has occurred, after irradiation with protons or alpha particles of equal LET. Areas represent the total yield of dsb per Gy per dalton, as reported in Figure 1.***

can also be seen from Figure 1 that cl occurs also below what will be a convenient threshold value of energy depositions in DNA segments to obtain acceptable results.

In order to understand track structure features determining the dependences of the yield of dsb and cl on LET and particle type, the contribution of primary ion and of secondary electrons tracks in inducing dsb and complex lesions were analyzed separately.[32] Results obtained (see Fig. 3) indicated that the LET and particle dependence of the yields of cl is mainly due to energy deposition from the primary ion combined with secondary electrons at short distances from the ion track (that is, mainly low energy electrons). Such a component is predominant at high LET, but it becomes comparable with the contribution of secondary electrons below ≈ 15 keV/μm. This is consistent with experimental results showing similar RBE for high energy heavy particle and sparsely ionizing radiation such as X-rays, where only electrons contribute to DNA damage.

These results are in agreement with conclusions obtained by other groups with more phenomenological approaches (i.e., which focus on the agreement with experimental results and do not make specific hypotheses on the nature of relevant DNA damages), based on microdosimetric calculations[33] or on specific analysis of track structure features. [34,35]

3 COMPARISONS BETWEEN SIMULATIONS AND EXPERIMENTS

The comparison between the yields of dsb can be made directly, with the caution suggested by the intrinsic limits of both simulated and measured data. [21]

The question is more complicated if one wants to compare theoretical predictions with experimental data regarding DNA complex damages. No selective measurements of damage complexity are in fact currently available. Good agreement was found between measured RBE for inactivation and simulated cl yields vs LET, by assuming that cl are lethal damages and by taking into account stochastic aspects as the number and the length of cell nucleus traversals, cell geometry, etc. Since

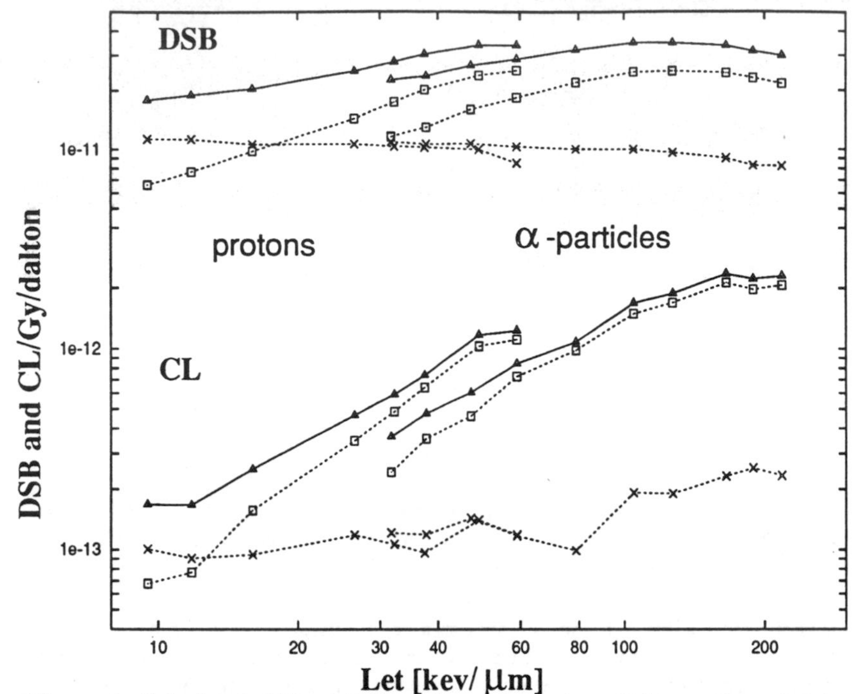

Figure 3 *Calculated yields of DSB and complex lesions (CL) per Gy and dalton (—△—-) induced by protons and α-particles as a function of LET. The yields of DSB or CL induced by secondary electron tracks only (- - -×- - -) and with the contribution of the energy deposited by the primary ion and by correlated ion-electron events (- - -□- - -) are also reported. Lines are guides for the eye only.*

the relevance of clustered damage is strictly related to its resistance to repair, a more direct comparison can be done between yields of cl and unrejoined dsb. There is large amount of evidence of the dependence of the fraction of unrejoined dsb on LET.[36] This dependence is in very good agreement with the trend of the ratio cl/dsb.[21] Another possible approach is to compare the ratio cl/dsb (simulated) vs LET, with the ratio lethal lesions/dsb (measured) (see Figure 4). The general trend as a function of LET of the two sets of results is very similar. Also the values are similar, even if no great significance should be given to this aspect. Indeed, other laboratories found the same trend of experimental data, but different values.[38]

4 CONCLUSIONS, OPEN QUESTIONS AND NEEDS

The results presented here have focused on the complexity of DNA damage. Other models give more importance to the proximity of elementary DNA lesions:[39] this aspect should be particularly relevant in modelling chromosome aberration[40,41,42] and in explaining shouldered survival curves. These two assumptions should obviously not be regarded as mutually exclusive, and it is a fundamental goal for the future to assess the relative importance of lesion complexity and lesion proximity in relation with different biological end-points.

From the theoretical point of view, further investigation on this issue requires the development of detailed DNA models including DNA cross sections, higher order

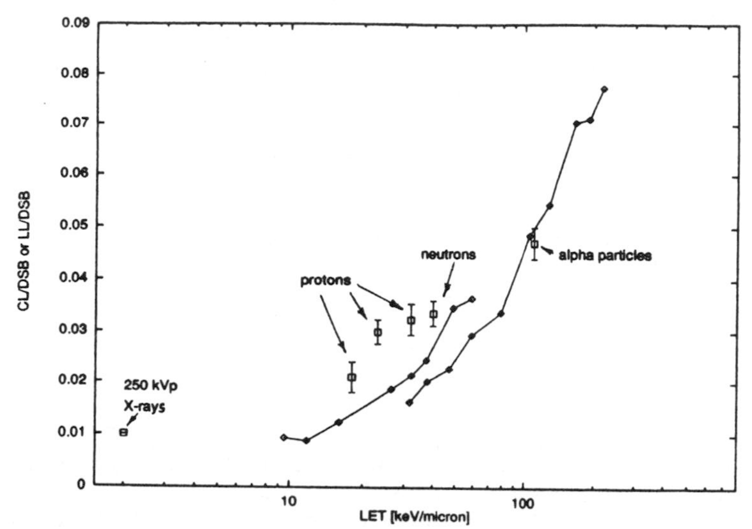

Figure 4 *Ratio of cl/dsb (diamonds, simulated, from Figure 3) and of lethal lesions/dsb (squares, experimental, from Prise*[37]*) as a function of LET.*

structures, hydration shell, etc., models of the chemical and structural modification of the DNA molecule following radiation insult, and models of the conversion of DNA damages into biological end-points.

On the experimental side, more data are probably needed on radiation quality dependence of DNA fragmentation, rejoining studies as indirect measurements of "complex damage" (including small fragments), radiation quality dependence of repair/misrepair of DSB, and, more generally, on different endpoints at low doses, possibly measured in the same experimental conditions.

Finally, we need a better understanding of the characteristics and role of clusters of energy deposition events in inducing biological damage. If the importance of clusters is confirmed, it becomes of great interest to develop devices such as cluster counters,[43] and to investigate the possibility of assuming the number of energy deposition clusters with specific characteristics as a new reference quantity, more reliable than dose for predicting radiation effects.

Acknowledgments. This work was partially supported by the EU (Contract No. F14P-CT95-0011). We would like to thank all the people of the two European consortia supported by EU on *Biophysical models for the induction of cancer by radiation* and on *Experimental data related to the induction of cancer by radiation of different qualities* for the important common discussions. In particular we are indebted to M. Belli, G. Simone and A. Tabocchini for the fundamental help in the analysis of the techniques for dsb measurements.

References

1. R. Katz, *Radiat. Res.*, 1994, **137**, 410
2. H.G.Paretzke, 1994 *Physics in Medicine and Biology*, **39a**, 362
3. J.F. Ward, *Radiat. Res.*, 1985, **104**, S-103.
4. U. Hagen, *Radiat. Environ. Biophys.*, 1994, **33**, 45.
5. M. Frankenberg-Schwager, *Radiat. Environ. Biophys.*, 1990, **29**, 273.
6. D.T. Goodhead, *Int. J. Radiat. Biol.*, 1994, **65**, 7.
7. M. Löbrich, P.K. Cooper, B. Rydberg, *Int. J. Radiat. Biol.*, 1996, In press.

8. W.R. Holley, A. Chatterjee, *Radiat. Res.*, 1996, **145**, 188.
9. W. Friedland, P. Jacob, H.G. Paretzke, T. Stork *This volume*
10. D.E. Lea, 'Action of radiation in living cells', Cambridge University Press, London, 1946.
11. J.J. Butts and R. Katz, *Radiat. Res.*, 1967,**30**, 855.
12. H.G. Paretzke, 'Kinetics of Nonhomogeneous Processes', Wiley, New York, 1987, p. 89.
13. H. Nikjoo, M. Terrisol, R.N. Hamm, J.E. Turner, S. Uehara, H.G. Paretzke, D.T. Goodhead, *Radiat Prot Dos*, 1994, **52**, 165.
14. A.M. Kellerer, H.H. Rossi, *Radiat. Res*, 1978, **75**, 471.
15. M. Zaider, *Radiat. Res.*, 1993, **134**, 1.
16. D.J. Brenner, J.F. Ward, *Int. J. Radiat. Biol.*, 1992,**61**, 737.
17. V. Michalik, *Int. J. Radiat. Biol.*, 1992, **62**, 9.
18. D.T. Goodhead, H. Nikjoo, *Int. J. Radiat. Biol.*, 1989,**55**, 513.
19. D.T. Goodhead, J. Thacker, R. Cox, *Int. J. Radiat. Biol.*, 1993, **63**, 543.
20. D.E. Charlton, H. Nikjoo, J.L. Humm, *Int. J. Radiat. Biol.*, 1989, **56**, 1.
21. A. Ottolenghi, M. Merzagora, L. Tallone, M. Durante, H.G. Paretzke, W.E. Wilson, 1995, *Radiat. Environ. Biophys.* **34**, 4, 239.
22. H. Nikjoo H, P. O'Neill, M. Terrisol, D.T. Goodhead, *Int. J. Radiat. Biol.*, 1994, **66**, 453.
23. M. Zaider, M. Bardash, A. Fung, *Int. J. Radiat. Biol.*, 1994, **66**, 459.
24. J.E. Turner,R.N. Hamm, H.A. Wright, R.H. Ritchie,J.L. Magee , A. Chatterjee, W.E. Bolch, *Radiat Phys Chem*, 1988, **32**, 503.
25. M. Terrisol, A. Beaudré, *Radiat Prot Dosim*, 1990, **31**, 175.
26. I.G. Kaplan, V.Y. Suchonosov, *Radiat. Res.*, 1991, **127**, 1.
27. M. Terrisol, M. Demonchy, E. Pomplun, *This volume.*
28. J.M. Goodfellow, L. Cruzeiro-Hansson, O. Norberto de Souza, K. Parker, T. Sayle, Y. Umrania, *Int. J. Radiat. Biol.*, 1994, **66**, 471.
29. P. O'Neill, E.M. Fielden, *Adv. Radiat. Biol.*, 1993, **17**, 53.
30. T. Melvin, S.W. Botchway, A.W. Parker, P. O'Neill, *J.C.S. Chem. Commun.*, 1995, 653.
31. S.G. Swarts, D. Becker, M. Sevilla, K.T. Wheeler, *Radiat Res*,1996, **145**, 304.
32. A. Ottolenghi, M. Merzagora, M. Durante, H.G. Paretzke, W.E. Wilson, 'Radiation Research 1895-1995', Universitatsdruckerei H. Sturtz, Wurzburg, 1995, 112.
33. G. Leuthold and G. Burger, *Radiat. Environ. Biophys.*, 1988, **27**, 177.
34. D. Harder and E.R. Bartels, 'Radiation Research, A Twentieth-Century Perspective', Academic Press, 1992, 427.
35. D.E. Watt, L.A. Kadiri, S. Glodic, 'Biophysical Modelling of Radiation Effects', Adam Hilger, Bristol, 1992, p. 201.
36. D. Blöcher, *Int. J. Radiat. Biol.*, 1982, **42**, 317.
37. K.M. Prise, *Int. J. Radiat. Biol.*, 1994, **65**, 43
38. M. Belli, F. Ianzini, O. Sapora, M. A. Tabocchini et al, *Adv. Space Res.*, 1996, **18**, 1/2, 73.
39. D.J. Brenner, *Radiat. Res.*, 1990, **124**, S29.
40. A.A. Edwards, V.V. Moisseenko, H. Nikjoo, *Radiat Environ Biophys*,1996, **35**,25.
41. H.L. Wu, M. Durante, T.C. Yang, submitted for publication, 1996.
42. R.K. Sachs, D.J. Brenner, *Int. J. Radiat. Biol.*, 1993, **64**, 677.
43. F. Sauli, *Radiat. Prot. Dosim.*, 1995, **61**, 29-38.

IONISING RADIATION INDUCED CLUSTERED DAMAGE TO DNA - A REVIEW OF EXPERIMENTAL EVIDENCE

K. M. Prise

Gray Laboratory Cancer Research Trust, PO Box 100, Mount Vernon Hospital, Northwood, Middlesex, HA6 2JR. UK

1 INTRODUCTION

Understanding the interactions of ionising radiation with cellular DNA requires an appreciation of the relationship between the structure of radiation tracks as they cross the DNA and the molecular nature and distributions of the damage produced. Track structure modelling studies using Monte-Carlo methods have attempted to relate the energy deposition patterns of ionising radiation of differing linear energy transfer (LET) to subsequent biological effects[1]. These modelling studies have predicted the importance of clustered damage to DNA, and its relationship to radiation quality. Ideas on the clustering of radiation damage on DNA are not new (see for example Howard-Flanders, 1958[2]). The concept of clustered damage in the form of locally multiply damaged sites was originally derived by Ward[3] from studies on the radiation chemistry likely to be associated with localised radical formation on the DNA. In these situations, he specifically considered the situations where there was the probability of more than one radical being produced at the sites of spurs (<100 eV in 4 nm), or blobs (100-500 eV in 7 nm and short tracks (>500 eV) of energy deposition events in aqueous systems. The term locally multiply damaged sites (LMDS) refers to the likely clusters of damage produced at these DNA locations. Recently, both modelling[4] and experimental studies[5] have predicted the production of regionally multiply damage sites (RMDS) where association of damage is over much larger distances related to the higher-order structures of the DNA. This paper will briefly review the evidence for the production of clustered damage and give pointers to areas of future research.

2 ENERGY DEPENDENCE OF CLUSTERED DAMAGE

Track structure studies rely heavily on relating the size of the energy deposited in volumes of the DNA to the production of DNA damage. This requires, to some extent, assumptions regarding the dependence of lesion formation on energy deposited. Some of the Monte-Carlo studies have assumed threshold energies for the production of single-strand breaks (ssb) of approximately 17.5 eV and double strand breaks of around 50 eV[6]. Until recently there has been a lack of experimental information regarding the energy dependence of strand breakage frequencies in DNA from exposure to energies low enough to be of relevance in track structure studies. Despite the wide range of energy deposition events which occur in DNA when radiation interacts with it, the most frequent energy deposition event is around 25 eV[7]. We have carried out a study of the energy dependence of both ssb and double-strand break (dsb) formation in plasmid DNA irradiated with 7-150 eV photons. Due to the limitations of sources available for such studies (low photon flux) and the ranges of the photons considered, our preliminary work and studies by other groups[8] have been done in DNA samples irradiated under vacuum. Here, where most of the effects will be due to direct damage to the DNA, we have found that the quantum efficiency of both ssb and dsb production appears to decrease with energy, falling rapidly at energies below 10 eV. This is in line with the expected differences in the energies of the secondary electrons produced by these photon energies. A close correlation between the yields of ssb and dsb was observed over a wide range of photon energies. These studies suggest a common precursor for these lesions, i.e. that energy deposited by a photon in the helix and leading to the production of DNA ssb has a 1 in 20 probability of producing a dsb. This is similar to that predicted by Siddiqi and Bothe[9] for the involvement of a radical transfer mechanism in the production of dsb in DNA irradiated in dilute solution. Given that these energy dependence studies have been carried out with DNA irradiated under vacuum conditions where we are monitoring direct ionisation on the DNA itself, it is important to consider the role of hydration on both the mechanisms and efficiencies of damage formation.

The prediction of a radical transfer mechanism for dsb production is in contrast to differences in the clustering of radicals with low LET radiation we have observed in hydrated model and cellular systems (see section 3). It is now important to extend these studies to determine the energy dependence of lesion formation for more complex lesions with the development of assays for these (see section 4) and the careful determination of the role of the hydration state of the DNA on the quantum efficiencies. For complete information, required to input in to track structure studies, it is necessary to couple the data from these photon studies with those monitoring the effectiveness of low energy electrons. Our initial studies with low energy electrons (25e V - 4 keV) have suggested that 25 eV electrons may be much less efficient at inducing dsb in comparison to ssb. Although the most frequent energy deposition events by most ionising radiations is in the 20 eV region, high LET radiations are likely to produce unique high energy deposition events in the DNA or, more importantly, in regions of the chromatin[1]. Recent studies have suggested that high LET radiations may produce radical species which are not found with low LET radiations. In particular, phosphate radicals which could lead directly to strand breakage are produced[10]. The importance of direct photoionisation of the phosphorus atoms in DNA via K-shell excitation has been shown to lead to an increased dsb to ssb ratio[11] indicating a distinct role for energy absorbed in phosphorus in strand break induction.

3 CLUSTERING OF RADICALS ON DNA

Localised energy depositions close to DNA will have an enhanced probability of increasing the local concentration of free radicals and these subsequently forming clustered molecular damage. For example it has been calculated that 2 - 5 radical pairs can be generated within a radius of 1-4 nm[12]. The reactivity of these radicals and their subsequent lifetimes is critically determined by the presence of chemical modifiers such as oxygen and thiols at the time of irradiation. Several groups, including our own, have utilised these factors to use chemical modifiers as probes for understanding the underlying multiplicity of the radicals produced on DNA and relating this to the yields of clustered damage produced. Using fast reaction techniques, which allow us to measure the rates of both the fixation of free-radical damage by oxygen and its chemical repair by agents such as thiols, we have been able to follow the reaction rates of the radicals involved in leading to both ssb and dsb in model plasmid and intact cellular systems. The kinetics of these processes can be explained on the basis of the free radical precursors of a dsb consisting of two radicals which are produced as a single ionising radiation track crosses the DNA helix[13]. Steady-state studies by other groups[14] have confirmed the importance of a bi-radical intermediate in the formation of dsb under conditions designed to mimic those of the cellular environment. Our studies of the kinetics of these processes in cellular systems have shown that the chemical repair kinetics of the free radical precursors of dsb follow those for the precursors of lethal events, and more importantly, are consistent with the two radical model. We are now extending these studies to monitor the kinetics of the free radical origin of lesions in DNA irradiated with high LET radiations. Our studies to date are consistent with both interactions between radicals and increasing radical multiplicity being important determinants of the effectiveness of high LET radiations[15].

4 CLUSTERING OF MOLECULAR LESIONS

Attempts to provide direct molecular evidence for the existence of clustered damage have proved difficult, principally because of the lack of suitable assays which give information on the distributions of the different types of lesions ionising radiation induces in DNA. Ward[16] has defined three variables which are important in defining the molecular structure of a LMDS. They are 1) the ratio of base damages to strand breaks; 2) the size of the LMDS i.e how many base-pairs of DNA it involves and 3) the number of individual lesions per LMDS. An important tool for assessing these parameters is to use radiations of differing LET. Much of the existing data infers the clustering of molecular lesions on the basis of the changes in the relative yields of ssb and dsb with radiations of different LET. For example, our own studies have shown that the lethality of a dsb increases with radiation quality, such that dsb induced by α-particles are up to 4 times more likely to kill a cell than low LET induced dsb[17,18]. Subsequent studies by Weber and Flentje[19] have extended this approach to show that dsb lethality reaches a maximal value at around 200 keV/μm and decreases beyond this. These changes in lethality are matched by studies which have shown that the ability of a cell to rejoin dsb depends on the LET of the radiation. In particular, dsb induced by high LET radiation are repaired at a slower rate and the overall level of residual or unrejoined breaks is higher[20].

Studies in model plasmid systems, where extracts of cellular repair enzymes have been used to mimic repair processes, have shown that the ability of cell extracts to repair radiation damage decreases as the complexity of the damage increases[21]. By creating model sections of DNA with specific lesions, known base-pairs apart, it has been shown that the efficiency of certain repair enzymes is critically related to lesion complexity[22]. Other studies, using enzymes which recognise specific base lesions produced by radiation and converts these to strand breaks, have shown that for high LET radiations the ratio of ssb to dsb decreases after enzyme treatment. This would be expected if the complexity of the lesions is increasing[23]. The use of concatenated DNA systems where two supercoiled rings of DNA are interlinked to each other has shown an increased yield of the linear form in one ring accompanied by a ssb or dsb in the other ring after high LET radiation consistent with an increased localisation of multiple radical attack over two DNA helix molecules[24].

Despite these studies, no clear molecular signatures have been obtained for clustered damage. Several groups are developing techniques for the end-labelling of strand breaks with a view to the detection of any small fragments which may be induced in the 1-15 bp regions where these lesions are likely to occur. Other groups are developing techniques, based on sequencing technology, to allow precise determination of the frequency and location dependence of clustered damage in model DNA systems[25]. These techniques can also give limited information regarding the structures of the break ends produced, which are known to change in response to the degree of direct versus indirect damage and the presence of oxygen.

5 CLUSTERING RELATED TO CHROMATIN STRUCTURE

More recently the concept of regionally multiply damaged sites (RMDS) has been defined to classify distributions of damage which may occur over larger, kilobase pair (kbp), distances on the DNA[5]. In particular, studies have suggested that the distribution of damage induced by high LET radiation tracks as they cross a cell nucleus may be related to the higher-order structures of the chromatin within the cell nucleus. Rydberg[5] has shown that there is an enhancement of fragments in the 1-2 kbp size range in cells irradiated with high LET iron and nitrogen ions in comparison to X-rays. He observes an additional 0.2 fragments per Gbp per Gy for X-rays of this size in comparison to an extra 0.8 fragments per Gbp per Gy for nitrogen ions with an LET of 97 keV/μm. This suggests that 20% of the breaks induced in these cells by a high LET nitrogen track have another break within 1-2 kbp. These fragments are likely to be caused by breakage of the 30 nm solenoid structure of the DNA as the particle tracks interact with it. Interestingly, the kinetics of rejoining of these fragments appears to be no different from the rejoining of total cellular DNA under these circumstances. Our own studies indicate that α-particles have an increased probability of inducing fragments in the 8-300 kbp region in comparison to X-rays and that these fragments are caused by a single-hit mechanism rather than a two-hit random breakage[26]. These fragment sizes are in the range of the higher-order chromatin loop structures (10 - 200 kbp) known to be present in cells and related to transcriptional activity.

Given the evidence that the higher-order structure of the cell nucleus and chromatin appears to determine important features of the distribution of damage in the cellular genome, a careful assessment of the relationship between gene activity and chromatin structure needs to be considered. Preliminary studies with X-rays, monitoring the fragmentation of actively transcribed *c-myc* in comparison to fragmentation in the whole genome, have suggested that the distribution of breaks within the gene are not random and regions close to matrix attachments sequences (MAR's) may be protected from radiation damage[27]. By digesting cellular DNA after irradiation with restriction endonucleases and probing with Southern blotting techniques, coupled to pulsed-field electrophoresis, it is possible to follow the correct and incorrect rejoining of dsb in the fragments produced[28,29]. These studies have shown that the level of incorrect rejoining increases with LET. Although non-random fragmentation of the genome may be occurring due to high LET radiations, the localised clustering of damage at the sites of these breaks may be influencing the reparability of these lesions.

6 CONCLUSIONS

In conclusion, a range of studies has demonstrated the importance of the concepts of damage clustering on irradiated DNA. In particular, two types of clustered damage need to be considered. Firstly, the LMDS structures of Ward, which occur at a single DNA helix and are restricted to 1-15 bp in size. Secondly, RMDS, which occur where radiation tracks cross more than one helix at a time and can give lesions 0.1-200 kbp apart, but non-randomly distributed in the genome. The development of molecular assays for the determination of molecular signatures of clustered damage is an urgent requirement to consolidate the importance of this mechanism and allow input of the molecular information of the distributions and structures of these lesions to be included in track structure models. At a more fundamental level, the exact energy dependence of the production of simple and more complex lesions in DNA for both photons and electrons is required to further strengthen track structure models of radiation action. The energy dependence of lesion formation and the influence of the DNA microenvironment may provide fundamental information on the mechanisms of dsb formation.

Acknowledgement

This work was supported by the Cancer Research Campaign and the Radiation Protection Research Action Programme of the European Community. The author is grateful to Prof. B.D. Michael for comments on the manuscript.

References

1. D.T. Goodhead, *Int. J. Radiat. Biol,* 1994, **65,** 7.
2. P. Howard-Flanders, *Adv. Biol. Med. Phy.*, 1958, **6,** 553.
3. J.F. Ward, *Radiat. Res.,* 1981, **86,** 185.
4. W.R. Holley and A. Chatterjee, *Radiat. Res.,* 1996, **145,** 188.
5. B. Rydberg, *Radiat. Res.,* 1996, **145,** 200.
6. D.E. Charlton, H. Nikjoo and J. L. Humm, *Int. J. Radiat. Biol,* 1989, **56,** 1
7. A.M. Rauth and J.A. Simpson, *Radiat. Res.,* 1964, **22,** 643.
8. K. Hieda, *Int. J. Radiat. Biol,* 1994, **66,** 561.
9. M. A. Siddiqi and E. Bothe, *Radiat. Res.,* 1987, **112,** 449.
10. D. Becker, Y, Razskazovskii, M.U. Callaghan and M.D. Sevilla, *Radiat. Res.,* 1996, **146,** 361.
11. C. LeSech, H. Frohlich, C, Saint-Marc and M. Charlier, *Radiat. Res.,* 1996, **145,** 632.
12. D.J. Brenner and J.F. Ward, *Int. J. Radiat. Biol,* 1992, **61,** 737.
13. K.M. Prise, S. Davies and B.D. Michael, *Radiat. Res.,* 1993, **134,** 102.
14. J.R. Milligan, J.Y-Y. Ng, C.C.L. Wu, J.A. Aguilera, R.C. Fahey and J.F. Ward, *Radiat. Res.,* 1993, **134,** 102.

15. B.D. Michael, K.M. Prise, M. Folkard and B. Vojnovic, *Radiat. Prot. Dosim.,* 1994, **52,** 277.
16. J.F. Ward, *Prog. Nucl. Acid. Res.,*1988, **35,** 95.
17. K.M. Prise, M. Folkard, B. Vojnovic, S., Davies, M.J. Roper and B.D. Michael, *Int. J. Radiat. Biol,* 1990, **58,** 261
18. K.M. Prise, *Int. J. Radiat. Biol,* 1994, **65,** 43.
19. K.J. Weber and M. Flentje, *Int. J. Radiat. Biol,* 1993, **64,** 169.
20. T.J. Jenner, C.M. deLara, P O'Neill, and D.L. Stevens. *Int. J. Radiat. Biol,* 1993, **64,** 265.
21. P.S. Hodgkins, M.P. Fairman, and P O'Neill, *Radiat. Res.,* 1996, **145,** 24.
22. M.A. Chaudhry and M. Weinfeld, *J. Mol. Biol.,* 1995, **249,** 914.
23. R. Roots, W. Holley, A. Chatterjee, M. Irizarry and G. Kraft., *Int. J. Radiat. Biol,* 1990, **58,** 55.
24. J.R. Milligan, C.L. Limoli, C.C.L., Wu, Y.-Y., Ng and J.F. Ward. 'Radiation damage in DNA: Structure/function relationships at early times', A.F. Fuciarelli and J.D. Zimbrik (eds) Battelle Press, Columbus, Ohio, 1995, p.165.
25. V. Isabelle, C. Prévost, M. Spotheim-Maurizot, R. Sabattier and M. Charlier, *Int. J. Radiat. Biol,* 1995, **67,** 169.
26. H.C. Newman, K.M. Prise and B.D. Michael, *submitted*
27. A. Sak, M. Stuschke, N. Stapper, and C. Streffer, *Int. J. Radiat. Biol,* 1996, **69,** 679.
28. M. Löbrich, B. Rydberg and P.K. Cooper, *Radiat. Res.,* 1994, **139,** 142.
29. M. Löbrich, B. Rydberg and P.K. Cooper, *Proc. Natl. Acad. Sci. USA,*1995, **92,** 12050.

HIGHER-ORDER CHROMATIN STRUCTURES AS POTENTIAL TARGETS FOR RADIATION-INDUCED CELL DEATH

K. G. Hofer, X. Lin and M. H. Schneiderman

Institute of Molecular Biophysics (KGH, XL)
Florida State University
Tallahassee, Florida 32306-3015

Department of Radiation Oncology (MHS)
University of Nebraska Medical Center
Omaha, Nebraska 68198-1050

1. INTRODUCTION

Traditionally, the majority of cellular radiation studies have been performed with external radiation sources which irradiate the entire cell volume. External exposures are well suited for studying the overall effects of ionizing radiations because the total dose, dose rate, and duration of exposure can be accurately controlled. However, irradiation of the entire cell induces random damage in all cell compartments. For this reason, studies using external radiation sources to clarify the mechanisms and intracellular targets for radiation-induced cell death have so far yielded mostly inconclusive results.

An elegant method for inducing preferential damage in selected cellular targets is to incorporate radionuclides into different cell compartments. The subsequent radionuclide decay results in highly localized irradiation of the labeled cell organelles, particularly in experiments performed with Auger electron emitting radionuclides such as ^{125}I, ^{111}In, or ^{67}Ga.[1-4] Auger emitters decay by electron capture and/or internal conversion. This process is accompanied by complex atomic vacancy cascades and copious emission of low-energy electrons. In the case of ^{125}I, about 20 electrons are emitted per decay, most with energies below 1 keV. Such low-energy electrons have a very short range in biological matter, and their collective action leads to highly localized energy deposition in the immediate vicinity of the decay site with little or no overlap exposure of distant sites.[5-7]

Early work with Auger emitters revealed that biological damage varied greatly with radionuclide location. For example, in one experiment 60 ^{125}I decays at the DNA were sufficient to cause 50% cell death, but with ^{125}I attached to the plasma membrane ~20,000 decays/cell were required to produce the same effect.[8] Similar findings reported by a number of investigators confirm that Auger emitters in the cell nucleus are extremely cytotoxic, but identical decay events at extranuclear sites (plasma membrane, cytoplasm, cytoplasmic organelles) are relatively nontoxic to cells.[3,4,8] In particular, ^{125}I incorporated as ^{125}I-iododeoxyuridine ($^{125}IUdR$) into cellular DNA is highly radiotoxic and produces effects similar to those observed with alpha particles and other high-LET radiations.[1-4] Adverse biological effects include DNA strand breaks,[9-11] mutations,[12,13] chromosome aberrations,[14,15] malignant transformation,[16] division delay,[17,18] and cell death.[1-4]

2. PARADOXICAL EFFECTS OF ^{125}I DECAYS IN DNA

DNA-bound ^{125}I is known to produce large numbers of DNA double-strand breaks (DSB), perhaps as many as one DNA DSB/^{125}I decay.[9-11] Since DNA is presumed to be the primary target for radiation-induced cell death[19,20], it would seem logical to attribute the lethal consequences of ^{125}I to its high efficiency in causing DNA DSBs. However, recent findings in our laboratory suggest that DNA damage may not be the sole mechanism for cell killing and that damage to higher-order structures in the cell genome may contribute to (or modify) radiation-induced cell death.[21-25]

2.1 The "Chase Effect"

2.1.1 Experimental Results. Studies on ^{125}I-induced cell death are usually performed on asynchronous cell populations labeled with ^{125}IUdR for long periods of time to ensure uniform labeling of all cells. Under these conditions DNA-bound ^{125}I invariably produces high-LET-type cell death. The studies reported here were performed on Chinese hamster ovary (CHO) cells pulse-labeled with ^{125}IUdR. The experimental techniques are illustrated in Figure 1. CHO cells synchronized by mitotic cell selection were plated, resynchronized at the G_1/S boundary of the cycle with aphidicolin, allowed to progress into S phase for 30 min, and then labeled for 10 min with ^{125}IUdR. The cells were cultured (chased) in nonradioactive medium and cell aliquots were harvested after chase periods ranging from 15 min to 6 h. The cells were frozen and stored in liquid nitrogen for up to 5 weeks to accumulate ^{125}I decays. When the labeled cells had accumulated the desired number of decays, aliquots were thawed and cell survival was evaluated by colony formation.

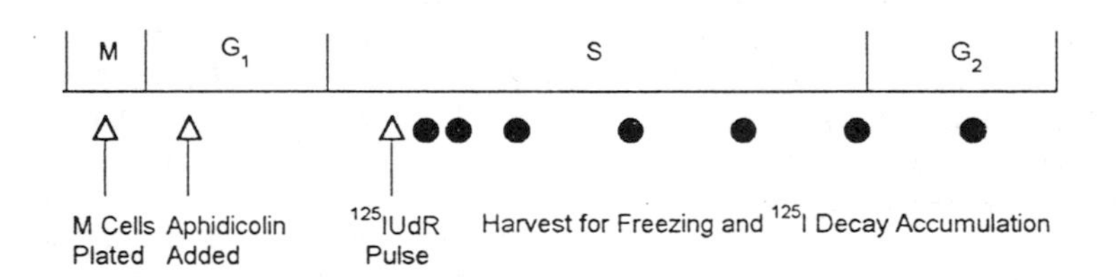

Figure 1 *Experimental design for synchronization and pulse-labeling of CHO cells*

The labeling technique described above ensured that the same "10-min segment" of DNA was labeled in all cells; only the length of the postlabeling chase period was varied. Nevertheless, the survival response of cells showed an unexpected low-LET to high-LET shift in radiation action (Figure 2). Cells frozen within 1 h after labeling yielded low-LET survival curves (large shoulder, D_0 135 decays/cell); cells harvested at hour 5 showed high-LET-type cell killing (no shoulder, D_0 40 decays/cell).[21,22] Similar experiments on irs-20 cells (a repair-deficient mutant of CHO) showed an even greater effect (Figure 3).

The striking shift in cell death might be explained by postulating that DNA damage per ^{125}I decay changes with time after labeling, either by increasing the amount of damage induced, or by decreasing damage repair. However, neutral filter elution studies did not indicate any change in the induction or repair of DNA DSBs.[25] In contrast, chromosome damage closely mirrored the pattern of cell death. Using micronucleus formation as an index for chromosome damage,[26,27] the response of cells frozen at different times after labeling showed a marked low-LET to high-LET shift (Figure 4). The correlation between micronucleus formation and cell survival was such that when survival data were plotted against micronucleus frequency rather than ^{125}I decays/cell, the widely separated survival values for early and late harvested CHO cell became superimposed.[24,25]

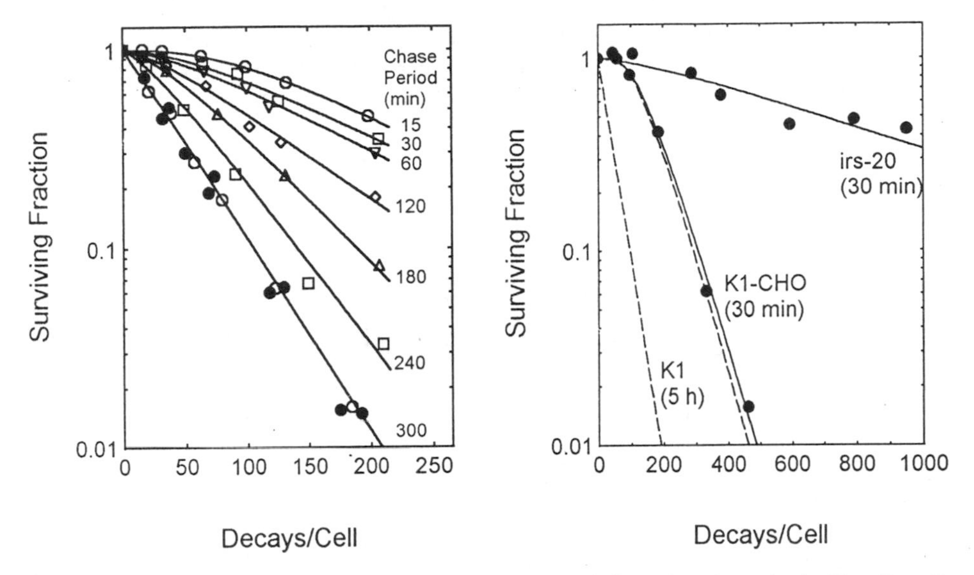

Figure 2 *Fractional survival of CHO cells harvested and frozen at different times after pulse-labeling with ^{125}IUdR*

Figure 3 *Fractional survival of irs-20 cells harvested and frozen 30 min after labeling with ^{125}IUdR*

2.1.2 Interpretation. If DNA is assumed to be the primary target for radiation death, the results indicate that DNA maturation increases damage induction or decreases damage repair. However, the shift in ^{125}I action with time after labeling is not accompanied by any detectable change in DNA DSB induction.[25] This is not surprising because with ^{125}I covalently bound to DNA the labeled DNA is obviously always located within the high-LET region around the decay site. As a result, DNA at that site invariably sustains high levels of DSBs, i.e., the induction of DSBs by ^{125}I decays would not be expected to show a low-LET to high-LET shift as the chase period is increased from 15 min to 5 h.

It could be argued that even if induction of DNA DSBs is not linked to cell death, repair may be more efficient in newly replicated DNA, resulting in enhanced cell survival after short postlabeling chase periods. DNA DSB repair studies using neutral filter elution do not support this interpretation,[25] but it is conceivable that the lethal effects of ^{125}I result from a small subset of DSBs that are not adequately resolved by neutral filter elution. However, irs-20 cells show an even bigger chase effect than wild-type CHO cells. Since irs-20 cells are deficient in DNA DSB repair,[28] the enhanced resistance of these cells after short postlabeling chase periods cannot be attributed to increased DSB repair.

An alternate, and internally more consistent, explanation would be to assume that higher-order genome structures (chromosome backbone, DNA loop domain, centromere, nuclear matrix) play a role in cell death by radiation. Using the chromosome backbone as an example, the chase effect could be interpreted as indicating that newly replicated DNA needs to undergo maturation before the DNA becomes associated with the chromosome backbone. During the maturation period ^{125}I decays would produce many DNA DSBs, but the backbone would remain outside the high-LET region around the decay site. Thus, although the DNA would be subject to high-LET effects at all times after labeling, the chromosome would sustain high-LET damage only after ^{125}I-labeled DNA segments have become associated with the backbone. As a consequence, there would be no direct link between ^{125}I-induced DSBs and cell lethality, but chromosome damage (as represented by micronucleus formation) would parallel the pattern of cell death (Figure 4).

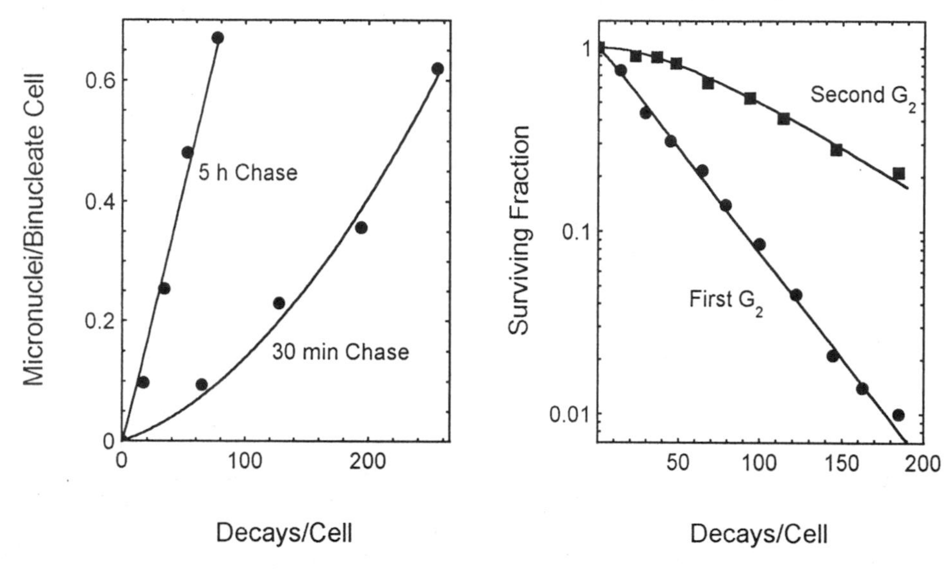

Figure 4 *Micronucleus frequency in CHO cells frozen 0.5 or 5 h after pulse-labeling with $^{125}IUdR$*

Figure 5 *Fractional survival of CHO cells subjected to ^{125}I decays during the first or second G_2 phase after labeling*

2.2 The G_2 Effect

2.2.1 Experimental Results. CHO cells were synchronized and pulse-labeled with $^{125}IUdR$ as shown in Figure 1. Cell samples were harvested and frozen either during the first G_2 phase after labeling, or during the G_2 phase of the next cycle. The results revealed an intriguing discrepancy in the radiation response of G_2 cells. ^{125}I decays during the first G_2 phase after labeling were highly toxic, yielding a shoulder-free survival curve (N = 1) with a D_0 of 40 decays/cell. In contrast, cells subjected to ^{125}I decays during the G_2 phase of the next cell cycle exhibited survival curves with a shoulder (N = 2) and a D_0 of 80 decays/cell (Figure 5). In short, identical decay events occurring in the same stage of cell cycle progression resulted in very different cell lethality depending on whether the decays were accumulated during the first or second G_2 phase after labeling.

Control experiments showed that when cells were subjected to ^{125}I decays during the first G_2 phase and then allowed to divide, the cells gave rise to daughter cell pairs where either both daughter cells were alive, or both daughter cells had sustained lethal damage. In contrast, cell exposed to ^{125}I decays during the second G_2 produced four different pairs of daughter cells: (1) both cells survived (live-live); (2) one daughter cell survived and the second died (live-dead); (3) one daughter cell died and the second survived (dead-live); (4) both daughter cells died (dead-dead). In other words, cells exposed during the first G_2 phase behaved as if they had a single target, whereas cells exposed during the second G_2 phase displayed dual-target kinetics of cell death. Lethal damage to one target did not affect the second target and both targets had to sustain lethal damage to produce a "dead-dead" pair of daughter cells, i.e., dual targets located within a single cell acted as independent entities as if already distributed between two separate daughter cells.[23]

The above experiments were repeated several times, always with the same results. CHO cells exposed to ^{125}I decays during the first G_2 phase after labeling showed single-hit kinetics of cell death, cells exposed during the second (or third) G_2 phase showed dual-hit kinetics. Moreover, the phenomenon was not limited to cell survival, it was also apparent in studies on mutation induction and micronucleus formation. The induction of HGPRT mutations during the first G_2 phase was twice as efficient as during the second G_2, and the same was true for micronucleus formation (data not shown).

These discrepancies could not be attributed to differences in radiosensitivity, repair capacity, or other cell cycle effects because both cell groups were exposed to ^{125}I decays during the same phase of the cell cycle. Also, no divergence in radiation action was noted in experiments where synchronized CHO cells were subjected to external X ray exposures during two successive G_2 phases (unpublished data). It was clear, therefore, that the shift in cell killing was somehow related to the unique dose distribution characteristics of ^{125}I and its location in newly replicated (daughter) DNA during the first G_2 phase, and in parent DNA during subsequent G_2 phases.

2.2.2 Interpretation. The ^{125}I studies on G_2 cells imply differential damage induction during the first and second G_2 phase. At first glance, this notion seems highly unlikely because G_2 cells during successive G_2 phases are structurally and functionally identical to each other. For instance, all G_2 cells have chromosomes with two sister chromatids and, according to the conventional view, each chromatid contains a single DNA double helix. Thus, if DNA is the primary target for radiation-induced cell death, an ^{125}I decay in DNA should either be able to damage only its own double helix, or it should simultaneously damage the double helices in both sister chromatids. Therefore, if DNA strands are indeed arranged as continuous, uniform DNA double helices, it would be impossible to envision a situation where the high-LET region around the site of ^{125}I decay could encompass both targets during the G_2 phase of the first cell cycle, while identical decays occurring during the next G_2 phase would damage only one of the two target structures.

This contradiction can be resolved by postulating that newly replicated DNA strands are organized differently from parental strands, at least in small regions of the genome. For example, if during G_2 short segments of the DNA double helix unwind and daughter strands become located in close proximity to each other, a single ^{125}I decay originating in either of the two labeled daughter strands could damage both strands and (possibly by enzymatic magnification) both DNA helices. During the next cell cycle the labeled strands would form part of the parental DNA, and ^{125}I decays originating in peripherally located parental strands could damage only one of the two DNA double helices.

The concept of single-stranded DNA regions in the mammalian genome is not new; open DNA regions have been detected at scaffold/matrix attachment sites by a number of methods.[29-33] Base unpairing in these regions seems to be important for DNA attachment to the nuclear matrix, and sequence mutations that diminish base-unpairing capability lead to reduced affinity of the DNA for nuclear matrix *in vitro.* Proteins such as nucleolin, a key nucleolar protein of dividing cells, have been shown to recognize and bind to both synthetic and naturally occurring DNA fragments with high base-unpairing potential.[34,35] As shown below, DNA topology at these specific regions of the genome may have major implications on target structure and mechanism of radiation action.

3. THE NATURE OF THE RADIATION TARGET: A SPECULATIVE MODEL

The chase experiments shown in Figures 2-4 provide strong circumstantial evidence for the involvement of higher-order structures in radiation-induced cell death. However, chase studies can never fully control for cell cycle effects. Different cell samples are harvested at different points in the cycle, so the possibility of altered damage induction and repair cannot be completely dismissed. Fortunately, these objections do not apply to experiments where cells are exposed to ^{125}I decays during successive G_2 phases because cells in one G_2 should obviously show the same radiobiological properties as cells in any other G_2.

A model that would account for both the chase and the G_2 data is shown in Figure 6. According to this model there are specific sites within chromosomes where DNA double helices unwind and newly replicated daughter strands are placed on the inside of a central structure, with parental strands located on the outside. The nature of the central structure is not specified. It could be a nuclear matrix attachment point for DNA, proteins forming the base of the DNA loop domain, or a higher-order chromatin structure at the chromosome backbone. In either case, the DNA would be arranged such as to permit close interactions between the two daughter strands. For clarity, the single-stranded DNA segments are depicted as straight stretches of DNA, but this does not exclude the possibility that the segments might form secondary configurations such as stem-loops.

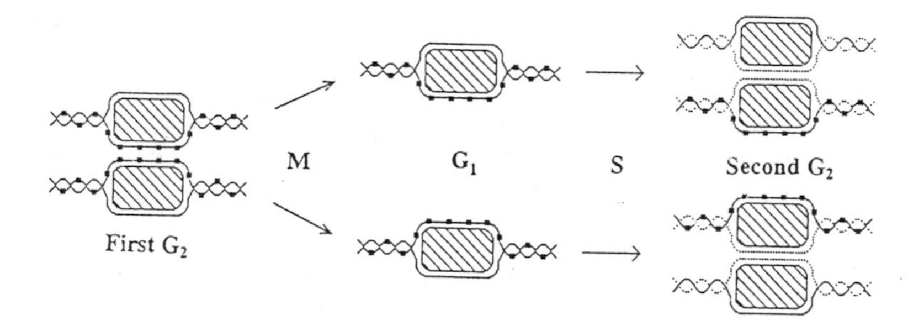

Figure 6 *Model of DNA organization at "single-stranded" chromosome regions during the first and second G_2 phase after ^{125}IUdR labeling (labeled strands indicated by solid squares). The model depicts the arrangement of newly replicated daughter DNA strands and parental DNA strands at specific sites where the DNA double helix unwinds and the DNA strands attach non-randomly to an unspecified structure (chromosome backbone, scaffold, matrix, nuclear envelope) in the cell nucleus. The configuration depicted here is hypothetical; the actual arrangement of the physical structure may be quite different.*

Regardless of the actual configuration of DNA at single-stranded DNA regions, this arrangement would ensure close proximity between the two daughter strands during the first G_2 phase after labeling. During mitosis the interaction would cease because the sister chromatids separate. The single-stranded configuration may or may not be retained during the G_1 phase, but after DNA replication the same structure would be observed during the second G_2 phase. However, the labeled DNA strands would now be parental strands and would therefore be located on the exterior of the central structure.

The proposed model would account for the marked shift in ^{125}I action from one G_2 phase to the next. During the first G_2 phase ^{125}I decays originating from either of the two daughter strands have an equal chance of simultaneously damaging DNA in both sister chromatids, but ^{125}I decays in parental strands during the next G_2 would damage DNA in only one chromatid. If each set of DNA targets requires 40 decays to sustain lethal damage, then cells accumulating ^{125}I decays first G_2 cell would require only 40 decays/cell to prevent colony formation. During the second G_2 phase again 40 decays/target would be required for cell inactivation, but with ^{125}I decays distributed randomly among two targets, and damage from each decay restricted to only one target, the D_0 would be 80 decays/cell. Moreover, since colonies could arise from surviving cells in "live-dead" and "dead-live" daughter cell pairs, the survival curve would acquire a shoulder ($N = 2$).

According to this model the chase effect shown in Figures 2-4 would be interpreted as indicating that newly replicated DNA is not immediately attached to the higher-order target structure, but becomes associated with the target during DNA maturation.

The model would also reconcile the contradiction between ^{125}I and X ray survival data for G_2 cells. With X rays the response of all G_2 cells would be identical because the radiosensitive target would be unchanged, that is, exactly the same DNA regions would be unwound. Thus, unlike the highly localized Auger electrons from DNA-associated ^{125}I, X ray photons would always encounter the same target configuration, so the X ray response of CHO cells would not be expected to vary from one G_2 phase to the next.

In conclusion, our findings indicate that DNA-bound ^{125}I shows a pronounced shift in radiation action with time after labeling and with cell progression from one G_2 phase to the next. To account for these paradoxical effects a model is proposed that invokes higher-order chromatin structures as potential radiation targets. The spatial organization of DNA in the target structure is such that a single ^{125}I decay originating in daughter DNA strands can simultaneously damage two targets during the first G_2 phase, but identical decays in parental DNA during the second G_2 can damage only one of the targets. It should be noted that the work described here could not be duplicated with any external radiation source, high-LET or low-LET, because external irradiation does not result in differential exposure of daughter DNA versus parent DNA in the genome. This demonstrates again the unique decay and dose distribution characteristics of Auger electron emitters and their utility for investigating molecular and cellular mechanisms of radiation action.

References

1. K. G. Hofer and W. L. Hughes, *Radiat. Res.*, 1971, **47,** 94.
2. L. E. Feinendegen, H. H. Ertl, and V. P. Bond, 'Biophysical Aspects of Radiation Quality', pp. 419 - 430. IAEA, Vienna, 1971, p. 419.
3. K. G. Hofer, C. R. Harris, and J. M. Smith, *Int. J. Radiat. Biol.,* 1975, **28,** 225.

4. A. I. Kassis and S. J. Adelstein, *Radiat. Res.,* 1980, **84**, 407.
5. K. G. Hofer, G. Keough, and J. M. Smith, *Cur. Top. Radiat. Res. Q.,* 1977, **12,** 335.
6. D. E. Charlton and J. Booz, *Radiat. Res.,* 1981, **87,** 10
7. L. S. Yasui, A. S. Paschoa, R. L. Warters, and K. G. Hofer, 'DNA Damage by Auger Emitters, Ed. K. F. Baverstock and D. E. Charlton, Taylor and Francis, London, 1988, p. 181.
8. R. L. Warters, K. G. Hofer, C. R. Harris, and J. M. Smith, *Cur. Top. Radiat. Res. Q.,* 1977, **12,** 389.
9. R. E. Krisch, F. Krasin and C. J. Sauri, *Int. J. Radiat. Biol.,* 1976, **29,** 37.
10. R. B. Painter, B. R. Young and H. J. Burki, *Proc. Nat. Acad. Sci. USA,* 1974, **71,** 4836.
11. R. F. Martin and W. A. Haseltine, *Science,* 1981, **213,** 896.
12. H. L. Liber, P. K. LeMotte and J. B. Little, *Mutat. Res.,* 1983, **111,** 387.
13. R. A. Gibbs, J. Camakaris, G. S. Hodgson and R. F. Martin, *Int. J. Radiat. Biol.,* 1987, **51,** 193.
14. G. Iliakis, G. E. Pantelias, R. Okayasu and R. Seanor, *Int. J. Radiat. Biol.,* 1987, **52,** 705.
15. P. C. Chan, E. Lisco, H. Lisco and S. J. Adelstein, *Radiat. Res.,* 1976, **67,** 332.
16. P. K. LeMotte, S. J. Adelstein and J. B. Little, *Proc. Nat. Acad. Sci. USA,* 1982, **79,** 7763.
17. R. L. Warters and K. G. Hofer, *Radiat. Res.,* 1977, **69,** 348.
18. M. H. Schneiderman, K. G. Hofer and G. S. Schneiderman, *Radiat. Res.,* 1990, **122,** 337.
19. R. B. Painter, 'Radiation Biology in Cancer Research', Ed. R. E. Meyn and H. R. Withers, Raven Press, New York, 1980, p. 59.
20. I. R. Radford, G. S. Hodgson, and J. P. Matthews, *Int. J. Radiat. Biol.,* 1988, **54,** 63.
21. K. G. Hofer, N. VanLoon, M. H. Schneiderman, and D. E. Charlton, 'Biophysical Aspects of Auger Processes', Ed. R. W. Howell, V. R. Narra, and K. S. R. Sastry, American Institute of Physics, New York, 1992, p. 227.
22. K. G. Hofer, N. VanLoon, M. H. Schneiderman, and D. E. Charlton, *Radiat. Res.,* 1992, **130,** 121.
23. K. G. Hofer, N. VanLoon, M. H. Schneiderman, and G. V. Dalrymple, *Int. J. Radiat. Biol.,* 1993, **64,** 205.
24. K. G. Hofer and S. P. Bao, *Radiat. Res.,* 1995, **141,** 183.
25. K. G. Hofer, X. Lin, and S. P. Bao, *Acta Oncol.,* 1996, **35,** (in press).
26. A. V. Carrano and J. A. Heddle, *J. Theor. Biol.,* 1973, **38,** 289.
27. A. Wakata and M. S. Sasaki, *Mutat. Res.,* 1987, **190,** 51.
28. M. A. Stackhouse and J. S. Bedford, *Int. J. Radiat. Biol.,* 1994, **65,** 571.
29. H. Probst and R. Herzog, *Eur. J. Biochem.,* 1985, **146,** 167.
30. T. Kohwi-Shigematsu and Y. Kohwi, *Biochem.,* 1990, **29,** 9551.
31. J. Bode, Y. Kohwi, L. Dickinson, T. Joh, D. Klehr, C. Mielke, and T. Kohwi-Shigematsu, *Biochem.,* 1992, **225,** 195.
32. J. H. Taylor, *BioEssays,* 1990, **12,** 289.
33. E. Palecek, M. Robert-Nicoud, and T. M. Jovin, *J. Cell Sci.,* 1993, **104,** 653.
34. L. A. Dickinson, T. Joh, Y. Kohwi, and T. Kohwi-Shigematsu, *Cell,* 1992, **70,** 631.
35. L. A. Dickinson and T. Kohwi-Shigematsu, *Mol. Cell Biol.,* 1995, **15,** 456.

Acknowledgement

This work was supported by Grant CA 21673 from the National Cancer Institute, NIH.

DUAL SPATIALLY CORRELATED NUCLEOSOMAL DOUBLE STRAND BREAKS IN CELL INACTIVATION

Anders Brahme[1], Björn Rydberg[2] and Patrik Blomquist[3]

[1]Dept. of Medical Radiation Physics, Karolinska Institutet, Box 260, S-171 76 Stockholm
[2]Life Sciences Div. Lawrence Berkeley Lab., University of California
[3]Dept. of Cell and Molecular Biology, Karolinska Institutet, S-171 77 Stockholm

1 INTRODUCTION

Most classical theories of cell inactivation have been focused on the role of single versus double strand breaks on a straight section of the elementary 2 nm wide DNA helix. In living cells only a small fraction of the DNA is organized in this way except possibly during the S phase of the cell cycle. During the G_0 G_1 and G_2 phases most of the DNA is strongly condensed first on histones to form the beads-on-a-string structure of about 11 nm crossection. The beads consist of the 8 histones that form a nucleosome around which the DNA helix is wound about two turns before it is linked to the next nucleosome etc. A linker histone contacts the DNA at the exit and entry of the nucleosome. This beads-on-a-string structure is further condensed to a 30 nm diameter fiber of super coiled DNA. It is clear that this more compact form of DNA will strongly influence the geometrical structure of radiation induced DNA damage sites.

Equally important as the structure of compacted DNA for the geometry of DNA damage sites is the geometrical arrangement of individual energy depositions along particle tracks. It is clear that independent of the particle species (electrons, photons, neutrons or heavy ions) most of the imparted energy is delivered by primary or secondary electrons. In photon beams nearly all the energy imparted is due to secondary electrons or positrons from photoelectric, inelastic or pair production events. In low kV x-ray and heavy charged particle beams most of the energy deposition is due to low energy δ-rays in the low to sub keV range. It is therefore of particular interest that such low energy δ-rays or "electron track ends" have a high probability to deliver multiple local or clustered energy depositions.

Recently a number of interesting Monte Carlo studies of the interaction of individual energy depositions along a particle track with compacted or opened DNA in the cell nucleus have been published[1-5]. This paper presents some new experimental results on the cellular and nucleosomal level and describes them in terms of a more simplified qualitative model of compacted DNA damage. This more qualitative model describes some of the most common events leading to cell inactivation as seen also in some of the more detailed Monte Carlo and other more qualitative approaches[6].

To illustrate the geometrical relationship between the energy depositions of different low energy electron tracks Monte Carlo results of Paretzke[7] were used, as shown in Figure 1. For simplicity the simple solenoidal model with 6 nucleosomes per turn of the 30 nm fiber was used to get a more realistic picture of the compacted DNA helix at distances somewhat larger than the interstrand distance of 2 nm. More or less randomly placed low energy electron tracks covering the energy range from 50 to 2000 eV are also shown superimposed in Figure 1 over the 30, 11 and 2 nm DNA fibers.

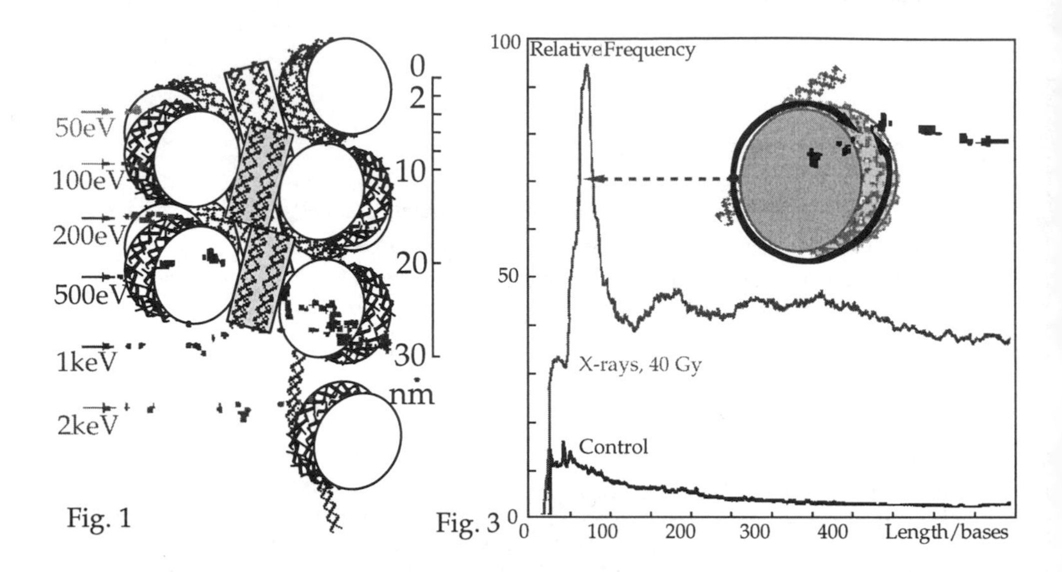

Fig. 1

Fig. 3

2 MATERIAL AND METHODS

Two somewhat different low LET radiation qualities were used in the present experiments. For the nucleosomal target 50 MV photons from the Racetrack accelerator at the Karolinska Hospital were used so that all four different DNA targets could be irradiated simultaneously to closely the same dose. The photon beam was quite narrow so the total irradiation time was less than 15 minutes even if the doses were quite high (10^2, 10^4, 10^5 and 10^6 Gy). For the fibroblast irradiation a therapeutic x-ray unit (RT 250) operated at 225kVp, 15mA with 0.35 mm Cu filtration was used at a dose rate of 6.5 Gy/min.

In the first experiment identical in vitro reconstructed nucleosomes were used to study the effect of radiation on nucleosomal DNA. The nucleosome system was primarily developed for quite a different purpose, namely to study the role of the histones in transcriptional regulation and functional consequences of different nucleosomal positioning of transactive factor response elements. The nucleosomes therefore happened to contain the gluticocorticoid receptor in a very accurate configuration on the histone octamer core of the nucleosome. Four different DNA targets were used all with the same nucleotide sequence. Either the top or the bottom strand of the 165 base pair sequences was labeled at one end and in each case. The DNA was either free in solution or mounted on a histone octamer to form nucleosomes. The DNA construct used in this study was 2No2[8].

To study the effect of conventional x-rays on DNA damage in a cellular system normal human fibroblasts were irradiated to a total dose of 40 Gy. The fibroblasts were embedded in agarose plugs in ice-cold PBS during irradiations and afterwards lysed and the fragments were end labeled as described elsewhere[9]. Since the protocol requires long incubations, in particular for the lysis steps and washings, small double stranded fragments are lost by diffusion out of the plugs. Separation of DNA fragments was performed on denaturing polyacrylamide gels that had enough thickness to allow direct loading of agarose pieces (plugs) on top of the gels.

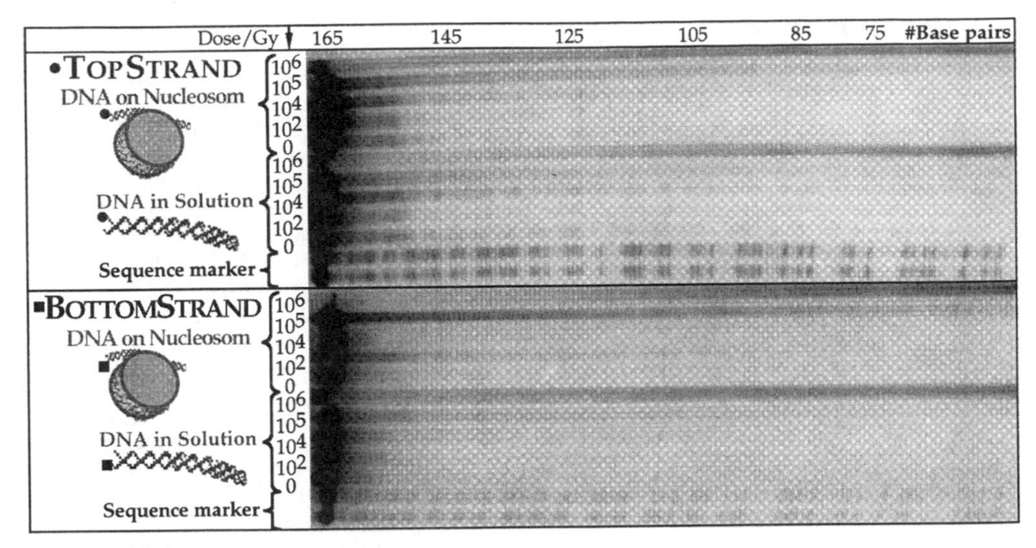

Fig. 2

3 RESULTS

The result of the nucleosomal irradiation is presented in Figure 2. The generated fragment distributions at stepwise increasing doses are presented both for the top and bottom strands and for the same DNA on a nucleosome and free in solution. It is seen that there is no clear preferential sites for the present DNA string neither on the nucleosome nor in solution. This indicates that the initial DNA damage is evenly distributed over the entire DNA sequence and there is a minimal difference between free and nucleosomal DNA. Since the DNA in each case is located outside the cell in an in vitro solution the results are completely independent of the repair enzymes that may be present in vivo. The results should however be representative also for the initial damage produced in vivo. It is also seen that at a dose of 10^6 Gy about one hit to each and every nucleosomal DNA segment is obtained in agreement with simple target calculations. Since the DNA in this case was labeled before irradiation at one end, only the fragments containing that end segment are seen in the figure.

In the irradiation of normal human fibroblasts, as shown in Figure 3, the labeling was done after irradiation so fragments from two single strand breaks on the same strand, or from a single and a double strand break are seen. It is striking to see in this case the very strong peak systematically falling at 78 base pairs. It is most likely that this peak is due to a single strong energy deposition event by a δ-electron somewhere on the periphery of the nucleosome. Due to the high geometrical probability of multiple close energy depositions by such electrons one of the DNA fibers may get a single or double strand break whereas the other fiber also gets a single or double strand break at more or less the same point on the periphery of the nucleosome. By the post irradiation end labeling the segment between these two damage sites will be detected. The interesting phenomenon is now that almost independent of where this hit cluster appears on the nucleosome periphery the segment will be equally long and approximately correspond to one turn around the histone-octamer or about 80 base pairs as shown in the insert of Figure 3. The reduction of the experimental peak to 78 bases as seen in Figure 3 could partly be caused by the loss of part of one base pair at each interaction site due to the breaks and the labeling procedure. The peak is thus most likely due to the increased geometrical probability caused by the independence of the exact location of the single clustered interaction site on the nucleosome periphery. The second much lower peak at about 180 base pairs is probably due to a similar hit cluster in the region where the helical DNA strings enter and leave the nucleosome or in other parts of the linker regions, since this is a less likely event.

4 CONCLUSIONS

It is clear that the damage inflicted by electron track ends on the periphery of a nucleosome may result in dual spatially correlated doubled strand breaks separated by about 80 base pairs. Obviously such a type of damage is hard to handle for the cell during the repair process when the histones are removed to allow full access for the repair enzymes. The fact that most of these strand breaks are blunt further reduces the probability to have a correct recombination and repair. In addition, there are also several possibilities for mis-repair such as loss of one turn or 80 bp of DNA or inversion of the DNA sequence in the same turn. Since a large part of the nuclear DNA is organized in this way this process may be a common lethal event leading to cell inactivation. As seen in Figure 3, damage sites causing a longer or shorter separation between the strand breaks are less probable, almost by a factor 3, but integrated over all possible separation distances they may contribute more, even though the probability for lethal rearrangements or losses is likely to decrease as the distance increases.

It is well known today that both the simpler classical single and double strand breaks are generally well repaired by approximately similar kinetics i.e. a fast and a slow repair rate constant of around 20 minutes and 3 hours, respectively. Therefore, due to the close relationship between this type of more severe DNA damage and the low energy electron track ends, this may be the main reason why ultra low energy x-rays and light ions have a high RBE. High energy electrons, photons and protons, on the other hand, produce much fewer low energy electron track ends per unit dose and thus fewer lethal events[10]. Since the fluence of low energy δ-electrons is closely proportional to the absorbed dose in all low LET beams the RBE is also close to unity. In summary it is possible that the relative fluence of low-energy δ-electrons in combination with the nucleosomal organization of the DNA explain the main characteristics of the direct radiation effect on cells.

ACKNOWLEDGMENTS

The authors are grateful to Bengt Lind and Peder Näfstadius for operating the Racetrack Accelerator and Örjan Wrange for providing the nucleosome reconstitution facilities. Support by the Swedish Cancer Foundation, 2222-B95-11XAC, 3035-B95-05XAB, 0037-B95-30XBC is gratefully acknowledged.

REFERENCES

1. H. Tomita, M. Kai, T. Kusama, Y. Aoki and A. Ito, *Int. J. Radiat. Biol.,* 1994, **66:6**, 669
2. H. Nikjoo and D. Goodhead, *Phys. Med. Biol.*, 1991, **36**, 229
3. V. Michalik, *Int. J. Radiat. Biol.*, 1992, **62**, 9
4. M. Terrissol, *Int. J. Radiat. Biol.*, 1994, **66:5**, 447
5. W.R. Holley and A. Chatterjee, *Rad. Res.*, 1996, **145**, 188
6. G.W. Barendsen, *Int. J. Radiat. Biol.*, 1993, **63**, 325
7. H. Paretzke, In 'Kinetics of Nonhomogeneous Processes', Ed. G.R. Freeman, John Wiley & Sons, 1987, USA, Chapter 3, p. 89
8. P. Blomquist, Q. Li and Ö. Wrange, *J. Biol. Chem.*, 1996, **271**, 153
9. B. Rydberg, *Radiat. Res.*, 1996, **145**, 200-209
10. A. Tilikidis and A. Brahme, *Acta Oncol.*, 1994, **33**, 457-469

REJOINING KINETICS OF DNA DOUBLE-STRAND BREAKS AND VARIATIONS IN RADIATION QUALITY

B. Stenerlöw, E. Höglund, E. Blomquist and J. Carlsson

Division of Biomedical Radiation Sciences
Uppsala University
Box 535
S-751 21 UPPSALA
Sweden

1 INTRODUCTION

Radiation-induced DNA double-strand breaks (DSBs) constitute a heterogeneous group of damages[1] and the biochemistry of those DSBs is complex and is assumed to depend on the radiation quality.[1,2] However, the induced level of DSBs is relatively insensitive to changes in the linear energy transfer (LET).[3-10] High-LET radiation is instead believed to induce more complex DNA damages than low-LET radiation, such as ^{60}Co photons, and an indirect way to probe the complexity of DSBs might be to follow the repair of these breaks.

For low-LET radiation the majority of the induced DSBs will be rejoined and usually two or more phases are identified; the fast kinetics has been proposed to represent repair mechanisms in non-coding sections of the genome[11] involving strand annealing between direct repeats, while the slow phase may reflect the high fidelity repair of damages at critical sites and it is possible that recombination events are operating.[12,13] DSBs induced by high-LET radiations are rejoined less efficiently than those induced by low LET radiation, and an increased fraction of unrejoined DSBs have been proposed as an important determinant in describing the molecular effects of high-LET radiations.[4,6,10,14,15] However, in some recent studies[7,10] different kinetics were seen for different LETs, but no differences were detected in the residual amount of DSBs after ≥ 6 h. Thus it seems like, at least in some cell systems, the molecular weight-distribution of DNA fragments could be restored but at a slower rate when the LET is increased.

In the following paper the DSB rejoining in three different cell types is described. Studies with nitrogen ions with LET of 125 keV/μm was, in relation to previous studies, extended to irradiations at four different depths in the Bragg peak of the beam, corresponding to LETs of 80, 125, 175 and 225 keV/μm. Our results indicate that DSBs induced by different radiation qualities are rejoined with a characteristic kinetics.

2 MATERIALS AND METHODS

Human glioma U-343MG, Chinese hamster V79 and low passage human skin fibroblasts GM5758 cells (Human Genetic Mutant Cell Repository, Camden, NJ) were grown as monolayers under standard conditions.[8] Cells for monolayer irradiation were plated in 3 cm culture dishes prior to the experiments. At the time of sample preparation the U-343MG and V79 cells were exponentially growing, whereas the GM5758 cultures were confluent and non-dividing in order to have a more homogenous cell population. Agarose gel plugs with ^{14}C-labelled cells were prepared either before (for DSB induction to prevent repair during plug preparation) or after (DSB rejoining) irradiation and the DNA was separated by pulsed-field gel electrophoresis, PFGE, as previously described.[10]

Following electrophoresis, each lane was cut at two positions corresponding to DNA sizes of 5.7 Mbp and 9 Mbp respectively and each gel slice was measured in a liquid scintillation counter.[10] The amount of DNA released from each gel section following electrophoresis was measured by the incorporated [^{14}C]thymidine activity and related to the total DNA content in the lane, giving the fraction activity released, FAR.

Cells were irradiated with photons (^{60}Co), 0.7-1.0 Gy/min, or nitrogen ions, >10 Gy/min, accelerated to energies between 32 an 45 MeV/u at the The Svedberg Laboratory in Uppsala. All irradiations were carried out on ice and the cells were placed on ice 30-60 minutes prior to irradiation. U-343MG and V79 cells were irradiated with nitrogen ions at LET = 125 keV/μm.[10] GM5758 cells for monolayer repair were irradiated at different positions of the Bragg peak of the nitrogen ion beam. The maximum range in water, after energy losses in scattering foil, mylar windows, air and transmission ionisation chamber was 4100 μm. By placing different absorbers (water equivalent) in front of the dishes the following LETs were calculated from established data on stopping power:[16] 80 keV/μm (510 μm), 125 keV/μm (2780 μm), 175 keV/μm (3460 μm) and 225 keV/μm (3740 μm).

3 RESULTS AND DISCUSSION

In human glioma U-343MG cells irradiated with 125 keV/μm nitrogen ions, two major differences, compared to photon irradiated cells, were obtained;[10] (1) a larger fraction of unrejoined DSBs after 6 h, and (2) less DSBs were rejoined by the initial fast kinetics (Figure 1). When rejoining was analysed 20-22 h after irradiation, the nitrogen ions gave 2.5-2.9 times more residual DSB than the gamma photons. In the same study, hamster V79 cells irradiated with 10 Gy-doses showed similar, but less accentuated, differences between the two radiation qualities up to 2 h after irradiation. No significant difference was found after >6 h. Thus it seems like some cells are able to rejoin all the high-LET induced DSBs, although with slower kinetics, while other cell types leave a significant fraction of DSBs

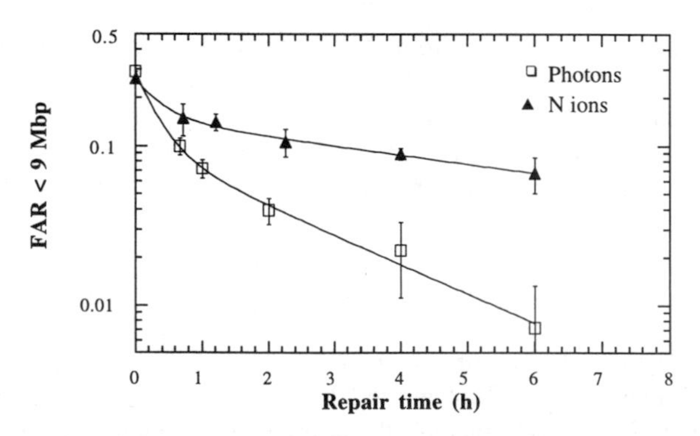

Figure 1 *Rejoining of DSBs in human glioma U-343MG cells after irradiation with 20 Gy of ^{60}Co photons or 125 keV/μm nitrogen ions. Data from Stenerlöw et al.*[10]

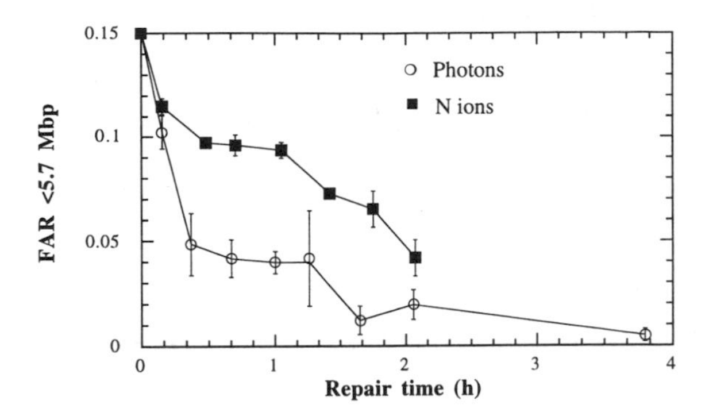

Figure 2 *Rejoining of DSBs in V79 cells after irradiation with 20 Gy of ⁶⁰Co photons or 125 keV/μm nitrogen ions. DSBs are expressed as the fraction activity released, FAR <5.7 Mbp. Photon data with SEM (n = 4-8) and ion data from duplicate samples (maximum errors)*

unrejoined. In both cases the relative biological effectiveness (RBE) for DSB induction were around 0.9. Regarding the initial amount of DSBs, it is also possible that correlated breaks lead to non-random distributions of small- and intermediate-sized DNA fragments,[9,*] thus increasing the RBE. These are generally not detected on standard PFGE gels.

In order to get a more detailed picture of the initial rejoining, V79 cells were irradiated with photons or 125 keV/μm nitrogen ions and the DSB rejoining was compared during the first 2 h (Figure 2). These results were in agreement with our earlier findings, and since a larger fraction of the ion-induced DSBs was rejoined by the slow kinetics, the curves for the two radiation qualities separate within the first 30 min. However, extrapolation of the curves up to 4-6 h will result in levels of damage (FAR < 5.7 Mbp) of the same order. Similar results have been obtained by Kysela *et al.*,[7] where DSBs induced by intermediate-LET neutrons were rejoined more slowly than the X-ray induced but no differences were detected in the residual amount of DSBs.

When an accelerated ion slows down, the ionisation density increases. To investigate whereas a systematic increase in LET resulted in a systematic change in the rejoining of DSBs, normal human fibroblasts were irradiated with 20 Gy-doses at different positions in the Bragg peak of the nitrogen ion beam. The results from the first experiments are summarised in Table 1. The same radiation dose gave slightly different initial FAR values, resulting in different RBE for DSB induction. When the initial yields were normalised (*i.e.*100% FAR remaining when t = 0 h) and the following rejoining was compared, the curves separated and the DSBs induced by the different radiation qualities rejoined with their characteristic kinetics. Note that the large difference in "FAR Remaining" between photons and 80 keV/μm nitrogen ions after 1 h, was almost disappeared at 6 h. These experiments are in progress and more time points will be measured. The stated errors in Table 1 are estimates of standard errors from curve-fits to several data points during 0-6 h.

* DNA-fragment distributions induced by ionising radiation: effects of linear energy transfer. B. Stenerlöw, E. Höglund and J. Carlsson. Manuscript in preparation.

Table 1 *Rejoining of DNA Double-strand Breaks in Human Fibroblasts GM5758 after Irradiation at Different LETs*

Radiation	*LET (keV/μm)*	*Induction RBE*	*% FAR Remaining* 1 h	6 h
γ (^{60}Co)	< 0.8	1	9 ± 3	2 ± 1
Nitrogen ions	80	≥1	38 ± 5	3 ± 1
	125	≤1	53 ± 4	17 ± 3
	175	0.8	56 ± 5	–
	225	0.6	67 ± 6	32 ± 6

Of those DSBs that are actually rejoined, the rejoining rate of the slow phase has been found to be independent of the LET.[5,8,10] Thus, a subgroup of the ion-induced DSBs may be irrepairable and permanent whereas the remaining breaks will be rejoined with nearly the same rate as the photon-induced DSBs. However, the extent of misrepair was unknown and by only measuring changes in the size-distribution of large DNA fragments, misrepair and loss of small fragments may be scored as rejoining.

From the above results we conclude that irradiation with increased LET does not always result in increased amounts of residual DSBs. On the other hand, clear differences can be detected within the first hours of rejoining.

References

1. D. T. Goodhead, *Int. J. Radiat. Biol.*, 1989, **56**, 623.
2. J. F. Ward, *Radiat. Res.*, 1985, **104**, S103.
3. J. Heilmann, G. Taucher-Scholz and G. Kraft, *Int. J. Radiat. Biol.*, 1995, **68**, 153.
4. D. Blöcher, *Int. J. Radiat. Biol.*, 1988, **54**, 761.
5. M. Frankenberg-Schwager, D. Frankenberg, R. Harbich and C. Adamczyk, *Int. J. Radiat. Biol.*, 1990, **57**, 1151.
6. T. J. Jenner, C. M. DeLara, P. O'Neill and D. L. Stevens, *Int. J. Radiat. Biol.*, 1993, **64**, 265.
7. B. P. Kysela, J. E. Arrand and B. D. Michael, *Int. J. Radiat. Biol.*, 1993, **64**, 531.
8. B. Stenerlöw, J. Carlsson, E. Blomquist and K. Erixon, *Int. J. Radiat. Biol.*, 1994, **65**, 631.
9. B. Rydberg, *Radiat. Res.*, 1996, **145**, 200.
10. B. Stenerlöw, E. Blomquist, E. Grusell, T. Hartman and J. Carlsson, *Int. J. Radiat. Biol.*, 1996, **70**, 413.
11. A. Price, *Semin. Cancer Biol.*, 1993, **4**, 61.
12. M. A. Resnick, *J. Theor. Biol.*, 1976, **59**, 97.
13. M. K. Derbyshire, L. H. Epstein, C. S. H. Young, P. L. Munz and R. Fishel, *Mol. Cell. Biol.*, 1994, **14**, 156.
14. M. A. Ritter, J. E. Cleaver and C. A. Tobias, *Nature*, 1977, **266**, 653.
15. J. Heilmann, H. Rink, G. Taucher-Scholz and G. Kraft, *Radiat. Res.*, 1993, **135**, 46.
16. J. F. Ziegler and J. M. Manoyan, *Nucl. Instrum. Meth.*, 1988, **B35**, 215.

CLUSTERING OF DNA BREAKS IN CHROMATIN FIBRE: DEPENDENCE ON RADIATION QUALITY

S.G. Andreev, I.K. Khvostunov*, D.M. Spitkovsky[#], V.Yu. Chepel[&]

Institute of Biochemical Physics, Kosygina str.4, 117977, Moscow, Russia,
*Medical Radiological Research Centre Korolev str. 4, 249020 Obninsk, Russia,
[#]Medical Genetics Centre, Kashirskoe shosse 1, Moscow, Russia,
[&]Universidade de Coimbra, Coimbra 3000, Portugal

1 INTRODUCTION

An explanation of difference in lethal effects between low and high linear energy transfer (LET) radiation is often associated with different complexity of locally (≤20-30 bp) damaged sites in DNA resulting from high local ionization concentration[1]. Alternative hypothesis suggests that clusters of DNA double strand breaks (dsbs) are more effective in causing lethal effects than single dsbs[2]. Monte Carlo track structure based models were developed to study complexity of dsbs and local damaged sites in linear DNA structure[3,4] for high LET radiation. Track structure analysis of DNA complex breaks and RBE-LET dependence of cell inactivation for protons and alpha particles showed that complex lesions-submicrodeletions up to 30 base pairs (bp) in DNA might correlate with lethality[5]. Complexity of the radiation damage may be expressed not only on the level of small DNA sites up to 20-30 bp but also over larger scales[6,7]. In present work we study the complexity of radiation-induced lesions expressed as clusters of DNA dsbs arising from DNA folding in higher-order chromatin structure. We present modelling evidence, based on the biophysical analysis of chromatin damage, that clusters of DNA dsbs produced by a single track in chromatin might be involved in reproductive cell death.

2 METHODS

The Monte Carlo code DeTrack[8] was used to generate stochastic track structure for different heavy ions in water gas of unit density for LET range from 10 to 10^4 keV/μm. Spatial structures of B-DNA, condensed chromatin fibre (solenoid model: 32 nm diameter, 7.4 nucleosomes per turn, 47 bp linker), decondensed fibre (extended solenoid and zig-zag models) approximated by DNA chain with nucleosomes, and 50 kbp coiled chromatin domains were generated as described previously[9]. Dsb and ssb (single strand breaks) induction was simulated in line with the model proposed by Charlton *et al.*[3] with modifications; algorithm has been published elsewhere[10]. Energy deposition inside the sugar-phosphate volume greater than E_{min} has been assumed to result in ssb. Any two ssbs on opposite strands within 30 bp were scored as single dsb.

3 RESULTS AND DISCUSSION

Figure 1 shows the frequency distribution of various number of dsbs per unit dose induced by alpha particle track of 102 keV/μm in 5.2 kbp chromatin solenoid segment compared with extended (zig-zag or solenoid) structure and linear DNA of the same molecular weight. Probability of none and one dsb in dense chromatin is lower than in decondensed structures but induction of multiple dsbs is more frequent than in decondensed fibre and linear DNA. Further information about multiple dsbs in chromatin could be derived from spectra of the DNA fragment lengths between pair of dsbs. Figure 2 shows the frequency distribution of alpha particle induced DNA fragments in chromatin computed for two extreme cases, i.e. condensed (solenoid) and decondensed (extended zig-zag and solenoid) conformations. Peaks at about 100 bp and 1200-1400 bp reflect DNA periodicity in nucleosome and solenoid respectively. The latter disappears for extended structures. These data generally suggest clustering of DNA dsbs at different distances within the 5 kbp chromatin segment although the shape of distribution and peak position depend on the fine structure of chromatin fibre.

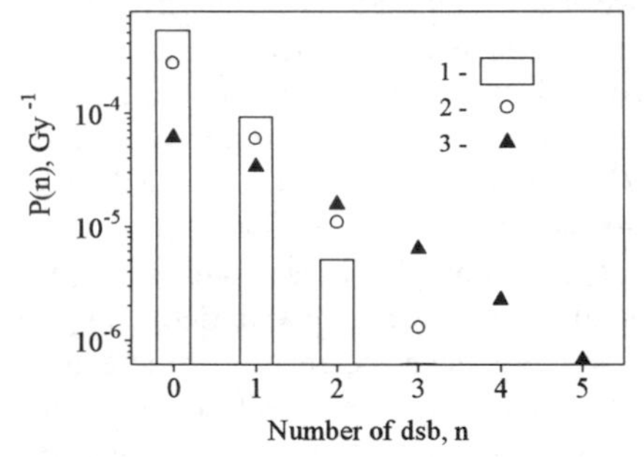

Figure 1 *Probability (per unit dose) of n dsbs induced by alpha particle of 102 keV/μm in linear DNA (1) <dsb>=0.17, extended fibre (2) <dsb>=0.25, solenoid dense fibre (3) <dsb>=0.82. <dsb> per single event.* E_{min} *= 20 eV.*

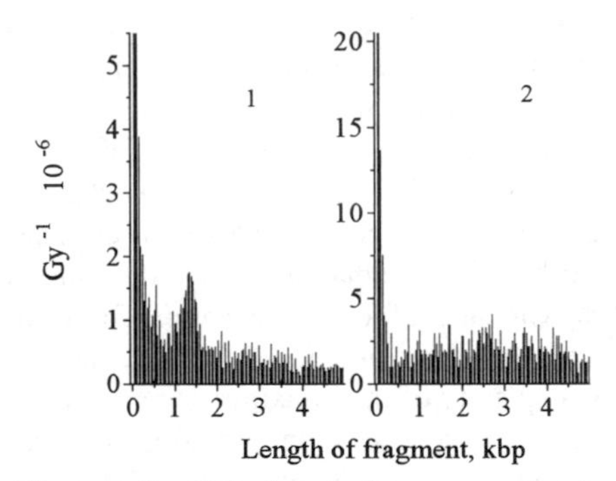

Figure 2 *Calculated frequency distributions (per unit dose) of DNA fragment lengths induced by alpha particle of 102 keV/μm in condensed (1) and extended 5 kbp chromatin fibre (2).* E_{min} *= 20 eV.*

Next level of chromatin folding in the cell is the 50-90 kbp domains. Calculated frequency distribution of DNA fragments induced by 102 keV/μm alpha particle in 50 kbp chromatin domain in compact coiled state is given in Figure 3. This result suggests that clustering of dsbs can occur also at distances of about 10^4 bp if conditions for high compaction of chromatin are realized in the cell. At present time it is little known whether these dense structures (solenoid, coiled domain) exist and if so, where is their localization in genome, in bulk chromatin or in specific areas of heterochromatin.

Simulation of data on dsb induction in the in vitro system modelling cellular chromatin is demonstrated in Figure 4. Experimental yield of dsbs in SV40 DNA irradiated intracellularly by different heavy ions (p, alphas, N, Ar, Xe)[11] is shown in comparison with yield predicted theoretically for SV40 minichromosome. This compact structure was modelled as 32 nm solenoid fibre of 5243 bp DNA length[12]. Data fit was obtained in the range of 10-10^4 keV/μm if one suppose that a fraction of SV40 DNA in closed circular form exists at the beginning of exposure. It suggests that viral DNA doesn't form a compact minichromosome with 100% effectiveness in the host cell. The following parameters were chosen: effectiveness of minichromosome formation is 0.7, E_{min} =17.5 eV.

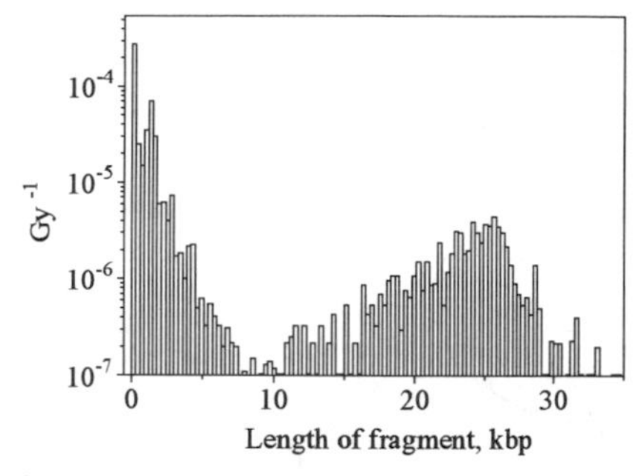

Figure 3 *Calculated frequency distribution (per unit dose) of DNA fragment lengths induced by alpha particle of 102 keV/μm in coiled 50 kbp domain.* E_{min} = *20 eV.*

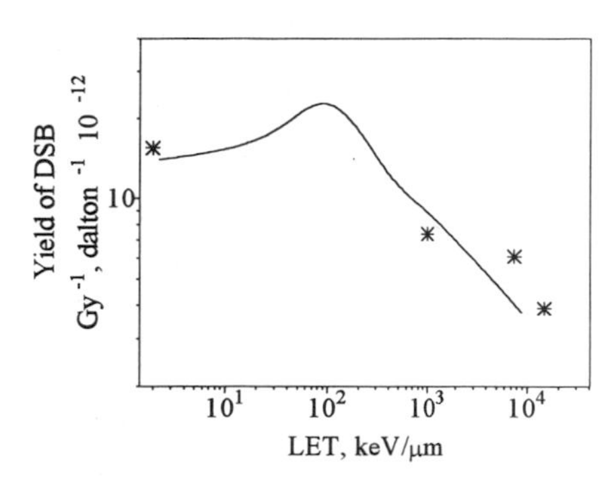

Figure 4 *Calculated (____) and experimental (**) yield of dsbs in SV40 DNA irradiated intracellularly by heavy ions of different LET*[11] .E_{min}=*17.5 eV.*

To verify the hypothesis that two dsbs formed in close proximity in chromatin are a lethal damage[2] we calculated probabilities of one, two, two and more dsbs *P(dsb=1), P(dsb=2),* $P(dsb\geq2)$ induced in 5.2 kbp chromatin segment (solenoid model) per unit dose as a function of LET. Figure 5a shows comparison of experimental RBE-LET relationship for cell inactivation presented by Barendsen[2] with probabilistic characteristics of alpha particle induced dsbs cluster. Theoretical curves are normalised at 10 keV/μm. Good correlation is seen in a wide range of LET for cluster of two dsbs located mainly within nucleosome (see Figure 2). However, one can not exclude the contribution of clusters with two and more dsbs (upper curve) in 5 kbp segment if one assumes that not all but fraction of these clusters would lead to lethal damage.

Further test of cluster hypothesis comes from analysis of data for relative efficiency of cell inactivation by protons and alpha particles. Figure 5b shows the results of fitting of RBE survival data[13] with computed probabilities of two-dsbs cluster formation as a function of LET for protons and alpha particles. Similarity for shape of theoretical and experimental curves is observed for alpha particles and minimal energy of inducing a ssb E_{min} = 20 eV. There is no direct relationship between dsbs and inactivation curves for protons if E_{min} = 20 eV, although qualitative agreement may be obtained for 14 eV. The simplest interpretation of this observation may be as follows. Clustered breaks induced by protons are repaired more efficiently than those induced by alpha particles. Since protons are efficient in inducing two dsb cluster in chromatin (E_{min} = 20 eV) only part of initial damage could become lethal whereas for alpha particles more clusters are converted to lethal event. Detailed interpretation of these data however will require further study and simulations. Under these assumptions results in Figures 5a,b are consistent with hypothesis that clusters of dsbs formed by a single track in chromatin might be linked to radiation-induced cell death.

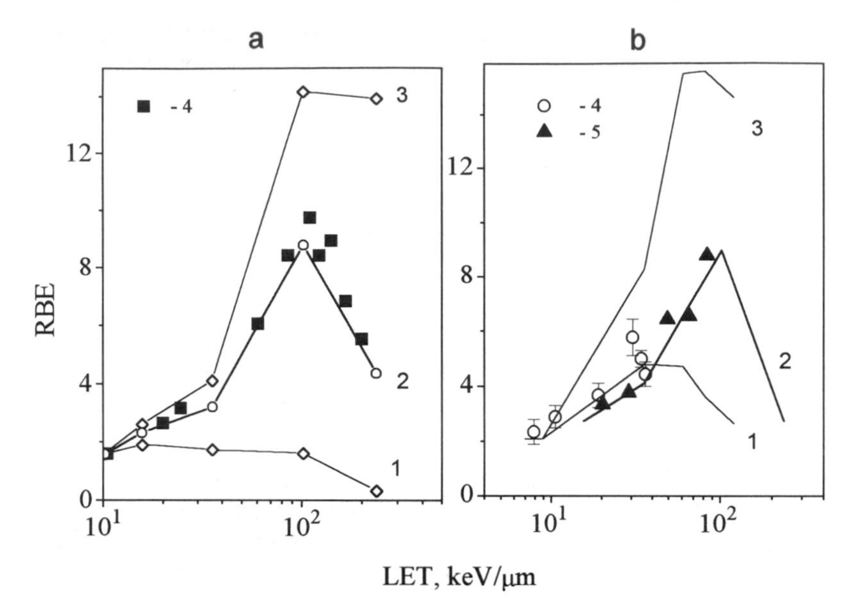

Figure 5 *Comparison of RBE data for cell inactivation with relative efficiency of dsbs cluster production vs LET. **a**. Calculations for alpha particles,* E_{min}*=20 eV, P(dsb=1) (1), P(dsb=2) (2), P(dsb≥2) (3). Experimental data (4) are from reference*[2]. ***b**. Calculations for protons,* E_{min}*= 14 eV, P(dsb=2) (1)* E_{min}*= 20 eV, P(dsb=2) (3) and alpha particles,* E_{min}*=20 eV, P(dsb=2) (2). Experimental data are from reference*[13]*, protons (4), alpha particles (5).*

References

1. J.F. Ward, *Int. J. Rad. Biol.*, 1994, **66**, 427.
2. G.W. Barendsen, *Radiat. Prot. Dosimetry*, 1990, **31**, 235.
3. D.E. Charlton, H. Nikjoo, J.L. Humm, *Int. J. Rad. Biol.*, 1989, **56**, 1.
4. V. Michalik, *Int. J. Rad. Biol.*, 1992, **62**, 9.
5. A. Ottolenghi, M. Merzagora, L. Tallone, M. Durante, H.G. Paretzke, W.E. Wilson, *Rad. Environm. Bioph.*, 1995, **34**, 239.
6. V.Yu. Chepel, I.K. Khvostunov, L.A. Mirny, T.A. Talyzina, S.G. Andreev, *11 Symposium on Microdosimetry*, Gatlinburg, USA, 1992, Book of Abstracts, p22.
7. K.M. Prise, M. Folkard, H.C. Newman, B.D. Michael, *Int. J. Rad.Biol.*, 1994, **66**, 537.
8. V.A. Pitkevich, V.G. Vidensky, V.V. Duba, *Atomic Energy*, 1982, **52**, 90.
9. V.Y. Chepel, I.K. Khvostunov, L.A. Mirny, T.A. Talyzina, S.G. Andreev, *Radiat. Prot. Dosimetry*, 1994, **52**, 259.
10. I.K. Khvostunov, S.G. Andreev, V.A. Pitkevich, V.Yu. Chepel, 'RADIATION RESEARCH 1895-1995. Congress Proceedings', Eds. U.Hagen, D.Harder, H.Jung, C.Streffer, Universitatsdruckerei H.Sturtz AG, Wurzburg, 1995, Vol. 2: Congress Lectures, p. 254.
11. G. Taucher-Scholz, J.A. Stanton, M. Schneider, G. Kraft, *Adv. Space Res.*, 1992, **12**, 73.
12. A.J. Varshavsky, V.V. Bakaev, S.A. Nedospasov, G.P. Georgiev, *Cold Spring Harbor Symp. Quant. Biol.*, 1978, **42**, 457.
13. M. Belli, F. Cera, R. Cherubini, D.T. Goodhead, A.M.I. Haque, F. Ianzini, G. Moschini, H. Nikjoo, O. Sapora, G. Simone, D.L. Stevens, M.A. Tabocchini, P. Tiveron, *Radiat. Prot. Dosimetry*, 1994, **52**, 305.

SURVIVAL OF V79 CELLS TO LIGHT IONS: AN ANALYSIS OF THE MODEL SYSTEM.

G.F. Grossi*, M. Durante*, G. Gialanella*, E. Mancini*, M. Merzagora°, F. Monforti°, M. Pugliese* and A. Ottolenghi°

* Dipartimento di Scienze Fisiche, Università "Federico II", and INFN, Sezione di Napoli, Mostra d'Oltremare Pad. 20, 80125 Napoli, Italy
° Dipartimento di Fisica, Università di Milano, and INFN, Sezione di Milano, Via Celoria 16, 20133 Milano, Italy

1 INTRODUCTION

Radiobiological properties of light ions are of great interest both for practical and theoretical reasons. Protons are the most important component of space radiation, and are the major cause of biological damage by neutrons. Besides, the use of protons in radiation therapy is rapidly increasing, because of the better dose localization that can be achieved with respect to photons. However, the RBE of low-energy protons is still under debate.

Most *in vitro* measurements of the RBE of low-energy light ions have been performed in Chinese hamster V79 fibroblasts (1-5), and found to reach values as high as 6-8 at low doses. It is still unclear whether these measurements apply as well to other biological systems, including human cells; relevance of these measurements for hadrontherapy and radiation protection has also to be proven.

One of the main problems in determining the efficiency of low energy ions is beam dosimetry. Measurements are not performed in track-segment conditions, because around the Bragg peak charged particle energy will quickly decrease inside the cell. Therefore adsorbed dose, and ultimately the biological response, will depend on target cell geometry. Measurements carried out with different cell geometry provided some different results. For example, Belli *et al.* (3) found that deuterons and protons had different efficiency in the inactivation of flattened V79 cells attached to mylar, but Folkard *et al.* (5) did not detect any difference in rounded V79 cells resting on polyvinylidine filters.

To determine factors affecting response of V79 cells to slow ions, we have performed experiments and Monte Carlo simulations in controlled conditions. Flattened cells have been exposed to protons through the thin mylar attaching surface. Cell thickness and shape has been measured by confocal microscopy. Variability in cell and nucleus geometry, piling up probabilities, and other biophysical parameters have been taken into account to perform the Monte Carlo simulation.

2 MATERIALS AND METHODS

2.1 Experiments

V79 cells were kindly donated by Dr. M. Belli and grown in α-MEM medium supplemented with 10% serum. Samples for irradiation were plated in glass wells with a 3 μm mylar base. Cell density at the time of irradiation was about 10^5 cells/cm^2. The well was closed and exposed at the Tandem TTT-3 accelerator, where the beam is extracted through an aluminized mylar window. Incident proton beam energy was between 0.6 and 4 MeV. Details of the radiobiological facilities at the Tandem accelerator have been published elsewhere (6). Following irradiation, cells were trypsinized and plated at low density.

Dishes were fixed and stained after one week. Colonies containing more than 50 cells were scored as survivors. Details of the survival curves will be published elsewhere.

2.1.1 Dosimetry. Radiation dose was calculated as:

$$D=k \cdot F \cdot LET_{inc} \quad (1)$$

were F is the beam fluence (protons/cm^2) measured by silicon detectors and CR-39 plastics, and LET_{inc} is the LET of the beam as incident on the cell monolayer (i.e., immediately after traversal of the mylar base). The LET is calculated by computer programs from the beam energy, measured by silicon detectors in air in the position of the cell monolayer. The factor k takes into account the energy variation inside the cell. From confocal microscopy measurements (see below), we have estimated that average cell monolayer thickness was about 6 μm and approximated the track-average LET with the LET calculated at the cell midplane. Thus, $k= LET_{3\mu m}/ LET_{inc}$ in our experiments.

2.1.2 Confocal microscopy. Cell thickness and shape have been measured in living V79 cells attached on mylar by using the Zeiss laser scanning confocal microscope in the Swiss Institute for Experimental Cancer Research (Lausanne). The medium was stained by fluorescein, and the cells appeared black on a white background. Image reconstruction capability of the microscope was employed to perform measurements on transversal sections. Average cell thickness and shape were measured. At the cell density used for irradiation, about 10% of the cells were piled up. It is still unclear whether such piling up is a general feature of the cell line, or is produced by specific agents (serum, mycoplasma contamination, etc.).

2.2 Modeling

A distribution of cells of different shape and volume has been exposed to simulated proton and deuteron beams by means of specially constructed computer codes.

2.2.1 *Cell model.* Flattened cells are modeled as suggested from confocal and electron microscopy images. Numerical parameters are the cytoplasm thickness between the mylar base and the nucleus, and nucleus volume and shape. Average values and standard errors were used to generate truncated gaussian distributions, and were obtained by our microscopic observations or by data reported in the literature. About 10^5 cells randomly chosen from the distributions have been selected for each data point. Piling up of the cells (about 10% of the total population) has also been simulated.

2.2.2 *Simulated irradiation.* For a given dose and energy, the average number of particle traversals is calculated by the nuclear cross-sectional area. The actual number of traversals is calculated assuming Poisson distribution. For Monte Carlo calculations of the energy inside the cell, each nucleus was divided into thin layers and energy inside each layer was calculated. Calculations of the k-factor in equation 1 have been performed with this method. Ion beam is assumed to be monoenergetic.

2.2.3 Biological response. DNA is assumed to be distributed uniformly inside the cell nucleus. The induction of DNA damage was simulated using the track structure code MOCA15 (7). Monte Carlo simulations suggest that clustered DNA double-strand breaks (2 or more breaks within 30 bp) are the crucial lesions leading to cell killing, but not all of the complex lesions are actually lethal (7). Here we have assumed that a clustered damage has a 0.5 probability to be lead to cell death.

3 RESULTS AND DISCUSSION

Output of Monte Carlo simulations is the number of cells killed for a given dose. After re-normalization based on experimental RBE at low LET, we can compare the RBE-LET curves with our experimental data. A dose of 10 cGy has been used for simulations, while experimental RBEs have been calculated as the ratio of the initial slopes of the proton and γ-ray survival curves. Results are shown in Figure 1. Our parameter-free simulation is in broad general agreement with experiments.

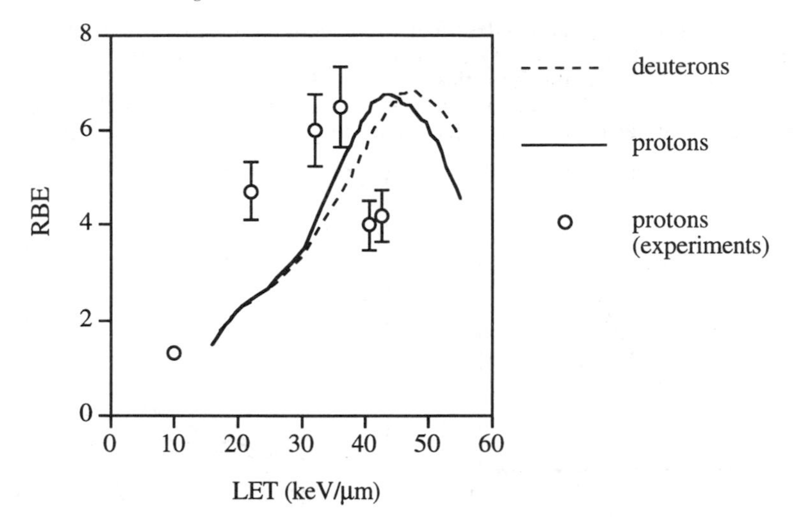

Figure 1. *Calculations of RBE of protons and deuterons at different LETs and comparison with measured values. LET is evaluated at a depth of 3 μm.*

Efficiency of deuterons and protons is similar at low LET, but some difference is predicted at high LET.

Data in Figure 1 do not take into account the cellular piling up that has been observed in these V79 cells under our growth conditions. Piling up can have an important effect on cell survival, as shown in Figure 2, where survival curves simulated for deuterons and protons at a LET (evaluated at a depth of 3 μm) of 40 keV/μm are compared with proton experimental data points at the same LET. Piling up probability improves the agreement between calculations and experiments. Protons become less efficient than deuterons at this energy, especially at high doses.

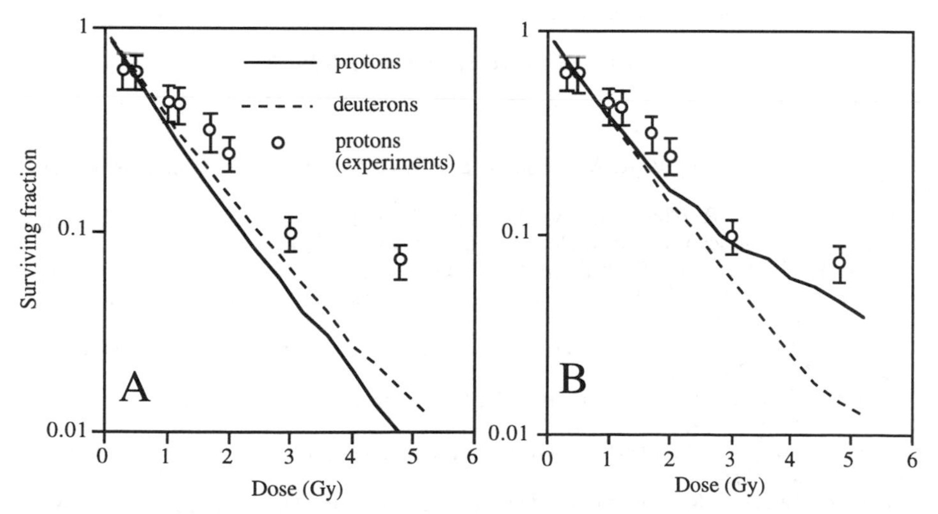

Figure 2. *Simulated survival curves of V79 cells exposed to deuterons or protons at the same LET of 40 keV/μm. A) No cell piling up; B) 10% of the cells are piled up. Points are average values from three separate experiment with protons at the same LET.*

All doses and LET have been calculated as described in Methods (equation 1): radiation LET reported in Figure 1 does not take into account the variability in cell geometry and shape. However, the Monte Carlo simulation show that the factor k varies sharply at high LET, especially for protons (Figure 3). This simulation demonstrate that calculation of the dose in low-energy ion experiments can be seriously affected by the geometry of the cell target.

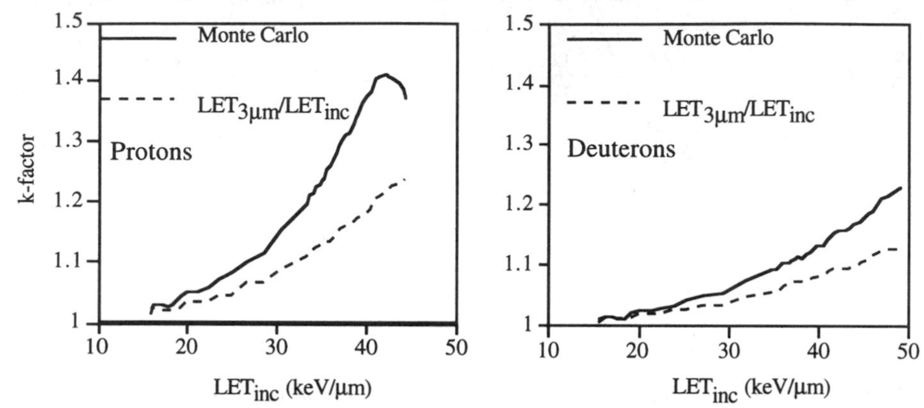

Figure 3. *Calculation of the k-factor (ratio D/F·LET_{inc}) with a Monte Carlo simulation at different LET of protons (left) and deuterons (right). The k value used in experimental dosimetry, evaluated by the LET at the cell midplane (3 μm), is also shown.*

In conclusion, Monte Carlo simulation of V79 cell survival to low-energy light ions gave results in broad agreement with experimental results. These results confirm the importance of clustered lesion for the induction of lethal effects. The agreement between experiments and calculations could be improved taking into account cellular piling-up, variations of the k-factor, and the actual beam energy spectrum as incident on the cells. The calculations demonstrate that variability in cell geometry, shape, volume, and piling-up in the cell monolayer play a critical role in determining the observed effect when experiments cannot be performed in track-segment conditions.

This work was partially supported by EU (Contract No. F14P-CT95-0011).

References

1. A. Perris, A. A. Pialoglu, A. A. Katsanos and E. G. Sideris, *Int. J. Radiat. Biol.*, 1986, **50**, 1093.
2. M. Belli, F. Cera, R. Cherubini, A. M. Haque, F. Ianzini, G. Moschini, O. Sapora, G. Simone and A. Tabocchini, *Int. J. Radiat. Biol.*, 1993, **63**, 331.
3. M. Belli, F. Cera, R. Cherubini, D. T. Goodhead, A. M. Haque, F. Ianzini, G. Moschini, H. Nikjoo, O. Sapora, G. Simone, D. Stevens and A. Tabocchini, *Radiat. Prot. Dosim.*, 1994, **52**, 305.
4. M. Folkard, K. M. Prise, B. Vojnovic, S. Davies, M. J. Roper and B. D. Michael, *Int. J. Radiat. Biol.*, 1989, **56**, 221.
5. M. Folkard, K. M. Prise, B. Vojnovic, H. C. Newman, M. J. Roper and B. D. Michael, *Int. J. Radiat. Biol.*, 1996, **69**, 729.
6. M. Napolitano, M. Durante, G. F. Grossi, M. Pugliese and G. Gialanella, *Int. J. Radiat. Biol.*, 1992, **61**, 813.
7. A. Ottolenghi, M. Merzagora, L. Tallone, M. Durante, H. G. Paretzke and W. E. Wilson, *Radiat. Environm. Biophys.*, 1995, **34**, 239.

Chromosome architecture and aberrations

NUCLEAR ARCHITECTURE AND ITS ROLE IN RADIATION-INDUCED ABERRATIONS

C. Cremer (1,2), Ch. Münkel (3), S. Dietzel (4), R. Eils (2), M. Granzow (4), H. Bornfleth (1), A. Jauch (4), D. Zink (5), J. Langowski (3,2), T. Cremer (5,2)

(1) Institute of Applied Physics, University Heidelberg, D-69120 Heidelberg
(2) Interdisciplinary Center of Scientific Computing, University Heidelberg, D-69120 Heidelberg;
(3) German Cancer Research Center, D-69120 Heidelberg;
(4) Institute of Human Genetics, University Heidelberg, D-69120 Heidelberg
(5) Institute of Anthropology and Human Genetics, University Munich (LMU), D-80333 München.

1 ABSTRACT

In recent years the mammalian cell nucleus has emerged as a highly compartmentalized structure, containing distinct chromosome territories subdivided in chromosome arm territories, R- and G-band domains, and numerous multiprotein complexes. These interact specifically with DNA and RNA and fulfill distinct functional roles in transcription, splicing, replication and repair. We discuss implications of such a territorial organization of the cell nucleus for the induction of radiation-induced chromosome exchanges.

2 INTRODUCTION

In electron micrographs of mammalian cell nuclei, the individual chromosomes cannot be discerned; the fine structure seen appears to be compatible with extended chromatin fibers, resulting in a "nonterritorial" organization of chromosomes.[1] In such nonterritorial models the euchromatic parts of interphase chromosomes may be regarded as long, flexible polymer fibers that extend throughout the entire cell nucleus and intermingle strongly with the fibers from other chromosomes. Under such circumstances it appears reasonable to expect that translocation rates of individual chromosomes induced by ionizing radiation are proportional to their DNA content. This hypothesis may be regarded as a valid first interpretation of the experimentally found frequency of breaks involved in translocation events in human lymphocytes.[2,3] There are, however, exceptions, concerning several autosomes and the sex chromosomes.[2,4] If the break frequency is proportional to the DNA content, one would also expect that in different species having the same total DNA content, the relative dicentric yields under else similar conditions should be similar. Strong deviations from this rule, however, have been observed in a number of mammalian species. Instead, the dicentric yields were found to increase linearly with the effective chromosome arm number. To explain these findings, a contact interface model of chromosome exchanges was proposed.[5] The model suggested a fundamental role of a territorial organization of interphase chomosomes in the formation of chromosome aberrations. It predicted (i) that chromosome arms occupy distinct nuclear domains and (ii) that the only relevant target volume for radiation lesions leading to interchange events is a region of thickness h (the "rejoining distance") which includes the surfaces of chromosome arm domains. In the following, the different (explicit or implicit) assumptions of contact interface (and related) models, and recent experimental evidence for them will be discussed.

3 "CONTACT INTERFACE" MODELS AND THEIR IMPLICATIONS FOR ABERRATION FORMATION

3.1 Assumption I: Chromosomes Occupy Distinct Territories in the Cell Nucleus

Two decades ago, the mammalian cell nucleus was still viewed by many cell biologists as a bag with chromatin fibers from different chromosomes intermingling in the nucleoplasm like spaghetti in a soup. Laser-UV-Microbeam experiments performed in the late 1970s and early 1980s indicated that in the nucleus of living Chinese hamster cells, the chromatin fiber of an individual chromosome was not extended over the entire nucleus but was constrained to a nuclear subvolume. When a small part of the nucleus of a living cell was microirradiated and cells were followed to the subsequent mitosis, chromosomal alterations were restricted to a few chromosomes.[6-8] These results supported a territorial organization of chromosomes in the cell nucleus as proposed at the turn of this century.[9-11]

Isotopic and nonisotopic in situ hybridization experiments with human genomic DNA as a probe made it possible for the first time to visualize entire human chromosome territories directly in somatic hybrid cell nuclei.[12-14] The decoration of entire human chromosomes by chromosomal in situ suppression hybridization ("chromosome painting") with chromosome-specific DNA obtained from flow sorted chromosomes[15-17] confirmed the existence of chromosome territories in nuclei of both normal and tumor cells.[18-20]

3.2 Assumption II: Chromosome Territories are Mutually Exclusive

The existence of chromosome territories is compatible with a variety of models, such as the "random walk/giant loop" model,[21,22] the "spherical subdomain" model,[23,24] or other modifications (C. Münkel, G. Kreth, unpublished results). Such models allow different degrees of intermingling of chromatin fibers from neighbouring chromosome territories. In an effort to study this problem experimentally, we have fixed and hybridized human cell nuclei under conditions which preserved their 3D-structure as much as possible. Two color chromosome painting, laser confocal serial sections and 3D-reconstructions of neighbouring/overlapping chromosome territories revealed that apparent intermingling of chromatin was limited to a border zone of the two adjacent territory surfaces.[24,25]

3.3 Assumption III: Chromosome Territories Are Composed Of Mutually Exclusive Chromosome Arm Domains, R- And G-Band Domains

Chromosome arm and band specific DNA probes[26] were used to paint the respective chromosomal subregions. These experiments demonstrated that chromosome arms and bands occupy distinct domains within chromosome territories[24] (Dietzel et al., submitted). In other experiments, DNA was differentially labeled in early and late S-phase with chlorine and iodine modified nucleotides (CldU, IdU). The cells were further cultured for several cell cycles until fixation. Immunocytochemical visualization of incorporated nucleotides was performed with two antibodies that distinguish between CldU and IdU and demonstrated the segregation of labeled chromosome territories with differentially visualized R-bands (early replicating DNA) and G-bands (late replicating DNA). 3D confocal microscopy and 3D-image analysis showed that the overlap volume observed between R-and G-band domains was in the order of a few percent only (D. Zink et al., manuscript in preparation).

3.4 Assumption IV: Chromosome Territories Have Characteristic Enveloping Surfaces: Most Of The Chromosomal DNA Is Localized Within This Enveloping Surface

Multifluor fluorescence in situ hybridization was used to simultaneously visualize specific DNA sequences, such as chromosome specific alphoid sequences, chromosome specific subtelomeric sequences and various genes, together with their respective chromosome territories in 3D-preserved human cell nuclei[27-29] (S. Diezel et al., unpublished data). Light optical serial sections were obtained for the different fluorescence emissions ("colors") by confocal laser scanning microscopy. These 3D-data stacks were used a) to obtain the enveloping surface and volume of a "painted" chromosome territory by Voronoi tesselation algorithms and appropriate thresholding procedures,[30] and b) to obtain the localizations (barycenters) of specific DNA sequences relative to the enveloping surface of the territory. The localization of the specific sequences was noted in most cases as distinct signals within or very close to the enveloping surface of the chromosome territory, but rarely remote from it. These findings support the notion that giant chromatin loops which would extend from the surface of one chromosome territory and penetrate to a substantial degree into a neighbouring chromosome territory are not common, at least in the cell types and under the conditions studied so far.

3.5 Assumption V: The Formation Of Radiation Induced Interchanges Occurs In A Small Volume At Risk Between The Surfaces Of Adjacent Territories: The Interchromosome Domain (ICD) Compartment

According to the hypothesis first proposed by Savage and Papworth,[5] neighbouring chromosome arm domains should act on each other via chromatin localized in boundary zones. They can be operationally defined and experimentally determined as the enveloping, three-dimensional surface area of a given arm domain which is shared with the respective surface of another arm domain. An increase in the average adjacent surface area between two chromosome arm domains in a population of irradiated nuclei is expected to increase the overall probability of exchange events between the two arms. The same considerations apply to translocation events between two neighbouring chromosome territories.[24,31] More specifically, we hypothesize that the illegitimate coupling of DNA-strands required for the induction of exchange events takes place within an interchromosomal domain (ICD) space. The hypothesis of an ICD-compartment was based on present knowledge of nuclear architecture and the biophysical properties of chromatin and proposed as a general model for the structural and functional compartmentalization of the cell nucleus.[27,32-34] The ICD-compartment model assumes a hypothetical, three-dimensional network of intranuclear channels that contain the machineries for transcription, splicing, DNA replication, and preferential repair, as well as the protein complexes involved in the formation of exchange events. The ICD-channels supposedly start at nuclear pores and expand between chromosome territory surfaces. Branches of the channel network may extend from the territory periphery into the territory interior. Interchromatin channels have been suggested repeatedly[35-42] but evidence is still circumstantial. Thc ICD-compartment model provides a physicochemical rationale for the formation and topology of such a channel network. On the basis of experimental evidence obtained for isolated metaphase chromosomes[43] we assume that chromosome territories and chromosomal domains in the nucleus of living cells are also negatively charged, resulting in repulsive electric forces between adjacent chromatin surfaces. The width of the resulting channels can be modulated by local differences in the size of the repulsive forces, by movements of chromosomal domains and their respective surfaces, and by the enrichment of macromolecules and macromolecule complexes with negative electric net charge within the ICD-space, including proteins which enter this space via the nuclear pore, as well as RNA transcribed and released at chromosomal domain surfaces. The assumed physical properties of the ICD-space allow for the preferential transport of negatively charged proteins and RNAs by channeled diffusion and/or via matrix core filaments located in this space. The model is compatible with but does not require the existence of an in vivo nuclear matrix[32]. Although it should be clearly pointed out that the existence of such an interchromatin domain channel network is still entirely speculative, the predicted physical effects seem plausible according to electrophysiological measurements of the potential difference between chromatin and non-chromatin,[44] (H. Koester et al., manuscript in preparation) and calculations based the on the Debye-Hückel and Donnan potential theory performed under highly simplified assumptions (C. Cremer, unpublished) and justify further experimental tests of the ICD-compartment model.

3.6 Assumption VI: Experimental Data Available For Radiation-Induced Interchromosome Exchanges Require A Pronounced Variability Of Chromosome Territory Structure And Territory Arrangements

As outlined above, interchanges should occur only at boundary zones formed by adjacent chromosome territories. In mitotically active human cell types, such as human amniotic fluid cells, fibroblasts and PHA stimulated lymphocytes, we have noted that the spatial distribution of chromosome territories in the cell nucleus is quite variable.[45-47] Nonetheless, the distribution of specific chromosome territories cannot be fully explained by models taking into account geometrical constraints provided by the various sizes and shapes of chromosome territories (see below[23] and unpublished data). The probability that the same reciprocal translocation occurs independently in two cells is rather small, if the respective double strand breaks are randomly arranged within the nuclear space.[37] Geometrical constraints, however, might favor (or reduce) the possibility of double strand breaks within the respective chromosome bands to come sufficiently close to each other for illegitimate recombination. Such constraints could result from specific patterns regarding the 3D-structures and intranuclear distributions of chromosome territories and arm domains, respectively (see below). The question to which extent chromosome exchange patterns vary as a result of cell type specific differences in the suprachromosomal organization of cell nuclei can still not be answered satisfactorily. Such comparisons require the development of models of chromosome territory organization and distribution that will allow quantitative predictions on the formation of chromosome aberrations.

4 PREDICTIONS OF CHROMOSOME ABERRATION FORMATION FROM QUANTITATIVE MODELS OF CHROMOSOME TERRITORY ORGANIZATION AND DISTRIBUTION

4.1 Simple Geometrical Models

The general dependence of the interchange frequency per chromosome (t-rate) on the surface area of chromosome territories can easily be calculated for a few highly simplified scenarios. For example, in the scenario considered by Savage and Papworth[5] chromosome arm domains were represented by hexagons with constant cross sections. This model can be further quantitated by assuming that the length of each hexagon is proportional to the DNA content of each modeled chromosome arm domain and that a certain fraction of the hexagon surface is attached to the nuclear envelope preventing this surface fraction from participation in exchange events. This model (and various other versions of the contact interface/ ICD-compartment model) would predict an approximately linear increase of the t-rate with the chromosomal DNA-content. If chromosome territories, however, would be represented by equivalent spheres with a volume proportional to DNA-content, the surface area (and thus the t-rate) would be proportional to (DNA content)$^{2/3}$. Cytogenetic observations[2,3] indicate a considerable deviation from the monotonous increase predicted from such simple geometrical models.

4.2 The Random Walk/Giant Loop Model

The observation that chromosomes can form intrachanges, e.g. rings, between sites that are many megabase pairs (>100 Mbp) apart from each other, requires a high flexibility of chromosome territory folding. This high flexibility was confirmed by distance measurements in 3D-preserved cell nuclei between DNA-sequences physically mapped along a chromosome of interest[22] (Dietzel et al., manuscript in preparation). The results of the measurements performed by Yokota et al. were explained by a random walk/giant loop model of chromosome territories with two organizational levels at scales larger than 100 kbp.[21] Giant loops in the order of 3 Mbp were simulated by a random walk of 300 nm long and 30 nm wide segments. The giant loops were considered to be attached to a randomly folded backbone structure. This model predicts a largely variable geometry of the chromosome territory as a whole. Furthermore, one should expect that almost any section of a chromosome territory would have a substantial probability to be exposed at the peripheral territory surface adjacent to other chromosome territories; thus, the model is compatible with breakpoints for chromosomal exchanges that occur both in R- and G-bands. In addition, the model predicts a substantial amount of intermingling between giant loops from two chromosome arm domains of the same territory and by the same token between adjacent territories (Münkel et al). These expectations of the random walk/giant loop model seem to be in conflict with a) observed differences in the intrachromosomal breakpoint frequencies;[48] b) with the apparently small amount of intermingling observed between adjacent chromosome territories and chromosome arms.[24,25] (Dietzel et al., manuscript in preparation), as well as between R-and G-band regions of the same chromosome territory (D. Zink et al).

4.3 The Spherical Subdomain Model

Relative chromosome territory surface area estimates were obtained by computer simulations[23,24] of entire human cell nuclei using a "spherical subdomain" model: In the starting configuration, the 46 territories of a male human nucleus were simulated as compact, globular structures of tiny spherical elements hold together by attractive forces. Each sudomain element represented approximately 1 Mbp and the number of elements used for the simulation of a given chromosome territory was proportional to its DNA content. The so far globular, small starting configurations were placed in an initially random position into a sphere representing the nuclear volume and then allowed to grow to their final size assuming that each chromosome territory occupies a coherent nuclear subvolume proportional to its DNA content. During this process all subdomain elements were allowed to move according to the actual forces (repulsion between elements belonging to different territories) until a minimum of free energy was reached. Starting configurations which were incompatible with subsequent growth of all chromosome territories to their final size were eliminated. In the end, the final position, form, surface size and average contact surface of chromosome territories in 384 possible configurations were calculated. These configurations showed a considerable variability of territory shapes. The probability of a more interior or exterior territory positioning in the model nucleus was dependent on DNA content. Finally, the probability was determined that two territories hit by a randomly placed straight line were adjacent to each other. The resulting relative mean "translocation rates" for "territories" with a given number of elements predicted relatively small deviations from the results obtained by simple geometrical models. Nonetheless, the spherical subdomain model has several advantages: It takes into account a) that chromosome territories and chromosome arm domains are not simple geometrical structures (rods, spheres, hexagons etc.), but have variable shapes; b) evidence for a small degree of chromatin intermingling in "border zones" providing a volume at risk between adjacent territories. The spherical subdomain model is also compatible with the prediction that an inter-chromosomal domain (ICD)-space (see above) with variable width is formed in these border zones, where repair complexes involved in the illegitimate rejoining of dsb's from two different chromosomes may be located.

5 PERSPECTIVES: THE FORMATION OF TUMOR SPECIFIC CHROMOSOMAL ABERRATIONS MAY CRITICALLY DEPEND ON THE TOPOLOGY OF CHROMOSOME TERRITORIES IN NORMAL PRECURSOR CELLS

Present experimental evidence indicates substantial differences in the intrachromosomal aberration frequencies (e.g. Folle and Obe[48]). In the framework of the contact interface/ICD-compartment model this is compatible with a nonrandom distribution of DNA-sequences in the territory, i.e. a specific topology. Apart from its fundamental role in the functional architecture of the cell nucleus, the topology of chromosome territories might play an important role in the formation of cancer related chromosomal translocations induced by ionizing radiation. A consideration of possible topological constraints in the formation of the translocation t(9;22)(q34;q11) may be helpful to exemplify their role in the genesis of chronic myelogenous leukemias (CML). The probability of such a translocation in normal (non-leukemic) precursor cells may increase or decrease to a great extent by these constraints. Let us assume that double strand breaks (dsb's) have been induced in the break point cluster regions of both the ABL and BCR genes[49] of a normal precursor cell. As long as the two breakpoint cluster regions of chromosome territories 9 and 22 are sufficiently distant from each other, the fusion of the ABL and BCR regions by illegitimate recombination is prohibited. Location of the ABL and the BCR break point regions at the surfaces of two neighbouring chromosome 9 and chromosome 22 territories could result in a topologically "exposed" configuration, where the ABL and BCR regions come sufficiently close to each other to bind to the same repair complex and thus can be joined together. In topologically restricted configurations the ABL and BCR regions are prevented from a simultaneous binding to the same repair complex (localized in the boundary zone). For example, a fusion will be prevented as long as the ABL and BCR regions are buried within the two territories and thus are not accessible to the same repair complex. These considerations suggest that for cells with a given rejoining efficiency, the probability of an erroneous coupling of two dsb's induced in the ABL and in the BCR breakpoint regions, respectively, is not simply a function of the euclidean distance between the two dsb's but of a variety of topological parameters. The frequency of exposed configurations might differ in different cell types and possibly show even interindividual variations. It seems possible that a specific topological configuration facilitated the formation of the translocation t(9;22)(q34;q11) in normal precursor cells of a patient with CML, but not in other somatic cell types. Alternatively, such a translocation may not be favored by a specific chromatin topology and occur in many cell types with the same frequency.

Multifluor-FISH[50,51] in combination with advanced forms of far-field fluorescence 3D-microscopy of intact nuclei should allow us to study experimentally the topology of chromosome territories in general and of cancer related subregions in more detail. Recent microscopical studies[52-54] and our unpublished results indicate that under suitable conditions it is possible to localize fluorescent targets at the surface and in the interior of cell nuclei with a 3D-precision in the order of about 20 - 30 nm. By labeling of neighbouring sites with fluorochromes of different "spectral signature" and separate image registration of them, the present limits of 3D-resolution (about 200 nm laterally,

700 nm axially) may be overcome with the respect to 3D distance determinations. This is a direct consequence of the fact that in multi-channel fluorescence 3D-microscopy, the spatial positions of each spectrally different target can be measured independently. After careful correction of the chromatic shifts induced, a high spectral distance precision microscope should allow, in principle, 3D-distance measurements with a *"resolution equivalent"* between differently colored targets down to 20 to 30 nm (corresponding to 2-3 nucleosome diameters). In such approaches, topological studies of chromosome territories and subregions will no longer be limited by the limits of 3D light microscopic resolution, but by artefacts introduced by the observation of fixed, permeabilized and denatured nuclei. To overcome these most critical limitations, methods for the in vivo visualization of specific chromatin regions are urgently required.

Acknowledgements

This work was supported by grants from the Deutsche Forschungsgemeinschaft and the European Community (Biomed 2) to C.C. and T.C. We are particularly grateful for stimulating discussions to Drs. J.A. Aten (Amsterdam), U. Hagen (Munich), M. Hausmann and A. Esa (Heidelberg), A.T. Natarajan (Leiden), G. Obe (Essen), R. Sachs (Berkeley), J.R.K. Savage (Didcot) and L. Trakhtenbrot (Tel Hashomer).

References

1. D.E. Comings, *Am. J. Hum. Genet.*, 1968, **20**, 440.
2. K. Tanaka et al., *J. Radiat. Res.*, 1983, **24,** 291.
3. J. Lucas et al., *Int. J. Radiat. Biol.*, 1992, **62,** 53.
4. M.C. Mühlmann-Diaz and J.S. Bedford, ‘Chromosomal Aberrations, Origin and Significance’, G. Obe and A.T. Natarajan, Springer-Verlag, Heidelberg, 1994, p.125.
5. J. R. K. Savage and D.G. Papworth, *Mutat. Res.*, 1973, **19,** 139.
6. C. Zorn et al., *Exp. Cell Res.*, 1979, **124,** 111.
7. C. Cremer et al., *Hum. Genet.*, 1980, **54**, 107.
8. T. Cremer et al., *Hum. Genet.*, 1982, **60**, 46.
9. C. Rabl, *Morphologisches Jahrbuch (Leipzig)*, 1985, **10**, 214.
10. E. Strasburger, ‘Die stofflichen Grundlagen der Vererbung im stofflichen Reich.’ Gustav Fischer, Jena, 1905.
11. T. Boveri, *Arch. Zellforschung,* 1905, **3,** 181.
12. L. Manuelidis, *Hum. Genet.*, 1985, **71,** 288.
13. M. Schardin et al., *Hum. Genet.*, 1985, **71**, 281.
14. D. Pinkel et al., *Proc. Natl. Acad. Sci.*, 1986, **83,** 2934.
15. K.E. Davies et al., *Nature,* 1981, **293,** 374
16. C. Cremer et al., *Cytometry,* 1984, **5,** 572.
17. L.L. Deaven et al., *Cold Spring Harb. Symp. Quant. Biol.*, 1986, **51**, 159.
18. T. Cremer et al., *Hum. Genet.*, 1988, **80**, 235.
19. P. Lichter et al., *Hum. Genet.*, 1988, 80, 224.

20. D. Pinkel et al., *Proc. Natl. Acad. Sci.*, 1988, **85**, 9138.
21. R. Sachs et al., *Proc. Natl. Sci.*, 1995, **92**, 2710.
22. H. Yokota et al., *J. Cell Biol.,*1995, **130**, 1239.
23. C Münkel, Ph.D. Thesis. Faculty of Physics, University Heidelberg (1995); C. Münkel et al., *Bioimaging,* 1995, **3,** 108.
24. C. Cremer et al., *Mutat. Res.*, 1996 in press.
25. R. Eils et al., *J. Cell Biol.,*1996 in press.
26. P.S. Meltzer et al., *Nature Genetics*, 1992, **1**, 24.
27. T. Cremer et al., *Cold Spring Harb. Symp. Quant. Biol.*, 1993, **58,** 777.
28. R. Eils et al., *Zoological Studies (Taiwan),* 1995, **34**, Suppl. I, 7.
29. A. Kurz et al., *J. Cell Biology*, 1996, in press.
30. R. Eils et al., *J. Microscopy,* 1995, **177**, 150.
31. K. Tanaka et al., *Int. J. Radiat. Biol.*, 1996, **70**, 95.
32. T. Cremer et al., 'Kew Chromosome Conference IV', P.E. Brandham and M.D. Bennett, Royal Botanic Gardens, Kew , 1995, p.63.
33. T. Cremer et al., 'Radiation Research 1895 - 1995', U. Hagen et al., Würzburg, 1995, Vol. 2: 459.
34. R. Zirbel et al., *Chromosome Res.*, 1993, **1,** 93.
35. G. Blobel , *Proc. Natl. Acad. Sci*, 1985, **82**, 8527.
36. L.S. Chai and A.A. Sandberg, *Cell Tissue Res.* 1988, **251**, 197.
37. J.R.K. Savage, 'Mutation and the Environment, part B', Wiley Liss, 1990, p.385.
38. D. Spector, *Proc. Natl. Acad. Sci.* 1990, **87**, 147.
39. Z. Zachar et al., *J. Cell Biol.*, 1993, **121**, 729.
40. J. Kramer et al., Trends in Cell Biology 1994, **4**, 35.
41. D.G. Wansink et al., *Molecular Biology Reports* 1994, **20**, 45.
42. S.V.Razin and I.I. Gromova, *Bio Essays* 1995, **17,** 443.
43 F.F. Bier et al., *Electrophoresis* 1989, **10,** 690.
44. H.B. Oberleitner et al., *Eur. J. Physiol.* 1993, **423,** 88.
45. P. Emmerich et al., *Exp. Cell Res.* 1989, **181,** 126.
46. S. Dietzel et al., Bioimaging 1995, **3**, 121.
47. S. Popp et al., *Exp. Cell Res.* 1990, **189,** 1.
48. G.A. Folle and G.Obe, *Int. J. Radiat. Biol.* 1995, **68**, 437.
49. D.C. Tkachuk et al., *Science* 1990, **250,** 559.
50. M. Speicher et al., *Nature Genetics* 1996, **12,** 368.
51. E. Schröck et al., *Science* 1996, 273, 494.
52. U. Kubitschek et al., *Biophys. J.* 1996, **70**, 2067.
53. B. Rinke et al., *Proc. SPIE,* 1996 in press.
54. J. Bradl et al., *SPIE Proc.*, 1996 in press.

MODELLING OF CHROMOSOME EXCHANGES IN HUMAN LYMPHOCYTES EXPOSED TO RADIATIONS OF DIFFERENT QUALITY

V.V.Moiseenko[+], A.A.Edwards[*], H.Nikjoo[#] and W.V.Prestwich[+]

[+] McMaster University, Department of Physics and Astronomy, 1280 Main Street West, Hamilton, Ontario L8S 4K1, Canada
[*] National Radiological Protection Board, Chilton, Didcot, Oxfordshire, OX11 0RQ, UK
[#] Medical Research Council, Radiation and Genome Stability Unit, Harwell, Oxfordshire, OX11 0RD, UK

1 INTRODUCTION

Substantial efforts have been made to model radiation induced chromosome aberrations in mammalian cells. Our approach has concentrated on the radiation quality dependence of dicentric yields in human lymphocytes in an attempt to better understand the mechanisms involved.[1] To model the statistical fluctuations of energy deposition, the production of double strand breaks (DSB) and the competition between their restitution and misrepair, Monte Carlo simulation on a cell by cell basis has been used. We have concentrated on dose effect relationships for four different radiations: ^{60}Co, 250 kVp x-rays, 8.7 MeV protons and 23.5 MeV ^{3}He ions. The dicentric yields Y for these radiations conform to the equation $Y=\alpha D+\beta D^2$ where D is the dose in Gy and α and β are fitted coefficients whose values are shown in Table 1. The basic mechanism investigated is that dicentrics are formed by pairwise interaction between DSB and that the probability that any two breaks will interact is distance-dependent.

With this model we have found that the α coefficient is determined by intra-track DSB interaction and increases with LET. The inter-track coefficient β does not change with LET at least up to values of 23 keV μm^{-1} corresponding to the helium ions used experimentally, provided that the RBE for DSB production remains at 1.0. If DSB are produced by clustered ionizations then the RBE for DSB production increases with higher LET and the consequence is that the predicted β coefficient increases as LET increases.[1] This paper investigates further the effect of clustering of ionizations on the eventual production of exchange aberrations by using the superior k-means method[2] rather than the associated volume method used previously. Inter- and intra-chromosome exchanges have also been modelled by incorporating chromosome domains within the model.

2 METHODS

Methods of modelling of energy deposition and partitioning of ionizations into clusters were described in detail elsewhere[1,3]. Computer generated tracks in water vapour including secondary electrons were used.[4,5] Ionizations only have been used to describe the energy deposition pattern. When ionizations were partitioned into clusters by the k-means method an extra requirement that each cluster could be enfolded by a sphere with a diameter corresponding to DNA dimensions 2.3 nm has been applied. Clusters of 3 or more ionizations have been considered as candidate events to be converted into DSB.

Table 1 *Theoretical and experimentally derived coefficients of the linear-quadratic dose-response curve $\alpha D+\beta D^2$ for dicentrics.*

Radiation type	Theoretical without correction*		Theoretical with correction*		Experimental, ± SE	
	α, Gy^{-1}	β, Gy^{-2}	α, Gy^{-1}	β, Gy^{-2}	α, Gy^{-1}	β, Gy^{-2}
^{60}Co	0.020	0.066	0.019	0.062	0.018±.003	0.060±.006
X-rays	0.040	0.105	0.035	0.096	0.037±.005	0.067±.007
Protons	0.052	0.134	0.046	0.120	0.044±.008	0.058±.006
^{3}He ions	0.22	0.215	0.18	0.175	0.40 ±.04	-

The chromatin fibre was modelled as a hollow cylinder with external diameter 30 nm and internal diameter 9 nm. Each chromosome was confined to a sphere of 2.5 μm within the 6 μm diameter cell nucleus to represent domains. A looping structure to each chromosome was additionally imposed similar to the scheme used by Sachs et al.[6]

By using Monte Carlo techniques the ionization cluster pattern was overlaid on the chromatin structure pattern and clusters of 3 or more ionizations within chromatin became candidate DSBs. The final conversion to DSB assumed that for ^{60}Co, 50 DSB per cell per Gy were produced on average. This criterion meant that for x-rays, protons and ^{3}He ions the RBE values for DSB production were 1.3, 1.4 and 1.8 respectively. Breaks closer than 2 kb were then considered as one break.[7] This correction made little difference to the RBE for DSB; the RBE for helium ions with respect to ^{60}Co reduced from 1.8 to 1.6. Calculations for the yield of exchanges were performed with and without this correction. For each radiation a simulated dose of 0.5 Gy was used and 100,000 cells with unique DSB distributions were generated. This was sufficient to establish the linear and quadratic coefficients through intra- and inter-track exchanges respectively. The ratio of the probabilities of exchange to repair, R, was assumed to follow an inverse power law $R=k/r^n$. The separation distance of any two breaks is r in μm, k and n are parameters fitted to obtain the dicentric yield curve for ^{60}Co. Half of inter-chromosome exchanges were assumed to be dicentrics and half of intra-chromosomal inter-arm exchanges, centric rings. Adjusted values of the constants were k=0.00025 and n=0.95.

3 RESULTS

Figs. 1 and 2 show the frequency distributions of intra-track cluster pairs for the four radiations considered in this paper. Fig. 1 shows the distributions over the range from 200 nm up to the 6 μm nuclear diameter chosen. Fig. 2 shows the detail for separation distances up to 200 nm.

Table 1 shows the derived α and β coefficients for dicentrics with and without the correction for close DSB on the same chromosome. The major discrepancies between predicted and observed coefficients are that the linear coefficient for ^{3}He ions is underestimated and the quadratic terms for protons and ^{3}He ions are overpredicted. The

* Correction means that DSB separated by 2 kb or less have been considered as one DSB.

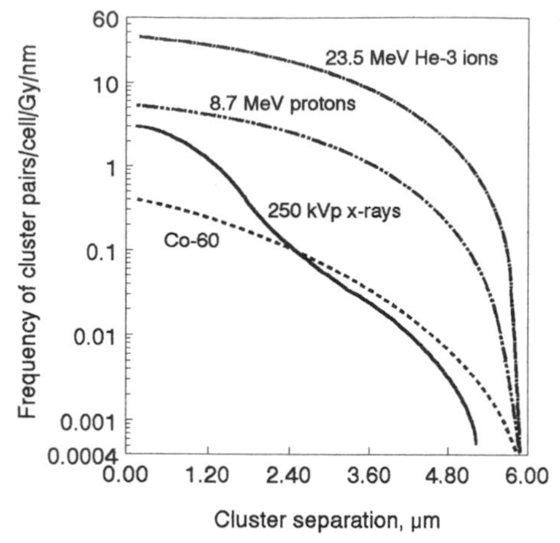

Figure 1 *Frequency distributions of cluster pairs for separation distances from 0.2 to 6 μm*

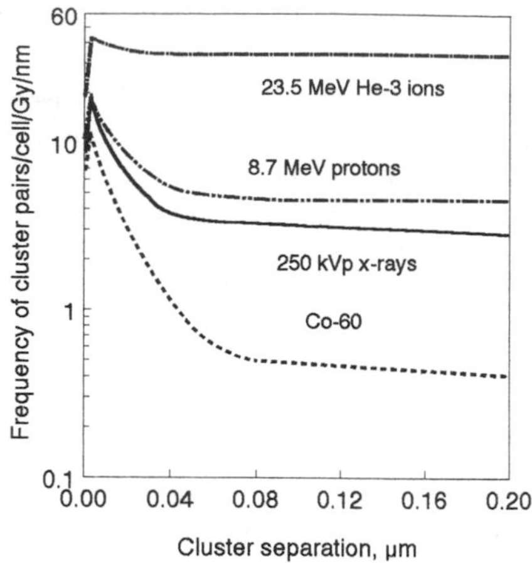

Figure 2 *Frequency distributions of cluster pairs for separation distances up to 200 nm*

increase in the predicted quadratic term is proportional to the square of the RBE for DSB which follows naturally from the assumption that chromosome exchanges result from pairwise interaction of DSB.

F values (ratio of dicentric to centric ring yields) have been separately estimated for the intra- and inter-track components of dose-response data for each radiation. For inter-track exchanges there is no dependence of F on radiation quality which fell in the range from 30 to 35. For intra-track interactions this value is radiation-dependent with a continuous decline from 25 for ^{60}Co to 10 for ^{3}He. Predicted F values are higher than usually quoted in literature.[8] These values however are sensitive to the chosen domain dimensions and repair/exchange function.

4 DISCUSSION

The assumption that chromosome exchanges result from pair-wise interaction of DNA double strand breaks leads to a prediction that the dicentric dose-response data for radiations with LET of 20-30 keV μm^{-1} (i.e. LET values requiring several particles per nucleus in the dose range of interest) would show a curvature as large as that observed for γ-rays. Allowing RBE>1 for DSB would make this curvature even stronger increasing as the square of the RBE for DSB. This contradicts experimental observation in human lymphocytes.[9] The assumption that close DSB would behave as one break when exchanges are considered leads to a reduced quadratic term for ^{3}He ions. The reduction however is small and it seems unlikely that the correction for the close DSB can explain the very much reduced β term for ^{3}He ions.

One further possibility remains. It is known that high LET radiations produce more complex breaks than low LET radiations.[10] If we further propose a separate mechanism for aberration production from these severe breaks, it might be possible to reduce the β-term for ^{3}He ions. This mechanism needs to predict a yield linear with dose, e.g. an interaction between the severely broken ends and undamaged DNA.[11] We have not as yet carried out specific modelling to test this hypothesis but inspection of clustering distributions in electron and ^{3}He tracks suggest that the ratio of simple to complex ionization clusters, which determine the relative importance of pairwise exchange and the postulated linear mechanism, does not alter sufficiently to produce a large reduction in β.

The model predicts that the F value depends on radiation quality but only for the intra-track term. This is because DSB exchange probability is separation dependent and for rings the distance is also restricted. Inter-track DSBs are random throughout the nucleus independent of quality whereas intra-track separations are quality dependent. By contrast, Savage and Papworth[12] assume that DSB interactions are confined to adjacent chromosome arms. This causes dicentrics and rings to have the same interaction criterion and hence F is independent of quality.

5 CONCLUSIONS

We have investigated the hypothesis that chromosome exchange aberrations are caused by pairwise exchange between DSBs. The model works well at low LET, but some problems arise. One is that current theories that ionization clusters cause DSBs leads to an RBE greater than 1.0 for DSB caused by higher LET radiations. This leads to a quadratic term for exchange aberrations which increases with LET contrary to observation in human lymphocytes. Attempts to adjust the DSB production mechanism have failed to solve the problem. If the concept of pairwise exchange is to be retained because it explains the linear-quadratic response at low LET so well, another linear mechanism which is of more importance at high LET needs to be postulated.

References

1. V.V.Moiseenko, A.A.Edwards, H.Nikjoo, W.V.Prestwich, Rad.Res., 1996 (in press).
2. J.A.Hartigan, 'Clustering Algorithms', John Wiley & Sons, New York, 1975.
3. V.Moiseenko, A.J.Waker and W.V.Prestwich, 'Energy Deposition Patterns from Tritium and Different Energy Photons - A Comparative Study', COG Report, COG-96-281-I, RC-1670, 1996.
4. S.Uehara, H.Nikjoo and D.T.Goodhead, Phys.Med.Biol., 1993, **38**, 1841.
5. W.E.Wilson, J.H.Miller and H.Nikjoo. PITS: A Code System for Positive Ion Track Simulation. In: Computational Approaches in Molecular Biology. Edited by M.N.Varma and A.Chatterjee, Plenum Press, New York, 1994.
6. R.K.Sachs, G. van den Engh, B.Trask, H.Yokota, E.Hearst, Proc.Natl.Acad.Sci. USA, 1995, **92**, 2710
7. W.R.Holley and A.Chatterjee, Rad.Res., 1996, **145**, 188.
8. D.J.Brenner and R.K.Sachs, Radiat.Res., 1994, **140**, 134
9. A.A.Edwards, D.C.Lloyd, J.S.Prosser, P.Finnon and J.E.Moquet, Int.J.Radiat.Biol., 1986, **50**, 137.
10. D.T.Goodhead, Int.J.Radiat.Biol., 1994, **65**, 7
11. K.H.Chadwick, H.P.Leenhouts, Int.J.Rad.Biol., 1978, **33**, 517.
12. J.R.K.Savage and D.G.Papworth, Rad.Res., 1996, **146**, 236

MODELING LOW AND HIGH LET FISH DATA ON SIMPLE AND COMPLEX CHROMOSOME ABERRATIONS

A. M. Chen†*, P. J. Simpson§, C. S. Griffin§, J. R. K. Savage§, D. J. Brenner‡ J. N. Lucas*, and R. K. Sachs†,

†Dept. Mathematics, University of California, Berkeley, CA 94720. §MRC, Radiation and Genome Stability Unit, Didcot, OX11 0RD, UK. ‡Center for Radiological Research, Columbia University, New York, NY. *Lawrence Livermore National Laboratory, Livermore, CA.

1 INTRODUCTION

With fluorescent *in situ* hybridization (FISH) many different categories of chromosome aberrations can be scored. The spectrum of aberration frequencies indicates aberration formation mechanisms and reflects radiation quality[1–4]. Analyzing the implications of observed yields requires a model, explicit or implicit. There is evidence[4] that: (a) the classic random breakage and reunion model[5] is appropriate; and (b), proximity plays a role, i.e. free ends from DSBs initially formed far apart are less likely to undergo illegitimate reunion than free ends from DSBs initially close together. Chen et al.[3] developed a Monte-Carlo computer implementation of the random breakage and reunion model, modified to incorporate proximity effects by assuming the cell nucleus is divided into interaction sites. It was assumed all DSB free ends eventually rejoin ("completeness"). They analyzed FISH data on chromosome aberration yields in human lymphocytes after acute low LET irradiation. The model has two adjustable parameters: the number of interaction sites per cell nucleus and the average number of reactive DSBs per Gy. Reasonable fits were obtained to data on a considerable number of different aberration types. The present paper extends the model of Chen et al. to high LET and applies it to published FISH aberration data[2] for fibroblasts subjected to x-ray or ^{238}Pu α-particle radiation.

2 METHODS

In the picture to be used here, the first step of aberration formation is that ionizing radiation induces multiple DSBs. The next step is that most of the DSBs are systematically restituted, while the other DSBs become "reactive", i.e. the two free ends move apart instead of restituting. In the third and final step free ends of reactive DSBs gradually undergo reunion, and unless a free end happens to undergo reunion with its own partner ("accidental restitution"), the reunion is illegitimate, causing a chromosome aberration.

The model to be used incorporates proximity effects into the random breakage and reunion model. For low LET, the method is described by Chen et al.[3]. In brief, a cell nucleus is regarded as divided into S sites, where S is an adjustable parameter and typically has values in the range 5-25. Monte-Carlo computer simulation of an experiment is performed, as follows.

In each simulated cell, the computer assigns the 46 chromosomes at random to the sites. It then inflicts reactive DSBs at random, i.e. the probability a reactive DSB is on a particular chromosome arm is proportional to the length, in bp, of that arm. The average number of reactive DSBs/cell is assumed proportional to dose, and the average number of reactive DSBs/cell per Gy, which we denote by δ, is the second and last adjustable parameter. After reactive DSBs are inflicted, free ends from reactive DSBs in the same site rejoin at random, but free ends from different sites cannot interact. After Monte-Carlo simulation of ~1,000,000 cells according to these rules, the computer adds up the number of aberrations of various kinds, adopting whatever FISH painting method and/or scoring criteria are used in the experiment to be simulated, and computes aberration frequencies.

The random infliction of reactive DSBs when simulating low LET experiments implies that the number of reactive DSBs per cell, and the number of reactive DSBs on a particular arm of a particular chromosome, are both Poisson-distributed random variables. For modeling α-particle irradiation, it is appropriate to modify the simulation procedure so that the distribution of reactive DSBs, instead of being Poisson, is determined microdosimetrically from the site geometry.

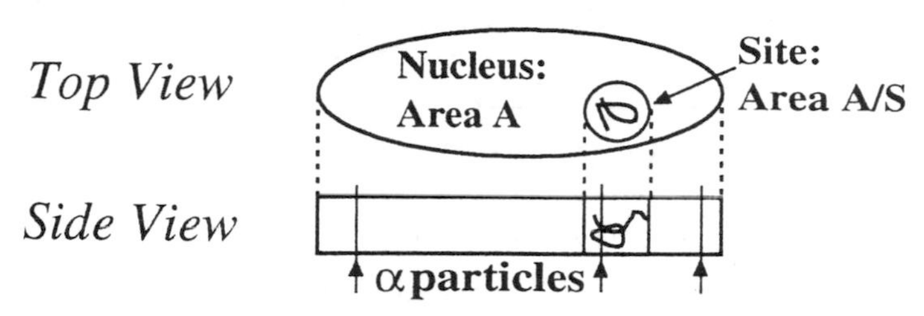

Figure 1 *Geometric configuration.*

Approximating the geometry of the flattened fibroblasts, we regard the cell nucleus as a right cylinder (Figure 1) with a (roughly ellipsoidal) cross section having area A=230 μm^2, the observed average[2]. The nucleus is assumed to consist of S sites, each a right cylinder with cross-sectional area A/S. In this formulation the height and the detailed shape of the various cross sections does not enter directly into the calculation. Other, perhaps slightly more realistic, geometries have been considered, but give similar results. In all cases the key point is the following: many sites are missed entirely, receiving specific energy 0; others are hit and among these some get a specific energy much larger than the dose.

For the case of right-cylindrical sites (Figure 1), and using LET approximation[6], the probability $P(n \mid x)$ that a site containing a fraction x of the genome has n reactive breaks is "Poisson-Poisson" conditioned on x, i.e.

$$P(n \mid x)=\sum_{j=0}^{\infty} \frac{(\tau D)^j}{j!}\exp[-\tau D]\times\frac{(j\delta x/\tau)^n}{n!}\exp[-j\delta x/\tau]. \qquad (1)$$

Here τ is the average number of tracks per Gy per site, determined by the geometry and the LET as $\tau \approx 11.9/S$ Gy^{-1}. Any one term in the sum (1) is the probability that j α-particle tracks make n reactive DSBs in the site. Apart from using (1), instead of a Poisson distribution, to determine the probabilities for reactive DSBs, the Monte-Carlo computer simulation proceeds just as in the low LET case.

3 RESULTS

The model was applied to published data[2] on x-ray or α-particle irradiated human fibroblasts. It was found by trial and error that appropriate values of the adjustable parameters are the following: for x-rays, S=13 sites per cell nucleus and δ=2.4 reactive DSBs per Gy; for α-particles, S=25 and δ=9.5 Gy^{-1} (corresponding to an average of ~0.8 reactive DSBs per α particle track and to an RBE for reactive DSBs of 9.5/2.4≈4). Figure 2 compares observed relative yields for chromosome 1 to the model with these parameters.

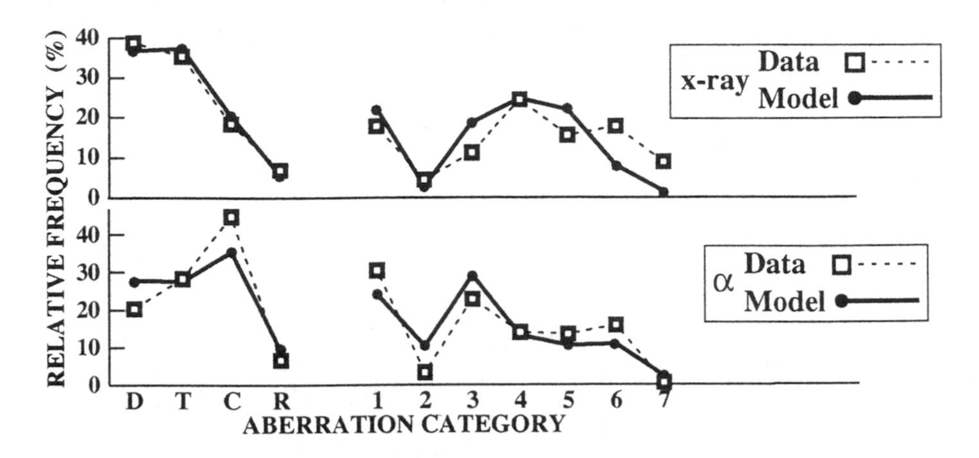

Figure 2 *Observed and predicted relative aberration yields.*

Values are connected by straight lines to guide the eye. On the left of the figure, the data is from Griffin et al.[2], Table 1. The aberration types are apparently simple dicentrics (D), apparently simple translocations (T), visibly complex exchange aberrations (C), and apparently simple centric rings (R); the doses (4 Gy of x-rays or 0.68 Gy of α particles) give comparable cell survival for the two radiations[2]. On the right, details on relative percentages for the visibly complex exchange aberrations are shown; the data is from Griffin et al.[2], Table 2. 1-7 specify aberration groups[1]. For example, group 1 consists of insertions, and groups 4 and 5 consist of two different kinds of three-way interchanges.

The data in Figure 2 concern *relative* frequencies of different aberration *types*. As will be reported elsewhere, acceptable results were also obtained with the same parameter values when modeling: (a) observed average aberration numbers per irradiated cell; and (b) observed relative frequencies of aberrations involving different painted chromosomes, of different lengths.

Once parameter values are determined, other quantities of interest, even if not directly observed, can be simulated. For example, using the parameters above, the simulated frequency of cells with non-transmissible[7] aberrations (painted or not) after α radiation is 36% at 0.41 Gy, 52% at 0.68 Gy, and 69% at 1.00 Gy. Similarly, predictions are obtained for inversions and acentric rings, painted or not.

4 DISCUSSION

It is seen in Figure 2 that the model, using only two adjustable parameters for each radiation quality, reproduces the main trends for many different aberration categories. Despite differences in cell type and radiation quality, the optimum parameters for x-rays are remarkably (indeed implausibly) similar to the values (S=13 sites, δ=2.3 Gy^{-1}) found earlier for γ-irradiated lymphocytes[3].

In agreement with the data, the model predicts that visibly complex aberrations are relatively more frequent at high than at low LET (C, Figure 2). The intuitive reason for this difference is that at high LET fluctuations give some sites high specific energy, which, especially in view of the large value of δ, acts like a high dose of low LET radiation, tending to make large numbers of complex aberrations. The larger site number, S=25, found when simulating the α-particle experiments may point to a weakness of the model. However it is also possible that track-structure effects potentiate proximity effects, i.e. that bunched DSBs tend to interact over shorter distances, corresponding to smaller sites and thus more sites per cell.

The present approach can be applied to more detailed models of α-particle track structure, to other geometries, or to other radiations by generalizing equation (1), using a 1-event specific energy distribution function[6]. Simulations are then possible for per-cell averages (e.g. average number of dicentrics per cell) but not always for distributions (e.g. the number of cells which have, respectively, 0, 1, 2, ... dicentrics). The reason for this limitation is that correlations among neighboring sites are being neglected, which is not always an acceptable approximation when considering high LET distributions[8]. The limitation is not relevant to our present discussion, both because the data modeled consists only of averages and because the particular geometry assumed can justify neglecting correlations.

5 CONCLUSION

The classic random breakage and reunion model, modified to incorporate proximity effects and implemented by Monte-Carlo computer simulations, adequately interprets detailed data on simple and complex aberrations produced by x-rays or α-particles. For the minority of DSBs which are reactive, the RBE was found to be ~4.

Acknowledgements

Research supported in part by AWU LAB COOP Graduate Fellowship (AMC), by LLNL under US DOE contract W-7405-Eng-48 (JNL and AMC), by NIH grant ES07361 (DJB), and by NSF grant DMS-93-02704 (RKS).

References

1. P. J. Simpson and J. R. K. Savage, *Int. J. Radiat. Biol.*, 1995, **67,** 37-45.

2. C. S. Griffin, S. J. Marsden, D. L. Stevens, P. J. Simpson and J. R. K. Savage, *Int. J. Radiat. Biol.*, 1995, **67,** 431-439.

3. A. M. Chen, J. N. Lucas, D. J. Brenner, F. S. Hill and R. K. Sachs, *Int. J. Radiat. Biol.*, 1996, **69,** 411-420.

4. R. K. Sachs, A. M. Chen and D. J. Brenner, *Int. J. Radiat. Biol.*, 1997, in press

5. D. E. Lea 'Actions of Radiations on Living Cells', Cambridge University Press, London, 1946.

6. A. M. Kellerer, 1985, in 'The Dosimetry of Ionizing Radiation, vol. I'. Edited by: K. Kase, B. Bjarngard and F. Attix, Academic Press, Orlando, pp. 77-162.

7. J. R. Savage, *Mutation Research,* 1995, **347,** 87-95.

8. D. J. Brenner and R. K. Sachs, *Radiat. Prot. Dosimetry,* 1994, **52,** 21-24.

INTER-CHROMOSOMAL HETEROGENEITY IN THE FORMATION OF RADIATION INDUCED CHROMOSOMAL ABERRATIONS

A. T. Natarajan, S. Vermeulen, J. J. W. A. Boei, M. Grigorova and I. Dominguez
MGC Department of Radiation Genetics and Chemical Mutagenesis
Leiden University & J.A. Cohen Institute, Wassenaarseweg 72, Leiden
The Netherlands

1. INTRODUCTION

It is generally believed that ionizing radiation induces DNA lesions randomly among the genome and the ensuing repair occurs also in a random manner. The DNA double strand break (DSB) is considered to be the most important lesion contributing to radiation induced chromosomal aberrations[1]. In simplistic terms, the induced DSBs are mostly repaired correctly and a small fraction is mis-repaired or remains unrepaired leading to exchanges and fragments respectively. Some earlier as well as recent studies have indicated that radiation induced chromosomal lesions and their repair are not random. In Chinese hamster primary embryonic cells, chromosomes which are mainly euchromatic sustain more damage than heterochromatic chromosomes though the repair kinetics appears to be reverse, i.e., faster in euchromatic in comparison to heterochromatic regions. In a detailed analysis of break points of radiation induced chromosome aberrations, distinct differences between Chinese hamster chromosomes on the distribution of deletions and exchanges have been observed[3].

The introduction of fluorescent in situ hybridization (FISH) technique using chromosome specific DNA libraries has made it possible to identify translocations and dicentrics with great accuracy and ease[4,5]. In addition to human chromosome specific DNA libraries, mouse and Chinese hamster chromosome specific libraries have become available[4,6,7,8,9]. Recent studies, using these chromosome specific probes, have revealed that there is a heterogenous response of different human and Chinese hamster chromosomes with regard to radiation induction of exchange type of aberrations. Human chromosome arm specific DNA libraries have been generated[10] which allows detection of pericentric inversions (exchanges between the arms of the chromosome) and the frequencies of such exchanges are much higher than one would expect on the basis of their DNA content indicating that proximity between chromosomes plays a significant role in the origin of exchanges. Some of the results obtained in our laboratory in this area are presented below and discussed.

2. MATERIAL AND METHODS

Freshly drawn human blood samples or splenocytes from Chinese hamsters were irradiated with different doses of X-rays and cultured in vitro in a medium containing foetal calf serum, antibiotics and mitogens (phytohaemoglutinin in the case of human lymphocytes and concanavalin A in the case of Chinese hamster splenocytes). Cytological preparations were made in a routine way and used for FISH. Human chromosome specific DNA libraries for # 1 and 4, arm specific human chromosome libraries for # 1 and 3, as well as Chinese hamster chromosome specific libraries for # 2 and # 8 were used for FISH. The methodology for FISH has been described earlier[11]. Human chromosome specific probes were generated from bluescribe plasmids obtained originally from Dr. J.W. Gray. Chinese hamster chromosome specific libraries were generated in our laboratory by flow sorting of chromosomes followed by PCR methods[7,9].

Multi-colour fluorescence was used to detect exchange aberrations involving specific chromosomes. Human chromosome arm specific probes were a gift from Dr. Meltzer (NIH, Bethesda, U.S.A). The two arms were differently painted, red and green by means of digoxyginin or biotin labelling and visualized with TRITC or FITC respectively. The slides were scored with Zeiss Axioplan microscope using a triple filter combination.

3. RESULTS AND DISCUSSION

3.1. *Human lymphocytes.*

Frequencies of translocations based on painting with DNA libraries of three or four chromosomes have been used routinely to estimate the frequencies for the whole genome, assuming that all chromosomes respond with equal frequencies expected on the basis of their DNA content. However, there have been indications that not all human chromosomes respond with the frequencies of radiation induced chromosome exchanges expected [12]. Recently, we completed a study on the relative induction of dicentrics and translocations involving chromosomes #1 and # 4 following X-irradiation (2 Gy) of human blood samples from two donors. About 2250 first division cells for each donor were analysed for the presence of dicentrics and reciprocal translocations using dual colour FISH. Chromosome #1 and # 4 constitute about 8.1% and 6.2% of the human genome. While the frequencies of dicentrics and reciprocal translocations were about equal, i.e., 2.12 and 2.36 respectively for chromosome # 1, those involving chromosome # 4 were 1.36 and 2.97 respectively, i.e., more reciprocal translocations than dicentrics[13]. The frequencies of dicentrics were close to the expected values based on the DNA content of these two chromosomes. However chromosome # 4 was involved in many more reciprocal translocations (about a factor of 2) than expected. We have suggested that the increased radiosensitivity of chromosome # 4 may be related to the low concentration of transcribing genes, as indicated by the paucity of CpG islands in this chromosome[14]. Recently, a detailed study on the participation of 12 different human chromosomes (# 1,2,3,4,6,7,X,8,9,10,12 and 14) in exchange aberrations following irradiation of human lymphocytes with 3 Gy of X-rays has been carried out[15]. There were considerable differences between the chromosomes in their involvement in exchanges. The most striking ones were, chromosomes #1 and #3, which participated in exchange aberrations below the expected levels. Similar to our observations, chromosome # 4 responded with much higher frequencies of translocations and expected frequencies of dicentrics. These data indicate that there must be inherent differences in the formation of dicentrics and translocations[16]. They also found that chromosome # 9 responded with higher frequencies of both dicentrics and translocations. Further detailed analysis of break points involved in the exchanges in these chromosomes may throw some light on the factors leading to the observed heterogeneity.

3.2. *Chinese hamster splenocytes.*

The karyotype of Chinese hamster (2n=22) is characterised by distinct meta-centric, submeta-centric and acrocentric chromosomes and all the chromosome pairs can be easily identified[17].

Based on the chromosome size and the relative GC/AT ratio of each pair they can be flow sorted and chromosome specific DNA libraries can be generated using DOP-PCR method[8]. We have used these libraries and FISH to estimate the frequencies of dicentrics and translocations involving chromosomes # 2,3,X,Y and 8 following different doses of X-irradiation of splenocytes of Chinese hamster. Compared all other chromosomes studied, chromosome # 8 was involved in more dicentrics and translocations than expected on the basis of its size. Compared to chromosome # 2 (16.7% of the genome), chromosome # 8 (4.5 % of the genome) responded with an increase of dicentrics (1.5 fold) and translocations (about 2.5 fold)[18]. In a recent study, involving scoring of more than 2000 cells per point, we found that Chromosome # 8 participates in exchange aberrations about 2 to 3 fold higher than chromosome # 2, confirming our earlier observations[19]. Chinese hamster chromosomes are characterised by the presence of large blocks of interstitial telomeric repeats near the centromeres, in addition to the terminal ones. We have shown earlier in Chinese hamster cells, that regions with intercalary telomeric repeats are more often involved in induced and spontaneous chromosome aberrations[20,21]. Chromosome # 8 is very rich in intercalary telomeric repeats and this feature is most probably the cause for the increased participation of this chromosome in exchanges[18].

3.3 *Inter- vs intra-chromosomal exchanges in human lymphocytes.*

Inter-chromosomal exchanges induced by ionizing radiation are dicentrics and translocations and intra-chromosomal exchanges are centric rings and pericentric inversions. The frequencies of inter-chromosomal exchanges are higher than intra-chromosomal exchanges and this ratio is termed as F values. F values have been reported to vary and in some cases, high F values are obtained following high LET radiation in comparison to low LET radiation[22]. It has also been proposed that the F values can be used as "finger prints" to identify the nature of past radiation exposures, especially high LET ones[23]. The F values appear to vary very much in different experiments, both for high and low LET radiations. Values from 3.2 to 27 for neutrons and 4.5 to 18.6 have been reported in the literature[24,25,26].

Using arm specific painting probes for chromosomes # 1 and # 3 and dual colour FISH, we determined the frequencies of intra-chromosomal exchanges (pericentric inversions and centric rings) and inter-chromosomal exchanges (dicentrics and translocations) in X-irradiated (2.5 Gy) human lymphocytes[27]. The ratio between pericentric inversions and rings was found to be 1, indicating that these two events occur with equal probability. For intra-changes to inter-changes the ratio (F value) was found to be between 6 and 9. We employed the "paint" nomenclature[28] to determine the number of colour junctions involving the two arms of the painted chromosomes Based on the total number of junctions, it was found that exchanges between the arms of the same chromosome occur 6.4 to 8.7 times more than inter-chromosomal exchanges calculated on the basis of the DNA content of chromosomes and random induction of aberrations in the total genome. Chromosomal organization in interphase nucleus appears to promote the formation of more intra-changes than inter-changes following X-irradiation, most probably due to close proximity of the two arms of the chromosomes. The frequencies of exchanges between the painted chromosomes (homologous) were very similar to the values obtained for exchanges between painted and unpainted chromosomes, indicating no preference for homologous exchanges and the homologous chromosomes are not proximaly situated in the interphase nucleus of G0 lymphocytes.

4. Conclusions

There is clear evidence for inter-chromosomal heterogeneity for induction of aberrations by low LET radiation. Increased sensitivity observed in some of the chromosomes, could be attributed to the level of condensation in interphase, the content of transcribing genes (CpG islands) and the distribution of interstitial telomeric sequences. The proximity of two arms of a metacentric chromosomes in interphase nucleus promotes formation of more intra-changes than inter-changes calculated on the basis of their DNA content.

Acknowledgements. The study was supported by EU Nuclear Fission Safety Programme.

References:

1. A.T. Natarajan and G.Obe, *Chromosoma*, 1984, **30**, 120.
2. P. Slijepcevic, P and A. T. Natarajan, *Mutation Res.*, 1994, **323**, 113.
3. P. Slijepcevic, P and A.T.Natarajan, *Int. J. Radiat. Biol.* 1994, **66**, 747.
4. D. Pinkel, T. Straume and J.W. Gray, *Proc. Natl. Acad. Sci. U.S.A.* 1986, **83**, 2934
5. A.T. Natarajan, R.CVyas, F. Darroudi, and S. Vermeulen, *Int. J. Radiat. Biol.* 1992, **61**, 199
6. J. J. W. A.Boei, A.S. Balajee, P.de Boer, W. Rens, J.A. Aten, L.H.F. Mullenders and A.T. Natarajan, *Int. J. Radiat. Biol.* 1994, **65**, 583.
7. A. S. Balajee, I. Dominguez and A. T. Natarajan, *Cytogenet.Cell Genet.*1995, **70**, 95
8. Y. Xiao, P. Slijepcevic, G. Arkesteijn, F. Darroudi and A. T. Natarajan, *Cytogenet, . Cell Genet* 1996 (in press).
9. Y. Xiao, F. Darroudi, A.G.J. Kuipers, J.H. de Jong, P. de Boer and A. T. Natarajan, *Cytogenet. Cell Genet.* 1996 (in press).
10. X-Y.Guan, H. Zhang, M. Bitner,Y. Jiang, P. Meltzer and J.Trent, *Nature Genet.*1996, **12**, 10.
11. J. J. W. A. Boei, S. Vermeulen and A.T. Natarajan, *Mutation Res.* **349**, 127
12. S. Knehr, H. Ziizelberger, H. Braselman and M. Buchinger, *Int. J. Radiat. Biol.* 1994, **65**, 683.
13. J. J. W. A. Boei, S. Vermeulen and A. T. Natarajan (unpublished results).
14. A. T. Natarajan, A.S. Balajee, J.J.W.I.Boei, F. Darroudi, I. Dominguez, M. Hande, M. Meijers, P. Slijepcevic, S. Vermeulen and Y. Xiao. *Mutation Res.* (in press)
15. S. Knehr, H. Zitzelberger, H. Braselman, U. Nafrstedt and M. Bauchinger, *Int. J. Radiat. Biol.* 1996, (in press).
16. A. T. Natarajan, A.S. Balajee, J.J.W.A. Boei, S. Chatterjee, F. Darroudi, M. Grigorova, M.M. Noditi, H.J. Oh, P. Slijepcivic and S. Vermeulen, *Int. J. Radiat. Biol.* 1994, **66**, 615.
17. A. T. Natarajan and W. Schmid. *Chromosoma,* 1971, **33**, 48.
18. I. Dominguez, J. J. W. A. Boei, A. S. Balajee and A. T. Natarajan, *Int. T. Radiat. Biol.* 1996, **70**, 199
19. M. Grigorova and A. T. Natarajan (unpublished results)
20. A. S. Balajee, O.H. Oh and A. T. Natarajan, *Mutation Res.* 1994,**307**, 307.
21. P. Slijepcevic, Y. Xiao, I. Dominguez and A. T. Natarajan, *Chromosoma,* 1996,**104**, 596.
22. L. R. Hlatky, R.K. Sachs and P. Hahnfeldt. *Radiat. Res.* 1992. **129**, 304.
23. D. J. Brenner and R.K. Sachs, *Radiat. Res.* 1994.**140**, 134.
24. D. Scott, H. Sharpe, A.L. Batchelor, H.J. Evans and D.G. Papworth, *Mutation Res.*1970, **8**, 367.
25. E. Schmid, M. Bauchinger and O. Hug. *Mutation Res.*1972,**16**, 307.
26. Bauchinger, M. E. Schmid, G. Rimpl, and H. Kuhn, *Mutation Res.*1975,**27**, 103.
27. A. T. Natarajan, J.J.W.A. Boei, S.Vermeulen and A.S. Balajee, *Mutation Res.*(in press)
28. J. D. Tucker, W.F. Morgan, A.A. Awa, M. Bauchinger, D. Blakey, M.N. Cornforth, G.L. Littlefield, A. T. Natarajan and C. Shassere, *Cytogenet. Cell Genet.* 1995, **68**, 211.

EFFECTIVENESS OF ULTRASOFT X-RAYS AT INDUCING COMPLEX EXCHANGES IN HUMAN FIBROBLASTS

Carol S. Griffin, Mark A. Hill, David L.Stevens and John R.K.Savage
MRC,Radiation and Genome Stability Unit
Harwell
OX11 ORD,UK

1 INTRODUCTION

The mechanism by which many chromosomes come together to interact to form multiple exchange aberrations, when they occupy discrete domains in interphase, is difficult to explain. From the high frequency of multiple chromosome exchanges that have been observed after low doses of high LET radiation and high doses of low LET radiation,we have postulated that either many chromosomes are found together at preexisting sites, or damaged DNA may be interacting with undamaged DNA.[1,2]

In order to resolve further this problem,we have used Cu-L and C-K ultrasoft x-rays with track lengths of 40 and 7nm respectively. We report that, in spite of average track lengths the majority of which will be less than the dimensions of a DNA chromatin fibre, complexes are induced with a high efficiency and at a frequency approaching that obtained after high LET α-particle irradiation.

1.1 Ultrasoft x-rays

The electrons generated by ultrasoft 0.28keV C-K and 0.96keV Cu-L x-ray photons deposit their energy in tracks of length ≤7nm and 40nm respectively and provide an ideal tool for analysing the spatial distribution of breaks and misrepair processes (Figure 1). We have undertaken the analysis of chromosome structural changes, produced by ultrasoft x-rays, in HF12 untransformed human fibroblasts in G1 phase, using Fluorescence in situ hybridization (FISH). Mean absorbed doses were calculated from a cell thickness of 5μm.

An RBE of 1.6 for cell killing was calculated for Cu-L x-rays relative to 250kVp x-rays and is comparable to an RBE of 1.7, reported previously for cell killing after Al-K x-rays.[2] Multicoloured, chromosome specific DNA probes, for chromosomes 1 and 2 and an α-satellite pan-centromeric probe, were used to examine in-vitro radiation-induced chromosome type exchange aberrations. After mean absorbed doses of 0.37-2.22Gy of Cu-L and 1.1Gy of C-K x-rays, the relative frequencies of complex exchanges derived from 3 or more breaks in 2 or more chromosomes,were 21-44% (Figure 2).

There was found to be no difference between Al-K and Cu-L x-rays, in the overall frequency of chromosome aberrations. Cu-L x-rays were more effective at producing complex exchanges, but less effective at producing simple exchanges,at mean doses below 2Gy. Simple exchanges involve the interaction of only 2 chromosomes and are believed

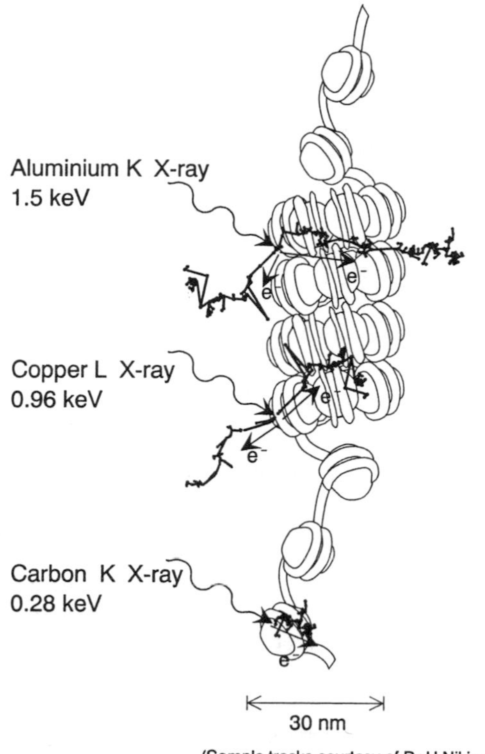

Figure 1. *Ultrasoft x-ray tracks in relation to the dimensions of a chromatin fibre*

to result from the passage of just one track.[3] The reduced efficiency of Cu-L x-rays for producing simple exchanges suggests that a track length of greater than 40nm is required for efficient production of simple interactions involving 2 chromosomes. At doses above 2Gy a high yield of excess fragments was observed after Cu-L x-rays, which was not seen after Al-K x-rays.

Recent studies using PCC[4,5] show different kinetics for each kind of chromosome aberration, simple exchanges being formed rapidly, and having a large linear dose component, and complex exchanges forming more slowly, and having a large quadratic component. 1-2 hours after exposure to 15Gy of Al-K and Cu-L x-rays, 40% of dsbs were found to be unrejoined compared with <20% after 15Gy of 250kVp x-rays[6] and 58% after 10Gy of α-particles.(C.de Lara per.comm.).

We believe that there is considerable movement of DNA during interphase, allowing interactions between lesions originally separated in space. Slow rejoining breaks may contribute to the formation of complex exchanges to result in the relatively high frequencies observed after ultrasoft x-rays. The different kinetics shown by simple and complex exchanges may also indicate a different DNA target. Neary in 1967[7] proposed 2 targets for cell killing, one independent of LET, and one sensitive to LET, and Goodhead[8] later proposed one repairable lesion requiring ≥100eV, dominant at low LET, (transcribing DNA, simple exchanges?) and one requiring ≥300eV and dominant

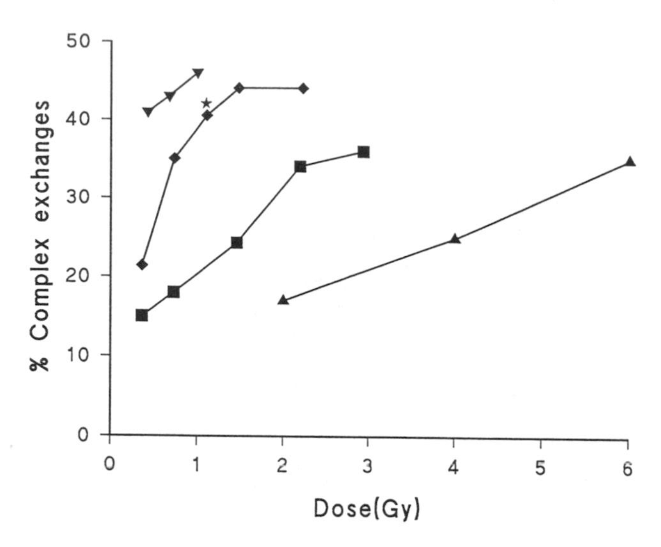

Figure 2 *Complex exchanges observed in chromosome 1 as a percentage of total exchanges.(▼)α-particles,(◆)Cu-L x-rays,(★)C-K x-rays,(■)Al-K x-rays,(▲)250kVp x-rays from Simpson and Savage* (3).

at high LET (non-transcribing DNA,complex exchanges?).

The involvement of lesions other than double strand breaks (dsb) may be needed to account for the high frequency of complexes. A mean absorbed dose of 1Gy of Al-K x-rays or Cu-L x-rays deposited in a nucleus of approximately 900μm^3 will generate 4000 and 6500 tracks respectively. Estimated dsb/Gy/cell together with tracks/cell calculations for Al-K and Cu-L x-rays give values of 2% of tracks producing 1dsb, and only 0.02% of tracks producing 2dsbs. Other possible lesions include base damage, which may be slower to repair than dsb.

Another important factor which may contribute to the high frequency of complexes after ultrasoft x-rays is the attenuation factor. The rapid attenuation of the low energy x-ray photons through the cell results in an exit dose of only 8.3% and 6.1% of the entry dose for Cu-L and C-K x-rays respectively. The consequence of this attenuation is that a much higher proportion of the dose is received by the first μm segment of the cell (attached to the Hostaphan) than the last μm segment of the cell. The deposition of the majority of the tracks into the DNA closest to the nuclear membrane may result in a high frequency of chromosome interactions, particularly if the DNA is concentrated around the nuclear pores, or at transcription sites close to the nuclear membrane.

In conclusion we report that, in spite of very short track lengths, Cu-L and C-K x-rays produce a relatively high frequency of complex exchanges, compared to 250kVp x-rays. This high frequency may be due to multiple chromosome associations formed close to the nuclear membrane, together with slow rejoining lesions moving to specific repair sites for interaction.

Acknowledgements
This work was supported in part by CEC contracts F14P-CT95-0011 and PL00950004.

References

1. C. S. Griffin, S. J. Marsden, D. L. Stevens, P. Simpson and J.R.K. Savage, *Int.J.Radiat.Biol.*, 1995, **67**, 431.
2. C. S. Griffin, D. L. Stevens and J. R. K. Savage, *Rad. Res.,* 1996. **146,** 138.
3. P. Simpson and J. R. K. Savage, *Int.J.Radiat.Biol.,* 1996, **69,** 429.
4. R. Greinert, E. Detzler, B. Volkmer and D. Harder, *Rad. Res.,*1995, **144,** 190.
5. M. Durante, K. George, H. Wu and T. C. Yang, *Rad. Res.,* 1996, **145,** 274.
6. S.Botchway, PhD thesis, 1996.
7. G. J. Neary, R. J. Preston and J. R. K. Savage, *Int.J.Radiat.Biol.,* 1967, **12,** 317.
8. D. T. Goodhead, R. J. Munson, J. Thacker and R. Cox, *Int.J.Radiat.Biol.,* 1980, **37,** 135.

ON THE NATURE OF OBSERVED CHROMATID BREAKS

John R K Savage and Alison N Harvey

MRC Radiation and Genome Stability Unit,
Harwell,
OX11 0RD, UK

'BREAKAGE-AND-REUNION' AND REVELL'S 'EXCHANGE' THEORIES

It is a basic assumption of Classic breakage-and-reunion theory that the open chromatid breaks, seen at metaphase, are the unrejoined (or unrejoinable ?) residue of primary breaks, most of which have restituted, and the rest rejoined, illegitimately, with other breaks close in time and space, to form aberrations [1, 2]. In other words, we are actually seeing the progenitors of the chromosomal structural exchanges.

Now, the primary break of the Sax/Lea theories was envisaged as a complete severance of a chromosome "backbone" (the "*chromonema*") having a diameter ~0.1μm, and this yielded two open ends which could move around, independently, meeting up with other ends to form exchanges.

When this theory was formulated, the nature and amount of DNA, its complex packing and integration with proteins, and its vital role in chromosome integrity, were unknown. Today, we recognise the importance of DNA double-strand breaks (dsb) in the formation of (or, as a trigger for) aberrations. However, these breaks occur in structures many orders of magnitude below the hypothetical "chromonema", and it is clear, from work with restriction endonucleases introduced into cells, that many dsb can be produced in chromosomes without leading to an obvious breakdown of chromosome integrity. It is therefore a rather questionable practice (although regularly done) to transfer, directly, the Sax/Lea breakage-and-rejoining concepts, on the basis of the equation: primary break = dsb.

In the late 1950's, the first serious challenge to the breakage-and-reunion theory was launched by Revell [3]. Briefly, he argued that all open breaks ("chromatid discontinuities") seen at metaphase were not residual primaries, but were secondary, the result of partial failure in an exchange process (i.e. they were incomplete intra-arm intrachanges). Thus, we never see the progenitors of structural changes, and we have no real evidence that they exist.

It is not the purpose of this paper to discuss, in detail, the arguments and associated controversy which arose from this challenge.These have been reviewed many times [4-6]. Rather, we wish to present a curious and consistent observation that has arisen from our extensive chromatid aberration work with Chinese hamster cells over several years [7-9], and which has a bearing on this topic.

COLOUR-JUMP "BREAKS"

In order to increase the accuracy of chromatid aberration classification, it is our regular practice to irradiate G2 cells where the chromosomes have incorporated BrdU and the sister-chromatids may be subsequently differentiated by Giemsa staining ("harlequin" staining; TB dark, BB light or, occasionally, TT dark, TB light). The open breaks seen may be in the dark, or the light chromatid, but always there is a proportion (~20%) which show a colour-jump (like a SCE) at the point of discontinuity.

Revell proposed [3] certain numerical relationships between various aberration types, based upon the premiss that there were four, equally likely, simple intrachange types, two involving interactions of lesions between sister chromatids (inter-arm, Types 1 and 4) and two involving interactions between lesions within an arm (intra-arm, Types 2 and 3). As two lesions are involved, there are 8 possible incomplete forms. Of these, two (4a,4b) will be seen as incomplete isochromatid deletions, and one (2b) as a minute deletion, The remaining 5 (1a,1b,2a,3a,3b) will all be scored as "breaks" when the chromosomes condense at metaphase. Now, two of these breaks (1a, 1b) come from inter-chromatid interactions, and, in harlequin chromosomes, will always be accompanied by a colour-jump (*c* breaks*) at the discontinuity, whilst the remaining three will show no jump. If Revell's original contentions were correct, we would expect 40% of breaks to be *c**. The proportion is much less, which has led some to dismiss the Exchange theory out of hand [10], for although some intrachange-derived breaks are allowed, breakage-and-reunion theory considers their contribution to be negligible. However, when we take into account the non-colour-jump contribution, it can be seen that such derived breaks are not negligible. We have considerable evidence from a number of organisms that the simple intrachange types are not equally likely, and in Chinese hamster cells, intra-chromatid interactions predominate. Consequently, the observed *c** proportion may take a variety of values.

Most workers accept the compromise that the low values come from a mixture of intrachange-derived and "simple" breaks. If this is so, it follows that the relative frequency of *c** must vary with dose, since, from Poisson considerations, one-track events will dominate at very low doses, but this dominance will diminish as dose rises, with the quadratic increase of two-track exchange events.

However, the accumulated results from our experiments show that, within the limits of sampling error, the proportion of *c** breaks is constant. From our work, the following facts have emerged:

1) The observed relative *c** frequency (expressed as a proportion of the total open breaks scored) is independent of the X-ray dose (and therefore of the absolute number of breaks present) up to 2Gy, and also of the sampling time with respect to G2 cells.

2) This proportion is not affected by the BrdU substitution method used to obtain sister-chromatid differentiation (TB/BB or TT/TB) nor by the concentration of BrdU in the medium within the range 0.5 - 20ug/ml.

3) "Spontaneous breaks", measured in unirradiated control cells occur at a very low frequency. Nevertheless, they show the same proportion of *c** breaks as found in the irradiated cells.

4) High LET radiation (^{238}Pu α-particles) also produce c^* breaks, but at a slightly higher relative frequency than X-rays. As with X-rays, their proportion also appears to be independent of dose [11].

5) We have investigated a range of different Chinese hamster cell lines, including V79-379A, CHO-K1 and its radiosensitive counterpart *xrs5*, which displays a higher absolute frequency of open breaks for a given radiation dose because of difficulties in dsb rejoining. The proportion of c^* breaks remains at the same constant level in all these lines.

6) Colour-jump breaks can also be seen to participate in *interchange* interactions to produce *intra-interchange* complexes [8]. This is to be expected, as the readily visible "tri-radial", an *isochromatid/chromatid interchange*, belongs to this group of complex exchanges.

7) Such breaks also occur in untransformed human fibroblasts (HF19) and JU56 wallaby cells (Harvey and Moore, unpublished).

RESTRICTION ENDONUCLEASE "BREAKS"

Recently [9], we have completed a series of time-course studies using restriction endonucleases (RE) in G2. RE introduced into cells produce dsb with clean, consistent ends at defined DNA sequences, and are extremely efficient at generating chromosomal aberrations. We anticipated that if any treatment is going to provide simple, unrejoined breaks, then RE would be most likely candidates.

Three RE were used ; *Alu* I, which cleaves the DNA to produce blunt-ended dsb, and the cohesive cutter *Sau3A* I and its isoschizomer, *Mbo* I. Under conditions which simulate the intra-cellular environment, *Alu* I and *Sau3A* I have been shown to be relatively short lived, with respect to cutting, and therefore represent relatively "acute" exposures. *Mbo* I, however, continues cutting the DNA for over 24h, which, by analogy with radiation, represents a "chronic" exposure [12]. Surprisingly, in spite of considerable differences in recovered aberration levels between the enzymes, the relative c^* frequency, in all cases, was as high as, or higher than, that observed in the ionizing radiation studies, suggesting that the observed open breaks are derived by a similar processing mechanism to that found for radiation and spontaneous damage.

INFERENCES FROM COLOUR-JUMP "BREAK" CONSTANCY

The most striking feature of our collected experiments, and one that demands explanation, is the *constancy* (within the limits of sampling error) of the proportion of breaks showing the jump. It matters not how the breaks are produced, or how many are recovered; the proportion derived from a two-lesion exchange interaction appears to be invariate. To account for this observation,we infer:

a) That there is probably a common processing mechanism behind the formation of visible open breaks, (i.e. they are *secondary*, or derived aberrations) regardless of the type of initiating lesion. The constant proportion of c^* then arises, either from some property of the process, or, more likely because a fixed proportion of the lesions are disposed in such a way that processing, in their case, leads to a colour-jump.

b) That if, in addition, there is any substantial presence of single-lesion primary breaks (of Lea/Sax type), these also must form a constant proportion, irrespective of mode of origin, otherwise the c^* proportion would fluctuate.

Now inference b) is very difficult to reconcile with the expectations of breakage-and-reunion theory. Since, from Poisson considerations, the frequency of 'sites' containing two or more lesions must be a function of the number (or density) of lesions initially induced, then the relative proportions of single and multi-break aberrations must change with dose, i.e. their ratio cannot be independent of dose. A constant ratio would require the shapes of the single and multi-break aberration dose-response curves to be identical, and this is contrary to to all basic theory.

Alternatively, if the contribution of simple primary breaks is always negligible, then so will any effect be on proportion of c^* breaks with dose, given a common processing mechanism.

Colour-jump break constancy poses no such problems for the Exchange theory. Revell proposed that *all* observed open breaks arise from the partial failure of a two-lesion processing mechanism, so a constant, dose- or agent-independent c^* subset can be expected, since all the intrachange types are inter-related [3]. The proportion recovered will be determined by the disposition of the initial lesions in relation to the configuration of the chromatid targets and to any preferences in interaction mode within the two types of *intrachange* site.

Thus, it would seem that observed chromatid "breaks" are just as much processed aberrations, as the familiar exchanges, often seen within the same cells, and not, as is usually assumed, the visible progenitors of such aberrations.

REFERENCES

1. K. Sax, *Genetics,* 1940, **25,** 41-68.
2. D. E. Lea, 'Actions of radiations on living cells', 1st edn., 1946, Cambridge University Press.
3. S. H. Revell, *Proc. Roy. Soc. B.*, 1959, **150,** 563-589.
4. H. J. Evans, *Int. Rev. Cytol.*, 1962, **13,** 221-321.
5. S. H. Revell, *Advan. Radiat. Biol.*, 1974, **4,** 367-416.
6. J. R. K. Savage, *Brit. J. Radiol.*, 1989, **62,** 507-520.
7. J. R. K. Savage and A. N. Harvey, *Mutation Res.*, 1991, **250,** 307-317.
8. J. R. K. Savage and A. N. Harvey, in: G. Obe, A.T.Natarajan (eds.) 'Chromosome aberrations: Origin and significance.' 1994, Springer-verlag, Berlin, pp 76-91.
9. A. N. Harvey and J. R. K. Savage, *Int. J. Radiat. Biol.*, 1996, (in press).
10. S. Wolff and J. D. Bodycote, *Mutation Res.*, 1975, **29,** 85-91.
11. C. S. Griffin, A. N. Harvey and J. R. K. Savage, *Int. J. Radiat. Biol.*, 1994, **66,** 85-98.
12. N. D. Costa, W. K. Masson and J. Thacker, *Somatic Cell Mol. Genet.*, 1993, **19,** 479-490.

MODELLING THE INDUCTION OF CHROMOSOMAL ABERRATIONS BY IONISING RADIATION

A.A. Edwards

National Radiological Protection Board
Chilton
Didcot
Oxon OX11 ORQ

1 INTRODUCTION

The variation of the yield of chromosomal aberrations in cells with dose and with radiation quality has formed a set of radiobiological data against which theories and models of their production may be tested. Early work, based on results of irradiations of Tradescantia pollen grains enabled Lea[1] to picture the formation of dicentrics and centric rings as exchange aberrations between a small proportion of primary breaks. Based on the observation that x-rays produced aberrations whose number was proportional to the square of dose whereas neutron induced aberrations were proportional to dose, Lea[1] was able to infer that interacting breaks were within about 1 μm when formed. These basic ideas still persist in many subsequent models and in present day theories. The objective of this paper is to review the more popular theories and point out specifically where they succeed and where they fail to reproduce quantitatively cytogenetic observations.

Much theory was originally invented and applied to both chromatid- and chromosome-type aberrations. It is the latter application that is envisaged for this paper.

2 MODELS

2.1 Breakage and Reunion or Exchange

Two basic mechanisms have been proposed. Firstly the chromosome breaks, which nowadays are considered to be double strand breaks, produce separated ends which are free to wander in the nucleus and join randomly with other free ends.[1] The second idea, expressed by Revell,[2] is that the primary events are not complete breaks but unstable points of damage on chromosomes which may wander in the nucleus and exchange in a pairwise fashion.

Neary[3] followed the mathematical consequences of the Revell hypothesis but at the same time stated that "the question of whether the primary lesion is an actual break or not makes little difference to the formal analysis". In other words his analysis applies equally well to the Lea or the Revell hypotheses. In his theory Neary demonstrated that the dose effect relationship for aberration production should follow the linear-quadratic equation and that the linear coefficient α should increase with LET to a maximum and then decrease. By contrast the square law coefficient β was constant at low LET and decreased as LET increased. Qualitatively these predictions agree with experimental observations but it was to be years before they could be tested quantitatively.

2.2 Microdosimetric Approaches

The theory of dual radiation action[4] was based on ideas of Lea, Revell and Neary but was not specific to the production of chromosomal aberrations in cells. It covered diverse biological effects such as cell killing, mutations, growth reduction, lens opacification and carcinogenesis in laboratory animals. The basic tenet was that a generalised biological effect was proportional to the square of the energy deposited in a small volume. Kellerer and Rossi showed that with this assumption the predicted behaviour of the linear term α with LET was similar to that deduced by Neary. The dose-squared component however was independent of LET. In order to fit published cellular data the volume of interest needed to be about 1 μm in diameter. The success of the theory was that, for the first time, an explanation of why 250 kVp x-rays were about twice as effective as cobalt-60 γ-rays in producing a biological effect, was forthcoming. This effect is seen with chromosomal aberrations in human lymphocytes as well as with other endpoints.[5] The failure occurred when particle ranges were very much smaller than 1 μm, eg. low energy particles[6,7] or for charged particle irradiations of LET well above that for peak effectiveness.[8] A generalised formulation of the theory of dual radiation action appeared later,[9] but this involved the specification of two functions, neither of which was given any analytical form. Attempts to give analytical form to these functions (one is dependent only on the radiation, the other dependent on the interaction between individual transfers of energy within the cell nucleus) have been made by Brenner and Zaider[10] but applied to cell killing experiments. Monte Carlo calculations were used to calculate the proximity function of the separation of energy deposition events. Numerical integration was used for the interaction part of the calculation. Edwards et al[11,12] and Moiseenko et al[13] have used a similar approach in an attempt to describe yields of chromosome aberrations in human lymphocytes. This will be referred to later.

2.3 Hit Size Effectiveness Approach

This is an approach[14,15] which has been tested principally at low doses where cells are crossed by either one particle or are left unexposed. The theory supposed that a single particle can cause the observed biological effect which may be cell killing, a chromosome aberration, a mutation or a cancer initiating event. The probability of the effect is a function only of the energy deposited by that particle.

The approach, at least at a cellular level, has been applied usually to cell killing although similar approaches using volumes in the region of 1 μm diameter have been considered for chromosomal aberrations.[16-18] This gives similar results to that of the theory of dual radiation action. Indeed one paper[16] calculated a "specific quality function" at a diameter of 6 μm which simulates a lymphocyte nucleus and this is identical to the hit size effectiveness function. While it is possible to define such a function for a limited range of radiations, the responses to low energy x-rays and higher LET particles, eg. 460 keV ^{20}Ne ions have been shown to be poorly predicted by this approach. In addition there is no specific biological mechanism which relates to the model.

2.4 Mechanistic Modelling

More recently, modelling has been based on the more mechanistic ideas of Lea, Revell and Neary and pairwise exchange between DSBs or free-exchange of DSB ends have been investigated. Sachs and co-workers[19] have followed the "breakage and reunion" model of Lea to explain how exchange aberrations are formed. The random formation and interaction of DSB ends were modelled using Monte Carlo techniques. They have considered low LET radiations only. By restricting the number of reactive DSBs per Gy and the sites over which they can exchange it is possible to explain the numbers of complex rearrangements seen and their increase with dose. Unfortunately the model does not explain the linear-quadratic behaviour up to doses of 4-6 Gy as measured with lymphocytes. It could explain the dose response where curvature is much less evident, eg. in V-79 cells.[20] No attempt has been made yet to account for alterations of dose response curves with radiation quality.

Edwards and co-workers[11-13] have investigated the consequences of Revell's exchange hypothesis where aberrations like the dicentric are thought to arise from pairwise exchange between damaged sites on chromosomes. The emphasis in this work was to explain the variation in the dose effect relationship with radiation quality. In order to reproduce the observed linear-quadratic curve, it was necessary to introduce the notion of proximity, that is, closer breaks had a greater chance of exchanging. This is in line with the theories of Neary and of generalised dual radiation action. Breaks which contributed to the linear intra-track term were closer than those contributing by the quadratic inter-track term. The model predicted that the close intra-track DSB exchanged more rapidly than the more distant inter-track exchanges and the rate of formation of exchanges was dependent on dose. However, the model was not able to predict the relatively large number of complex rearrangements now observed or the reduction in curvature of dose response seen at LET values of 20-30 keV μm^{-1}.

3 FUTURE DEVELOPMENTS AND CONCLUSIONS

It is judged that the most productive way forward in modelling is to use mechanistically based theories. The processes underlying physical variations of energy deposition in cells and in DNA, the production of lesions on chromosomes (double strand breaks) and, the repair and misrepair of those breaks all need modelling. These may be regarded as random processes and are best approached using Monte Carlo simulation on a cell by cell basis. If the objective is to explain the mechanism of production of the biological effect it is important to focus on a single well defined cell system. The data set concerning the induction of chromosomal aberrations in human lymphocytes is sufficiently well developed for this purpose.

Initially, it is necessary to explain the shape of the dicentric yield curve for one particular low LET radiation and explain how this is modified numerically at higher LET. The model must also be capable of predicting the rate of dicentric formation as described using premature chromosome condensation measurements. Following this the model may be extended to predict other aberrations such as centric rings, acentrics, translocations and the existence of cells containing complex rearrangements.

Acknowledgements

Partial funding from CEC contract No. F14P-CT95-0011 is acknowledged.

References

1. D. E. Lea, *Brit. J. Radiol.* Supplement 1, 1947, 75.
2. S. H. Revell, *Advances in Radiation Biology*, 1974, **4**, 367.
3. G. J. Neary, *Int. J. Radiat. Biol.*, 1965, **9**, 477.
4. A. M. Kellerer and H. H. Rossi, *Curr. Topics in Radiat. Res. VIII,* 1974, 85.
5. V. P. Bond, C. B. Meinhold and H. H. Rossi, *Health Phys.*, 1978, **34**, 433.
6. R. P. Virsik, C. H. Schäfer, D. Harder, D. T. Goodhead, R. Cox and J. Thacker, *Int. J. Radiat. Biol.*, 1980, **38**, 545.
7. A. A. Edwards, D. C. Lloyd and J. S. Prosser, *Radiat. Prot. Dosim.*, 1990, **31**, 265.
8. A. A. Edwards, P. Finnon, J. E. Moquet, D. C. Lloyd, F. Darroudi and A. T. Natarajan, *Rad. Prot. Dosim.*, 1994, **52**, 299.
9. A. M. Kellerer and H. H. Rossi, *Radiat. Res.*, 1978, **75**, 471.
10. D. J. Brenner and M. Zaider, *Radiat. Res.*, 1984, **99**, 492.
11. A. A. Edwards, V. V. Moiseenko and H. Nikjoo, *Int. J. Radiat. Biol.*,1994, **66**, 633.
12. A. A. Edwards, V. V. Moiseenko and H. Nikjoo, *Radiat. Environ. Biophys.*,1996, **35**, 25.
13. V. V. Moiseenko, A. A. Edwards and H. Nikjoo, *Radiat. Environ. Biophys.*,1996, **35**, 31.
14. V. P. Bond and M. N. Varma, CEC Report EUR 8395 EN, Proc. of the Eighth Symposium on Microdosimetry, Jülich, 27 September - 1 October 1982, Eds. Booz and Ebert, 1983, 423.
15. M. N. Varma and V. P. Bond, ibid 439.
16. M. Zaider and D. J. Brenner, *Radiat. Res.*, 1985, **103**, 302.
17. A. A. Edwards, D. C. Lloyd and J. S. Prosser, *Rad. Prot. Dosim.*, 1990, **31**, 265.
18. M. N. Varma and M. Zaider, Biophysical Modelling of Radiation Effects, Eds. K. H. Chadwick, G. Moschini and M. N. Varma, Adam Hilger, 1992, 145.
19. A. M. Chen, J. N. Lucas, F. S. Hill, D. J. Brenner and R. K. Sachs, *Int. J. Radiat. Biol.*, 1996, **69**, 411.
20. J. Thacker, R. E. Wilkinson and D. T. Goodhead, *Int. J. Radiat. Biol.*, 1986, **49**, 645.

FURTHER EVIDENCE FOR THE ASSOCIATION OF CHROMOSOME ABERRATION YIELD COEFFICIENT α WITH "FAST, SHORT-RANGE" AND OF COEFFICIENT β WITH "SLOW, LONG-RANGE" PAIRWISE INTERACTION BETWEEN DNA LESIONS

R. Greinert, E. Detzler, E. Bartels. K. Schulte, O. Boguhn, C. Thieke, R. P. Virsik-Peuckert and D. Harder

Institute of Medical Physics and Biophysics, University of Göttingen, Germany

Introduction

Coefficient α of the equation describing exchange-type chromosome aberration yield in eukaryontic cells, $y = \alpha D + \beta D^2$, increases with LET, and coefficient β decreases with dose fractionation or protraction. These characteristics have already lead D. Lea to suggest that the α component is due to intratrack and the β component to intertrack pairwise interaction between radiation-induced chromosomal lesions. The surprisingly high α value for carbon characteristic X rays in spite of their very short photoelectron range[10] and the disappearance of β with increasing LET (possibly due to a kinetic competition advantage of the α over the β reaction pathway) later led to the suggestion that the α component is due to a "fast, short-range" lesion interaction, while the β component appears to originate from a "slow, long-range" lesion interaction[11]. Today it is known that the β component originates from pairwise interaction between radiation-induced lesions located on *both* of the interacting chromosomes[3], and these two lesions have been identified as *fast-repaired DNA double-strand breaks*[5]. The more difficult problem of the α component is discussed in this paper.

Delayed-fusion PCC experiments: Kinetic difference between α and β components

A fundamental difference between the α and β components of the exchange-type chromosome aberration yield has been discovered by observing the *temporal development* of these aberrations in the cell[4]. Human T lymphocytes irradiated in G_o phase with 4 Gy of 150 kV X-rays were fused with mitotic CHO cells at graded delay times, t_D, after irradiation. The time needed for cell fusion and chromosome condensation was, in most experiments, 60 min, but shortening to 30 min gave very similar results. By observing the prematurely condensed interphase chromosomes thereby formed, the stages of development of the exchange-type aberrations could be studied as a function of post-irradiation time. Using C-banding to identify the dicentric chromosomes, it was observed that their yield started with a finite value at $t_D = 0$. This was followed by an S-shaped rise of the yield curve, reaching saturation at about $t_D = 10$ h (fig. 1a). The dose-yield curve of the fast component, obtained at $t_D = 0$, was strictly linear, whereas the dose-yield curve obtained in the time region of saturation ($t_D = 8$ h) was linear-quadratic and approached the known first metaphase values[4]. This proved that the lesion interaction leading to the β component was much slower than that associated with the α component.

When experiments were performed using chromosome exchanges made visible by combination of the PCC and FISH techniques, with "painting" of chromosome 4, a similar slow kinetics was now seen for the formation of the *complete exchanges*, i.e. of reciprocal translocations and of dicentrics with associated bi-coloured fragment. The yield of complete exchanges was very small at $t_D = 0$, but increased in an S-shaped fashion, reaching saturation near 20 h (fig. 1b). At $t_D = 8$ h, their dose-yield relationship was linear-quadratic.

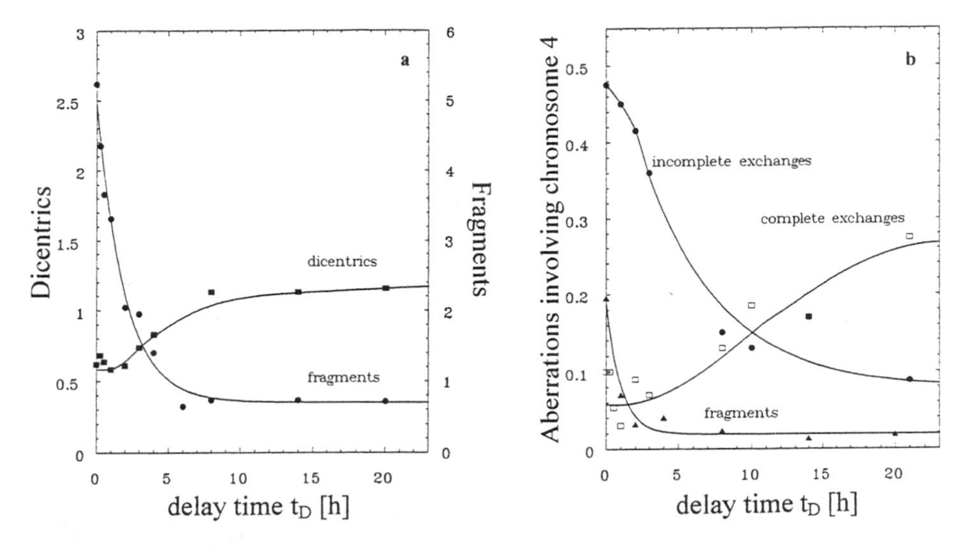

Figure 1. Kinetics of chromatin rearrangements seen in human T lymphocytes under delayed-fusion PCC conditions after 4 Gy of 150 kV X-rays. Abscissa: Post-irradiation time before cell fusion. Ordinate: yield per cell. 1a) Dicentric chromosomes (each with one fragment) and excess acentric fragments. Giemsa plus C-banding. 1b) Complete exchanges (bicoloured chromosome pairs), incomplete exchanges (colour changes in single bicoloured chromosomes) and painted fragments, all involving chromosome #4.

The fast decline of the un-attached *chromosome fragments* (fig. 1b) agrees with that of the fast-repaired DNA double-strand breaks in lymphocytes from the same donor, observable by pulsed-field gel electrophoresis[9], whereas the slow decline of the *incomplete exchanges* agrees with that of the slowly repaired dsb. The dose-yield relationship of the incomplete exchanges was *strictly linear* at $t_D = 0$ as well as $t_D = 8$ h, suggesting that they originate from *single-track traversals*, i.e. are early forms of the α component. Their gradual disappearance in the course of several hours indicates that they are the residue of not yet stable *intermediate exchange products* (*IEP*) of chromatin fibres, distorted by PCC stress. Their transitory existence until their development into stable chromosome exchanges or towards restitution can be regarded as a key to further understanding, e.g. of the phenomenon of potential lethal damage to cells. These observations with FISH-marked chromosomal exchanges, among which the dicentrics are subgroups, agree with the direct observation of the dicentrics in fig. 1a.

For explanation of the surprising similarity between the temporal profiles of the PCC fragments and incomplete exchanges with the time-dependent decline of the number of DNA lesions observable as double-strand breaks, one may consider the molecular forces to which the chromatin fibre is subjected in cell preparation either for dsb measurement or for premature chromosome condensation. Since both treatments will result in DNA scission at unstable sites of the chromatin, we suggest that the well-known dsb repair kinetics observable under PFGE or gradient centrifugation conditions and the gradual disappearance of the fragments and incomplete exchanges observable under PCC conditions *are only two different manifestations of the same phenomenon, the gradual disappearance of unstable sites in the chromatin.* This means in particular, that the hitherto unidentified slowly repaired and dose-proportional dsb component seen under PFGE or gradient centrifugation conditions and the slowly disappearing incomplete exchanges seen under PCC conditions are merely two different, preparation-dependent manifestations of the same unstable chromatin sites arising from single-particle traversals.

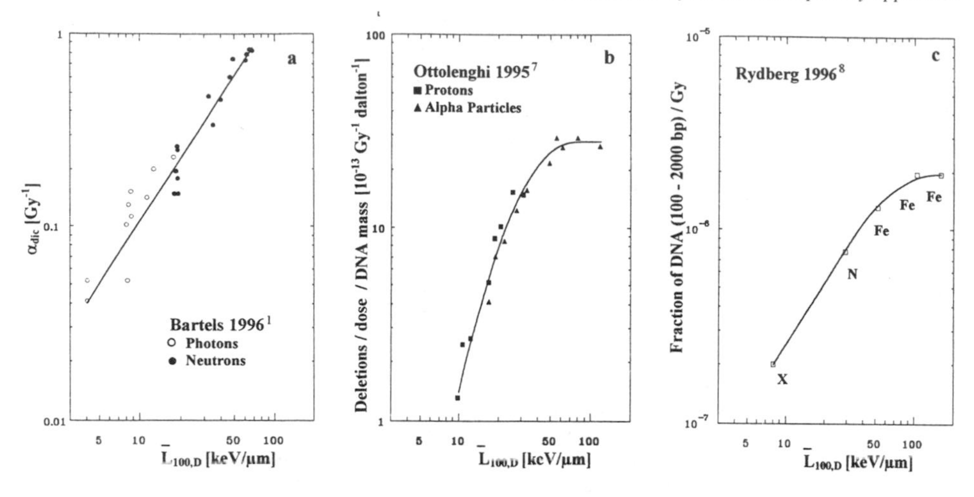

Figure 2. Dependence on dose-mean restricted LET of the yield coefficients a) for dicentric chromosomes, observed in first post-stimulation mitosis of human T-lymphocytes, b) for DNA "deletions", defined by the occurence of two dsb less than d_{del} = 30 bp apart (Monte-Carlo result) c) for DNA fragments of length 100 to 2000 bp.

Proximity of dsb in DNA: Nature of the α component

For dicentric chromosomes in human T lymphocytes α_{dic} was found to be *proportional to dose-mean restricted linear energy transfer* $\bar{L}_{100,D}$ [11], and this has been confirmed in many experiments with various radiations up to $\bar{L}_{100,D}$=70 keV/μm (for photons and neutrons see fig. 2a). The proportionality of the dose-related yield with $\bar{L}_{100,D} = \overline{L_{100}^2}/\bar{L}_{100}$ means *pair-wise lesion interaction along the particle track*. Another typical example, not shown here, is the proportionality of coefficient α for cell inactivation with dose-mean restricted LET, holding for protons, deuterons, He-3 ions and alpha particles[2].

For sufficiently large electron and proton ranges the two interacting lesions have often been assumed to be a pair of dsb *on two different chromatin fibres*. However, this model is hardly applicable in the case of the C_K photo-electrons, whose practical range is only 5 nm, but who show the same proportionality of α_{dic} with dose-mean restricted LET as the other radiations[11]. This difficulty can be solved by assuming that the α component is determined by another type of intra-track lesion interaction, namely of a pair of dsb on proximate DNA sections *within the same chromatin fibre*. This pair of dsb can occur on proximate DNA turns upon the surface of the nucleosome, on proximate nucleosome windings of the 30 nm chromatin fibre or at the bases of chromatin loops. Such "clustered" DNA damage has been studied in Monte-Carlo simulations e.g. by Ottolenghi et al.[7] and by Holley and Chatterjee[6], and a unique dependence of the dose-related yield of pairs of dsb separated by less than 30 bp upon $\bar{L}_{100,D}$ has been obtained, see fig. 2b. (The steepness of this curve would decrease on the assumption of a larger interaction distance, since at $\bar{L}_{100,D}$ values less than about 20 keV/μm the maximum delta-ray range already exceeds 100 nm .) If this explanation of the essential intratrack interaction would be true, the yield of short DNA fragments in the range 100 to 2000 bp released at the sites of such dsb pairs should also be proportional to dose-mean restricted LET. Our plot of this yield, measured by Rydberg (1990), versus dose-mean restricted LET in fact shows the expected proportionality over a large range of $\bar{L}_{100,D}$ down to 225 kV X-rays (fig. 2c).

Hence there is already much evidence that clustered damage to DNA represents the lesion from which the α component of the exchange-type chromosome aberrations originates. Since the second chromosome involved in the exchange should then be an *intact* chromosome, a basic difference in mechanism as compared to the β component, whose formation requires the interaction between two dsb placed on two different chromosomes, has to be anticipated.

Conclusions

Delayed-fusion PCC experiments with human lymphocytes have disclosed *different reaction kinetics* of the α and β aberration yield components. It was found that incomplete exchanges are the early manifestation of the α component, and it was concluded that slowly repaired dsb are the manifestations of the same *intermediate exchange products (IEP)* of the chromatin fibres which under PCC conditions appear as incomplete chromosomal exchanges. The linear dependence of yield coefficient α for dicentrics in human lymphocytes upon dose-mean restricted LET has been used to suggest, by comparison with Monte-Carlo-simulated clustered DNA damage and with experimental short-fragment records, that the formation of *a pair of DNA double-strand breaks on proximate sections of the DNA* by a single particle traversal is the event from which the α component of the exchange-type aberrations originates.

Acknowledgments
The financial support given to us by Deutsche Forschungsgemeinschaft, Bundesamt für Strahlenschutz and Commission of the European Communities is gratefully acknowledged.

References

1. E.R. Bartels and D. Harder, Rad. Prot. Dosim. 1990, **31,** 211-215

2. F. Cera, R. Cherubini, M. Dalla Vecchia, S. Favaretto, G. Moschini, P. Tiveron, M. Belli, F. Ianzioni, L. Levati, O. Sapora, M.A. Tabocchini, G. Simone; this Symposium

3. M.N. Cornforth, Radiat. Res. 1990, **121**, 22-27

4. R. Greinert, E. Detzler, B. Volkmer and D. Harder, Radiat. Res.1996, **144**, 190-197

5. R. Greinert, B. Volkmer, R.P. Virsik-Peuckert and D. Harder, Int. J. Radiat. Biol. 1996, **70**, 33-43

6. W.R. Holley and A. Chatterjee, Radiat. Res. 1996, **145**, 188-199

7. A. Ottolenghi, M. Merzagora, L. Tallone, M. Durante, H.G. Paretzke and W.E. Wilson, Radiat. and Environm. Biophys., 1995, **34**, 239-244

8. B. Rydberg, Radiat. Res. 1996, **145**, 200-209

9. K. Schulte, Biophysikalische Analyse der Reparatur strahleninduzierter DNA-Schäden mittels Puls-feld-Gelelektrophorese. Diploma Thesis, Georg-August-Universität Göttingen 1996

10. R.P. Virsik, C. Schäfer, D. Harder, D.T. Goodhead, R. Cox, J. Thacker, Int. J. Rad. Biol. 1980, **38**, 545-557

11. R.P. Virsik, R. Blohm, K.-P. Hermann, D. Harder: Fast, short-ranged and slow, distant-ranged interaction processes involved in chromosome aberration formation. 7th Symp. Microdosimetry 1980 (Oxford), Ed.: H.G. Ebert, Harwood Publ. 1981.

IMPLICATIONS OF MICRODOSIMETRY FOR A STATE VECTOR MODEL OF CHROMOSOMAL ABERRATIONS AND CELLULAR TRANSFORMATION: THE CASE OF MULTIPLE PATHWAYS TO EFFECTS

D. Crawford-Brown[1,2], W. Hofmann[1], M. Nösterer[1] and P. Eckl[3]

[1]Institute of Physics and Biophysics, University of Salzburg, A-5020 Salzburg, Austria
[2]Institute for Environmental Studies, University of North Carolina, Chapel Hill, NC 27599-7400, USA
[3]Institute of Genetics and General Biology, University of Salzburg, A-5020 Salzburg, Austria

1 INTRODUCTION

There are several pathways by which radiation may produce cellular damage, all with the same endpoint of transformation and, eventually, cancer. These pathways involve alterations of the DNA, leading to either initiation or promotion. In initiation, the result is an aberrant cell with an altered phenotype but with relatively intact controls on growth. The pathway to initiation may take place through any one of at least three mechanisms involving either point mutations or larger structural changes in DNA. The radiation may:
(1) cause one or both of two double strand breaks (DSBs), allowing translocation of a suppressor region to a point on the DNA where it becomes inactive or allowing translocation of the oncogene itself to a point where it no longer is affected by the suppressor;
(2) cause a single-strand DNA break (SSB) located sufficiently close to oxidative damage to allow interaction during replication; this may result in either a DSB after replication or to production of a point mutation or shift in the sequence of the gene due to mis-repair. The oxidative damage may also caused by radiation induced radicals which are converted to reactive oxygen species capable of interacting with DNA.
(3) produce a DSB resulting in the deletion of a tumor suppressor gene or the supressor region for an oncogene;

2 STATE VECTOR MODELING

The three pathways may be written as a state vector model, all resulting in initiation of the irradiated cell. In this model, the first pathway requires two DSBs[1,2]. One is a specific DSB associated with an oncogene or with the suppressor region. The second DSB is non-specific, simply producing a place for translocation of the relevant portions of the DNA from the specific DSB. These two breaks then must interact, producing a translocation, and mitosis must occur prior to any rejoining or repair if the alteration is to be made permanent.

The second pathway requires that radiation produces either a SSB, oxidative damage, or both, either directly or through chemical reactions (oxidative damage). If this SSB is located sufficiently close to a site of oxidative damage, the two forms of damage will produce a DSB during replication of the DNA. The yield of DBSs, point mutations, etc. produced via this pathway will depend on repair of SSBs and oxidative DNA damage, and the expression of gene products reducing the rate of oxidative damage (adaptive response).

The third pathway results from a single radiation-induced transition in which the DSB sustains a deletion. Mitosis then must occur prior to rejoining or repair if the damage is to be fixed and the alteration made permanent.

Radiation may induce any of the transitions, although the specific DNA break in the first pathway appears to be due more to background events[1,2]. If $P_k(D)$ is the probability that a cell is initiated through pathway k, the total probability of initiation, $P_I(D)$, is:

$$(1) \qquad P_I(D) = 1 - \prod(1 - P_k(D))$$

where the product is from k equal to 1 to 3. For transformation, $P_I(D)$ must be multiplied by the probability of promotion, $P_P(D)$. This yields:

$$(2) \qquad P_T(D) = \{1 - \prod(1 - P_k(D))\} \times P_P(D)$$

where $P_T(D)$ is the probability of transformation. If the transitions described above are independent of those for cell killing, equation 2 may be modified as:

$$(3) \qquad P_T(D) = \{1 - \prod(1 - P_k(D))\} \times P_P(D) \times P_S(D)$$

where $P_S(D)$ is the cell survival probability at dose D. It is unlikely, however, that survival and the transitions associated with chromosomal damage are independent. Those cells with the greater probability of sustaining damage that leads to transitions should also be the cells sustaining the greatest degree of chromosomal damage, eventually resulting in a greater probability of cell death. This can be dealt with most directly by incorporating microdosimetric information on the distribution of cellular doses in the irradiated tissue.

Let dP(Z)/dZ be the probability density function for cellular doses Z (where Z is the mean value of the specific energy distribution in a particular cell nucleus). Then.

$$(4) \qquad P_T(D) = \int [dP(Z)/dZ]\, P_T(Z)\, dZ =$$

$$\int [dP(Z)/dZ] \times \{1 - \prod(1 - P_k(D))\} \times P_P(D) \times P_S(D) \times dZ$$

The three pathways to initiation may be described by differential equations and solved through numerical methods. The advantage of using numerical methods is that the model may incorporate stochastic methods, allowing a probabilistic formulation. The resulting differential equation for any state j in any particular pathway is:

$$(5) \qquad dN_j(t)/dt = \sum k_{i,j}(t) \times N_i(t) + k_{j+1}^{Rep}(t) \times N_{j+1}(t) - k_j^{Rep}(t) \times N_j(t)$$

where $N_j(t)$ is the number of cells in state j at time t (equivalent to dose D under constant dose-rate); $k_{i,j}$ is the probability-per-unit-time of a transition from state i to j (units of time^{-1}); and k_j^{Rep} is the rate constant for repair or rejoining of the induced damage for state j (units of time^{-1}). The value of $k_{i,j}$ has both a background and radiation-induced component:

$$(6) \qquad k_{i,j}(t) = k_{i,j}^{b}(t) + k_{i,j}^{r} \times DR(t)$$

where $k_{i,j}^{b}$ is the background rate constant (units of time^{-1}); $k_{i,j}^{r}$ is the radiation-induced rate constant (units of dose^{-1}); and DR is the dose-rate. Note that $P_S(D)$ in equation 4 is developed empirically from dose-response data on cell killing. Comparison of model predictions with dose-response data for chromosomal alterations and transformation given later are expressed as the probability of effect per surviving cell, so $P_S(D)$ is implicitly 1.0.

3 ADAPTIVE RESPONSE

Radiation damage, particularly to active genes, results in a cell responding through changes in several characteristics relevant to the second pathway (but probably not the other two pathways). These changes are a form of adaptive response[3], which is a feedback system in which the cell lowers the rate of further damage.

First, the radiation damage increases the concentration of antioxidant enzymes, thereby preventing further damage. This depresses the transition rate constant associated with production of oxdiative damage and, potentially, SSBs in the second pathway. Thus, $k_{i,j}$ for both of these transitions is a function of the cumulative dose at time t (since this is related to the total production of chromosomal damage). In particular, $k_{i,j}^b$ and $k_{i,j}^r$ decrease with increasing dose, until the adaptive mechanism is saturated. This does not, however, imply that the total $k_{i,j}$ decreases, since this also is a function of the dose-rate (equation 6)

Second, the damage induces an increase in repair enzymes when an active gene is damaged. This increases the rate constant for k_j^{Rep} and decreases the probability that a damaged chromosome enters mitosis and, hence, has the damage fixed. The net result of both forms of adaptive response is that accumulations of dose result in a lowering of the ability of radiation to induce further chromosomal changes. This has been shown in experiments where delivery of a small dose (the conditioning dose) prior to a larger dose (the experimental dose) lowers the yield of chromosomal aberrations from the second dose[3]. Presumably, this also lowers the transformation frequency, although this has not been tested.

4 IMPLICATIONS FOR MODELING

Eckl et al.[3] have studied the dose-response relationship for chromosomal aberrations in primary rat hepatocytes following *in vitro* gamma irradiation (Figure 1). The results shown in Figure in dicate a plateau in the relationship above 1 Gy. This same plateau has also been noted in studies of *in vitro* transformation of cells by low LET radiation[4]. Within the model presented here, there are two explanations for this plateau (existing experimental data make it impossible to select between these two explanations).

(1) Radiation acts primarily by the first pathway. A subpopulation of cells sustains the specific DNA damage prior to irradiation and the radiation affects only the transition associated with the non-specific damage. When all of this subpopulation of cells have sustained the non-specific DSB, the dose-response curve plateaus. Earlier modeling[1,2] showed that the transition rate constant for the non-specific damage required to reproduce the transformation data was approximately 4 per Gy, consistent with measurements of the probability of non-specific DSBs. If this explanation is correct, then the relevant nanodosimetric information is the probability that a sensitive volume on the order of 10 to 50 nm must sustain sufficient ionization to induce a DSB. It also implies that the spatial location of DSBs in a cell will be an important determinant of the transition associated with interaction of the specific and non-specific DSBs, with a closer spacing yielding a higher transition rate constant for the interaction.

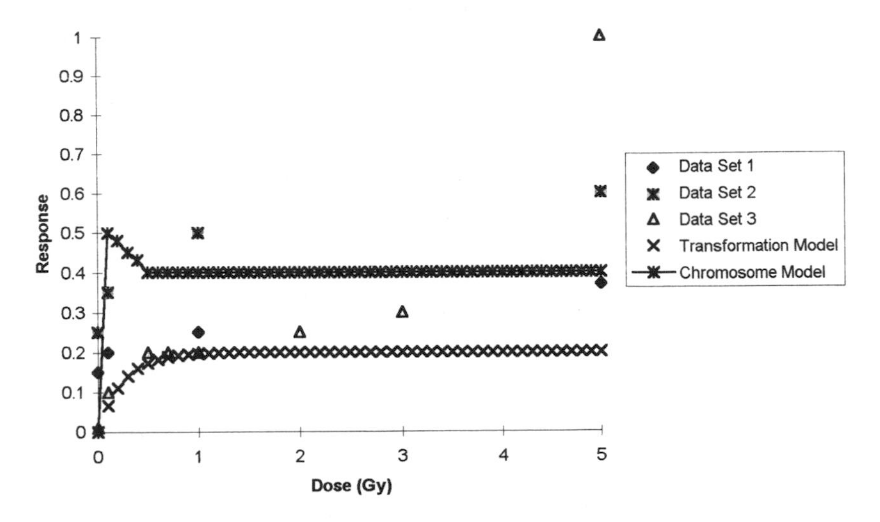

Figure 1 *Comparison of in vivo X-ray data and model predictions for chromosomal aberrations and transformation. Data are for total chromosomal aberrations without pre-irradiation (1), for total chromosomal aberrations with pre-irradiation of 2.5 mGy (2), and cellular transformation, in units of transformations per 1000 surviving cells (3).*

(2) Radiation acts primarily by the second pathway. As dose and, hence, chromosomal damage accumulate, the transition rate constants for the SSBs and oxidative damage decrease and the repair rate constant increases. If this is simulated in the model (second pathway), with the repair rate proportional to the rate of DNA damage and with $P_P(D)$ set to 0.01 for transformation, the result is an approximate plateau (Figure 1). If this explanation is correct, then the relevant nanodosimetric information is the probability that a sensitive volume on the order of less than 10 to 50 nm must sustain sufficient ionization to induce a SSB or oxidative damage, but not to produce a DSB (or else the other pathways would dominate). It also implies that the spatial location of damage within the nucleus is not important (since there is no interaction of DSBs required), but rather the spatial location of SSBs relative to sites of oxidative damage will be the important parameter.

Acknowledgments

This research was supported in part by the CEC, Contract FI4P-CT95-0025.

References

1. D. Crawford-Brown and W. Hofmann, *Int. J. Rad. Biol.*, 1990, **57**, 407.
2. D. Crawford-Brown and W. Hofmann, *Math. Biosci.*, 1993, **115**, 123.
3. P. Eckl, A. Karpf, W. Stöß, S. Knasmüller and R. Schulte-Hermann, IAEA/UNEP Seminar, Nairobi, IAEA-SR-184-10, 1993.
4. R. Miller, E. Hall and H. Rossi, *Proc. Nat. Acad. Sci, USA*, 1979, **76**, 5755.

Radiation quality and biological effectiveness

UNDERCOUNTING OF PARTICLE IRRADIATION-INDUCED DNA DOUBLE-STRAND BREAKS BY CONVENTIONAL ASSAYS

M. Löbrich

Strahlenzentrum der Justus-Liebig-Universität Giessen
Leihgesterner Weg 217
35392 Giessen, Germany

1 INTRODUCTION

It is a general consensus that DNA double-strand breaks (dsbs) are responsible for most biological effects from ionizing radiation in mammalian cells although base damages, DNA protein crosslinks and other lesions also contribute. The assays which are currently used for measuring dsbs in mammalian cells include low-speed sucrose gradient centrifugation[1], neutral filter elution[2] and pulsed-field gel electrophoresis[3,4] (PFGE). In order to minimize the doses used, the largest possible DNA fragments as defined by the limits of sensitivity of the assays are usually analyzed. For PFGE for example, the amount of DNA below a certain exclusion size, typically in the order of several megabasepairs (Mbp), is quantified as a function of radiation dose. In order to correlate this FAR-value (fraction of DNA radioactivity released from the gel plug into the gel) with the number of breaks induced, assumptions concerning the distribution of breaks throughout the genome and thus concerning the size distribution of radiation-induced DNA fragments are necessary. It is generally assumed that the breaks are randomly distributed throughout the genome and, if radiations of different quality are compared, that the distributions yielding the same FAR-values are identical for the radiations under investigation.

The induction of breaks has been found to depend only slightly on the LET of the radiation when assay were employed which measure dsbs on a Mbp-level. Although older work reported RBEs for mammalian cells significantly bigger than unity in the LET-range around 100 keV/μm[5,6], more recent investigations found no significant increase in this LET-range and a decline above several hundred keV/μm[7-9]. This RBE-LET relationship for dsb induction is in contrast to strong dependencies observed for complex biological endpoints like mutation induction and cell inactivation[10]. In an attempt to explain the LET effect on the basis of dsbs being the relevant lesions, theoretical considerations[11] and experimental evidence[12] have demonstrated clustering of breaks after particle irradiation in the DNA size range from about 100 bp to 2 kbp due to the organization of DNA in the 30 nm chromatin fiber. Such a clustering of dsbs results in an undercounting of breaks by the assays mentioned above and therefore in RBE-values for the induction of dsbs which are too low. Additionally, such clustering may increase the lethality and biological significance per individual break and thus may account for the high biological efficiency for high-LET radiation.

In addition to clustering in the size range from 100 bp to 2 kbp, a correlation of breaks may also be expected on larger distances due to the interaction of the particles with higher order DNA structures like, for example, the DNA loop structure[13]. The present report describes experiments designed to detect DNA fragments in the size range above 10 kbp.

2 EXPERIMENTAL PROCEDURE

Primary human fibroblasts GM38 were radioactively labeled with ^{3}H-thymidine for several days prior to the experiment and irradiated in the G_0-stage of the cell cycle[14]. The cells were lysed with proteinase K and the DNA was separated in an agarose gel using PFGE. Two different electrophoretic conditions were used to separate DNA-fragments between 10 kbp and about 1 Mbp[14]. After electrophoresis, the gels were reproducibly cut at the positions of size markers (chromosomes of Saccharomyces cerevisiae and DNA lambda-ladders) and the radioactivity inside each size zone was counted by liquid scintillation counting. The number of fragments in a size zone is proportional to the amount of radioactivity in the zone and inversely proportional to the DNA size of the fragments in the zone.

To calculate the number of breaks per unit of radiation dose and unit of DNA length, the fraction of the total amount of radioactivity in each size zone of a certain dose sample was multiplied by the DNA length unit and divided by the average DNA size of the corresponding size zone. This number of fragments per DNA length unit was then summed up for all sizes and divided by the radiation dose of the corresponding sample. For size regions for which no fragment measurements were made, the number of fragments was calculated assuming a random distribution of breaks with a dsb induction rate obtained by the FAR assay. This procedure yielded the dsb induction rate for the dose sample evaluated so that the number of breaks per unit of radiation dose and unit of DNA length could be determined from the average value of all samples of a given experiment. A detailed description of the evaluation procedure is given elsewhere[14].

3 RESULTS AND DISCUSSION

Table 1 summarizes the results of a typical experiment with 225 kVp X-rays and N particles with an energy of 29 MeV/u and LET at target of 97 keV/μm. The percentages of radioactivity in 8 different zones covering the size range from 9 to 1120 kbp are given for 50, 100, and 150 Gy of X-rays and for 63, 126, and 189 Gy of N particles (the error can be estimated to be +/- 10% for a typical experiment). The FAR-values were determined with the FAR assay in the same experiment and show approximately the same values for the 3 X-ray doses compared to the 3 N doses indicating a lower induction frequency for N particle irradiation.

The percentages of radioactivity in the different zones were divided by the widths of the zones (maximum minus minimum) and drawn in the upper part of figure 1 versus the average size value of the zones (approximated by the mean of maximum and minimum). These weight distributions for 189 Gy of N particles (squares) and 150 Gy of X-rays (circles) are compared to the theoretically expected weight distribution for 150 Gy of X-rays (diamonds) based on the assumption of a random distribution of breaks and a dsb induction rate of 5.8×10^{-3} breaks/Mbp/Gy as determined elsewhere[15]. The lower part of figure 1 shows the corresponding number distributions which were obtained by dividing the weight distributions by the fragment sizes.

Table 1 *Percentages of radioactivity in different size zones and FAR-values for samples irradiated with 3 different doses of X-rays and N particles*

	X-rays			N particles		
Size zone (kbp)	50 Gy	100 Gy	150 Gy	63 Gy	126 Gy	189 Gy
780-1120	4.62	9.63	13.35	4.57	9.65	13.45
365-780	5.41	13.58	21.69	6.85	14.91	23.18
225-365	1.20	3.85	6.52	1.91	5.55	7.88
145-225	0.53	2.05	2.51	0.88	2.85	3.43
97-145	0.10	0.38	1.14	0.29	0.86	1.95
48-97	0.05	0.38	0.81	0.38	0.81	1.78
23-48	0.05	0.09	0.22	0.12	0.26	0.52
9-23	0.04	0.10	0.22	0.13	0.28	0.57
FAR-value	38%	66%	83%	42%	70%	84%

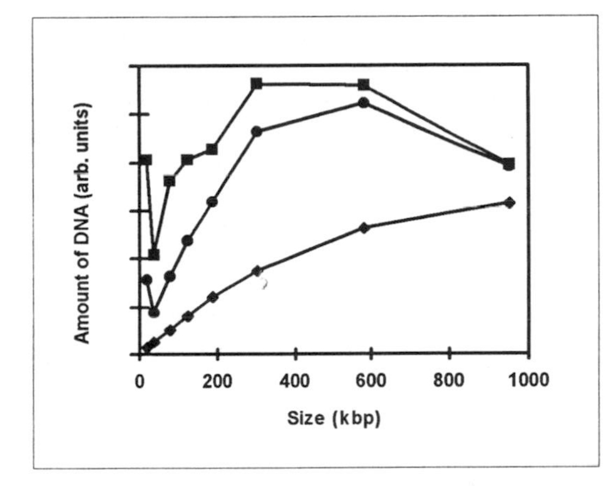

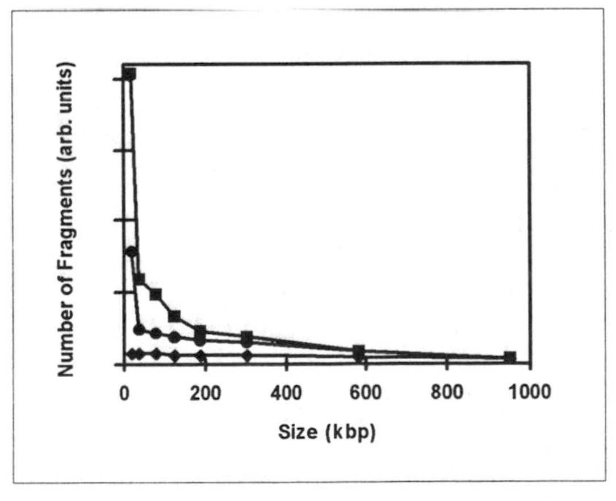

Figure 1 *Weight (upper part) and number (lower part) distributions of DNA fragments generated by 189 Gy of N particles (squares) and 150 Gy of X-rays (circles) compared to theoretically expected distributions for 150 Gy of X-rays (diamonds).*

While the experimentally determined distributions agree quite well with the theoretical distribution in the high molecular weight range (around 1000 kbp), a profound difference can be observed for smaller DNA fragments. As demonstrated in the lower part of figure 1, this difference is most significant for fragments below about 100-150 kbp. Comparing the experimental distributions for 150 Gy of X-rays and 189 Gy of N particles, it can be seen that although both radiations are at these two doses equally efficient in inducing megabasepair-spaced dsbs (as indicated by almost the same FAR-values of 83 and 84% and by an almost identical amount of DNA in the largest size zone), N particles are considerably more efficient in producing smaller DNA fragments particular in the size range below 100-150 kbp. This difference between low-LET and high-LET radiation is reflected in changing RBE-values for total dsbs. While the RBE for N particles was determined with the FAR assay from 3 independent experiments to be about 0.7, the corresponding value obtained by analyzing the entire fragment distribution (see section 2) changed to 1.1. The increase in RBE for 900 MeV/u Fe particles with an LET of 150 keV/μm was even more significant, changing the RBE from 0.64 to 1.8 if the non-random break distribution is taken into consideration[14]. Considering these results, it is obvious that the RBE-LET relationship for dsb induction as determined by conventional assays is misleading and falsely indicates a decreased efficiency in the LET range around 100 keV/μm. The RBE-values obtained by analyzing the entire fragment distribution in the total range of 0.1 kbp to 10 Mbp show an increasing induction efficiency for increasing LET and agree with theoretical calculations which predicted an increase in dsb yield with LET up to a few hundred keV/μm[16,17].

The observed clustering of dsbs arises from a non-random distribution of energy deposition by ionizing radiation and is presumably influenced by the highly organized structure of genomic DNA. A possible candidate of DNA organization which could affect clustering of breaks in the size range between 10 and 200 kbp is the DNA loop structure. However, it should be considered that loop domains of the chromatin fiber are not required to explain the observed effect because a random-coil model of the chromatin is expected to produce correlated breaks as well. The suggested clustering of breaks may have important biological implications. Dsbs which are induced in a correlated manner may be more likely to be mis-rejoined than breaks induced randomly, simply due to the greater proximity of incorrect ends. In addition, rejoining could proceed with loss of the DNA fragments between the correlated breaks, possibly causing deletion mutations. The non-random distribution of dsbs might therefore be reflected in an altered deletion spectrum for high-LET radiation compared to X-rays.

Acknowledgements

It is a great pleasure to acknowledge Drs B. Rydberg and P.K. Cooper for their support.

References

1. D. Blöcher, *Radiat. Res.*, 1982, **42,** 317.
2. M. O. Bradley and K. W. Kohn, *Nucl. Acids Res.*, 1979, **7,** 793.
3. D. T. Stamato and N. Denko, *Radiat. Res.*, 1990, **121,** 196.
4. D. D. Ager, W. C. Dewey, K. Gardiner, W. Harvey, R. T. Johnson, and C. A. Waldren, *Radiat. Res.*, 1990, **122,** 181.
5. G. Kampf and K. Eichhorn, *Studia Biophys.*, 1983, **93,** 17.
6. D. Blöcher, *Int. J. Radiat. Biol.*, 1988, **54,** 761.
7. J. Heilmann, H. Rink, G. Taucher-Scholz, and G. Kraft, *Radiat. Res.*, 1993, **135,** 46.
8. K. J. Weber and M. Flentje, *Int. J. Radiat. Biol.*, 1993, **64,** 169-178.
9. B. Rydberg, M. Löbrich, and P. K. Cooper, *Radiat. Res.*, 1994, **139,** 133.
10. J. Thacker, "Advances in Radiation Biology", O. F. Nygaard, W. K. Sinclair, and J. T. Lett, Academic Press, San Diego, CA, 1992, Vol. 16, pp. 77.
11. W. R. Holley and A. Chatterjee, *Radiat. Res.*, 1996, **145,** 188.
12. B. Rydberg, *Radiat. Res.*, 1996, **145,** 200.
13. J. W. Bodnar, *J. Theor. Biol.*, 1988, **132,** 479.
14. M. Löbrich, P. K. Cooper, and B. Rydberg, Int. J. Radiat. Biol., in press.
15. M. Löbrich, B. Rydberg, and P. K. Cooper, *Proc. Natl. Acad. Sci. USA*, 1995, **92,** 12050.
16. A. Chatterjee and W. R. Holley, *Int. J. Quantum Chem.*, 1991, **39,** 709.
17. W. R. Holley, A. Chatterjee, and J. L. Magee, *Radiat. Res.*, 1990, **121,** 161.

CELL INACTIVATION, MUTATION AND DNA DAMAGE INDUCED BY LIGHT IONS: DEPENDENCE ON RADIATION QUALITY

F. Cera, R. Cherubini, M. Dalla Vecchia, S. Favaretto, G. Moschini and P. Tiveron
Laboratori Nazionali di Legnaro-INFN, Via Romea, 4, I-35020 Legnaro, Padova, Italy

M.Belli, F.Ianzini, L. Levati, O.Sapora, M.A.Tabocchini
Istituto Superiore di Sanità and INFN-Sezione Sanità, Viale Regina Elena 299, I-00161 Roma, Italy

G.Simone
Istituto FRAE-CNR, Via P.Gobetti 101, I-40129 Bologna and INFN-Sezione Sanità, Viale Regina Elena 299, I-00161 Roma, Italy

1 INTRODUCTION

Over the recent years an increasing interest in radiation biology research is being devoted to the attempt of relating the initial energy deposition events to the observed biological effects. It is common knowledge that each individual charged particle travelling a material produces, along its path, a track of free electrons and ionizations characterized by a proper diameter and a local energy deposition, which mainly depend on the specific energy of the particle. Accelerated charged particles offer unique opportunities for experimental studies on cellular and molecular effects in cultured cells as a function of 'radiation quality'. By choosing particles of different mass, charge and energy, the macroscopic and microscopic distribution of energy deposition can be varied over a wide range and, therefore, correlated to different biological end points (cell killing; mutation induction; induction of DNA strand breaks; chromosomal aberration; etc.). Increased knowledge of the biological effects induced by radiations can provides constraints for identifying the radiation physical characteristics relevant to the biological action and for testing and validating proper biophysical models.

We have undertaken since many years a systematic comparative analysis of the biological effectiveness of different charged particles as a function of LET in inducing cell inactivation, *hprt* mutation and DNA dsb (considered as crucial lesions) in V79-753B Chinese hamster cells.

In our early works[1,2] we have shown, giving the first experimental evidence, that in the LET range 7 - 38 keV/μm low-energy protons are more effective than alpha particles at the same LET for inactivation and *hprt* mutation induction in V79 cells. Such results have subsequently been confirmed by independent experiments carried out at different European laboratories[3-5].

Parallel experiments on the induction of DNA damage, namely double-strand breaks (dsb), not gave, in contrast, any indication of difference in effectiveness for the different radiation considered (X- , gamma-rays, protons and alpha particles)[6].

Subsequently, comparing inactivation and mutation induced by protons and deuterons of the same LET, we have found that deuterons appear less effective than protons for the same biological end-points, at LET values lower than 30 keV/μm [7]. These findings result in contrast with very recently published data[5].

To extend our study to other heavier ions and to confirm the RBE anomalies found for the hydrogen isotopes we have recently undertaken systematic comparative experiments by using helium-3 and helium-4 ions in the LET range 40-150 keV/μm.

In the present work a progress report of the results obtained for V79 cell inactivation, *hprt* mutation induction and DNA damage is presented, together with those previously obtained for protons and deuterons for an overall direct (in-house) comparison.

2 MATERIALS AND METHODS

2.1 Cell Inactivation and Mutation Induction

Log-phase Chinese hamster V79-753B cells were irradiated as monolayer, through the mylar foil used as basis of the especially designed Petri dishes. Irradiations with 7-150 keV/μm monoenergetic light ion (protons, deuterons,$^3He^{++}$ and $^4He^{++}$) beams in air have been performed at the radiobiological facility set up at the 7MV Van de Graaff CN accelerator of the INFN-LNL (Legnaro - Padova, Italy)[8]. Experimental conditions concerning irradiation facility, beam dosimetry, inactivation and mutagenesis tests and cell culture methods have been described in detail elsewhere[1,2]. Primary energies of $^3He^{++}$ and $^4He^{++}$ ions were chosen in such a way to cover a wide LET range in order to match, where possible, the already investigated proton/deuteron LET values. The dose response curves were obtained in the range of 0.5-4.0 Gy using a dose-rate of about 1Gy/min. X-rays (200kV, 15mA with 0.2 mm Cu filter) were used as reference radiation.

2.2 DNA Damage

Log-phase Chinese hamster V79-753B cells, labelled with ^{14}C-thymidine, were irradiated as monolayer with monoenergetic protons, $^3He^{++}$ and $^4He^{++}$ ions in the LET range 11-104 keV/μm, at 0°C and a dose-rate of 20 Gy/min. Dose response curves were obtained in the dose range 5-120 Gy. X-rays and γ-rays (^{60}Co) were used as reference radiations.

3 RESULTS AND DISCUSSION

3.1 Cell Inactivation and Mutation Induction

The beam characteristics and the biological effectiveness (expressed as α/α_X) as well as cross section for the survival and mutation induction curves are reported in Table 1.

Table 1 *Beam characteristics, RBE and cross section for the biological end points studied*

	BEAM CHARACTERISTICS			INACTIVATION		MUTATION		DSB DAMAGE		
	Energy at cell midplane (3μm) (MeV)	LET (keV/μm)	Range (μm)	$RBE_{inactivation}$ α/α_x±SE	Cross section/$_{in}$ δ±SE (μm²)	$RBE_{mutation}$ α/α_x±SE	Cross section/$_{mut}$ δ±SE (x10⁻⁵μm²)	RBE_{DSB} $\alpha/\alpha_{r.r.}$ (reference radiation) ±SE	Lethal lesions/DSB ±SE	Mutagenic lesions/DSB ±SE
PROTON	0.56	37.8	9.8	4.4±0.4	3.4±0.1	-	-	-	-	-
	0.64	34.6	11.8	5.1±0.5	3.6±0.1	-	-	-	-	-
	0.76	30.5	15.7	5.8±0.6	3.6±0.1	5.9±0.9	17.3±0.7	0.87±0.15° 0.86±0.05°	0.060±0.001 0.070±0.003	0.333±0.021 0.333±0.020
	1.41	20.0	42.9	3.7±0.4	1.5±0.1	4.3±0.7	8.2±0.3	0.84±0.09	0.045±0.006	0.246±0.027
	3.20	10.9	170.6	2.9±0.4	0.6±0.1	3.5±0.5	3.7±0.1	1.24±0.10* 0.93±0.17° 1.05±0.06°	- 0.028±0.002 0.028±0.003	- 0.158±0.012 0.160±0.008
	5.00	7.7	370.3	2.4±0.3	0.4±0.1	2.6±0.4	1.9±0.1	-	-	-
DEUTERON	0.62	57.0	8.3	7.3±0.7	8.6±0.2	9.2±1.6	50.5±4.1	-	-	-
	0.80	48.0	11.7	6.4±0.7	6.4±0.3	-	-	-	-	-
	1.06	39.5	17.8	5.8±0.6	4.7±0.2	4.6±0.8	17.4±1.3	-	-	-
	1.51	30.8	30.7	5.1±0.5	3.2±0.1	4.2±0.7	12.5±0.8	-	-	-
	1.90	26.3	44.1	4.1±0.4	2.2±0.1	-	-	-	-	-
	3.16	18.4	103.1	2.2±0.4	0.8±0.1	2.6±0.4	4.6±0.2	-	-	-
	4.87	13.4	213.8	1.9±0.3	0.5±0.1	-	-	-	-	-
3HE	6.41	58.6	66.7	6.9±0.7	8.4±0.2	-	-	1.21±0.24°	0.052±0.006	-
	7.73	51.1	91.4	6.0±0.6	6.3±0.2	-	-	0.97±0.08*	-	-
	8.99	45.7	117.9	5.7±0.6	5.4±0.2	-	-	-	-	-
	10.21	41.5	146.1	5.2±0.5	4.4±0.1	-	-	0.98±0.19°	0.048±0.005	-
4HE	2.16	150.5	11.8	3.3±0.3	10.1±0.6	-	-	-	-	-
	3.79	104.2	25.2	6.3±0.6	13.6±0.5	-	-	1.03±0.09*	-	-
	6.17	74.0	52.8	4.8±0.5	6.9±0.4	-	-	-	-	-
	8.87	56.8	94.8	4.7±0.5	5.4±0.3	-	-	-	-	-
	10.12	51.5	118.2	4.7±0.5	5.0±0.3	7.5±1.3	37.1±2.5	-	-	-

* Costant Field Gel Electrophoresis, CFGE (damage evaluated as Fraction of Activity Released from the well)

° Low speed sedimentation tecnique

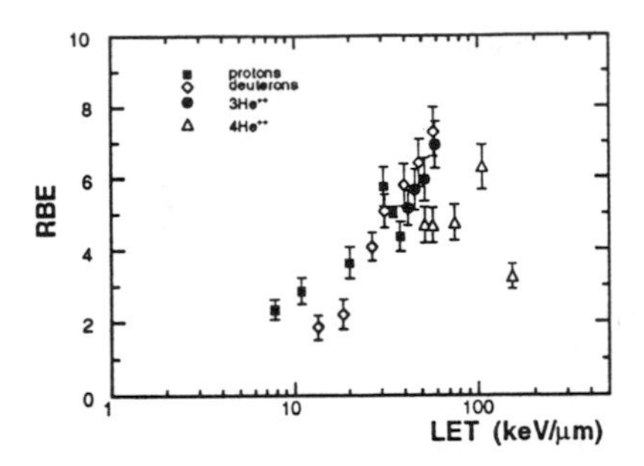

Figure 1 *RBE-LET relationship for cell inactivation induced by protons, deuterons and He ions*

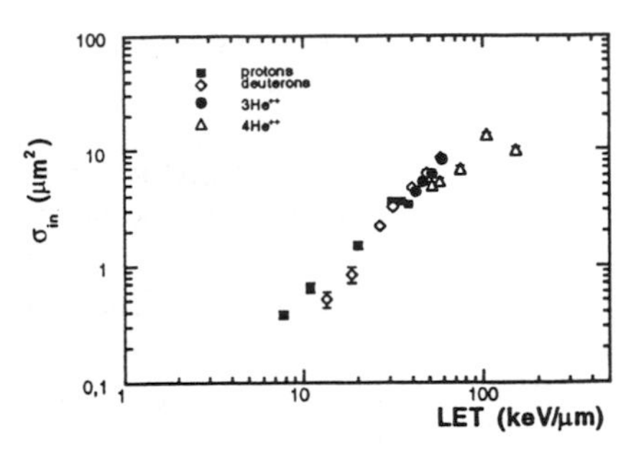

Figure 2 *Inactivation cross section as a function of LET for protons, deuterons and He ions*

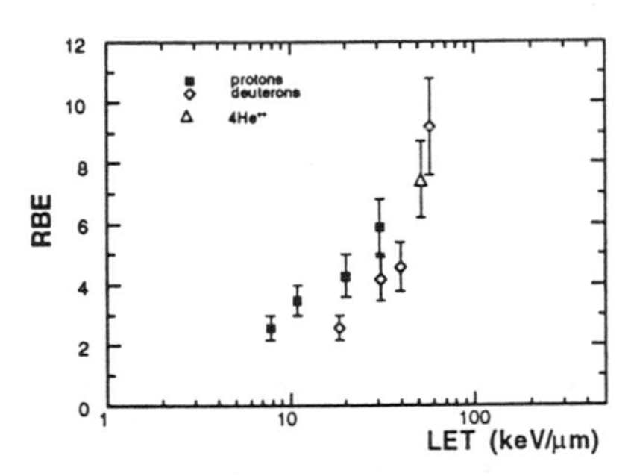

Figure 3 *RBE-LET relationship for hprt mutation induction induced by protons, deuterons and 4He++ ions*

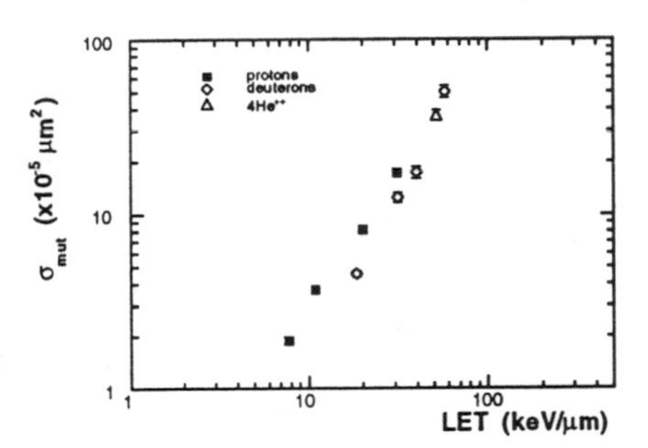

Figure 4 *hprt mutation cross section as a function of LET for protons, deuterons and 4He++ ions*

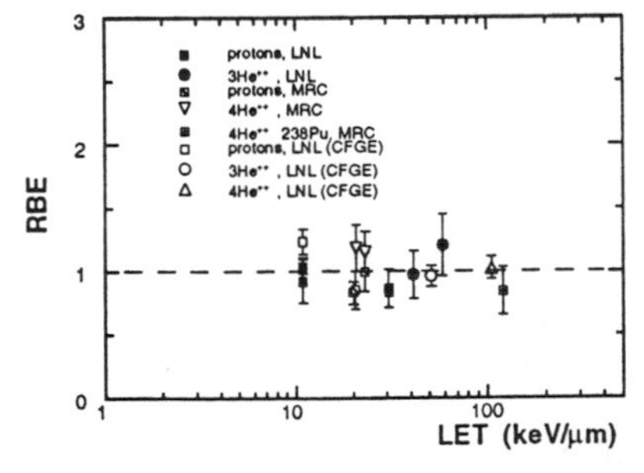

Figure 5 *RBE-LET relationship for dsb production induced by protons and He ions*

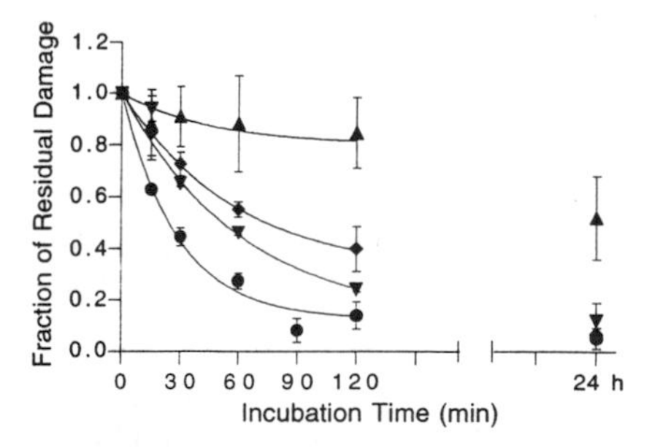

Figure 6 *Rejoining kinetics of DNA dsb induced after irradiation with (•) γ-rays, (▾) 11keV/μm protons, (♦) 52 and (▲) 104 keV/μm 4He++ ions*

Figures 1 and 2 show the RBE-LET and σ_{in} -LET relationships, respectively, for cell inactivation. It appears that the effectiveness for cell inactivation induced by $^{3}He^{++}$ ions in the LET range 40-60 keV/μm increases with the LET, while for $^{4}He^{++}$ ions rises as the LET increases up to about 100 keV/μm and then decreases, in agreement with literature data. Moreover, a different biological effectiveness can be observed between $^{3}He^{++}$ and $^{4}He^{++}$ ions: the RBE values for $^{4}He^{++}$ ions are in fact lower than those obtained for $^{3}He^{++}$ ions of comparable LET. These findings parallel what previously reported by us for protons and deuterons, but contrast with the data recently found by Folkard et al.[5].

The preliminary results for *hprt* mutation induced by 51 keV/μm $^{4}He^{++}$ is reported in Figures 3 and 4 along with the proton and deuteron data previously found[9]. Further experiments are planned to confirm and to extend the studies to higher LET values and to $^{3}He^{++}$ions.

The results here reported indicate that careful and more sophisticated particle track structure description would be necessary for a better understanding of the radiation action in biological matter for the identification of adequate parameters representative of the "radiation quality".

3.2 DNA Damage

Linear dose response curves for initial yield of DNA dsb were found for protons, $^{3}He^{++}$ and $^{4}He^{++}$ as well as X- and γ-rays. RBE values obtained with low speed sedimentation technique[10] together with those obtained with Constant Field Gel Electrophoresis (5%) are reported in Table 1 and Figure 5. It appears that the RBE produced by light ions with LET values up to about 100 keV/μm is not significantly different than unit, independently on the dsb detection technique used.

Figure 6 shows the rejoining kinetics of DNA dsb induced by γ-rays, proton and $^{4}He^{++}$ ions after exposure of V79 cell to a dose of 40 Gy. It appears that there are significant differences in the kinetics depending on both radiation type and LET. Moreover, the amount of dsb left unrejoined after 120 min incubation increases with LET, paralleling what already found with sedimentation technique for 11 and 31 keV/μm protons[10].
Further experiment are planned to extend the investigation to the other ions and LET values.

The results reported here give indirect experimental evidence that changing the type of ionizing radiation, from sparsely to densely, and increasing the LET there is an increase in the "complexity" of the induced damage. The lesion complexity affects the repair capability and, as a consequence, the final cellular effects. All this underlines a "damage quality" dependent on the "radiation quality".

References

1. M. Belli et. al, *Intern. J. Radiat. Biol.*, 1989, **55**, 93.
   M. Belli et al., *Intern. J. Radiat. Biol.*, 1991, **59**, 459.
2. M. Belli et al., *Intern. J. Radiat. Biol.*, 1993, **63**, 331.
3. M. Folkard et al., *Intern. J. of Radiat.. Biol.*, 1989, **56**, 221.
4. D.T. Goodhead et al., *Intern. J. of Radiat. Biol.*, 1992, **61**, 611.
5. M. Folkard et al., *Intern. J. of Radiat. Biol.*, 1996, **69**, 729.
6. M. Belli et al., *Intern. J. Radiat. Biol.*, 1994, **65**, 511.
7. M. Belli et al., *Radiat. Prot. Dosim.*, 1994, **52**, 305.
8. M. Belli et al., *Nucl. Instr. and Meth in Phys. Res.*, 1987, **A256**, 576.
9. R. Cherubini et al., Proc. of the 5th Workshop on Heavy Charged particles in Biology and Medicine, Darmstadt, Germany,1995. GSI-Report-95-10, ISSN0171-4556,(1995)73.
10. G. Simone et al., Proceedings of 10th ICRR, Wurzburg, Germany, (1995), *in press.*

THE SEPARATION OF SPATIAL AND TEMPORAL EFFECTS OF HIGH LET RADIATION

D. L. Stevens , S. J. Marsden, M. A. Hill, I. C. E. Turcu[*] , R. Allott[*] and D. T. Goodhead

MRC Radiation and Genome Stability Unit, Harwell, Didcot, Oxon, OX11 0RD, UK.
[*] Rutherford Appleton Laboratory, Chilton Didcot, Oxon, OX11 0QX, UK

1 INTRODUCTION

The question that this research aims to address is are radiobiological differences between high and low LET radiations due entirely to spatial differences in track structure, as is usually assumed, or are they due in some instances to temporal differences? The development of a very high intensity pulsed X-ray source has now made such investigations possible on a picosecond timescale.

High-LET alpha-particles traverse mammalian cells on timescales of the order of picoseconds, depositing a dose of the order of 0.2 Gy in a linear heavily ionising track, with little lateral spread due to delta rays. Single tracks are generally biologically effective, showing the characteristic quantitative and qualitative features of high-LET radiation with little or no increase in effectiveness per particle due to inter-track effects for multiple traversals. Thus, the physical conditions that determine the biological outcome are established before the timespan of the free-radical chemistry (10^{-12} to 10^{-9}) seconds or the subsequent cellular biochemistry.

By contrast, low LET radiations such as X - and γ-rays deposit small doses (of the order of mGy) per electron-track event, distributed over the whole cell. Biological effects of single tracks are usually not detectable. Larger doses show characteristic features of low-LET radiations, often including curvature of dose-response and dose-rate dependence from which substantial inter-track action can be inferred. In conventional low-LET irradiations doses equivalent to those received from single high LET particles are delivered over times that are long compared to the free radical chemistry and while biochemical/biological processes are already occurring.[1] Nevertheless, it is universally assumed that the characteristic responses from high-LET radiation are due to the spatial aspects of their track structure rather than these massive temporal differences.

With the High Brightness Picosecond X-ray Source[2] developed at the Rutherford Appleton Laboratory a dose equivalent to one alpha-particle per cell can be given in approximately 10 ps. So the high brightness picosecond X-ray source is unique in that it can irradiate samples with the usual spatial aspects of low-LET radiations but in a time period approaching that of high-LET radiations. If, surprisingly, biological differences dependent on LET were found to be due to the differing time scales of energy deposition, rather than spatial differences, it would radically alter our understanding of the mechanisms of radiation action and have implications for practical problems including radiotherapy.[3]

2 RADIATION SOURCE

The Rutherford Appleton Laboratory X-ray source has been described fully by Turcu *et al*[2] and Seifamirhosseini[4] and so only a brief description will be given here. 7 ps 248 nm KrF laser pulses were focussed onto a moving Copper tape target in a chamber filled with Helium at atmospheric pressure. Copper plasma L-shell X-rays, of energy 1.0 to 1.4 keV, were produced. UV and lower energy X-rays, also produced by the plasma, were removed from the beam with a Aluminium/parylen-CH membrane. A silicon PIN X-ray diode positioned adjacent to the irradiated dish was used for dosimetric measurements. The source produced 7 ps X-ray pulses delivering a surface dose to the cells of 0.2 Gy at a rate of 20 Hz.

3 RESULTS AND DISCUSSION

As a first test of the capabilities of the source and the irradiation methodologies, experiments have been carried out to measure the oxygen enhancement ratio (OER) for cell inactivation by picosecond X-ray pulses. It is well established that mammalian cells are less sensitive when irradiated under hypoxic conditions, and that the dose-modifying factor (i.e. OER) is approximately 2-3 for conventional X-rays and γ-rays. This is believed to be due to the absence of the oxygen that normally competes with the fast endogenous chemical repair and creates DNA damage substrates for subsequent biochemical repair and misrepair. For low velocity high-LET particles the OER approaches unity and this is commonly explained as being due to the initial molecular damage being too severe for adequate endogenous repair even in the absence of oxygen fixation . The severity may be ascribed to the local spatial properties, ionisation clustering, of the high LET tracks on the scale of DNA.

Therefore OER has a clear dependence on radiation quality, and provides one system for testing the question of spatial versus temporal differences. The conventional expectation for this endpoint is clearly that the OER for the picosecond X-rays would be similar to that for X-rays delivered over usual timescales of seconds to minutes, because the radiation timescale should not affect the local fast sulphydryl/oxygen chemistry around the DNA.

Three survival experiment were carried out on V79 cells under normal gassing conditions (air with 5% CO_2), and four under hypoxic conditions. For the latter, the oxygen levels in cells, surrounding biological medium and gas flowing in sealed biological dishes was minimised by continuously flushing with 5%CO_2 in nitrogen. The irradiation assembly as used by R Meldrum et al[5] was adapted to hold our sealed irradiation dishes and to allow gassing to take place whilst over the irradiation chamber.

The data obtained show that the picosecond X-ray pulses are much less effective at cell inactivation under hypoxic conditions and lead to an OER of about 2.5 (Figure 1).This value of OER is comparable to that obtained in our laboratory using similar biological conditions for 1.5 keV ultrasoft X-rays delivered over minutes from a continuous cold cathode source (2.0 ± 0.1) The values are consistent also with those reported by Raju et al [6] that showed only a slightly reduced OER for low energy X-rays. By contrast, we have obtained an OER of unity for 3.2 MeV alpha particles using similar biological methods.

These results suggest that, in accordance with conventional expectations, it is the spatial distribution of energy deposition rather than any temporal aspects of the radiation that distinguish between high and low LET characteristics of protection by hypoxia.

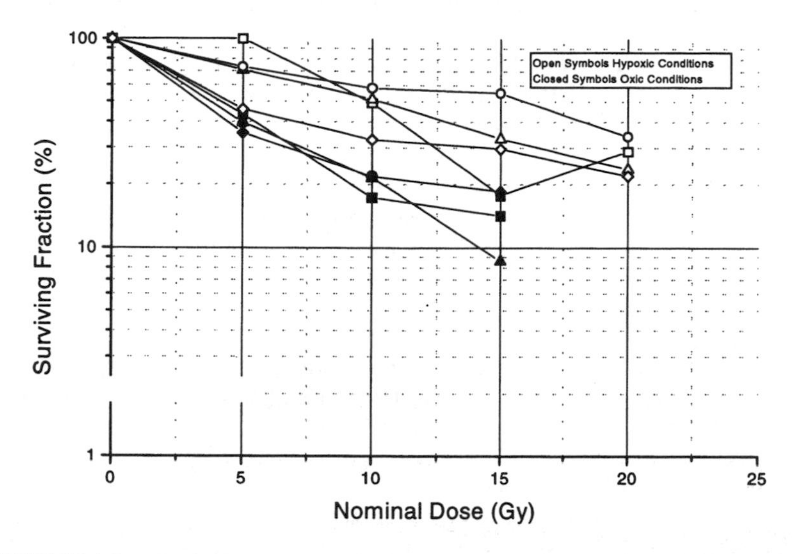

FIGURE 1 Survival of V79-4 cells irradiated with multiple soft X-ray pulses of approximately 0.2 Gy each, under hypoxic and oxic conditions.

4 CONCLUSION

These experiments have demonstrated that the picosecond X-ray source can be applied in single-pulse mode to quantitative experiments with monolayers of mammalian cells. The first such experiment supports the conventional expectation that the disappearance of an oxygen effect for high-LET radiation particles is due to their spatial properties rather than the very short time-scale of each traversal. For other effects, expectations may be less clear cut, although current interpretations are based almost entirely on spatial properties. Future work intends to extend the test to other biological effects. Particularly intriguing may be induced genomic instability[7] and cell signalling processes triggered by DNA or non-DNA targets.

5 ACKNOWLEDGEMENTS

The study was partly funded by the Commission of the European Union (Contract No F14P-CT95-0011).

References

1. P O'Neill, EM Fielden 'Molecular aspects of DNA damage and its modification', In 'Radiation Carcinogenesis and DNA Alterations', Editors FJ Burns et al (eds.), Plenum Press, New York., 1986

2 I.C.E Turcu, I.N Ross, P. Trenda, C.W Wharton, R.A. Meldrum, H. Daido, M.S. Schulz, P. Fluck, A.G. Michette, A.P. Juna, J.R. Maldonado, H. Shields G.J. Tallents, L. Dwivedi, J. Krishnan, D.L. Stevens, T.J. Jenner, D. Battani, H. Goodson,. *SPIE,* **2015**, 243.

3. DT Goodhead. *Health Physics*,. 1988, **55 No2**, 231

4. A. Seifamirhosseini *M.Sc. Thesis,* Kings College London, 1994

5. RA Meldrum, J Edgerton, W Meaking, CW Wharton, A Damerell. 'The kinetics and mechanism of the repair of soft x-ray damage in mammalian cellular DNA'. Science and Engineering Research Council, Central Laser Facility, Annual Report, 1992, p.200

6. M.R. Raju,S.G. Carpenter, J.J. Chemielewski, M.E. Schillaci, M.E. Wilder, J.P. Freyer, N.F. Johnson, P.L.Schor, and R.J. Sebring, *Radiation Research*, 1987,**110**,396

7. M.A.Kadhim, D.A. Macdonald, D.T. Goodhead, S.A. Lorimore, S.J. Marsden & E.G. Wright, *Nature*, 1992, **355**, 738.

THE RBE OF ACCELERATED NITROGEN-IONS FOR APOPTOSIS IN HUMAN PERIPHERAL LYMPHOCYTES EXPOSED *IN VITRO*

A. E. Meijer [a,b], U.-S. E. Kronqvist [c], R. Lewensohn [b,c] and M. Harms-Ringdahl [a,b]

[a]Dep. of Radiobiol., Stockholm University, S-106 91 Stockholm, Sweden
[b]Swedish Radiation Protection Inst., S-171 16 Stockholm, Sweden
[c]Unit of Med. Radiobiol., Dep. of Oncol./Pathol., Karolinska Inst., S-171 76 Stockholm, Sweden

1 INTRODUCTION

For a better understanding of the mechanisms involved in radiation induced apoptosis studies on different radiation qualities and doses are required.

In 1978 Hedges and Hornsey studied the RBE of 7 MeV neutrons with pycnotic cells as an end point of the radiation effect, in untransformed lymphocytes, and found an RBE of 1.0[1]. Also, they investigated the response of transformed lymphocytes involving changes in morphology and DNA-synthesis. A biphasic dose response was seen for both tests, indicating at least two sub-populations of lymphocytes giving RBE values in the range 1.95 to 2.45[1].

A study on the induction of apoptotic fragments in mouse thymocytes exposed to fast neutrons (4 or 62.5 MeV) showed an RBE of 1.0 six hours after irradiation. In this report the authors suggest that the principal site of damage is largely independent of LET[2].

In a series of experiments using fast neutrons (14 - 600 MeV), RBE values between 3 to 4 were obtained from studies of intestinal crypts[3]. In these experiments it was established that the time course of appearance and disappearance of apoptotic fragments was similar after high- and low-LET radiation, and also that the same plateau level was reached.

Although, there already are a few reports on the induction of apoptosis after high-LET irradiation, further investigations are needed for better understanding of the underlying mechanisms and the kinetics of the apoptotic processes. Therefore, we investigated the effect of radiation induced apoptosis in G_0-phase human peripheral lymphocytes exposed to radiation of different LET. In addition we studied the induction of apoptosis at different post-irradiation times.

2 MATERIAL AND METHODS

Peripheral lymphocytes from four healthy non-smoking individuals were obtained from freshly collected buffy coats after separation in Ficoll-paque. The cells were resuspended in basic medium (BM, containing 40% foetal calf serum) and kept in G_0-phase during the whole experiment.

The lymphocytes were irradiated (0, 1, 2 or 3 Gy) in BM at room temperature either with gamma-rays from a ^{137}Cs-source (0.6 Gy/min), or by accelerated nitrogen-ions

(140 keV/μm, 10-20 Gy/min) at the The Svedberg Laboratory, Uppsala, Sweden. Due to the small lateral size and penetration depth of the nitrogen-ion beam, specially designed plastic chambers were used for the high-LET irradiation. After irradiation the lymphocytes were resuspended in fresh BM ($5x10^6$ cells/ml), transferred to 12 well-plates ($10x10^6$ cells/well), and incubated (37°C, 5% CO_2, 80% humidity) up to 72 hours.

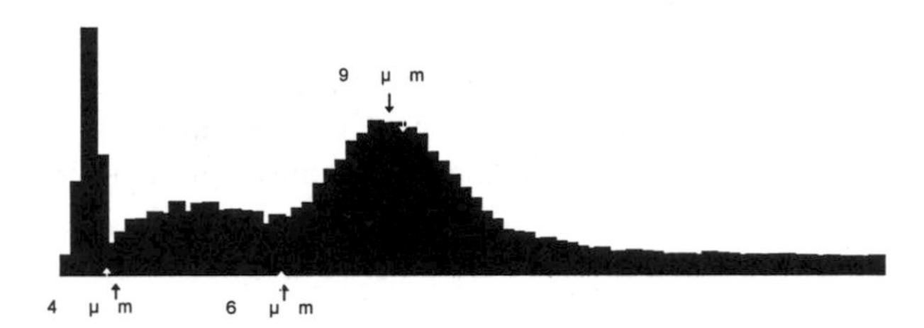

Figure 1 *Cell size measurements of G_0-phase lymphocytes. Apoptotic fragments are considered to be < 4 μm, apoptotic bodies 4 - 6 μm and cells of normal size >6 μm in diameter.*

Analysis of apoptotic cells, bodies, fragments and DNA-"ladders" were performed at different times after irradiation (0, 24, 48 and 72 hours) using three different techniques, 1) living, apoptotic and necrotic cells were scored by measuring morphological changes after dual staining in acridine orange (AO) and ethidium homodimer-1 (EthD-1), 2) normal cells, apoptotic bodies and fragments were scored (Figure 1), using a Coulter counter with a Channelyzer, measuring the cell size and distributions, and 3) DNA-"ladders" where DNA was prepared from $5\text{-}10^6$ cells by the salt solution method[4] and analysed by gel electrophoresis using 1.4% agarose gels.

3 RESULTS

The RBE values, calculated from the dose response data obtained from morphological and cell size studies, are presented in Table 1. These values were calculated from the linear component of the estimated dose response curves as illustrated in Figure 2.

The percentage remaining living cells, scored after morphological studies are presented in Table 2. These results corresponds very well to the detected DNA-"ladders" in Table 3.

Table 1 *RBE with standard errors for radiation induced apoptosis*

Time after irradiation (hours)	*Apoptotic cells morphological analysis*	*Apoptotic bodies cell size analysis*
0	1.0	1.0
24	2.0 ± 1.26	2.3 ± 0.31
48	2.7 ± 1.78	3.0 ± 0.53
72	1.3 ± 0.18	1.9 ± 0.34

Table 2 *Percentage remaining living cells with standard errors at different post-irradiation times*

Time after irradiation (hours)	*0 Gy*		*1 Gy*		*2 Gy*		*3 Gy*	
	Low LET	*High LET*	*Low LET*	*High LET*	*Low LET*	*High LET*	*Low LET*	*High LET*
0	100	100	100	100	100	100	100	100
24	80 ± 2	80 ± 2	69 ± 4	51 ± 6	62 ± 8	50 ± 4	63 ± 6	54 ± 8
48	80 ± 2	83 ± 6	60 ± 5	35 ± 3	55 ± 2	27 ± 1	46 ± 8	14 ± 4
72	80 ± 3	77 ± 1	46 ± 3	28 ± 6	35 ± 4	15 ± 4	30 ± 4	10 ± 3

Table 3 *Detectable DNA-"ladders" at different post-irradiation times and doses*

Time after irradiation (hours)	*Low-LET*	*High-LET*
0	-	-
24	-	2 and 3 Gy
48	3 Gy	1, 2 and 3 Gy
72	1, 2 and 3 Gy	1, 2 and 3 Gy

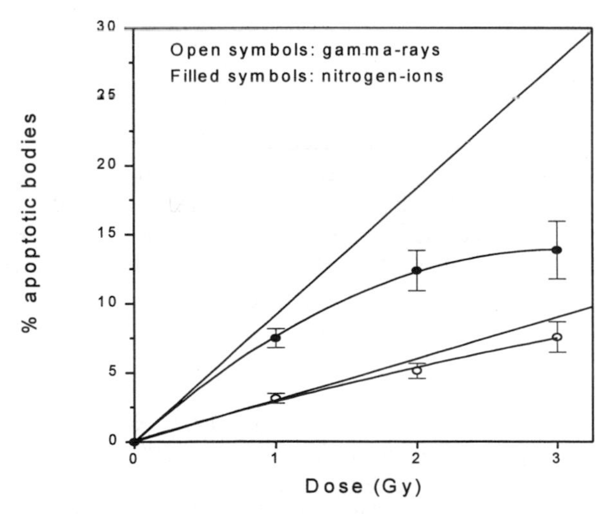

Figure 2 *Dose response curves on the formation of apoptotic bodies 48 hours after irradiation*

4 DISCUSSION

Our results show a significantly higher RBE for radiation induced apoptosis after high-LET as compared to low-LET irradiation at 24, 48 and 72 hours post-irradiation time. The results obtained after morphological analysis of apoptotic cells were in agreement with the results obtained from cell size analysis of apoptotic bodies, although these techniques detect different stages in the apoptotic process (Table 1).

The results we got are not in agreement with the RBE of 1.0 obtained after exposure to 7 MeV neutrons for interphase death in human peripheral G_0-phase lymphocytes[1]. However, in the study by Hedges and Hornsey, the results were obtained after exposures up to 15 Gy. According to our experience necrotic cell death will dominate in this cell system after exposures above 5 Gy.

In mouse thymocytes exposed to fast neutrons an RBE of 1.0 was obtained for the induction of apoptotic fragments six hours after irradiation[2]. This difference, as compared to our own results, may depend on a number of factors referring to the complexity of the radiation induced apoptotic response[5]. However, the short post-irradiation time before analysis of apoptotic fragments in the study by Warenius and Down may be the main reason for the RBE of 1.0. Analysis at later times might give a higher RBE even in this system.

There may be several targets for radiation induced apoptosis. Primary DNA-double strand breaks have been suggested to be the cause for induction of apoptosis in lymphoblastoid cell lines[6]. Also, membranes are likely to be involved in the total cellular response. With the staining technique applied in our study we also got information on membrane damage of apoptotic cells where our preliminary data show an increased incidence of membrane damage in apoptotic cells after high-LET as compared to low-LET irradiation (to be published).

Acknowledgements

We thank Margareta Lagerberg and Lena Rödin for excellent technical assistance. This work was financially supported by the Swedish Radiation Protection Institute, European Commission Program for Nuclear Fission Safety Contract no. FI 4PCT950001 and the Stockholm Cancer Society no. 95:120 and 96:153. Annelie E. Meijer was awarded a Young Investigators Award to attend the 12th Symposium on Microdosimetry.

References

1. M. J. Hedges and S. Hornsey, *Int. J. Radiat. Biol.*, 1978, **33**, 291.
2. H. M. Warenius and J. D. Down, *Int. J. Radiat. Biol.*, 1995, **68**, 625.
3. J. H. Hendry, C. S. Potten and A. Meritt, *Radiat. Environ. Biophys.*, 1995, **34**, 59.
4. L.-H. Zhang and D. Jenssen, *Mut. Res.*, 1991, **263**, 151.
5. M. Harms-Ringdahl, P. Nicotera and I.R. Radford, *Mut. Res.*, 1996, **366**, 163.
6. I. R. Radford, I.R., *Int. J. Radiat. Biol.*, 1986, **49**, 611.

A VERSATILE MAMMALIAN CELL IRRADIATION RIG FOR LOW DOSE RATE ULTRASOFT X-RAY STUDIES

M. A. Hill, D. L. Stevens, D. A. Bance, and D. T. Goodhead.

MRC Radiation and Genome Stability Unit,
Harwell, Didcot,
Oxon, OX11 0RD,
UK.

1 ABSTRACT

The design of a mammalian cell irradiation rig is described in which up to 20 dishes can simultaneously receive individually prescribed doses with irradiation times variable from minutes to days. Temperature and gassing arrangements allow the cells to be kept at a constant 37°C or 5°C under aerobic or anaerobic conditions. The rig is designed to be used for both ultrasoft x-ray and alpha irradiations.

In its first application V79-4 cells were irradiated using aluminium K ultrasoft x-rays. An increase in survival was observed for cells irradiated at a reduced dose rate compared to acute doses of about 8Gy but little increase was seen at 4Gy.

2 INTRODUCTION

Ultrasoft x-rays 0.1 - 5 keV provide a unique tool for investigating the energy and spatial requirements for specific types of radiation damage. They produce isolated tracks of electrons with small well defined low energies and very short tracks, comparable to that of critical structures of the cell, such as DNA, nucleosomes and chromatin fibres. Many publications have shown these tracks to be efficient at inducing biological damage in acute irradiations[1]. The electrons produced by ultrasoft x-rays are similar to the numerous electron ‘track ends’ produced in the slowing down spectrum of more energetic electrons produced by low LET radiations such as ‘hard’ x-rays; they contribute approximately 30% to the absorbed dose.

Aluminium K characteristic x-rays (1.487keV) interact mainly with the oxygen atoms within the cell producing photo- and Auger-electrons (956eV and 516eV respectively) with a combined range of <70nm. The RBE for inactivation of V79-4 cells compared to ^{60}Co γ-rays is ~2.2 at 10% survival[2].

Irradiation at low dose rates will separate these isolated tracks in time so reducing the possibility of interactions between separate tracks or of saturation of repair processes. Thus the capabilities of single, isolated, short tracks should be revealed.

3 THE IRRADIATION RIG

The irradiation rig (figure 1) can be used with various irradiation facilities including the MRC cold-cathode ultrasoft x-ray source[1], the MRC plutonium 238 alpha source[3] and

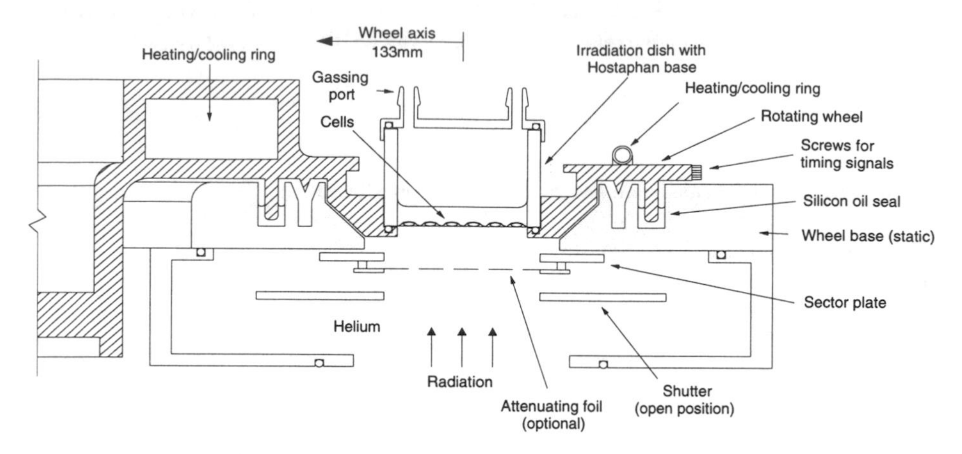

Figure 1 *Schematic of irradiation rig.*

the Rutherford Appleton Laboratory's high brightness picosecond laser induced ultrasoft x-ray source[4]. Up to 20 Hostaphan-based dishes containing cells can be mounted in the horizontal wheel. This can be used in static mode to irradiate dishes individually or rotated at 3 revs per minute with the dose to individual dishes determined by the opening and closing sequence of the shutter. Temperature and gassing facilities allow optimum growth conditions to be maintained during protracted irradiations. Alternatively the temperature can be reduced to $5^{O}C$ or below to stop repair processes or cells may be gassed with $N_2 + CO_2$ to study the effects of hypoxia.

The sector plate incorporates both an aperture larger than the dish for static irradiations and an open sector subtending 12° at the wheel centre for rotating irradiations. The sector defines the edge of the exit beam and ensures that all positions on the irradiation dishes are exposed for an identical length of time. A choice of smaller sector plates are available down to 1.5^{O}. For ultrasoft x-ray experiments the instantaneous dose rate to the cells can be reduced by the addition of an attenuating foil beneath the sector.

The shutter can be computer controlled to vary the dose received by individual dishes for the rotating wheel. Timing signals for the shutter sequence are from cap head screws on the perimeter of the wheel detected by an inductive sensor. Dummy dishes are used at positions coinciding with the shutter opening or closing.

Dish temperature is controlled using a Churchill chiller thermo circulator pumping water plus antifreeze mix around the heating/cooling rings on the wheel either side of the circle of dishes. The connection to the rotating wheel is made using multiple O-ring seals to a central stationary boss.

The dishes can be flushed with air + CO_2 or $N_2 + CO_2$ during the experiment. The gas is filtered and passed through a bubble pot containing distilled water, preventing the dishes from drying out. Connections to the wheel are made via rotating seals on the central boss and passed serially from dish to dish through ports in the dish lids.

An air path would significantly attenuate ultrasoft x-rays or significantly reduce the range of alpha particles. For this reason helium is flushed (~200ml/min) between the irradiation source and the base of the dishes. The gas seal between the rotating wheel and the base is maintained by silicon oil wells just inside and outside the circle of dishes.

4 EXPERIMENTAL

Al_K ultrasoft x-rays (1.487keV) were produced by the MRC cold cathode discharge tube[2]. Typical dose rates of 2.25Gy/min at the cell surface were obtained with a cathode potential of 3kV and a discharge current of 1mA. The bremsstrahlung background was <1% at the sample position. Dosimetry measurements were made before and after each set of irradiations, with an air filled ion chamber positioned 1mm behind an empty Hostaphan dish[2].

A comparison was made between acute irradiations (~2.25Gy/min) and extended irradiations (~0.075Gy/min) for given doses. Static acute irradiations were made before and after the extended irradiation to check for variation in the survival response with time. The extended irradiations utilised the rotating wheel with a 12° sector plate and a computer operated shutter controlling the dose given to each dish, including the option for zero dose. Two dishes were irradiated per dose point and the cells pooled.

V79-4 Chinese hamster cells were recovered from liquid N_2 storage and maintained in exponential growth for several days. The cells were replated in Hostaphan based irradiation dishes 40-46 hours before irradiation, forming a monolayer of fully flattened cells. A confocal laser scanning microscope was used to make thickness measurements on live attached cells in the irradiation dishes[5]. An average thickness of 4.5µm was obtained with no change over the duration of the irradiations, during which all the dishes including the controls were kept on the wheel at ~36.5^{O}C and gassed with 95% air 5% CO_2 mixture. On removing from the wheel a 26mm diameter disc at the centre of each dish was cut out and the cells harvested and a colonogenic assay performed.

5 RESULTS AND DISCUSSION

The survival curves obtained for pre-acute, post-acute and extended irradiations of V79 cells with Al_K x-rays is shown in figure 2. Due to the attenuation of the x-rays through the cell, the mean dose received is a factor of 0.74 less than the dose incident on the cell surface, assuming a cell thickness of 4.5µm.

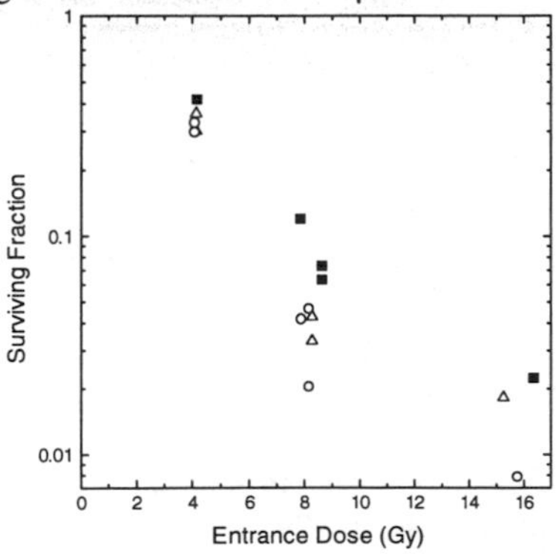

Figure 2 *V79 survival for Al_K irradiation (Δ pre-acute, ■ extended, O post acute).*

Pre and post acute irradiations give similar survival curves, as expected. The extended irradiation data show an increase in the level of survival at intermediate to high doses, but remain rather effective at low doses suggesting single-track action. This is consistent with an earlier split dose experiment[6]. From the limited data, both the acute and extended survival curves appear to exhibit a shoulder. This would indicate the potential for a further increase in survival for extended irradiations if the dose rate were reduced. (The dose points above 15Gy are indicating the start of a tail due to the presence of a small number of clumped cells for that particular experiment, resulting in the partial shielding of a small fraction of the cells.)

Although ultrasoft x-rays are more effective per unit dose than hard x-rays they appear to produce damage characteristic of low LET radiations in general, showing features such as shouldered survival curves and increased survival at low dose rates or fractionated doses. This implies that low energy secondary electrons are largely responsible for the lethality of low LET radiations such as hard x-rays and γ-rays[1].

Assuming a cylindrical nucleus of thickness 4.5μm and area[7] of 134μm^2 the total number of tracks in the nucleus per Gy is 2730. Acute irradiations produce tracks at the rate of 6150 tracks/min and extended irradiations at a rate of 205 tracks/min corresponding to 68 tracks per revolution with a mean nearest neighbour distance of 1.15μm. This distance will increase when lower dose rates are used, further isolating the tiny tracks.

6 SUMMARY

The irradiation rig provides a versatile facility for the irradiation of mammalian cells with ultrasoft x-rays or alpha particles. The ability to change the temperature, gassing and dose rate enables a wide range of experiments to carried out, including investigation of the ability of isolated electron tracks to produce biological damage.

Acknowledgements

The authors would like to thank the mechanical and electronic workshops at the MRC Harwell for their expertise in the construction of the rig. This work was partially supported by EC contract F14P-CT95-0011.

References

1. D. T. Goodhead and H. Nikjoo, *Radiat. Prot. Dosim.*, 1990, **31**, 343.
2. D. T. Goodhead and J. Thacker, *Int. J. Radiat. Biol.,* 1977, **31,** 541.
3. D. T. Goodhead, D. A. Bance, A. Stretch and R. E. Wilkinson, *Int. J. Radiat. Biol.,* 1991, **59**, 195.
4. I. C. E. Turcu et al, *SPIE,* 1994, **2015**, 243.
5. K. M. S. Townsend, A. Stretch, D. L. Stevens and D. T. Goodhead, *Int. J. Radiat. Biol.,* 1990, **58**, 499.
6. D. T. Goodhead, R. Cox and J. Thacker, 'Proc. 7th Symp. on Microdosimetry' Harwood Academic, London, 1980, p. 929.
7. K. M. S. Townsend and S. J. Marsden, *Int. J. Radiat. Biol.*, 1992, **61**, 549.

MICRODOSIMETRY OF CELLS *IN VITRO* IRRADIATED BY ALPHA PARTICLES

W. B. Li W. Z. Zheng X. Zhang Y. F. Gong J. Li D. C. Wu

Institute of Radiation Medicine, Beijing
27 Taiping Road, 100850
Beijing, P. R. China

1 INTRODUCTION

The radiobiological effects and dosimetry to mammalian cells *in vivo* and *in vitro* of alpha particles emitted from ^{239}Pu and ^{238}Pu have been studied earlier in our laboratory, including experimental measurements of the microdosimetric spectra of alpha particles from ^{241}Am, ^{210}Po and ^{239}Pu and microdosimetry calculations with the aid of image analyzer.[1] The awareness increased recently of the potential health effects from radon inhalation and other alpha emitting radionuclides internal deposition in human have attracted our attention to continue previous research work on dose delivered to a single cell from radon progeny and the biological effects induced on the cellular level.[2,3] The project of mechanism of radiation carcinogenesis proposed to study the cellular and molecular carcinogenic and mutagenic effects of alpha emitters demands an alpha particle exposure apparatus to give an accurate measurement energy distribution and even the ion track structure spatial profiles in the biological tissues on the order of DNA dual-helix dimensions. Reviewed the exposure systems used in several laboratories,[4 8] a planar alpha source irradiation system was designed and constructed to meet the further research on different cell systems.[9] In this paper, microdosimetry calculations and other parameters of C3H10T1/2 cells irradiated by ^{238}Pu alpha particles are compared with experimental results

2 MATERIALS AND METHODS

2.1 Exposure System

The major components of exposure system are shown in figure 1. The sources are six sealed electroplated ^{238}Pu on a 45 mm diameter stainless steel disk which can be horizontally rotated round its axial in variable speeds, from 200 to 300 rpm, its activity range from 3.27 to 200 MBq; the honeycomb collimator, which 15 mm in height and 100 mm in diameter, can move circular motion without rotation, its honeycomb diameter is 3 mm and the distance between any two channel centers is 3.2 mm; the shutter disc is operated by a computer controlled step motor that can be programmed for required exposure time; in exposure period, the incubation chamber is filled with 5% CO_2 and kept at 37.3 ℃. An alpha particle traverses a 41.5 mm of He and exit windows with 1.93 μm Hostaphan film into a 2 mm air then passes through the cell culture dish base(1.93 μm Hostaphan film) which its inner diameter is 45 mm before it reaches the cell surface.

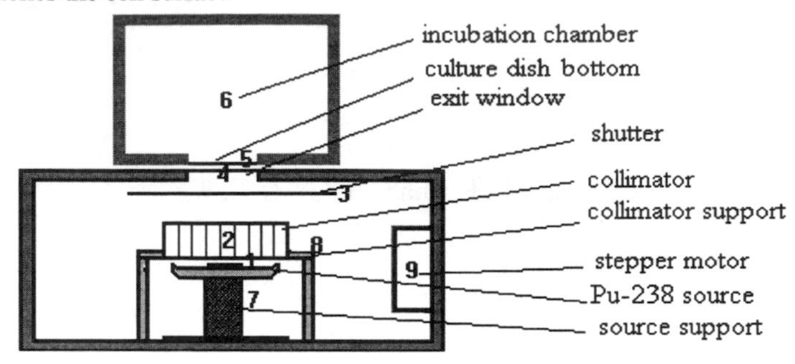

Figure 1 *Schematic diagram of the alpha particle exposure apparatus*

2.2 Theoretical Calculations

2.2.1 Geometry Evaluation. The uniform distribution planar source, activity be A_0, should be equally divided into small enough area which can be considered as a point source, activity a_i. The mean number, *m*, should equal the product of a_i and the solid angle subtended by the nucleus. Selected one cell nucleus in the center of the culture dish for convenient calculation.The local source area that may irradiate the cell nucleus through the collimator channel was evaluated by analytical method, and turned out to be a circular area 8.15 mm^2 right down the nucleus and other 18 small areas surround the source disc center in three different distances.There are 6 equal areas in the same distance; their areas are 5.225, 0.992, 0.083 mm^2 respectively.

2.2.2 Methods Used in Calculation. Microdosimetry calculation methods[10-11] were used to calculate the probability density in specific energy, *f(z)*, in irradiating the cells for one minute and then obtain the mean specific energy, $\bar{z}$, i.e., the desired dose rate, $\dot{D}$, to a single cell nucleus.

$$F\{f(z)\} = \exp[-m(1 - F\{f_1(z)\})] \tag{1}$$

where *F{}* denotes a Fourier transform, $f_1(z)$ is the probability density in specific energy, *m* is the mean number of energy deposition events in the cell nucleus.

$$\dot{D} = \bar{z} = \int_0^\infty z f(z) dz \tag{2}$$

The method developed by Wilson[12] based on Monte Carlo simulation track structure results evaluated by Paretzke[13] was used to calculate the $f_1(z)$. Since the track structure calculation was for liquid H_2O an approximation biological tissues, the He and air layer length should be converted to that in tissues. The microdosimetry computation code, MICRODOSE1, was used to calculate *f(z)*.[14]

3 RESULTS AND DISCUSSIONS

3.1 Results

Converted into tissues, the distances between the 19 small areas sources and the cell nucleus center in the He and air are 15.8, 15.9, 16.0 and 16.2 μm. There is no different on the tissue distance interval 0.1 or 1.0 μm when evaluating *f(z)*,[1] only the single-event density at 16 μm was calculated assumed the cell nucleus was a sphere when calculating *f(z)* and the spectra was shown in figure 2 (a). The experimental and calculation LET in tissues at the distance 16 μm from the cell nucleus are 111 keV. μm^{-1} and 110.6 keV. μm^{-1} respectively. The calculating dose rate is 0.37 Gy.min^{-1} (corrected by the transparency of 0.802) compared with 0.20 Gy.min^{-1} in experiment. Table 1 shows the other valuable parameters both in calculations and experiments. The details of experiments on dosimetry such as energy distribution measured with a silicon surface barrier detector and track-etch, and cellular biological effects have been completed including some microdimetric parameters.[9]

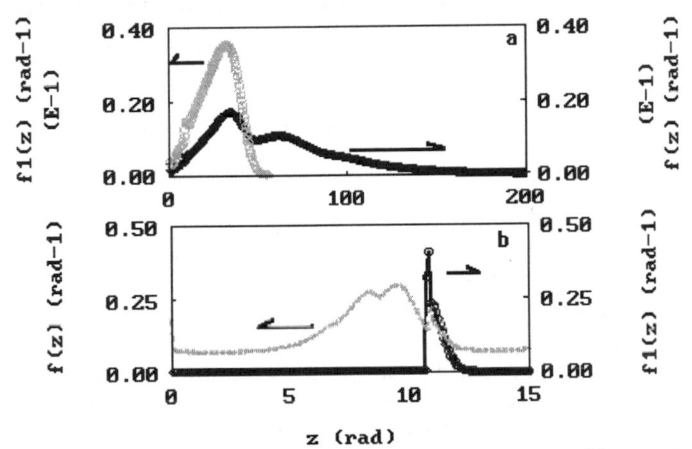

Figure 2 *Spectra for C3H10T1/2 cell nucleus irradiated by 5.25 MeV ^{238}Pu alpha particles. (a) single event density is calculated based on Wilson[12] assuming a sphere shape nucleus; (b) single event density is calculated based on simple Monte Carlo simulation assuming a cylinder shape nucleus(height/diameter = 2.54/7.44)*

The theoretical value 0.37 Gy.min^{-1} is larger than 0.20 Gy.min^{-1} may be due to a presumed sphere cell nucleus instead of a realistic oblate. A simple Monte Carlo simulation on an alpha particle traversal distribution in a cylinder with 2.54 μm height and 7.44 μm diameter which is approximated to the oblate cell nucleus is performed in the basic principle of Borak[15] without considering the secondaries. The calculation of *f(z)* is shown in figure 2 (b). The dose rate based on Monte Carlo calculation is 0.18 Gy.min^{-1}, slightly smaller than experimental results,which perhaps is due to the simple simulation and the approximation of the oblate.

Table 1 *Some Valuable Parameters Obtained by Experiments and Calculations*

parameters	experiment	calculation(sphere/oblate)
no hit probability		0.44/0.18
mean traversals/cell nucleus	5.8	1.3[a] /5.4

a) if corrected, the value will be 3.2(ratio of realistic cell nucleus areas to spherical circular areas)

3.2 Discussions

f(z) for different cell shapes can differ greatly in curve shape and distribution, it will impact the mean specific energy. The approximation of a plated oblate cell to a sphere will contribute error to the mean dose of cell.To obtain more accurate energy distribution and value in specific energy, perhaps not only the information of interactive between radiation with different media before it reaches the cell nucleus but also the energy deposition in different shapes of sites for individual particle and its secondaries should be studied in details. This will certainly prompt us to study ion track structure in nanometer site.

Microdosimetry calculation methods developed by Kellerer[10] and Roesch[11] can resolve the probability density in specific energy in cellular nucleus level and inactivation of cells induced by radiation when combining the probability densities with cellular inactive model replacing D and D_0 by z and z_0 in the case of single event densities for different radiation are available. However, cellular biological effects are consequences of the changes of genes, DNA target model combine with ion track structure can describe the initial physical process of the radiobiological effects for understanding the mechanism of radiation carcinogenesis.[16,17]

ACKNOWLEDGMENT

This work is supported by NSFC of China with contract No. 35900034.

REFERENCES

1. D. C. Wu and C. Q. Ye,'The Injurious effects of Inhaled Plutonium-Collected Papers of Research Theses', Institute of Radiation Medicine, Beijing, 1992.
2. D.C.Wu,J.Shou,'Proc.Tenth ICRR Cong.Vol.1'U.Hagen,et al,Ed,Wurzberg,Germany, 1995, pp85
3. W. B. Li and W. Z. Zheng, *Chinese J. Radiol. Med. Prot.*, 1995, **15**, 229. (in Chinese)
4. W.C.Inkret,Y.Eisen,W.F.Harvey,A.M.Koehler,M.R.Raju,*Radiat.Res.*,1990, **123**, 304.
5. H. Roos and A. M. Kellerer, *Phy. Med. Biol.*,1989, **34**, 1832.
6. D.T.Goodhead, D.A.Bance, A.Stretch, R.E.Wilkinson,*Int. J. Radiat. Biol.*,1991, **59**,195.
7. M. Napolitano, M. Durante, G.F.Grossi, M. Pugliese, G. Gialanella, *Int. J. Radiat. Biol.*, 1992, **61**, 813.
8. T.E. Hui, A. C. James, R. F. Jostes, J. L. Schwartz, K.L.Swinth, F.T.Cross, *Health Phys.*, 1993, **64**, 647.
9. X. Zhang, W.Z. Zheng, Y.F.Gong, W.B. Li, Z.H.Yang, *Radiat. Prot.*,1996,**16**,192. (in Chinese)
10. A.M.Kellerer,'Proc.2thSym.Microdosim',H.G.Ebert,Ed,Eurotom,Brussels,1970,pp.107
11. W. C. Roesch, *Radiat. Res.*, 1977, **70**, 494.
12. W.E.Wilson,Richland,WA:Pacific Northwest Laboratories,BNWL-2254 UC-34a, 1977
13. H.G.Paretzke, 'Proc. 4th Sym.Microdosim',Wiley, New York, H.G.Ebert,Ed, Eurotom, Brussels, 1973, pp.5122
14. W. B. Li and W. Z. Zheng, *Radiat. Prot. Dosim.*,1996, **67** (in press)
15. T. B. Borak, *Radiat. Res.*,1994, **137**, 346
16. H.G. Paretzke,' Kinetics of nonhomogenous Processes',G. R. Freeman, Ed, Wiley, New York, 1987, pp.87
17. D. T. Goodhead, *Raidiat. Environ.Biophys.*,1995, **34**, 67.

PREDICTION OF CELLULAR EFFECTS OF HIGH- AND LOW-LET IRRADIATION BASED ON THE ENERGY DEPOSITION PATTERN AT THE NANOMETER LEVEL

R. W. M. Schulte

Department of Radiation Medicine
Loma Linda University Medical Center
Loma Linda, CA 92354, USA

1 INTRODUCTION

There is increasing evidence that ionizing radiation causes a wide spectrum of DNA lesions with different grades of complexity. The most abundant lesions, i.e. single strand breaks, can be easily repaired by the intracellular repair system and are probably of no significance for any cell. On the other hand, complex DNA lesions with multiple strand breaks located within a few nanometers on either side of the DNA double strand are difficult to repair. Although not proven, these complex breaks may lead to cell death.

Most existing radiobiological dose response models do not take into account the existence of DNA lesions of different qualities. A new biophysical model, the two compartment (TC) model of cellular radiation response, is proposed: the model relates the amount of energy deposited at the DNA level to DNA lesion complexity and cell survival.

2 MATERIALS AND METHODS

2.1 Model Theory

2.1.1 Probability of Low- and High-Energy Deposits in DNA Sites. The critical *site* of the TC model is a cylindrical DNA segment of about 60 nucleotide pairs (20 nm). Passage of a site by a source particle of ionizing radiation, e.g. a high-energy photon, neutron, or ion, within the reach of its secondary particles is called an *event*. The energy deposited in a site during one event is a stochastic quantity that is subject to the probability density distribution $f_1(E)$.

For doses up to several hundred Gy, the probability that an individual site receives energy from more than one event is negligible. Thus, the probability density distribution for k events, $f_k(E)$, can be written as

$$f_k(E) = k\, f_1(E) \tag{1}$$

Given that the expected number of events for an absorbed dose D is n and the mass of a site is m, the probability density function can be expressed as

$$f(E,D) = nf_1(E) = Dmf_1(E)/\int_0^\infty Ef_1(E)dE \tag{2}$$

The TC model assumes that only energy deposits above a threshold energy E_{low} are of biological significance. Furthermore, it distinguishes between *two compartments* of energy deposition: (1) energy deposits between E_{low} and E_{high} lead to DNA damage that can be repaired by most cells but occasionally results in lethal damage; (2) energy deposits above E_{high} lead to damage that is mostly irreparable and typically results in lethal events.

The probabilities of a low-energy and a high-energy deposit in a particular DNA site can be expressed as

$$P_{low}(D) = \int_{E_{low}}^{E_{high}} f(E,D)dE = (1-q)\eta_{low}D \tag{3a}$$

and

$$P_{high}(D) = \int_{E_{high}}^{\infty} f(E,D)dE = q\eta_{high}D \tag{3b}$$

respectively, where we have introduced the following three model parameters:

partitioning parameter $\quad q = \int_{E_{high}}^{\infty} Ef_1(E)dE / \int_0^\infty Ef_1(E)dE \quad$ (4a)

low-energy-deposit efficiency param. $\quad \eta_{low} = \int_{E_{low}}^{E_{high}} f_1(E)dE / \int_0^{E_{high}} Ef_1(E)dE \quad$ (4b)

high-energy-deposit efficiency param. $\quad \eta_{high} = \int_{E_{high}}^{\infty} f_1(E)dE / \int_{E_{high}}^{\infty} Ef_1(E)dE \quad$ (4c)

2.1.2 Cell Survival. At this stage of modeling, only the residual (fixed) damage after repair is completed is considered. If s denotes the probability that a particular DNA site does not receive energy deposits that lead to lethal damage, we can write

$$s(D) = \left[1 - c_{\alpha,high}P_{high}(D) - c_{\alpha,low}P_{low}(D) - c_{\beta,low}P_{low}^2(D)\right] \tag{5}$$

where $c_{\alpha,\text{high}}$, $c_{\alpha,\text{low}}$, and $c_{\beta,\text{low}}$ are cell-specific constants. Eqn. (5) implies that lethality from low-energy deposits can result either from single events (linear term) or from a combination of two events (quadratic term), while lethality from high-energy deposits is always the consequence of a single event.

The probability of cell survival is equivalent to the probability that none of a total number of N_{crit} DNA sites receives energy deposits that lead to a lethal event. Inserting Eqn. (3) into Eqn. (5) and renaming the constants, the following expression for the probability of cell survival can be derived:

$$S(D) = \exp\left[-\left\{\alpha_{high}\nu_{high}qD + \alpha_{low}\nu_{low}(1-q)D + \beta_{low}\nu_{low}^2(1-q)^2D^2\right\}\right] \tag{6}$$

In this equation, the high- and low-energy-deposit efficiency parameters were normalized to their low-LET limit: $\nu_{high,low} = \eta_{high,low}/\eta_{high,low}(\text{LET} \to 0)$. Note that in the low-LET limit Eqn. (6) becomes the ordinary linear-quadratic cell survival formula.

2.2 Model Fit

Previously published frequency distributions for energy deposits in DNA segments of 18.4 nm length by alpha particles and protons of various energies and LETs, and by 20 keV electrons, were used to obtain numerical parameter estimates.[1] Using the least-square method, Eqn. (6) was fitted to experimental V79 cell-inactivation data, which were obtained after irradiation with alpha particles, protons, and X rays.[2] Systematic variations of E_{low} and E_{high} were tested to find the best combination for these parameters.

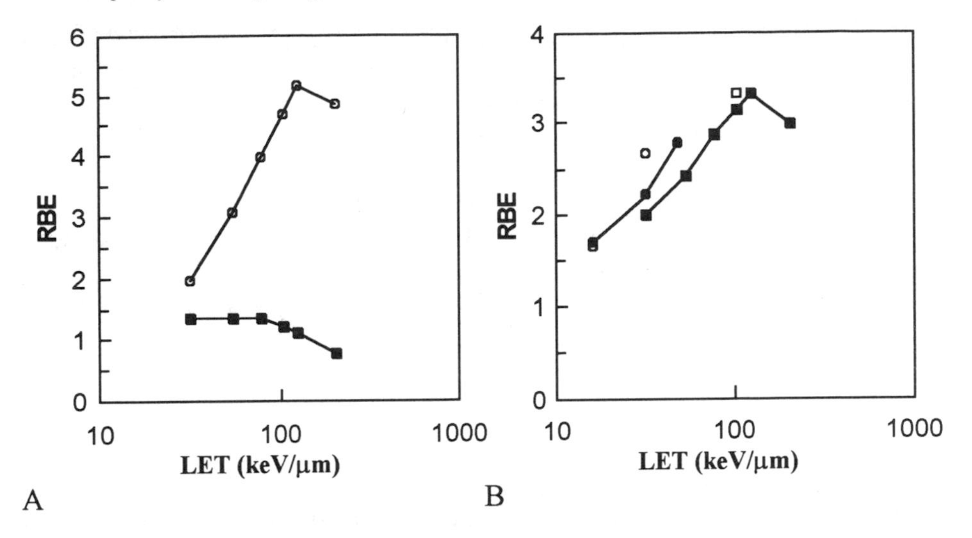

Figure 1 RBE predictions of the TC model. (A) LET dependence of the RBE of alpha particles for all energy deposits (■) and high-energy deposits (□). (B) LET dependence of the RBE for alpha particles (■) and protons (●) for cellular inactivation; open symbols represent measured RBE values.

3 RESULTS

3.1 Model Parameters

The best fit of Eqn. (6) to the experimental data was obtained for the threshold energy combination E_{low} = 50 eV, E_{high} = 140 eV. For this combination of threshold energies, the optimum cell-specific parameters for V79 cells were $\alpha_{low} = 0.12$ Gy^{-1} , $\beta_{low} = 0.027$ Gy^{-2}, and $\alpha_{high} = 2.9$ Gy^{-1}.

3.2 Relative Biological Effectiveness

Figure 1A demonstrates the LET dependence of the relative biological effectiveness (RBE) of alpha particles (relative to low LET 20 keV electrons) for the total number of energy deposits and the number of high-energy deposits in DNA segments. The RBE for the total number of energy deposits remains close to unity up to about an LET of 100 keV/μm and declines for higher LET values. The RBE for high-energy deposits by alpha particles increases steeply with LET and reaches a peak value for an LET of about 100 keV/μm.

Figure 1B shows the LET dependence of the RBE of alpha particles and protons (relative to low LET irradiation) for cellular inactivation. The RBE curve for alpha particles is similar to the RBE curve for high-energy deposits (Figure 1A), showing a maximum around 100 keV/μm. The RBE curve for protons is different from the curve for alpha particles, in that for the same LET value protons are more effective than alpha particles.

The TC model predicts that the RBE for cellular inactivation is cell specific. The maximum RBE for inactivation has the following dependency on the intrinsic radiosensitivity α_{low} :

$$RBE_{max} = (1-q)\nu_{low} + q\nu_{high}\alpha_{high}/\alpha_{low} \quad (7)$$

Thus, for the same radiation quality, higher RBE values are predicted for cells with lower radiosensitivity. In particular, if the parameter α_{high} is independent of the cell type, the TC model predicts a linear relationship between RBE_{max} and the reciprocal low-LET radiosensitivity.

4 DISCUSSION

A new biophysical model of cellular radiation action has been presented. The model differs from previous models by assuming two different categories of DNA lesions, i.e. reparable and irreparable DNA damage. Furthermore, it is assumed that these categories are related to a low- and a high-energy deposit compartment, respectively. The critical site of this model is a DNA segment of about 20 nm in length, thus stressing the importance of the energy deposition at the nanometer level.

The assumptions of the TC model are certainly oversimplified. It is likely that some of the damage caused by energy deposits beyond the threshold energy E_{high} is still reparable, although with a higher frequency of misrepair. Furthermore, energy that is deposited in bulk water surrounding the DNA molecules will contribute to the damage. This was not taken into account in Charlton's calculation of energy deposition frequencies in DNA,[1] on which the computation of the TC model parameters was based. Despite these potential shortcomings, it appears that the concept of threshold energies and different lesion qualities leads to valid predictions, such as peak RBE values around an LET of 100 keV/μm for alpha particles, and more effectiveness for protons than for alpha particles at the same LET. These effects have been verified experimentally.[3,4] Other TC model predictions, such as the linear RBE_{max} dependence on the reciprocal low-LET intrinsic radiosensitivity, remain to be proved.

5 REFERENCES

1. D.E. Charlton, H. Nikjoo, and J.I. Humm, *Int. J. Radiat. Biol*, 1989,. **56**, 1.
2. K.M. Prise, M. Folkard, S. Davies, and B.D. Michael, B.D., *Int. J. Radiat. Biol.*, 1990, **58**, 261.
3. D.T. Goodhead, M. Belli, A.J. Mill, D.A. Bance, L.A. Allen, S.C. Hall, F. Ianzini, G. Simone, D.L. Stevens, A. Stretch, M.A. Tabocchini, and R.E. Wilkinson, *Int. J. Radiat. Biol.*, 1992, **61**, 611.
4. G.W. Barendsen, *Radiat. Res.*, 1994, **139**, 257.

Risk extrapolation and cancer

THRESHOLD DOSE FOR CARCINOGENESIS: WHAT IS THE EVIDENCE?

C. Streffer

Institut für Medizinische Strahlenbiologie
Universitätsklikum Essen
45122 Essen
Germany

1. INTRODUCTION

A linear dose effect relationship without a threshold dose has been used by international committees for the estimation of stochastic radiation risks during the last decades[1-3]. Such dose effect curves have been derived from epidemiological studies on irradiated human populations and radiobiological experiments. Through the epidemiological studies the induction of cancer (including leukemia) has been observed in a dose range of around 0.1 to 4 Sv after whole body and partial body irradiation[1-3]. In order to estimate the radiation-induced cancer risk in the lower dose range an extrapolation of the dose effect curve has been performed in the above mentioned mode.

This procedure has often been criticised. It has especially been brought forward that no radiation effects occur or can be observed in the low dose range and therefore a threshold dose has to be assumed under these conditions. The Health Physics Society has stated recently "the Health Physics Society recommends against quantitative estimation of health risk below an individual dose of 5 rem in one year or a lifetime dose of 10 rem in addition to background radiation. Risk estimation in this dose range should be strictly qualitative accentuating a range of hypothetical health outcomes with an emphasis on the likely possibility of zero adverse health effects"[4]. On the basis of this statement Goldman has written an article published in Science which started with the sentence: "It is time to scientifically challenge the old tenet stating that cancer risk is always proportional to dose, no matter how small"[5].

The following arguments are usually brought forward for non-linearity and a possible threshold dose in radiation-induced carcinogenesis:

1) Epidemiological studies on irradiated human populations do not show increased cancer frequencies after radiation doses of the low dose range (< 100 mSv).
2) DNA repair reduces radiation-induced DNA lesions very efficiently and therefore the likelihood of cancers and genetic effects are also reduced.
3) Adaptive response which is induced by small radiation doses increases cellular radioresistance and may even lead to beneficial effects of a low level exposure.

4) Apoptotic processes remove damaged cells (cf. transformed and neoplastic cells) and therefore the probability of the development of cancer decreases.
5) For the development of a cancer several mutation steps probably in sequence are necessary. These phenomena reduce the probability for the induction of cancer and make a threshold in the dose effect curve probable.
6) For a dose effect relationship without a threshold cancer must develop from one damaged cell. Taking into account that cancer is induced through multiple mutation steps it is not clarified at which time clonal growth of malignant cells for the development of a cancer starts. Does this occur already with the cellular transformation processes briefly after a radiation exposure or does it only start when a preneoplastic lesion has developed probably several years after a radiation exposure?

2. EPIDEMIOLOGICAL STUDIES

The most important epidemiological studies for the evaluation of radiation-induced cancer risk are probably the studies on the survivors after the bombing in Hiroshima and Nagasaki. The data which have been published recently show a significant increase of total cancer mortality in a dose range of 50-100 mSv. The significant increase of leukemia occurs only after higher doses[6]. Other investigations which have studied the radiation effects in very sensitive human groups (cf. children) have found radiation effects in the same dose range as for total cancer in Hiroshima and Nagasaki[1,7]. In these studies where a dose effect curve can be established the data are usually described best by a linear or a linear-quadratic dose effect relationship without a threshold dose. However it can clearly been stated that no significant increase of cancer risk has been observed after radiation doses in the range of 10-50 mSv.

A number of studies have been undertaken in order to evaluate whether the cancer rates are higher in regions with a high natural radiation background in comparison to those regions where the exposure from natural radiation is low. Wei et al.[8] observed no increase of total cancer mortality when the mortality was measured in a region with a high background radiation (external radiation 2.2 mSv per year) and with a low level background radiation (0.8 mSv per year). These two regions were very similar with respect to the living style of the population which consisted mostly of farmer families and did not migrate over generations. Also there was not found any difference between these two regions with respect to the so-called radiosensitive cancers like leukemia, lung cancer or intestinal cancer. In general so far it was not possible to see any increased cancer induction after radiation exposures of humans in the range of 10-50 mSv.

On the other hand such a lack of a radiation effect in these dose ranges is not astonishing if one compares the radiation effect to be expected with the variability of "normal" cancer mortality which is caused by many factors like genetic predisposition and living style. Thus the cumulative cancer mortality until the age of 75 years was in Western Germany (population around 60 million) during the years 1970-1991 0.205 (around 20 % of all causes of death). The 95 % confidence interval (C.I.) for this figure was +/- 0.002. For the state of Bavaria (population: 10.7 million) the same figure was 0.200 and the C.I. +/- 0.005. If one calculates the radiation-induced cancer risk for mortality on the basis of a risk factor of 5×10^{-2} and a linear dose effect curve without threshold the risk would be after a radiation dose of 10 mSv 0.0005 and 100 mSv 0.005.

A comparison of these figures with the above C.I. shows that a radiation exposure of 100 mSv would just result in a cancer risk which is identical with the 95 % C.I. for "normal" cancer mortality of the Bavarian population. From these figures it is clear that even with very large populations it appears very plausible that the radiation effect for cancer mortality which is to be expected after a radiation dose of 10 mSv will disappear in the variability of the normal cancer mortality. There are other factors which modify the cancer mortality in a much stronger way than such a low radiation dose. Therefore it appears very improbable that the question whether a threshold dose exists or not in the low dose range can be solved by epidemiological studies only.

3. DNA REPAIR

DNA repair and its influence on the radiation-induced carcinogenic risk has been discussed extensively in many reviews[7]. DNA repair certainly can lower the cancer risk after exposures to ionizing radiation with low dose rate. However this effect appears to be less for carcinogenesis than for other radiation effects like cell survival and it appears also to be very different for various cancer entities[3,7]. If DNA repair lowers the risk, the steepness of the dose effect curve will be less. However these effects will not lead to thresholds.

4. ADAPTIVE RESPONSE

In 1977 Samson and Cairns[9] made the observation that a treatment of bacteria cells with a low dose of a mutagenic substance can lead to an induction of cellular repair functions. These authors incubated Escherichia coli cells with a very low concentration of the mutagen MNNG continuously and found that the frequency of mutations increased only during the first 60 minutes of incubation and then remained constant for seven days. If Samson and Cairns pretreated the cells with a low dose of MNNG (1 μg/ml) and then challenged the cells with a pulse of 100 μg/ml MNNG the pretreated (adapted) cells showed a far better survival and a lower mutation frequency than the non-adapted cells. The authors concluded that the low adapting dose of the mutagen induced an error-free repair pathway.

Similarly Olivieri et al.[10] discovered that a pretreatment of human lymphocytes in vitro with low concentrations of tritiated thymidine rendered the cells less susceptible to a subsequent high dose of X-rays. The authors measured the chromosomal aberrations and found a lower frequency of these aberrations in adapted cells in comparison to non-adapted cells. Similar results have been obtained by several other groups[11]. The assumption that DNA repair processes are induced by the adapting radiation dose was strengthened by the observation that this adaptive response could be inhibited by the drug aminobenzamide[12]. Since then many experimental groups have tried to find such adaptive response in various biological systems and for various biological endpoints. However it turned out that the results were very variable. They depended apparently on various genetic factors of the donors from which the cellular systems were obtained; a very narrow temporal window has to be used in order to find such adaptive response and the variability of various other factors is decisive[13].

The underlying mechanism of these phenomena is unclear up to now and it has not been possible to measure effects consistently. For carcinogenic effects such an adaptive response has not been shown up to now. Therefore there is no evidence that a threshold dose could be caused by such adaptive processes for carcinogenic risk or stochastic effects after radiation in general.

5. APOPTOSIS

Apoptosis is an active genetically programmed mode of cell death with characteristic morphological and biochemical features that distinguish this type of cell death from necrosis[14]. Apoptosis occurs to a certain degree in most tissues and leads to cell killing by physiological processes as well as after exposure to toxic agents. The morphological features are membrane blebbing, but no loss of membrane integrity, aggregation of chromatin at the nuclear membrane followed by cellular condensation and formation of membrane bound vesicles (apoptotic bodies). No desintegration of organells is included as it is seen in necrotic tissues. On the molecular level DNA fragmentation is observed. Cell death by apoptotic processes starts already early during prenatal development and is always connected with cell turnover.

Potten[15] has made the interesting proposition that differences in the site and incidence of apoptosis may contribute to the 100fold lower incidence of small intestinal cancer relative to colorectal cancer. Thus it has been observed that apoptotic cell death is much higher in the crypts of the small intestine than in the epithelium of the colon and rectum. In a number of reports the hypothesis has been brought forward that apoptotic processes may also reduce carcinogenesis after exposures to carcinogenic drugs and radiation[16]. However this has not been substantiated up to now by experimental evidence or other studies.

Norimura et al.[17] presented data that apoptosis suppresses radiation-induced teratogenesis. These authors studied the embryonic death and induction of abnormalities after X-ray exposures in p53 knock-out mice and wild-type mice. They observed that an X-ray dose of 2 Gy increased the rate of early death (death during the preimplantation period) much stronger in wild-type mice than in p53 knock-out mice. On the other hand the rate of fetuses with anomalies was higher in the knock-out mice in comparison to the wild-type mice. As p53 is thought to be one of the inducers for apoptotic death the authors concluded that radiation-induced lesions are not removed by apoptosis in the p53 knock-out mice and therefore a higher frequency of fetuses with abnormalities occured. This is not the case in the wild-type mice and therefore the early death is higher through apoptotic processes.

Although cancer induction has some common features with the induction of developmental abnormalities, there are on the other hand a number of differences in the involved processes and therefore it is certainly difficult to extrapolate from such studies to the induction of cancer after radiation exposure. Further it has to be taken into account that the apoptotic processes have to be induced and to occur in the neoplastic cells. However studies of apoptosis in tumor cells have shown that the degree of apoptosis is very different between tumor cell lines[18] and therefore it appears very questionable whether such processes can really reduce the cancer induction in a general way so that dose effect curves with a threshold result.

6. CARCINOGENESIS AS A MULTISTEP PROCESS

Carcinogenesis has been formulated as a multistep process already many years ago. These processes were devided into cell transformation, tumor promotion and tumor progression. During the last years it has become possible to identify the molecular events that underly the initiation and progression of human tumors[19]. Such molecular events have been studied on cell systems as well as on human tumors. It has been found that the induction of oncogenes and the suppression of tumor suppressor genes play a significant role. These processes are quite often initiated by mutation steps.

Fearon and Vogelstein[19] have proposed a genetic model for colorectal carcinogenesis under the inclusion of several mutations which follow each other stepwise. In patients with familial adenomatous polyposis (FAP) a mutation on chromosome 5q is inherited. This mutation is apparently responsible for the hyperproliferative epithelium present in these patients. The authors propose that in tumors arising in patients without polyposis the same region on chromosome 5 may be lost and/or mutated at a relatively early stage of carcinogenesis. From the hyperproliferating epithelium early adenoma are formed. It is further assumed that ras gene mutation (usually k-ras) occurs in the preexisting small adenoma and produces a larger and more dysplastic tumor through clonal expansion. These processes are followed by further loss of chromosomal material which includes the chromosomes 5, 17 and 18. On chromosome 17 the p53 gene is located and it appears that either deletions or mutations occur in this gene for the development of late adenomas to carcinomas. Fearon and Vogelstein[19] described that k-ras mutation is observed in about 60 % of the intermediate and late adenomas and in about 50 % of colorectal carcinomas. Loss of chromosome 17p was detected at a rate of about 10 % in intermediate adenomas, of about 30 % in late adenomas and around almost 80 % in carcinomas. This is strong evidence for the development of colorectal carcinomas from the preneoplastic stages of adenomas. Although many questions remain open this model has been widely accepted for the development of colorectal cancers.

One open question certainly is whether the sequence of these steps or the accumulation of these changes is most important. More than 90 % of the carcinomas have two or more of the above mentioned genetic alterations. However only 7 % of early adenomas had more than one of the genetic alterations and this percentage increased when the adenomas progressed to intermediate and late stages[19]. It remains completely open whether radiation-induced colorectal cancers develop in the same way via these preneoplastic stages. It is not clear when the first malignant cells develop which can then provoke the monoclonal growth of a cancer. Further it remains open whether radiation-induced lesions in one cell are sufficient in order to develop such a malignant cell. These are questions which have to be answered in order to get a definite answer whether the dose effect curve for the radiation-induced cancer processes have a threshold or not.

From experimental data it appears quite clear that only when these processes are developing from lesions in one cell then a dose effect curve without a threshold is scientifically valid. Studies of many groups on the formation of chromosomal aberrations after radiation exposure give good evidence that these effects occur through lesions in one cell and no interaction between cells has to take place. On the other hand all chromosomal aberration data especially with studies in the lower dose range show that these data can be best described by dose effect curves without a threshold[20]. In analogy to such experimental data it appears very plausible to accept such a dose effect curve also for genetic mutations in somatic and germ cells.

Further experimental studies on the induction of malformations after prenatal irradiation in a mouse strain with a genetic predisposition have nicely shown that the form of the dose effect curve depends on the question whether the radiation lesion was induced in one or several cells. In the mouse strain used specific malformations can be induced after a radiation exposure during the preimplantation period. The reason for this extraordinary effect is that this mouse strain has a genetic predisposition for this malformation[21]. It could be shown that the malformations were induced by radiation exposure to the one-cell stage as well as to the multicellular cell stages during the preimplantation period. However only a radiation exposure during the one-cell stage led to a dose effect curve without a threshold whereas the exposure to multicellular embryos always had a threshold in the dose effect curve[22]. This is a convincing example that only processes which are caused by an unicellular lesion lead to dose effect curves without a threshold. Therefore it appears necessary to solve the questions which have been formulated before.

Many studies on clinical cancers give evidence that a monoclonal growth of tumors really occur. This is documented by cytogenetic studies[7] by flow-cytometric studies in which usually only one clone of tumor cells is observed in a clinical cancer[23]. On the other hand there are examples where several clones have been found by flow-cytometric DNA measurements or by morphological observations. These processes may be due to the genomic instability which is induced after radiation exposure and during cancer development[24,25]. These phenomena and their consequences for cancer development have certainly to be investigated in more detail.

It further has to be taken into account that the biological processes of cancer development differ from tissue to tissue. Especially the tumor progression may be quite different. Therefore it appears very plausible that also the dose effect curve of radiation-induced cancer can differ in the various tissues and organs. While for most cancer entities a dose effect curve without a threshold appears plausible from the studies which have been performed up to now, it appears obvious that for the induction of osteosarcomas by bone-seeking radionuclides a dose effect curve with a threshold occurs[1]. These differences between various cancer entities certainly have to be included into the discussion under consideration of the complex biological processes which are involved in carcinogenesis.

7. CONCLUSIONS

Epidemiological studies on cancer risk after small radiation doses in the dose range of 10-50 mSv and lower will probably never show an increased cancer frequency as the radiation effect is so low that it disappears in the variability of "normal" cancer incidence. Therefore the argument that no increased cancer frequencies can be observed in this dose range is not scientifically valid for the discussion against an extrapolation of the dose curve without a threshold dose.

DNA repair may reduce the cancer rate after exposures with low dose rates. This effect may reduce the steepness of the dose effect relationship but will not introduce a threshold dose. Similar modifications of the dose effect curve may occur through adaptive response (although this has not been shown up to now with carcinogenesis) and apoptotic processes.

The development of a clinical cancer after radiation exposures in humans takes decades. The mechanisms of the involved processes are not solved up to now. However it is clear that several mutation steps are involved and preneoplastic lesions develop from which a cancer is formed. These very complex biological processes have to be clarified in order to give a definite answer in which way the extrapolation of cancer risk in the very low dose range has to be done.

The scientific data which are available are in favour of a dose effect curve without a threshold dose. One of the crucial questions is at which period of carcinogenesis the monoclonal growth of malignant cells starts in order to develop a clinical cancer.

References

1. UNSCEAR, Sources, effects and risks of ionizing radiation, United Nations, New York, 1988.
2. BEIR, Health effects to exposure to low levels of ionizing radiation, National Academy Press, Washington D.C., 1990.
3. ICRP, Recommendations of the International Commission on Radiological Protection, Pergamon Press, Oxford-New York-Frankfurt-Seoul-Sydney-Tokyo, 1991.
4. Health Physics Society, Position Statement: Radiation risk in perspective, HPS Newsletter, 1996, 3.
5. M. Goldman, Science, 1996, **271**, 1821.
6. D. A. Pierce, Y. Shimizu, D. L. Preston, M. Vaeth and K. Mabuchi, Radiat. Res., 1996, **146**, 1.
7. UNSCEAR, Sources and effects of ionizing radiation, United Nations, New York, 1993.
8. L. Wei, Z. Zongru and T. Zufan, China. J. Radiat. Res., 1990, **31**, 119.
9. L. Samson and J. Cairns, Nature, 1977, **267**, 281.
10. G. Olivieri, J. Bodycote and S. Wolff, Science, 1984, **223**, 594.
11. UNSCEAR, Sources, effects and risks of ionizing radiation, United Nations, New York, 1994.
12. J. D. Shadley, V. Afzal and S. Wolff, Radiat. Res., 1987, **111**, 511.
13. A. Wojcik and C. Streffer, Biol. Zent.bl., Gustav Fischer Verlag Jena, 1994, **113**, 417.

14. A. H. Wyllie, J. F. R. Kerr and A. R. Currie, Int. Rev. Cytol., 1980, **68**, 251.
15. C. S. Potten, Cancer and Metastasis Reviews, 1992, **11**, 179.
16. S. Kondo, Int. J. Radiat. Biol., 1988, **53**, 95.
17. T. Norimura, S. Nomoto, M. Katsuki, Y. Gondo and S. Kondo, Nature Medicine, 1996, **2**, 577.
18. N. J. Stapper, M. Stuschke, A. Sak and G. Stüben, Int. J. Cancer, 1995, **62**, 58.
19. E. R. Fearon and B. Vogelstein, Cell, 1990, **61**, 759.
20. U. Weissenborn and C. Streffer, Int. J. Radiat. Biol., 1988, **54**, 381.
21. C. Streffer and W.-U. Müller, Int. J. Dev. Biol., 1996, **40**, 355.
22. W.-U. Müller, C. Streffer and S. Pampfer, Radiat. Environ. Biophys., 1994, **33**, 63.
23. C. Streffer, D. van Beuningen, E. Gross, J. Schabronath, F.-W. Eigler and A. Rebmann, Radiotherapy and Oncology, 1986, **5**, 303.
24. N. W. Kim, M. A. Piatyszek, K. R. Prowse, C. B. Harley, M. D. West, P. L. C. Ho, G. M. Coviello, W. E. Wright, S. L. Weinrich and J. W. Shaw, Science, 1994, **266**, 2011.
25. S. Pampfer and C. Streffer, Int. J. Radiat. Biol., 1989, **55**, 85.

CARCINOGENIC RESPONSE AT LOW DOSES AND DOSE RATES: FUNDAMENTAL ISSUES AND JUDGEMENTS

Roger Cox

National Radiological Protection Board
Chilton
Didcot, Oxon, OX11 ORQ

EXTENDED ABSTRACT

Epidemiological and, indeed, animal studies have limited capacities to provide direct estimates of carcinogenic risk at low doses and low dose rates of ionising radiation. In the case of epidemiological studies on Japanese A-bomb survivors new evidence[1] suggests that excess cancer is present after doses of around 50 mGy while data from studies of *in utero* irradiated children and occupationally exposed nuclear workers imply that risk is present down to at least 20 mGy.[2] These direct human observations clearly limit the extent to which a threshold-type human carcinogenic response might apply after radiation and, for radiological protection purposes, it is reasonable to assume that human cancer risk rises as a simple function of dose without a dose-threshold.

Direct proof of this current assumption will, however, always remain elusive and for this reason it is increasingly important to take full account of the data that relate to the cellular/molecular mechanisms of carcinogenesis and the defences available to protect against cancer induction and development. These fundamental data, together with others relating to epidemiological and animal studies, have been summarised recently.[2]

In respect of fundamental considerations there is increasingly strong evidence that cancer develops via a complex multistage process driven by the stepwise accumulation of specific gene and chromosomal mutations.[2,3] Such mutational combinations in conjunction with influences from systemic and environmental factors serve to select cells having malignant phenotypes characteristic of the common and potentially fatal cancers.

In spite of this biological complexity the evidence available implies that in the majority of cases cancer develops, albeit at a low probability, as the consequence of an initial but specific mutation in single target stem-like cells in tissues.[3] Given current knowledge on the mechanisms of radiation-induced gene and chromosomal mutations in somatic cells it may be concluded, therefore, that a single track of ionising radiation traversing an appropriate cellular genome has a very low but finite probability of creating a mutation that will contribute towards cancer development. Thus, at the level of mutagenesis, no dose-threshold for cancer initiation should be expected.

As might be anticipated defence mechanisms against cancer development are, however, available. The first line of defence centres on DNA damage recognition and repair including the provision of cell cycle checkpoints to maximise repair efficiency and fidelity. These repair processes appear to operate in an essentially error-free fashion against simple DNA lesions occurring on one strand of DNA and probably correctly reconstitute the majority of simple co-incident double strand DNA lesions. In contrast the more complex double strand lesions known to be amongst those induced by ionising radiation are believed to be more prone to misrepair resulting in the gene/chromosomal deletions and DNA rearrangements that characterise mutational responses in somatic cells and early tumour-associated events in some animal systems.[4,5] Accordingly there is no reason to assume, even at the low abundance expected after low dose exposure, that all radiation induced DNA damage potentially associated with cancer induction will be repaired in the error-free fashion that is required for a DNA damage-related threshold response. Inducible error-free repair, a so called adaptive response, has been suggested as a possible mechanism for threshold-type responses. The evidence for this is however tenuous and not deemed sufficient to influence judgements in radiological protection.[6]

Post-DNA damage systems for cancer protection also exist and principal of these are terminal cellular differentiation and programmed cell death (apoptosis). In essence, if a cancer-initiated cell can be removed from the malignant pathway through differentiation to a non-dividing state or by the active process of apoptosis then risk will be negated. There can be little doubt that both these processes can efficiently remove a large fraction of potentially neoplastic cells prior to malignant commitment; the high frequency of specific gene mutations in tumours, eg. *p53*, that serve to block or bypass these defence processes testifies to their importance in the determination of cancer risk. Of far greater uncertainty is whether these processes which can operate throughout early phases of carcinogenesis exhibit a critical dependence upon initial dose such that risk is removed below say 50 mGy. Scenarios of this type cannot be formally excluded but find little support in the data available.

A further set of defences against malignant development may be provided by immune and non-immune cellular surveillance systems which act to actively eliminate neoplastic cells. It seems, however, that in the main these processes require the relatively strong expression of tumour-specific antigens and, accordingly, act principally against virally-associated neoplasms rather than the common radiogenic tumours. A major role for cellular surveillance in radiation carcinogenesis remains therefore somewhat uncertain. Even in the event that cellular surveillance were to be a significant determinant of radiogenic cancer risk the removal of that risk at low doses would require a strong dose-dependence for surveillance efficiency, ie. an enhancement at low doses. The evidence for this is weak and remains highly controversial.[6]

In summary, the weight of evidence from experimental studies relating to carcinogenic mechanisms and defences favours the view that at low doses and dose rates cancer risk overall will rise as a simple function of dose and that there is not a low dose interval within which risk may be discounted.

That is not to say that the fundamental data relating to judgements on this critical issue are wholly adequate. In addition to the need to more completely understand DNA damage repair and apoptosis at low doses together with their relationships with track structure, the whole question of the nature, consequences and fates of early pre-neoplastic events in irradiated tissues requires further resolution.[5] Although somewhat outside the field of radiation biology it is also important to gain a more clear view of the extent to which some tumours may not be of single cell origin[7] and resolve uncertainties on the role of cellular surveillance in cancer development.[8]

ACKNOWLEDGEMENTS

Work on radiation oncogenesis in the authors laboratory is supported in part by the Commission of the European Communities, contract F14P-CT95-0008.

References

1. D. A. Pierce, Y. Shimizu, D. L. Preston, M. Vaeth and K. Mabuchi, *Radiat. Res.*, 1996, **146**, 1.
2. R. Cox, C. R. Muirhead, J. W. Stather, A. A. Edwards and M. P. Little, Docs. NRPB, 1995, Vol. 6, No. 1.
3. UNSCEAR, Annexe E. IN 'Sources and effects of ionising radiation'. United Nations, New York, 1993.
4. K. H. Chadwick, R. Cox, H. P. Leenhouts and J. Thacker (Eds), 'Molecular mechanisms in radiation mutagenesis and carcinogenesis'. CEC EUR 1329, Luxembourg, 1994.
5. P. M. H. Lohman, R. Cox and K. H. Chadwick, *Int. J. Radiat. Biol.*, 1995, **68**, 331.
6. UNSCEAR, Annex B. IN 'Sources and effects of ionizing radiation', United Nations, New York, 1994.
7. M. R. Novelli, J. A. Williamson, I. P. M. Tomlinson, G. Elia, S. V. Hodgson, I. C. Talbot, W. F. Bodmer and N. A. Wright, *Science*, 1996, **272**, 1187.
8. T. Elliot, *Nature Genet.*, 1996, **13**, 140.

AN HSEF FOR MURINE MYELOID LEUKEMIA

*V.P. Bond, *E.P. Cronkite, *J.E. Bullis, +C.S. Wuu, +S.A. Marino and +M. Zaider
*Brookhaven National Laboratory, Upton, NY 1173, and +Department of Radiation Oncology, Columbia University, New York, NY 10032, USA

I. Introduction

In the past decade, a large amount of effort has gone into the development of hit size effectiveness functions (HSEFs), with the ultimate aim of replacing the present absorbed dose-RBE-Q system[1-3]. The reasoning has been that, while cancers are observable only at the organ level, they are in fact single-cell in origin. Because the absorbed dose can not provide information on single hits it is necessary to resort to microdosimetry. From this information, an HSEF can be derived.

To date there have been no sets of data available on animals exposed to radiations of several qualities, and for which microdosimetric data were available. Thus, it was possible to obtain HSEFs only for endpoints observable in the single cell, e.g. mutations or chromosome abnormalities, which were thought to have some relevance to the carcinogenesis process. In the present set of experiments large numbers of mice were exposed to radiations of several different qualities, and were observed throughout their entire lifespan for the appearance of myeloid leukemia. The HSEF developed for this neoplasm is presented and discussed.

II. Materials and Methods

Animals: All studies were performed on 12-16 week old male CBA/CaJ mice. The mice were maintained at BNL in AAALAC-approved quarters, on a twelve-hour light/dark cycle, and given acidified drinking water (pH 2.4) and Purina Lab Chow ad libitum.
Exposures: Photon exposures were to both X rays and ^{137}Cs. All animals were irradiated at 84 days of age. X-ray exposures were done in the BNL Medical Department, using 250 kVp x rays with 0.5 mm Cu + 1.0 mm Al filtration and a dose rate of 30 cGy/min. Gamma ray exposures were done at BNL using a dose rate of 50 cGy/min. All neutron exposures were done at the Radiological Research Accelerator Facility of Columbia University. The neutron energies were: 0.22; 0.44; 1.5; 6.0; and 14.0 MeV. Dose rates ranged from 5 to 60 cGy/hr.

All animals were checked twice daily, seven days per week. Those that appeared to be ill or moribund were euthanized. The body cavities were opened and the remains were placed in 10% formalin. Samples of the sternum, spleen, lymph nodes, thymus, lungs, liver, and kidney were taken.

III. Results and data analysis

The results consist of K=7 data sets, each set corresponding to a particular radiation type. Within each set, k, irradiations have been performed at J(k) different doses. Following exposure to dose D_{kj}, animals are observed for I consecutive time intervals, Δt, and for each interval, i, one records the number of animals, N_{kji}, that died during that time period and among these, the number of animals, M_{kji}, that have a tumor.

We use the relative risk model in which the cancer hazard rate, $\lambda(t,D)$, is:

$$\lambda(t,D) = \eta(D) \cdot \lambda_0(t) \tag{1}$$

The quantity obtainable from experiment, however, is the cumulative cancer rate, F(t,D):

$$F(t,D) = 1 - \exp\left[-\int_0^t \lambda(t',D)\,dt'\right], \tag{2}$$

From Eqs(1,2) one obtains:

$$F(t,D) = 1 - [1-F_0(t)]^{\eta(D)}, \tag{3}$$

where $F_0(t)=F(t,0)$. We further express $\eta(D)$ as a linear quadratic function of dose:

$$\eta(D) = 1+\epsilon(D) = 1+\alpha D+\beta D^2. \tag{4}$$

We shall treat $F_0(t)$ non-parametrically, that is assume no particular functional form for this quantity. The unknowns of the problem are then α_k, β_k, and $F_{0,i}$ with k=1,2,...,K; i=1,2,...,I. Estimators for these parameters may be obtained in the maximum likelihood sense. The log-likelihood function for this problem is[4]:

$$\log L = \sum_{k=1}^{K} \sum_{j=1}^{J(k)} \sum_{i=1}^{I} \left\{M_{kji}\log\left[1-(1-F_{0,i})^{\eta_{kj}}\right] + (N_{kji}-M_{kji})\,\eta_{kj}\,\log\left[1-F_{0,i}\right]\right\} \tag{5}$$

subject to positivity constraints on unknowns and the requirement that $F_0(t)$ is a monotonically non-decreasing function of t.

Fig.1 shows an example of the solution F(t,D) for 1.5 MeV neutrons (solid lines). Also shown (circles) is the fraction of animals with malignancies in each time interval. The α and β values thus obtained are given in the Table.

IV. Hit-size effectiveness function

The HSEF, h(z) is defined as:

$$\epsilon(D) = \int_0^\infty h(z)\,f(z;D)\,dz. \tag{6}$$

Here f(z;D) is the microdosimetric spectrum at dose D. At low doses[2]:

$$\epsilon(D) \approx D \int_0^\infty \frac{h(z)}{z_F} f_1(z)\,dz. \quad (7)$$

Here $f_1(z)$ is the *single-event* microdosimetric spectrum and z_F is its first moment. Thus

$$\alpha = \int_0^\infty \frac{h(z)}{z_F} f_1(z) = \int_0^\infty q(z)\,d(z)\,dz \quad (8)$$

q(z), which equals h(z)/z, has been termed specific quality factor[2]. Eq(8) can be used to obtain information on q(z) or h(z) as follows: in a series of experiments (such as those reported here), the initial slopes of the dose effect curves (α_i) as well as microdosimetric distributions, $d_i(z)$ are measured for a series of radiations, i. The integral equation, Eq(8), can be then converted to a system of integral equations and solved numerically[2]. In this analysis the α values have been converted to RBE (see the last column of Table 1), i.e. normalized to the value obtained for ^{137}Cs gamma rays.

The microdosimetric spectrum for the photon field was calculated using:

$$f(y) = \int_0^\infty \frac{N(E_\gamma)\,E_\gamma \mu(E_\gamma)}{y_F(E_\gamma)} f(y,E_\gamma)\,dE_\gamma \Bigg/ \int_0^\infty \frac{N(E_\gamma))\,E_\gamma \mu(E_\gamma)}{y_F(E_\gamma)}\,dE_\gamma\,. \quad (9)$$

Here $N(E_\gamma)$ denotes the photon energy spectrum and $f(y,E_\gamma)$ represents the single-event microdosimetric distribution in lineal energy that results from exposure to monoenergetic photons of energy E_γ; y_F is the frequency-averaged lineal energy and μ is the linear attenuation coefficient. Microdosimetric spectra for monoenergetic photons, $f(y,E_\gamma)$, were calculated using the computer code PHOEL[5] and our event-by-event transport codes[6]. Microdosimetric spectra for neutrons have been calculated using the concepts of crossers, stoppers, insiders and starters introduced by Caswell[7,8].

The function q(y) obtained in this analysis is shown in Fig.2; also shown here is the HSEF function h(z). They have been obtained by applying the Bayes theorem and maximum entropy principle[9] to the problem of solving Eq(8).

V. Conclusion

The accuracy possible in this analysis was limited because the population used was mice, with each observed leukemia representing but one cell that has been fully transformed malignantly. The sigmoid shape of the obtained function is reassuring and indicates that the HSEF - with complementary data - can be used to develop an internally consistent and coherent system for use in radiation protection practice. [Work supported by Grant CA12536 and by contract DE-FG02-88ER60631 from DOE].

Table

	250 kVp x	^{137}Cs	220-keV n	440-keV n	1.5-MeV n	6-MeV n	14-MeV n
α/Gy^{-1}	0.12	0.05	0.88	0.39	0.43	0.16	0.12
β/Gy^{-2}	0.	2.6 10^{-4}	0.	.0017	0.	0.	0.
RBE	2.4	1.	17.8	7.8	8.6	3.2	2.4

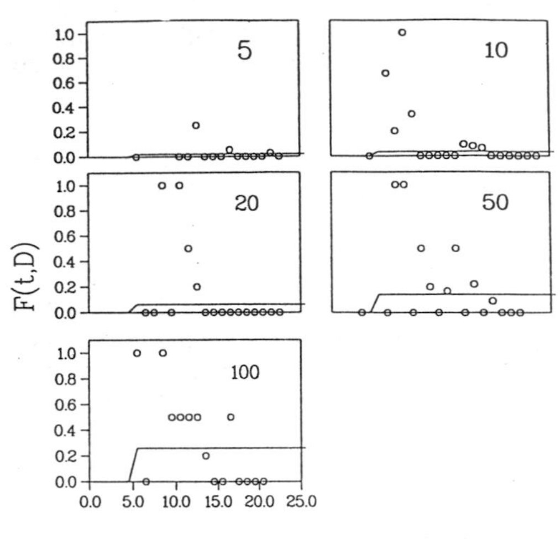

Fig.1: Solid lines: F(t,D) as a function of time at the doses indicated in each panel (in cGy) for 1.5-MeV neutrons. Circles: fraction of animals with malignancies in each time interval (Δt=50 days).

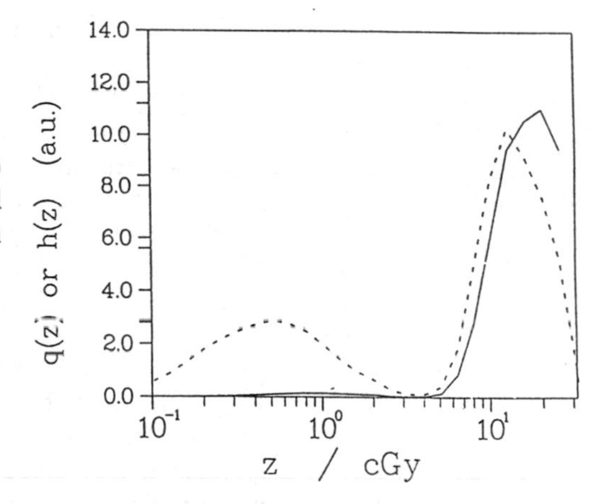

Fig.2: h(z) (solid) and q(z) (dashed) for the induction of myeloid leukemia in mice (the ordinate is in arbitrary units).

LITERATURE CITED

1. V.P. Bond and M.N. Varma, Proceedings of the Eighth Symposium on Microdosimetry. London: Harwood Academic, pp. 423-437 (1982).
2. M. Zaider, and D.J. Brenner, Radiation Research 103: 302-316 (1985).
3. C.A. Sondhaus, V.P. Bond, and L.E. Feinendegen, Health Phys. 59: 35-48 (1990).
4. D. Chmelevsky, A.M. Kellerer, J. Lafuma, and J.O. Chameaud, Radiat. Res. 91: 589-614 (1982).
5. J.E. Turner, R.N. Hamm, H.A. Wright, J.T. Modolo, and G.M.A.A. Sardi, Health Phys 39: 49-55 (1980).
6. M. Zaider, A.Y.C. Fung, and M. Bardash, In: Computational Approaches in Molecular Biology: Monte Carlo Methods (M.N. Varma and A. Chatterjee, Eds.). New York, Plenum Press, (1994).
7. R.S. Caswell, Radiat. Res. 27: 92 (1982).
8. M. Zaider, J.F. Dicello, and J.J. Coyne, Nucl. Instr. Meth. B40/41: 1261-1265 (1989).
9. M. Zaider and G.N. Minerbo, Phys. Med. Biol. 33: 1273-1284 (1988).

THE ESTIMATION OF NEUTRON QUALITY FACTORS: FUTURE PROSPECTS BASED ON FURTHER REVISIONS OF NEUTRON DOSES IN HIROSHIMA

P. R. Grimwood [1,2] and M. W. Charles [2]

[1] Nuclear Electric Ltd, Barnett Way, Barnwood, Gloucester. GL4 3RS.
[2] School of Physics and Space Research, University of Birmingham. B15 2TT.

1 INTRODUCTION

A unified system for evaluating the biological effects of mixed radiation fields by weighting the absorbed dose by a factor which depended upon the type of ionising radiation was first considered by Cantril and Parker in 1945[1]. Since then there has been much debate regarding the weighting values. Publication 60 of the International Commission on Radiological Protection (ICRP)[2] includes the latest recommendations of radiation weighting factors w_R to weight the absorbed doses for each type of radiation. The recommended value of w_R for all low-LET radiations is unity and between 5 and 20 (depending on energy) for neutrons. These values of w_R are said to be based on measured values of relative biological effectiveness (RBE) obtained from *in vitro* and *in vivo* laboratory experiments which have been reported in Publication 40 of the International Commission on Radiation Units and Measurements (ICRU)[3]. The range of estimates of neutron RBE is large and varies greatly for different biological endpoints in animals and cells. Sinclair[4] has recently pointed out that the endpoints most relevant to cancer induction could be used to support larger w_R values than those given by the ICRP and he alludes to possible future increases on this basis. With such difficulties of interpreting the radiobiological data it would clearly be more satisfactory if carcinogenic effects in man following neutron irradiation could be based directly on human data.

For many years, the Life Span Study (LSS) cohort of Japanese atomic bomb survivors has been used to estimate the biological effects of exposure to low-LET radiation. However, since the most recent major revision of the bomb dosimetry (DS86) the estimated neutron doses in Hiroshima and Nagasaki have been extremely small and so it has not been possible to estimate the risks from neutrons. The only reported discrepancy in the DS86 system was that there appeared to be an underestimate of the calculated compared to measured thermal neutron doses at distance between 1.0 and 1.6 km from the epicentre (the range where most survivors were situated). Recently Straume and colleagues[5] have collated contemporary measurements of neutron activation to show that beyond 1000 m in Hiroshima the thermal neutron doses are between two and ten times higher than given by the DS86 dosimetry. This paper attempts to evaluate what effect a revision of the neutron doses in Hiroshima (of the type suggested by Straume) will have on the use of the bomb survivor data to estimate the RBE of neutrons.

2 METHODS

This study makes use of the latest release of incidence data from the Life Span Study (LSS) of the Japanese survivors of the atomic bombings in Hiroshima and Nagasaki. The data was supplied by the Radiation Effects Research Foundation (RERF) and is the same data as used in the RERF's latest analysis of incidence data[6, 7].

This analysis is based on the relative risk (RR) model and uses solid tumour incidence as the endpoint of interest. The absorbed dose to the colon is used as a surrogate for organ dose. The data were stratified in terms of city, sex, age at exposure and follow-up interval and also divided into 7 exposure categories. In this study, as is usual, the population was restricted to those receiving less than 4 Gy whole body dose, since dose estimation is far less accurate for those survivors who are more highly exposed.

If i represents the level of stratum and j is the exposure category, then the number of deaths in the ij^{th} category can be defined as

$$M_{ij} = PYR_{ij} \cdot \lambda_i \cdot RR_{ij} \quad ,$$

where PYR_{ij} is the number of person years in category ij, λ_i is the mortality rate in stratum i assuming no exposure to radiation and RR_{ij} is the risk for the ij^{th} category, relative to the unexposed group in stratum i.

As a result of previous studies, a linear dose-response was assumed for both neutron and gamma radiation. This leads to the model for relative risk of

$$RR_{ij} = 1 + \alpha \cdot G_{ij} + \beta \cdot N_{ij} \quad ,$$

where G_{ij} and N_{ij} are the mean gamma and neutron colon doses in the ij^{th} category and α and β are the estimated excess relative risks per gray (ERR/Gy) of gamma and neutron radiation respectively. The estimate of RBE is then

$$\text{RBE} = \frac{\beta}{\alpha} \quad .$$

The inclusion of a dose-squared term for either gamma or neutron terms did not improve the fit to the data and did not alter the results significantly. However, this does not appear to be the case in analyses where leukaemia incidence was used as an endpoint. In that case, a G_{ij}^2 term does improve the fit significantly, although a linear dose-response for neutrons still seems to be most appropriate.

Data was fitted by maximum likelihood methods allowing α and β to be estimated. The ERR per gray of gamma rays and neutrons were evaluated for both the current dosimetry and for a revised DS86 dosimetry, which altered the neutron doses in Hiroshima by distance-dependent factors which were in line with the measured to calculated ratios quoted by Straume. Doses in Nagasaki were left unchanged.

Table 1 *Estimated Excess Relative Risks of Neutrons and Gamma Rays*

Dosimetry System	*Estimated ERR/Gy for Neutrons*	*Estimated ERR/Gy for Gamma Rays*
DS86	23 (-8.1, 56)	0.52 (0.30, 0.74)
Revised DS86	2.7 (-1.8, 7.2)	0.58 (0.41, 0.76)

Table 2 *RBE Estimates from the LSS of Japanese Atomic Bomb Survivors*

Dosimetry System	*Maximum Likelihood Estimate of RBE*	*Approximate 1-sided 95% Upper Confidence Bound*
DS86	46	180
Revised DS86	4.7	17

3 RESULTS

Table 1 shows the estimated excess relative risks per gray of neutrons and gamma rays, when both of these parameters were free to vary independently in the model. The ERR values were estimated for the current (DS86) dosimetry and for a the DS86 dosimetry which has been modified in line with Straume.

Table 2 shows the best estimates of the RBE of neutrons, as derived from the above model. The 90 % confidence intervals for these estimates all include zero.

4 DISCUSSION

The results from this study indicate that if the ERR/Gy is modelled separately for gamma and neutron radiation (this is equivalent to allowing the RBE to vary and thus be estimated), the low-LET risk appears to be insensitive to upward revisions of neutron doses. Thompson and colleagues analysis[7], which used a conventional relative risk analysis with the risk depending only on the dose equivalent and assuming a radiation weighting factor for neutrons of 10 estimated the ERR/Sv to be 0.63 for solid tumour incidence. Thus it seems likely that even if the dosimetry of the Japanese atomic bombs were revised in the manner suggested, the low-LET risks would remain virtually unchanged.

This would not be true for the case of neutrons. Using the current (DS86) dosimetry, the best estimate for the ERR per gray of neutron radiation is about 20. However, the uncertainties on the values are incredibly large, due to the very small neutron doses involved. If the dosimetry was revised in the manner suggested, there are indications that the best estimate of excess relative risk per gray will be around 3. This corresponds to an estimated RBE for neutrons compared to the atomic bomb gamma rays of about 5 and an upper bound of close to 20.

Tore Straume[8] has recently highlighted another complication in the use of radiation quality and radiation weighting factors. This concerns the choice of "reference" radiation. Straume pointed out, using data on dicentric chromosome aberrations in human lymphocytes, that nominally 'low-LET' radiations seem to have different RBEs, depending on energy, with higher energies being relatively less effective than lower energies. Since the mean energy of the gamma rays in Hiroshima and Nagasaki is between 2 and 5 MeV compared to a few hundred keV for photons typically encountered in the radiation workplace, there may be a need for further downward revision of RBEs from the Japanese data when applied to radiological protection.

5 CONCLUSIONS

It has been shown that the Life Span Study data of Japanese atomic bomb-survivors, with an upward revision of neutron doses based on recent neutron activation data in Hiroshima, may yet be the source of direct data on neutron cancer risks in man. Our results provide support for neutron RBEs which are no greater than the current ICRP neutron radiation weighting factors. Significantly lower values may in fact be more appropriate, particularly if a lower energy photon reference radiation, rather than that from weapons detonations, is used. A definitive analysis of risks from neutrons must however await the availability of revised dose estimates for each individual survivor. We hope that our results will encourage this revision process.

References

1. S. T. Cantril, H. M. Parker, 'The Tolerance Dose.' *Report MDDC-1100*, US Atomic Energy Commission, 1945.
2. ICRP Publication 60, '1990 Recommendations of the International Commission on Radiological Protection,' Annals of the ICRP, Pergamon Press, Oxford, 1991.
3. ICRU Report 40, 'The Quality Factor in Radiation Protection' *ICRU Report 40*, 1986.
4. W. K. Sinclair, *Health Physics*, 1996, **70**, 781.
5. T. Straume, et al., *Health Physics*, 1992, **63**, 421.
6. K. Mabuchi, et al., *Radiation Research*, 1994, **137**, S1.
7. D. E. Thompson, et al., *Radiation Research*, 1994, **137**, S17.
8. T. Straume, *Health Physics*, 1995, **69**, 954.

A NEW PARADIGM FOR RADIATION RISK ASSESSMENT

M. M. Elkind and R. L. Wells

Department of Radiological Health Sciences
Colorado State University
Fort Collins, Colorado 80523
U.S.A.

1 INTRODUCTION

Historically, radiation risk assessment has been based on the tacit assumption that *risk* can be predicted from a *dose-effect* curve. Because of the localization of radiation absorption events, this paradigm became supported by microdosimetric principles but also by animal studies and, in the instance of cancer, human data as from Hiroshima and Nagasaki. The need to account for cell killing (brief exposures) and the protraction of the dose became evident both usually having the effect of decreasing risk with dose. The shape of the dose-effect curve is still in contention (superlinear, linear, or sublinear), but the essential paradigm remains *risk* vs. *dose*.

2 NEOPLASTIC TRANSFORMATION, *in vitro* DATA

A new paradigm is suggested by studies of radiation-induced neoplastic transformation *in vitro*. Using the C3H mouse 10T1/2 cell system, the transformation of these cells was examined with X- and γ-rays (low-LET) and fission-spectrum neutrons, as a source of high-LET radiation, both at high-[1] and low-dose-rate exposures.[2,3] Figure 1 summarizes the results obtained with brief exposures;[1,3] when normalized for survival, the γ-ray and neutron (f-n) curves rise to plateaus apparently because cells that are induced to transform are as lethally sensitive as untransformed cells exposed to mid-to-large doses.

When low-LET exposures were protracted in time, as expected survival increased and concomitantly transformation decreased.[2] Our experience with protracted doses of neutrons was different, however. Survival was not changed with increasing time of exposure,[3] but surprisingly following small doses transformation was enhanced.[4-6] Figure 2 is a summary of the essential results. Normalized for cell killing at a high dose rate (D') the transformation curve rose to a maximum as in Fig.1. However, when the neutron dose was at a low D', in the region of small doses an enhanced transformation frequency was observed. As will be explained presently, the latter enhancement would result if a cell moiety sensitive to transformation is also a moiety sensitive to lethality. The enhancement shown in Fig. 2 at low D' was also observed with multifraction, high-dose-rate exposures.[6]

3 THE SENSITIVE WINDOW

To explain the anomalous results obtained with protracted doses of a high-LET radiation, a model was developed based on the assumption that a brief age-interval exists in the cell cycle when cells are sensitive to killing and transformation.[7] It was proposed that this window was in and around mitosis. The results that were obtained subsequently, Fig. 3, confirmed that the sensitive window consists of cells in late-G_2/M phase, an interval that is also the most sensitive to killing.[8]

4 CELL KINETICS & DEFICIENT REPAIR

Taken together, the results in Figs. 2 & 3 indicate that to observe enhanced transformation cells have to occupy the sensitive window, late-G_2/mitosis, during exposure. Cells that express contact inhibition of growth, as do untransformed 10T1/2 cells, progressively down-regulate their growth as they make cell-to-cell contact and accumulate out of cycle in G_o. In this respect, a down-regulated monolayer of 10T1/2 cells simulates many differentiated cell types *in vivo* which are non-cycling and, therefore, would not have cells in the sensitive window in contrast to actively cycling 10T1/2 cells, cell renewal systems *in vivo*, or cells which may be stimulated to cycle because of homeostatic controls following tissue injury, for example. In addition to adequate kinetics, repair deficiency is also required as illustrated by the observation that five daily fractions of high-D' neutrons were equally as effective in enhancing transformation by small total doses as were low-D' exposures.[6] Hence, the essential biophysical features of the model[7] that was proposed to explain the anomalous effect of protracted high-LET exposures are **kinetics** and **reduced** or **deficient repair**.

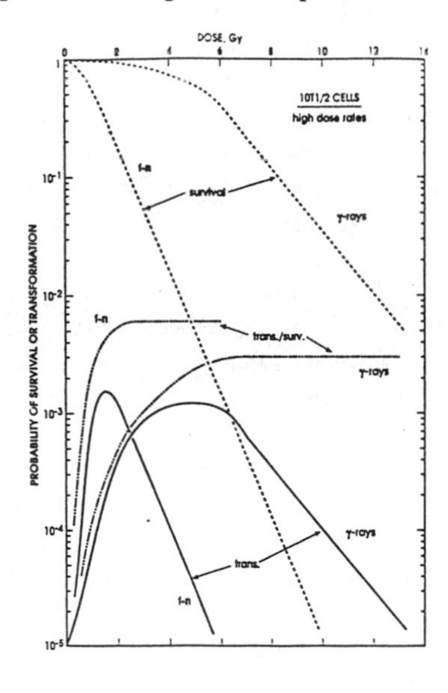

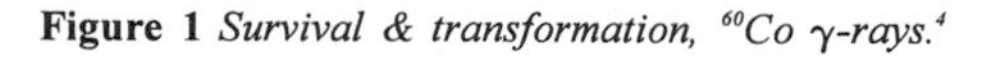

Figure 1 *Survival & transformation, ^{60}Co γ-rays.*[4]

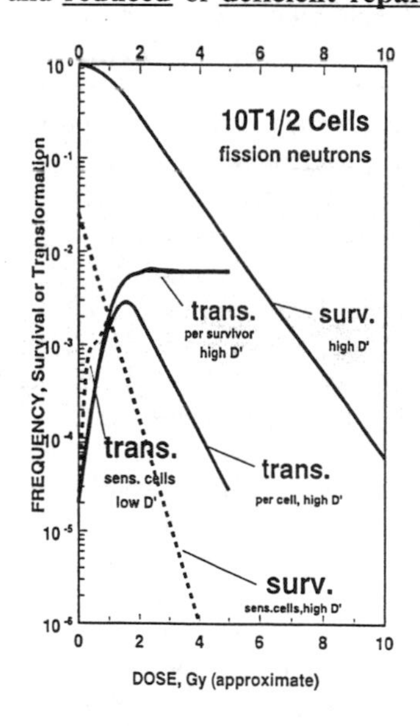

Figure 2 *Survival & transformation, high and low D'.*[4] *The dashed curves are for sensitive cells.*

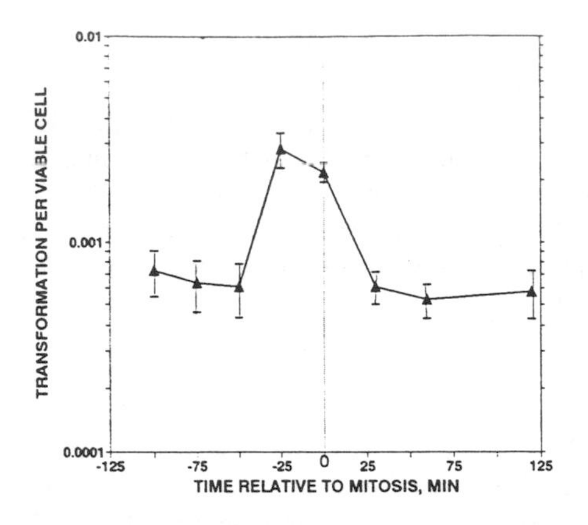

Figure 3 *Transformation window of sensitivity, late G_2/mitosis.*[7] *Dose = 3.5 Gy, ^{137}Cs γ-rays.*[8]

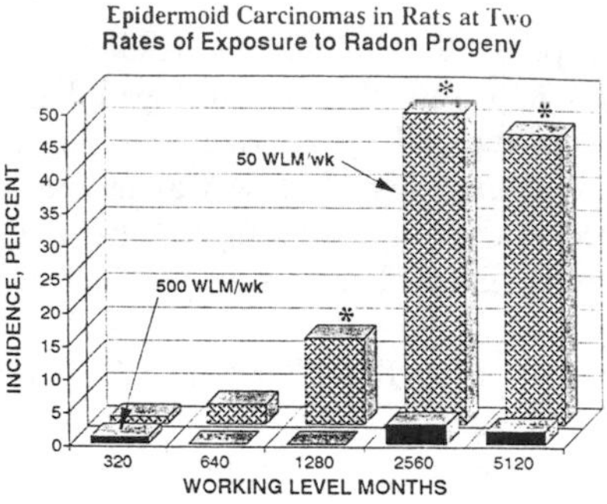

Epidermoid Carcinomas in Rats at Two Rates of Exposure to Radon Progeny

Figure 4 *Enhanced tumorigenesis in rats breathing radon at 50 vs. 500 WLM/week.*[9,10]

5 RADON & LUNG CANCER

The prediction of the model -- *enhanced transformation from protracted exposures* -- applies to results like those in Fig. 2 because of the reduced repair of neutron-induced damage. The high-LET α-particles emanating from the products of radon decay radiobiologically can be expected to act like fission neutrons leading to the expectation that enhanced neoplastic effects would be observed if the target cells in the lung were in or induced to cycle. Prolonged exposures and compensatory repopulation can be expected to occur if the degree of injury is great enough. An example of an enhanced yield of epidermoid carcinomas in rats breathing radon at two different rates of exposure, 50 & 500 WLM/week, is in Fig. 4. At exposures smaller than 320 WLM enhancement was not observed probably because the degree of injury and the periods of exposure were too small.[9,10] Enhanced tumorigenesis at reduced rates of exposure to radon have been reported for uranium and hard-rock miners.[11,12] (A number of other examples of the applicability of the model in the instance of protracted high-LET exposures have been cited.[13])

6 BREAST CANCER[13]

The formal requirements of the model are ***kinetics in the absence of repair***. *Kinetics* may depend on the cell-renewal character of the cells and tissue in question, homeostatic controls, lifestyle (as illustrated by tissue injury due to smoking), and other factors. *Absence of repair* even in repair-competent cells is a consequence of the nature of the damage produced by high-LET radiations. It can be expected that the model would also apply to protracted doses of low-LET radiations in those individuals or tissues whose target cells are deficient in repair. Most breast cancers arise in the ducts where the cyclic growth of the epithelium is under control of the periodic estrogen/progesterone secretions. Considerable cellular evidence exists to indicate deficient repair in a fraction of women too large to be accounted for by the frequency of ataxia telangiectasia heterozygotes in the population.[13] (Reference 13 contains only an incomplete list of these citations.) Hence, as predicted by the model, these women would be at an enhanced risk of radiation-induced breast cancer due to protracted exposures of low-LET radiation.

7 ELEMENTS OF A NEW PARADIGM

A simple, single-dose-effect paradigm for estimating risk is justified, aside from uncertainties in the data, if the target cells do not cycle, the dose is small, and the cells are competent in the repair of radiation damage. However, the latter stipulations may not always apply or may apply only in part. Protracted high-LET exposure is an example where repair competence would be moot. But most people are exposed to low-LET radiation over the course of a lifetime. Target cells may be naturally in cycle or induced to cycle depending on the dose, the inherent sensitivity of the cells (eg, in ataxia telangiectasia homozygotes repopulation may be sustained and extensive), and/or exposure to other cytotoxic agents. When a "hot spot" for an inductive process exists -- as illustrated for the neoplastic transformation of 10T1/2 cells, Fig. 3 -- a risk-assessment paradigm based on a *single dose-effect* becomes a limiting case whose justification may reflect limitations of the data more than an accounting for the essential biology.

A more general and therefore new paradigm would be one that accounts for ***cell kinetics, cell-cycle dependencies,*** **and** ***repair competence*** in addition degree of risk dependent on dose and the quality of the radiation. Not all of these elements may fully apply in a given case. But to account adequately for oncogenic changes in the molecular characterization of uncontrolled growth, the role(s) of inducing agents must be understood as in the instance of radiation-induced cancer.

Acknowledgements This work was supported by the U.S Department of Health & Human Services, Public Health Service, National Cancer Institute Grant No. CA47497, and U.S. Nuclear Regulatory Commission Grant No. NRC-04-94-103.

References

1. A. Han and M. M. Elkind, *Cancer Res.*, 1979, **39**, 133.
2. A. Han, C. K. Hill, and M. M. Elkind, *Int. J. Radiat. Biol.*, 1980, **37**, 585.
3. C. K. Hill, F. M. Buonagaro, C. P. Myers, A. Han, and M. M. Elkind, *Nature*, 1982, **298**, 67.
4. M. M. Elkind, *Int. J. Radiat. Biol.*, in press.
5. C. K. Hill, A. Han, and M. M. Elkind, *Int. J. Radiat. Biol.*, 1984, **46**, 11.
6. C. K. Hill, B. A. Carnes, A. Han, and M. M. Elkind, *Radiat. Res.*, 1985, **102**, 404.
7. M. M. Elkind, *Radiat. Res.*, 1991, **128**, S47.
8. J. Cao, R. L. Wells, and M. M. Elkind, *Int. J. Radiat. Biol.*, 1992, **62**, 191.
9. M. M. Elkind, *Int. J. Radiat. Biol.*, 1994, **66**, 649.
10. F. Cross, 'Radiation Research: A Twentieth-Century Perspective', Academic Press, San Diego, 1992.
11. S. C. Darby and J. M. Samet, 'Epidemiology of Lung Cancer', Marcel Dekker, New York, 1994.
12. J. H. Lubin, J. D. Boice Jr., R. W. Hornung, E. P. Radford, J. M. Samet, M. Tirmarche, A. Woodward, Y. S. Xiang, and D. A. Pierce, 'Radon & Lung Cancer Risk: A Joint Analysis of 11 Underground Miners Studies', U.S. National Cancer Institute, Bethesda, 1994.
13. M. M. Elkind, *Brit. J. Cancer*, 1996, **73**, 133.

COOPERATIVE BEHAVIOR OF IRRADIATED CELLS IN A THREE-DIMENSIONAL MODEL OF RADIATION CARCINOGENESIS

R. Bergmann[1], W. Hofmann[1], D. Crawford-Brown[1,2] and H. Oberhummer[3]

[1]Institute of Physics and Biophysics, University of Salzburg, A-5020 Salzburg, Austria
[2]Institute for Enviromental Studies, University of North Carolina, Chapel Hill, NC 27599-7400, USA
[3]Institute of Nuclear Physics, Technical University of Vienna, A-1040 Vienna, Austria

1 INTRODUCTION

The development of cancer in a living organism is not solely the result of a damage in the DNA-structure of individual cells, but communication between several cells can also have a major influence on the probability of the carcinogenesis process. Indeed, several studies[1-3] suggest that cells with existing damage can be inhibited by surrounding unaffected cells. Inhibition of communication between cells can have a major influence on the development of the cancer progress.[1-3] Therefore the development of cancer in a cell cluster might take a different course compared to cases where only independent cells are considered. For this study a Monte Carlo model was developed to simulate possible cooperative effects of spatially distributed cells in a multistage initiation-promotion model of radiation carcinogenesis.

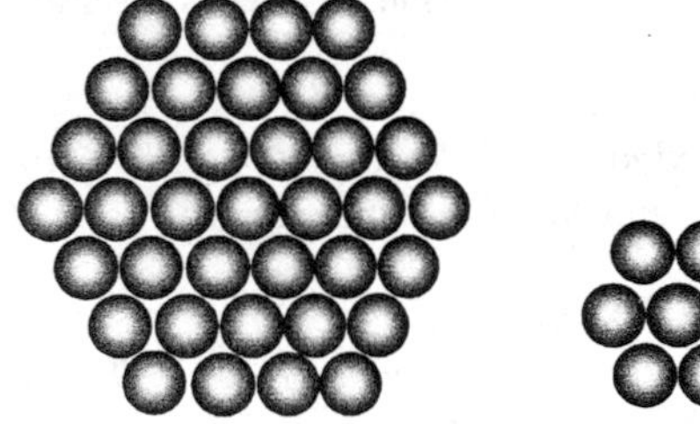

Figure 1 *Geometry of the cell cluster: One layer has 37 cells (left), and three layers are lying over each other (right)*

2 MODEL

Our model consists of a geometric cellular model (Figure 1), which simulates the dynamics of cellular growth and removal, and, a multistage cancer model for single cells to simulate the effect of radiation carcinogenesis for independent cells and cooperative cells, respectively.

The multistage model consists of four different states (Figure 2). It distinguishes between normal and initiated cells, also a reversible and irreversible promotional state of a cell was considered. This model is derived from the six stage cancer model of Crawford-Brown and Hofmann[4].

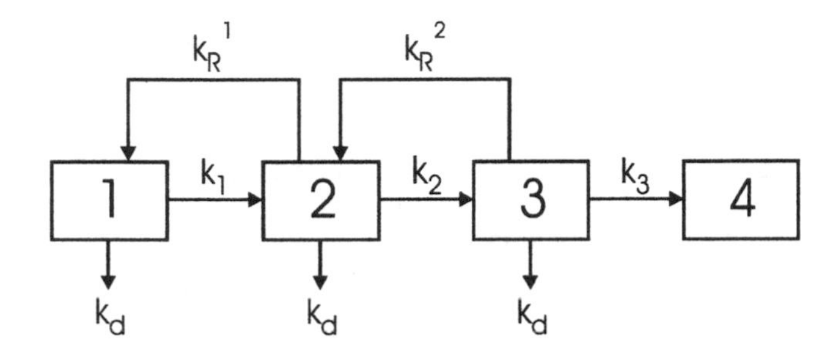

Figure 2 *Four-stage cancer model: state 1 - normal cell : state 2 - initiated cell : state 3 - reversibly promoted cell : state 4 - irreversibly promoted cell.*

The transition of a cell from one state to a new one for a given dose and time is determined by transition probabilities, which are derived from the six stage model.[4] In the cooperative part of our model, each cell is able to work as a signal receiver and emitter for the surrounding neighboring cells during one time cycle, where the sequence of communication between distinct cells is selected randomly. Communication between cells is assumed to occur via gap-junctions[5], therefore neighboring cells are primarily responsible for this effect and the cooperative effects in the model are simulated only by communication between neighboring cells. The cooperative effect is based on the idea of a loss of contact inhibition of initiated cells by surrounding dead cells. If an initiated cell communicates with a dead cell then an additional transition of the initiated cell to the reversible promotional state can occur. The transition probabilities for the two- or three dimensional cases are calculated such that an initiated cell is transformed to the reversible promotional state with a probability of one if all neighboring cells are dead.

The geometric cell structure was created of layers with 37 cells each, where each cell is surrounded by 6 cells in the two-dimensional case and by 12 cells in the three-dimensional case. The number of layers lying on each other can be selected. To maintain the symmetry in this model only the 19 inner cells in each layer can communicate with their neighboring cells. In this model cell death, cell removal and mitosis are also considered. Every time a cell is produced or removed, the cells are moved appropriatly in the cell cluster to simultate these mechanisms as realistically as possible. Therefore cells can move over time. Cells that are pushed out of the geometric boundaries are stored in a stack and re-enter the cell cluster, when a dead cell is removed by lysis. If the stack is empty, it is assumed that a normal cell enters the cluster. This mechanism is included to mirror the conditions beyond the cell boundaries.

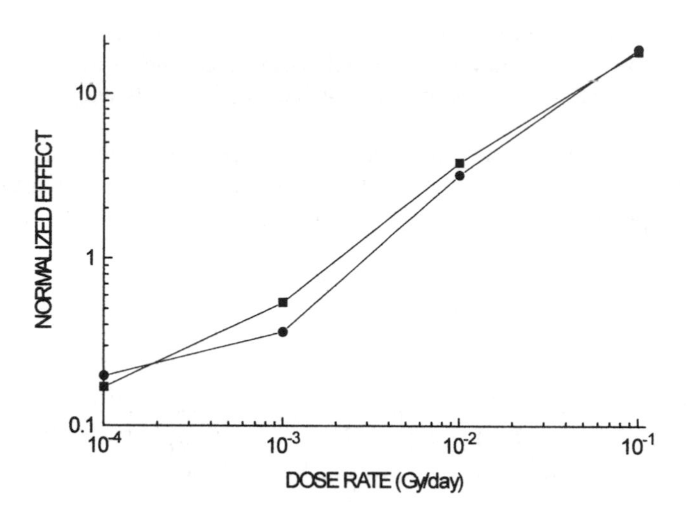

Figure 3 *Dose-rate effect curves for one layer of cells:*
■ *non-cooperating cells*
● *cooperating cells*

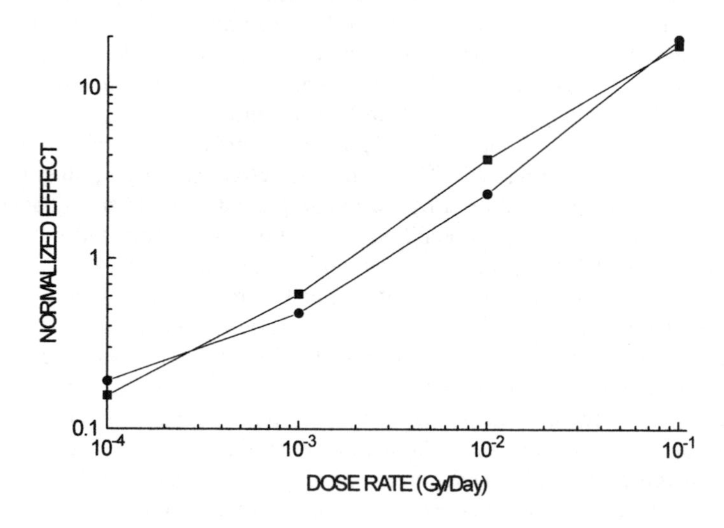

Figure 4 *Dose-rate effect curves for three layers of cells:*
■ *non-cooperating cells*
● *cooperating cells*

3 RESULTS

The behavior of this system was examined for one and three layers, that were irradiated with a constant X-ray dose rate. The simulations were stopped when the first irreversibly promoted cell occured (for 1 layer) or when three of them appeared (3 layers). This was assumed to normalize the occurence of an irreversibly promoted cell to the total number the cells. For every dose rate a distribution of simulation time was produced. The dose-rate effect in our model was defined to be inversely proportional to the time needed to reach a 60 percent probability of one (three) irreversibly promoted cells occurring. All curves were normalized to a total effect probability of one.

In Figures 3 and 4 the dose-rate effect curves in the cooperative case and in the case of independent cells for the one-dimensional and three-dimensional calculations are shown. For low doses, a slight decrease in the normalized effect was observed for the scenario of cooperating compared to not cooperating pairs of cells. There was no significant difference between the two- and three-dimensional calculations.

Despite the small effects shown here, preliminary calculations for other scenarios indicate that cell cooperation may be important. Obviously, cell killing and cell removal are very important factors for the evaluation of cooperative effects in cells.

4 CONCLUSIONS

The results presented here indicate that the cooperative effects between cells can lead to a delayed progress in cancer development and a nonlinear dose-response curve, predicting a smaller carcinogenesic risk at low doses than the commonly used linear dose-response model. Such a nonlinear dose-response function may have important implications for risk assessment purposes, for example in determining risk due to radon progeny in the human lung.

Acknowledgement

This research was funded by the CEC, Contract No. F14P-CT95-0025.

References

1. J.E. Trosko, C.C. Chang and B.V. Madhukar, *Radiat. Res.,* 1990, **123**, 241.
2. M.Z. Hossain et al., *Carcinogenesis*, 1989, **10**, 1743.
3. J.S. Bertram, *Radiat.Res.*, 1990, **123**, 252.
4. D.J. Crawford-Brown and W. Hofmann, *Math. Biosci.*, 1990, **115**, 123.
5. H. Mohamed, *Exp. Cell. Res.*, 1987, **168**, 422.

MODELLING ACUTE LYMPHOCYTIC LEUKAEMIA USING GENERALIZATIONS OF THE MVK TWO-MUTATION MODEL OF CARCINOGENESIS: IMPLIED MUTATION RATES AND THE LIKELY ROLE OF IONISING RADIATION

M.P. Little,[a] C.R. Muirhead[a] and C.A. Stiller[b]

[a]National Radiological Protection Board, Chilton, Didcot, Oxon, OX11 0RQ
[b]Childhood Cancer Research Group, 57 Woodstock Road, Oxford, OX2 6HJ

1 INTRODUCTION

One of the more frequently observed patterns in the age incidence curves for epithelial cancers is that the cancer incidence rate varies approximately as a power function of age. The multi-stage model of carcinogenesis of Armitage and Doll[1] (AD) was developed in part as a way of accounting for this approximately log-log variation of cancer incidence with age, at least for a variety of epithelial cancers in adulthood. A problem with this model is that it accounts less well for the patterns of age incidence for certain cancers of childhood, including retinoblastoma and leukaemia. The 2-mutation model developed by Knudson[2] to explain the incidence of retinoblastoma in children did so by taking account of the process of growth and differentiation in normal tissues. The stochastic 2-stage model of Moolgavkar and Venzon,[3] by taking account of cell mortality at all stages as well as allowing for differential growth of intermediate cells, generalized Knudson's model. In contrast with the AD model, there is a considerable body of experimental biological data supporting the initiation-promotion type of model.[4] Tan[4] has developed a number of generalizations of the 2-mutation model of Moolgavkar, Venzon and Knudson (MVK) and has documented much biological evidence in support of there being more than 2 rate-limiting mutations for certain cancer types. Recently Little[5 6] has presented a number of generalizations of the AD and MVK models and examined the behaviour of the excess risk when certain parameters are subject to small instantaneous perturbations. In this paper the generalized MVK model is fitted to UK acute lymphocytic leukaemia (ALL) incidence data.[7] ALL has a number of distinct subtypes, distinguishable by their immunophenotype. The CD10+ subtype accounts for most (> 60%) ALL cases.[7] The results of perturbing various of the parameters in the optimal CD10+ models are compared with what has been observed in radiation-exposed groups.

2 MATERIALS AND METHODS

The generalized MVK model supposes that at age t there are $X(t)$ susceptible stem cells, each subject to mutation to a type of cell carrying an irreversible mutation at a rate of $M(0)(t)$. The cells with 1 mutation divide into two such cells at a rate $G(1)(t)$; at a rate $D(1)(t)$ they die or differentiate. Each cell with 1 mutation can also divide into an equivalent daughter cell and another cell with a second mutation at a rate $M(1)(t)$. Similar processes take place at subsequent stages, until at the $(k-1)$th stage the cells with $(k-1)$ mutations acquire another mutation at a rate $M(k-1)(t)$ and become malignant. It can be shown[5] that the hazard function for the occurrence of the first malignant cell $h(t)$ may be written:

$$h(t) = -\int_0^t M(0)(s)\cdot X(s)\cdot \frac{\partial \Phi_1}{\partial t}[t,s]\, ds$$

where the functions $\Phi_i[t,s]$ $(1 \le i \le k)$ satisfy the equations:

$$\frac{\partial \Phi_i}{\partial s}[t,s] = [G(i)(s) + D(i)(s) + M(i)(s)]\cdot\Phi_i - G(i)(s)\cdot(\Phi_i[t,s])^2 - M(i)(s)\cdot\Phi_i[t,s]\cdot\Phi_{i+1}[t,s] - D(i)(s)$$

$$\frac{\partial^2 \Phi_i}{\partial t\,\partial s}[t,s] = [G(i)(s) + D(i)(s) + M(i)(s)]\cdot\frac{\partial \Phi_i}{\partial t} - 2\cdot G(i)(s)\cdot\Phi_i\cdot\frac{\partial \Phi_i}{\partial t} - M(i)(s)\cdot[\frac{\partial \Phi_i}{\partial t}\cdot\Phi_{i+1} + \frac{\partial \Phi_{i+1}}{\partial t}\cdot\Phi_i]$$

for $1 \le i \le k-1$ and for $s \le t$. The Φ_i also satisfy the boundary conditions: $\Phi_i[t,t] = 1$

$(1 \leq i \leq k\text{-}1)$; $\Phi_k[t,s] \equiv 0$; $\frac{\partial \Phi_k}{\partial t} \equiv 0$; $\frac{\partial \Phi_i}{\partial t}[t,s]\Big|_{s=t} = 0$ $(i < k\text{-}1)$; $\frac{\partial \Phi_{k-1}}{\partial t}[t,s]\Big|_{s=t} = -M(k\text{-}1)(t)$.
These equations are solved numerically to obtain h, which is fitted by maximum likelihood. In most fits the number of stem cells is assumed to be a constant, X_0, determined by the fit. For some fits (see Table 2) the number of stem cells is fixed, in order to better estimate the mutation rates M(i). Models are fitted in which the parameters are piecewise constant. In particular, models are fitted in which for some $T > 0$ the first mutation rate is given by $M(0)(t) = M(0)_-$ for $t \leq T$ and $M(0)(t) = M(0)_+$ for $t > T$, with similar expressions for the other parameters. In all fits the average age t is augmented by the average gestation length of 0.728 years (= 266/365.25). Further details on some of the model fits are given elsewhere.[8]

3 RESULTS

Table 1 *Fitted Parameters (+95% CI) for ALL 2-Mutation MVK Models, By Immunophenotype*

Parameter[a]	*Males*	*95% CI*	*Females*	*95% CI*
Null ALL				
T	12.2	(6.0, 17.2)	1.2	$(9.2 \times 10^{-1}, 3.2)$
$M(0)_+$	3.7	$(3.1 \times 10^{-1}, 1.6 \times 10^{5})$	3.3×10^{-2}	$(0.0, 9.2 \times 10^{-2})$
M(1)	3.1×10^{-3}	$(2.6 \times 10^{-9}, 1.0 \times 10^{-1})$	6.7×10^{-10}	$(1.1 \times 10^{-15}, 1.4)$
$G(1)_-$	11.2	$(3.9, 3.5 \times 10^{13})$	20.7	$(18.7, > 10^{15})$
$G(1)_+$	6.9×10^{-3}	$(< 10^{-15}, 7.2 \times 10^{-1})$	2.6	$(< 10^{-15}, > 6.1 \times 10^{2})$
CD10+ ALL				
T	3.2	(2.1, 6.1)	7.0	(6.2, 7.7)
$M(0)_+$	1.5×10^{-2}	$(8.4 \times 10^{-3}, 3.5 \times 10^{-2})$	2.5×10^{-2}	$(1.7 \times 10^{-2}, 3.6 \times 10^{-2})$
M(1)	9.7×10^{-4}	$(1.3 \times 10^{-5}, 1.1 \times 10^{-2})$	2.9×10^{-6}	$(3.8 \times 10^{-15}, 4.7 \times 10^{-2})$
$G(1)_-$	3.2	(1.5, 12.8)	10.5	(1.4, 15.0)
$G(1)_+$	3.3×10^{-1}	$(1.6 \times 10^{-1}, 6.1 \times 10^{-1})$	9.2×10^{-1}	$(1.3 \times 10^{-1}, 1.8)$
B-ALL				
T	6.1	$(1.6 \times 10^{-14}, 9.7)$	2.2	$(2.1 \times 10^{-5}, 6.2)$
$M(0)_+$	5.0×10^{-1}	$(0.0, 1.2 \times 10^{9})$	2.2×10^{-2}	$(6.8 \times 10^{-3}, 6.5 \times 10^{-2})$
M(1)	2.3×10^{-2}	$(1.8 \times 10^{-15}, > 10^{15})$	2.8×10^{-11}	$(1.8 \times 10^{-15}, 1.4 \times 10^{-1})$
$G(1)_-$	6.3	$(1.6, > 10^{15})$	20.8	(16.7, 23.4)
$G(1)_+$	8.1×10^{-2}	$(< 10^{-15}, 3.4 \times 10^{7})$	1.8	$(5.1 \times 10^{-1}, 4.5)$
T-ALL				
T	1.3	$(1.6 \times 10^{-5}, 2.1)$	23.2	(20.9, 27.2)
$M(0)_+$	2.0×10^{-2}	$(1.9 \times 10^{-7}, 3.6 \times 10^{-2})$	1.7×10^{5}	$(2.4 \times 10^{-1}, > 10^{15})$
M(1)	1.1×10^{-10}	$(< 10^{-15}, 9.8 \times 10^{-2})$	9.1×10^{-9}	$(3.8 \times 10^{-13}, 1.6 \times 10^{-2})$
$G(1)_-$	18.4	$(15.1, > 10^{15})$	11.9	$(1.9, 1.3 \times 10^{8})$
$G(1)_+$	1.2	$(7.6 \times 10^{-1}, 1.6)$	7.2×10^{-2}	$(< 10^{-15}, 1.0)$

[a]G(1), M(1) have units of $cell^{-1}$ $year^{-1}$; T has units of year; M(0) normalized so that $M(0)_- \equiv 1$.

As is shown elsewhere,[8] among various models considered, with between 1 and 3 stages, one version of the 2-mutation model provides an acceptable fit to each of the four ALL immunophenotypes. The model assumes that there is a step change in the first mutation rate (M(0)), or alternatively in the number of susceptible stem cells, and a change at the same age in the intermediate cell growth rate (G(1)). A similar 3-mutation model fitted each ALL immunophenotype not much worse than the above 2-mutation model.[8] This 3-mutation model also predicts a step change in the first mutation rate (M(0)), or alternatively in the number of susceptible stem cells, and a change at the same age in the two intermediate cell compartment growth rates (G(1), G(2)).[8] The 1-mutation model did not provide a satisfactory fit to any of the ALL immunophenotypes.[8] There is reasonable agreement of the time/age trends of the predicted excess in the fitted 2- and 3-mutation CD10+ models with those observed for ALL in the Japanese A-bomb incidence dataset[9], if one assumes that radiation acts to perturb the first mutation rate M(0) in either model (Figure 2). There are indications of lack of fit in the older age-at-exposure group, particularly if radiation acts to elevate mutation rates in either model for life, a possibility suggested by the work of Kadhim et al.[10] The small number of A-bomb ALL cases (32) implies that significance should not be attached to this inconsistency.

Figure 1 *Fitted and Observed CD10+ ALL Incidence in the LRF Study*[7]

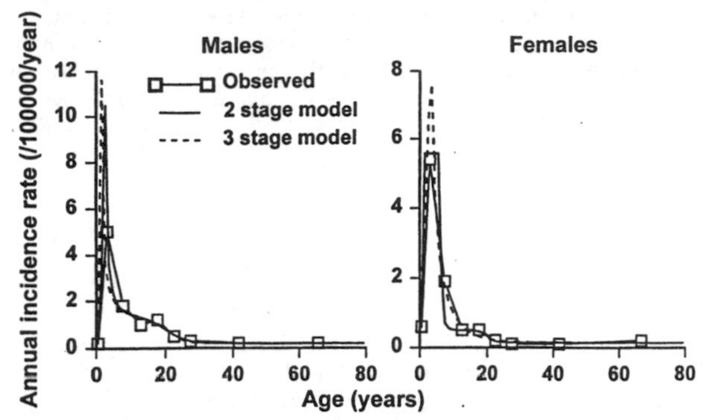

Figure 2 *Excess Male CD10+ ALL Incidence Resulting from Perturbation of Parameters in 2- and 3-Mutation Models, and Comparison with Excess in Japanese A-Bomb ALL Data*[9]

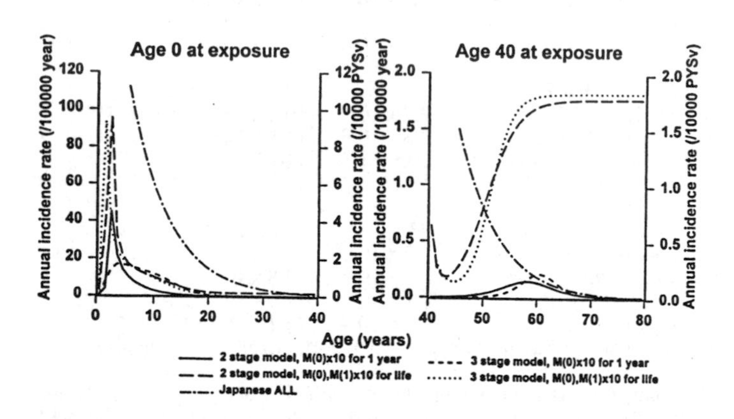

Table 2 *Estimated Mutation Rates for Optimal 2-Mutation MVK Model fitted to CD10+ ALL in the LRF Study*[7] *for Specified Values of Number of Stem Cells* (X_0)

Stem Cell Numbers ($=X_0$)	*Estimated Mutation Rates* ($= M(0)\_ = M(1)$) ($cell^{-1}$ $year^{-1}$) *Males*	*Females*
10^0	6.8×10^{-4}	6.2×10^{-4}
10^1	7.8×10^{-5}	6.4×10^{-5}
10^2	8.9×10^{-6}	6.5×10^{-6}
10^3	9.2×10^{-7}	2.1×10^{-7}
10^4	9.5×10^{-8}	2.8×10^{-8}
10^5	3.5×10^{-9}	2.8×10^{-9}
10^6	3.5×10^{-10}	2.7×10^{-10}
10^7	3.3×10^{-11}	2.7×10^{-11}
10^8	3.5×10^{-12}	2.8×10^{-12}
10^9	3.6×10^{-13}	3.1×10^{-13}
10^{10}	3.8×10^{-14}	3.0×10^{-14}

4 DISCUSSION

It appears that the common ALL subtypes can be reasonably well described by stochastic models which assume that 2 or 3 mutations are required to cause malignancy. Another notable feature is that marked age-variations in various of the model parameters are required to describe the age-incidence patterns of the respective ALL subtypes. The relatively small number of stages required by the optimal models for ALL is consistent with the observations in various radiation-exposed groups, in which excess leukaemias are manifest typically within a few years of exposure.[11] Little *et al.*[6 12] concluded from an analysis of the Japanese atomic bomb survivor data using MVK and AD models that the excess leukaemia incidence could best be explained by a model with 3 stages, with radiation acting to perturb the first and second mutation rates (M(0), M(1). The analysis of this paper suggests that ALL may be adequately described by 2- or 3-mutation models in which radiation acts on the first mutation rate M(0). Most of the optimal ALL models imply a significant reduction in the stem cell population or, equivalently, a reduction in the first mutation rate (M(0)). Sansoni *et al.*[13] document significant reductions in the blood concentrations of T- and B-lymphocytes with increasing age, as well as in the concentrations of various lymphocyte subpopulations (CD4+, CD8+). To this extent, the reductions that we found implied by the optimal CD10+ and B ALL models are plausible. The mutation rates implied by the optimal models are generally in the range 10^{-2} - 10^{-11} $cell^{-1}$ $year^{-1}$, albeit with very large confidence intervals. Mutation rates of the HPRT gene have been estimated to be of the order of 7×10^{-8} - 7×10^{-7} $cell^{-1}$ $year^{-1}$.[14] Rather larger mutation rates, 2×10^{-4} $cell^{-1}$ $year^{-1}$, can be inferred for stable chromosomal abnormalities.[15 16] Given the very wide confidence intervals on the mutation rates displayed in Table 1 none of the models yields mutation rates incompatible with those which have been measured. The analysis of Table 2 suggests that the stem cell population must be very small ($< 10^4$ cells) if the 2-mutation model is not to yield implausibly low mutation rates.

5 ACKNOWLEDGEMENTS

The CCRG is supported by the Department of Health and the Scottish Home and Health Department. This work was funded partially by the Commission of the European Communities under contract FI4P-CT95-0011.

References

1. P. Armitage and R. Doll, *Br. J. Cancer*, 1954, **8**, 1.
2. A.G. Knudson, *Proc. Natl Acad. Sci. U.S.A.*, 1971, **68**, 820.
3. S.H. Moolgavkar and D.J. Venzon, *Math. Biosci.*, 1979, **47**, 55.
4. W.-Y. Tan, 'Stochastic models of carcinogenesis', Marcel Dekker, New York, 1991.
5. M.P. Little, *Biometrics*, 1995, **51**, 1278.
6. M.P. Little, *J. Radiol. Prot.*, 1996, **16**, 7.
7. P.A. McKinney, F.E. Alexander *et al.*, *Leukemia*, 1993, **7**, 1630.
8. M.P. Little, C.R. Muirhead *et al.*, *Statist. Med.*, 1996, **15**, 1003.
9. D.L. Preston, S. Kusumi *et al.*, *Radiat. Res.*, 1994, **137**, S68.
10. M.A. Kadhim, S.A. Lorimore *et al.*, *Lancet*, 1994, **344**, 987.
11. M.P. Little, *J. Radiol. Prot.*, 1993, **13**, 3.
12. M.P. Little, M.M. Hawkins *et al.*, *Radiat. Res.*, 1992, 132, 207, 1994, **137**, 124.
13. P. Sansoni, A. Cossarizza *et al.*, *Blood*, 1993, **82**, 2767.
14. M.H.L. Green, J.P. O'Neill *et al.*, *Mutat. Res.*, 1995, **334**, 323.
15. E.J. Tawn, *Mutat. Res.*, 1987, **182**, 303.
16. J.D. Tucker, D.A. Lee *et al.*, *Mutat. Res.*, 1994, **313**, 193.

ANALYSIS OF LUNG CANCER AFTER EXPOSURE TO RADON USING A TWO-MUTATION CARCINOGENESIS MODEL

H P LEENHOUTS,[1] P A M UIJT DE HAAG[1] AND K H CHADWICK[2]

[1] RIVM, P.O. Box 1, 3720 BA Bilthoven, The Netherlands

[2] CEC, DG XII/F/6, rue de la Loi 200, Brussels, Belgium

1 INTRODUCTION AND OBJECTIVES

Radon is the largest contributor of the natural radiation dose to the population. The uranium miner data provides the most direct source of information on lung cancer risk due to radon. The analysis of lung cancer in uranium miners is important for estimating the radiation risk that indoor radon poses for the population. However, analysis of the data is complicated by the fact that lung tumours are also caused by smoking; moreover, smoking and radon may act synergistically.[1]

In our study a two-mutation carcinogenesis model[2,3] is used to analyse and explain the induction of lung tumours by radon and smoking. In an earlier study, Moolgavkar et al.[4] carried out a combined analysis of radon and cigarette smoking using a different two-stage carcinogenesis model. However, in the present investigation a somewhat different approach using fewer variables is chosen.

2 METHODS AND RESULTS

The two-mutation carcinogenesis model used to analyse the data is described by Leenhouts and Chadwick[2] and is shown schematically in Figure 1.

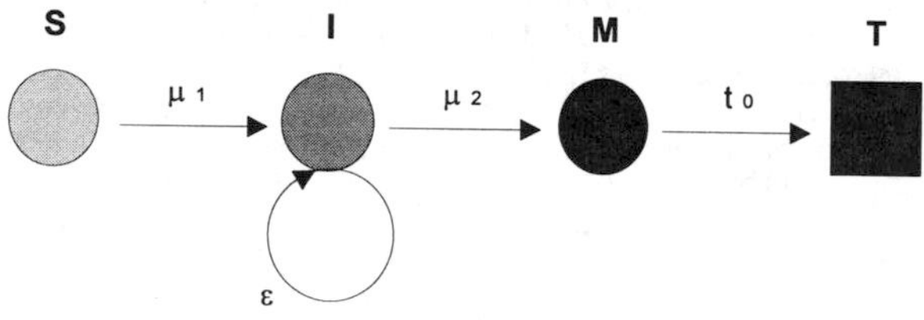

Figure 1. Schematic representation of the two-mutation carcinogenesis model.[2]

It is assumed that in a normal somatic (stem) cell (S), two mutations (μ_1 and μ_2) have to be induced to make the cell malignant. First, a normal cell enters in an intermediate stage (I) after one transformation; second, in the intermediate stage (I) it will participate in the cell division process and, in general, the number of intermediate cells will increase at a rate, ε, per unit time; third, the intermediate cell can undergo a second transformation and become malignant (M). After a given time period (t_e), a malignant cell will grow into a tumour (T), which can be detected, either as a visible tumour or as death due to the tumour.

The mathematics of the model used is given in Leenhouts and Chadwick.[2] The rate coefficients, μ_1, μ_2 and ε, are assumed to be independent of age. The number of stem cells (S) is assumed to increase proportionally with age during the first 20 years and be constant (10^7 per individual) for later years. The cell numbers I, M and T, assumed to be zero at the beginning of a lifetime, are calculated in 5-year steps.[2] For simplicity, it is assumed that both background mutation rates are equal, i.e. $\mu^{bg} = \mu_1^{bg} = \mu_2^{bg}$. Further, it is assumed that smoking and radon influence the mutation rates but do not change the expansion rate ε or the time t_e between the cell becoming malignant and tumour detection, which is assumed to be five years.

The working procedure of the analysis is to successively determine the coefficients of the model as described below.

2.1 Background lung tumour incidence

Population statistics on lung cancer incidence in non-smokers are used to determine the coefficients μ^{bg} and ε of the model for the background incidence of lung cancer. The data are from Peto et al.[5] Figure 1 shows the fit of the model to cumulative incidences in non-smoking males up to the age of 80.

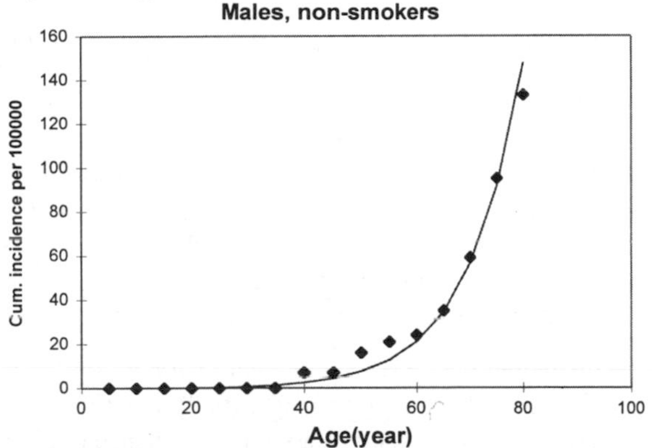

Fig. 2. Cumulative lung-tumour incidence with age in non-smoking males as presented by Peto et al.[5] The curve represents calculations using the model.

2.2 Effect of smoking

The influence of smoking is determined from epidemiological data on smoking in British radiologists by Doll and Peto.[6] Data for non-smokers and regular smokers constantly smoking since the age of 20, are used to determine the increase in the background mutation rates with smoking rate. The data are fitted using the model and assuming a linear dependence of the mutation rates on smoking rate σ equally for both mutations:

$$\mu_1 = \mu_2 = \mu^{bg} (1 + a \times \sigma) \quad (1)$$

where a is the increment of the mutation rate per unit smoking rate σ.

To fit the model to the data, μ^{bg} and ε are taken as determined (see section 2.1), and only σ is varied to find the best fit. To illustrate the results, Figure 3 shows cumulative incidence at age 80 as a function of smoking rate epidemiologically and as calculated by the model.

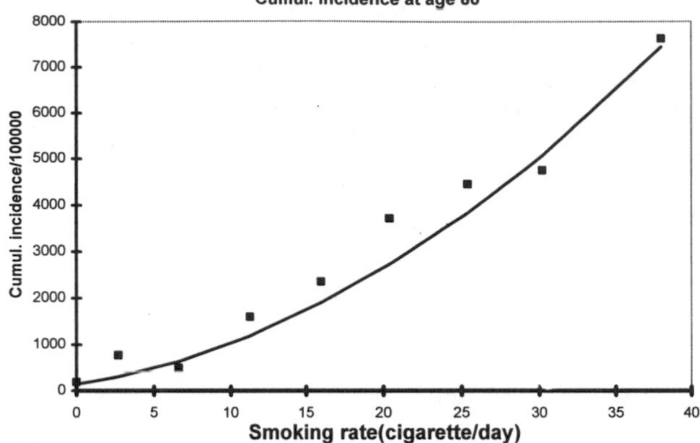

Fig. 3. Cumulative incidence of lung tumours for 80-year-old British doctors[6] as a function of smoking rate. The curve represents calculations using the model.

2.3 Effect of radon

The effect of radon is determined using the Colorado miner data, used several times in the past to analyse radon effects.[1,7,8] The data were obtained directly from the National Institute of Occupational Safety and Health (NIOSH, OH, USA). Only data for non-smokers and regular cigarette smokers are used in the present analysis. The total exposure in working-level month (WLM), as given by the data, is used as a basis for the accumulated individual exposure to radon. For the model it has been assumed that this exposure to the miners took place between the ages of 30 and 50. Three smoking groups are considered: non-smokers, smokers of one pack per day or less (average 0.76 pack per day), and smokers of more than one pack per day (average 1.14 pack per day).

The effect of radon on the mutation rate (i.e. at cellular level) is supposed to be proportional to the annual exposure, X, to radon (in WLM a^{-1}), and to the smoking rate σ. Further, the alpha radiation of radon is assumed to also have a chance of killing the exposed cells. The total effect on the mutation rate is given by:

$$\mu_i = \mu^{bg} (1 + a \times \sigma + b_i \times X) \exp(-s/q \times b_i \times X) \qquad (2)$$

where i is 1 or 2 and indicates the first or second mutation,

b_i is the increment of μ_i per unit radon exposure rate

and s/q is the ratio of the cell-killing and mutation-induction action of radon at the cellular level.[9]

For the fitting procedure, μ^{bg}, ε and σ are taken as determined (see section 2.1 and 2.2); only b_1, b_2, and s/q are varied to find the best fit. Figure 4 shows the cumulative incidence, at age 80, for non-smoking miners and miners regularly smoking 0.76 packs per day, and the fit of the model. Clearly, different radon effects in both groups can be observed, both in the epidemiological data and in the model calculations.

3 DISCUSSION

The modelling results show a consistent pattern of lung tumour incidence as a function of age in non-smokers, in smokers and in miners exposed to radon. Only six coefficients are assumed to be unknown and are used to find the best fit. The fitting of the model reveals a much larger radiation influence on the first than on the second mutation rate, possibly indicating a different nature of the two mutational steps.

Although the effects of smoking and radon at cellular level are assumed to be non-synergistic and only additive, their effects on tumour induction are more complicated. For example, using the model calculations for the miners, the excess *absolute* radon risk of 50 WLM radon can be calculated to be about 7 times larger for smokers than for non-smokers, but the excess *relative* radon risk is about 0.2 times smaller.

Although the model fitting may change when different assumptions are used for the analysis, the results show the importance of characterizing smoking habits for estimating the risks of radon. They also demonstrate the ability of the simplified two-mutation model to analyse tumour dependence on age, radon and smoking in epidemiological studies.

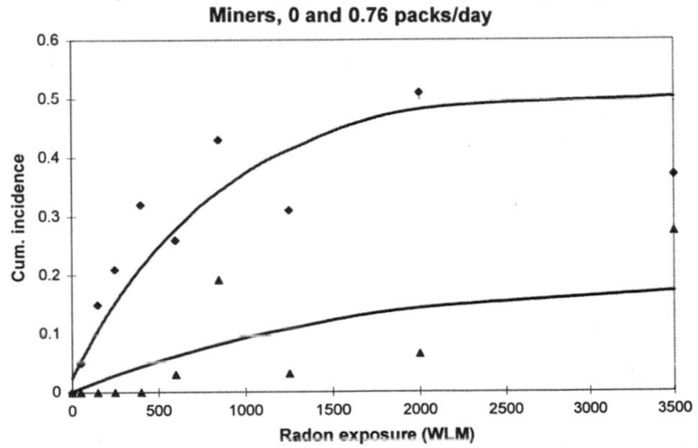

Fig. 4. Cumulative lung tumour incidence at age 80 for non-smoking miners (Δ) and miners regularly smoking 0.76 pack/day (◊) as a function of total exposure to radon. The curves represent calculations using the model.

References

1. J.H. Lubin et al., 'Radon and lung cancer risk: A joint analysis of 11 underground miners studies', NIH Publication No. 94-3644, 1994.

2. H.P. Leenhouts and K.H. Chadwick, *J. Radiol. Prot.*, 1994, **14,** 115-130.

3. H.P. Leenhouts and K.H. Chadwick, *Radiat. Prot. Dosim.*, 1994, **52,** 465-469.

4. S.H. Moolgavkar et al., *Epidemiology,* 1993, **4,** 204-217.

5. R. Peto et al., *Lancet,* 1992, **339,** 1268-1278.

6. R. Doll and R. Peto, *J. Epidem. Commun. Health,* 1978, **32,** 303-313.

7. R.W. Hornung and T.J. Meinhardt, *Health Phys.*, 1987, **52**, 417-430.

8. A.S. Whittemore and A. McMillan, *J. Natl.Cancer Inst.*, 1983, **71**, 489-499.

9. K.H. Chadwick and H.P. Leenhouts, 'The Molecular Theory of Radiation Biology', Springer Verlag, Heidelberg, 1981.

Microdosimetry applied to radiotherapy

MICRODOSIMETRIC CONSIDERATIONS IN THE TARGETED RADIOTHERAPY OF CANCER

T. E. Wheldon[1] and J. A. O'Donoghue[2]

[1]Departments of Radiation Oncology and Clinical Physics
Glasgow University and West Glasgow Hospitals NHS Trust
CRC Beatson Laboratories
Glasgow G61 1BD, UK

[2]Department of Medical Physics
Memorial Sloan-Kettering Cancer Center
New York, NY 10021, U.SA.

Summary
Targeted radiotherapy consists of the selective irradiation of tumour cells using radionuclides conjugated to tumour-seeking biomolecules. Beta-emitting and alpha-emitting radionuclides enable the irradiation of untargeted cells by 'cross-fire' from their radiolabelled neighbours. For these radionuclides, microdosimetric considerations of absorption of radiation energy suggest an optimal tumour size for radiocurability and have given rise to rationales for combined modality therapy of neuroblastoma and other tumours. Clinical evaluation of treatment strategies based on these rationales is now in progress. Auger electron emitters are generally considered to be radiotoxic only if targeted to DNA, with negligible cross-fire between adjacent cells. Microdosimetric considerations are important in selecting between alternative Auger emitters (eg ^{125}I or ^{123}I) and in gauging the efficacy of Auger emitters which are incorporated as part of the DNA structure (IUDR targeting) or which are bound to the DNA as part of the targeting adduct (radiolabelled hormones and gene targeting agents). Some recent evidence suggests a novel mechanism of Auger electron cell killing : by triggering apoptosis following irradiation of the cell membrane. If substantiated, this would create new possibilities for targeted radiotherapy, but will pose new microdosimetric questions on the energy absorption distribution at the level of the cell.

1 Introduction

Ionising radiation is presently one of the most generally useful modalities in oncology. Though variation of intrinsic cellular radiosensitivity between tumours may be of major importance in clinical outcome, the range of variability observed is much less than that for sensitivities to cytotoxic drugs. Also, the development of cellular resistance to radiation occurs less often than the emergence of clonal chemoresistance. However, conventional radiotherapy unavoidably entails irradiation of normal organs. It is the vulnerability of critical normal tissues, rather than the existence of absolutely radioresistant tumour cells, which limits the effectiveness of radiation treatment. Radiation is a good choice of cytotoxic agent for biologically selective targeting to tumour cells, for which a range of radionuclides are available (table 1).

Table 1: *Physical properties of radionuclides useful for targeted therapy*

Radio-nuclide	*Half-life*	*Emitted particle*	*Mean particle range*
^{90}Y	2.7 days	β	5 mm
^{131}I	8 days	β	0.8 mm
^{67}Cu	2.5 days	β	0.6 mm
^{199}Au	3.1 days	β	0.3 mm
^{211}At	7 hr	α	0.05 mm
^{212}Bi	1 hr	α	0.05 mm
^{125}I	60 days	Auger	~ 1 μm
^{123}I	15 hr	Auger	~ 1 μm

An important aspect of radiation therapy is the relatively advanced state of quantitative dosimetry and the existence of dose response relationships for tumour cell kill. Although dosimetry is less reliable for radionuclide therapy than external beam irradiation it nevertheless affords possibilities for the rational design of treatment strategies. Microdosimetry is especially relevant to this as tumour sterilisation requires consideration of radiation energy absorption on a cellular or a subcellular scale.

2 Targeting agents

2.1 Antibody targeting

Athough monoclonal antibodies now have demonstrable usefulness in pathological diagnosis, evidence of therapeutic efficacy is still very limited. Approaches to improve the therapeutic effectiveness of radiolabelled antibodies presently include antibody engineering techniques[1,2] and the exploration of local targeting strategies which might allow higher antibody uptake to be achieved [3].

2.2 Exploitation of biosynthetic peculiarities

Targeting agents other than antibodies are now attracting increasing attention. A promising example is meta-iodo-benzyl-guanidine (mIBG) which is preferentially taken up by catecholamine-synthesising cells. Malignant tumours of sympathetic nervous tissues, particularly neuroblastoma and phaeochromocytoma, often retain the property of high mIBG uptake, allowing this agent to be used for delivery of radio-iodine [4,5]. Similar approaches might be possible for melanoma[6] and other tumours.

2.3 New targeting agents

Molecular biology has provided a new class of agent. Some tumours, including gliomas[7] and squamous carcinomas[8] over-express the cellular receptor for epidermal growth factor (EGF). Administration of radiolabelled EGF should lead to at least partially selective radionuclide delivery to these tumours. Experimental studies have shown that it is possible to achieve preferential cell kill in vitro of cells over-expressing the EGF receptor by means of ^{131}I labelled EGF[9.10]. Preferential uptake of radiolabelled nerve growth factor (^{131}I-NGF) by neuroblastoma cells expressing the NGF receptor has also been demonstrated[11] A future challenge for targeted therapy is to exploit new molecular knowledge to effect tumour cell kill or gene inactivation by means of highly selective DNA targeting of radionuclides with sub-nuclear ranges of emission (Auger emitters) [12].

3 Dosimetry of targeted radiotherapy

The distribution of tumour dose as a result of radionuclide targeting is complex. Some tumour cells will not be accessible to the targeting agent some cells may not express the targeting receptor or may be in cell cycle phases or other phenotypic states which are less amenable to targeting by that agent.Therefore, the cellular distribution of radionuclide deposition will be non-uniform. How this affects the distribution of absorbed dose will depend on the energy spectrum of particles emitted and on the resultant distribution of particle path lengths. The cumulative dose distribution will also depend on the kinetics of uptake physical decay and biological clearance of the radionuclide. It is possible to derive some useful principles for the simple case of uniform uptake. This idealised scenario approximates to reality for small tumours (especially microtumours) treated with a small-molecule targeting but is a less good approximation for larger tumours and poorly diffusing agents.

3.1 *The microdosimetry of uniformly-targeted beta emitters*

Because beta particle path lengths (typically of mm dimensions) are many times greater than the diameter of a cell, these radionuclides deliver a large part of their dose by *cross-fire* ie the flux of beta particles to which a particular cell is exposed depends principally on the radionuclide concentration in its neighborhood). This is an advantage in averaging out small scale inhomogeneities in targeting, which diminishes with increasing scale of targeting inhomogeneities. However, very small microtumours or will receive less dose than larger tumours because of the relative absence of cross-fire. The fraction of the disintegration energy which is absorbed within uniformly radiolabelled spheres of differing size has been computed for variety of beta emitters. The absorbed fraction increases with sphere size until the sphere diameter exceeds the mean path length, then asymptotically approaches unity. Small microtumours whose diameter is less than that of the mean path length receive significantly less dose for any given activity of targeted radionuclide. Of course, larger tumours contain more clonogenic cells than smaller ones, so larger tumours would be less radiocurable than smaller ones, for any given dose level. The conflict of the cell number factor and the absorbed dose factor results in a peak radiocurability which occurs at the same tumour diameter for all activity levels. Detailed computation shows that the radiocurability peak occurs at a diameter which is somewhat greater than the mean path length of emitted beta particles. For ^{131}I the diameter for maximum radiocurability is around 3 mm corresponding to a sub-clinical micro tumour. Table 2 gives the corresponding results for a number of other beta emitters.

Table 2 : *Optimal tumour diameter for radiocurability by uniformly targeted beta emitters (abridged from data of O'Donoghue et al[13])*

Radio-nuclide	*Mean energy / disintegration (keV)*	*Optimal cure diameter (mm)*
^{199}Au	142.4	0.8
^{67}Cu	154.1	2.0
^{131}I	192.3	3.4
^{90}Y	939.1	34.0

Although the numerical results from this idealised model should not be taken too literally, the principle of the existence of an optimal diameter (ie that the very smallest tumours are not necessarily the easiest to cure) and the rank order of optimal diameters for various radionuclides are probably robust conclusions.

3.2 Combined modality treatments including targeted beta-emitters

Targeted radiotherapy is most often used clinically for patients who have disseminated malignant disease (ie a 'size spectrum' of malignant disease). Targeted beta-emitters are best suited to treatment of larger micrometastases (millimetre dimensions) but are not ideal for smaller micrometastases (down to single cells). This suggests a combination strategy in which non-interchangeable treatment modalities are deployed in combination to ensure that no particular tumour is likely to escape sterilisation by virtue of the size class to which it belongs . The argument has been developed mathematically[14] and it has been shown that the superiority of the combined modality approach, over any single treatment (for equal toxicity to normal tissues) appears to be a robust conclusion. Such strategies are being explored clinically for treatment of neuroblastoma, using ^{131}I-conjugated MIBG as the targeted radiotherapy component. In one approach, relapsed neuroblastoma patients receive ^{131}I-MIBG followed by external beam TBI (at reduced dose) then autologous bone marrow rescue and local radiotherapy to macroscopic tumours if necessary[15]. Recently, a UK multi-centre clinical study has commenced of 'first-line' ^{131}I-MIBG treatment followed by multi-agent chemotherapy (and surgery if appropriate) for newly presenting neuroblastoma patients with unfavourable biological markers ie indicating poor prognosis on conventional The approach may be generalisable to other tumours[17].

4 Targeted radiotherapy using Auger electron emitters

In contrast to the beta emitters, Auger emitters deliver radiation energy over ultra -short ranges (mm dimensions) . Auger emissions are effectively short-range high LET radiation for which long-range cross-fire makes almost no contribution. For targeted radiotherapy, this is a mixed blessing, allowing high-precision and high-efficiency killing of individual targeted cells but allowing untargeted cells to escape. DNA -targeting is considered mandatory for cell kill and this theoretical expectation is supported by a body of experimental evidence[18]. Possible exceptions to this will be considered below. The microdosimetry of Auger emitters has attracted considerable interest although it is recognised to be rather complex. Martin and Haselein[19] showed that for Auger electron emitters substituted into the DNA double helix (as ^{125}I-IUDR) DNA strand breakage was substantial within a few base pairs of the site of the substitution but fell off dramatically at longer distances. Auger emitters also differ somewhat in their efficacy of double strand breakage. Humm and Charlton[20] provided estimates for this and showed that an efficacy of about (or slightly less than) 1 double strand per disintegration was fairly typical. This allowed the cell killing efficacy of an Auger emitter to be expressed as the number of atoms originally incorporated in the DNA of each tumour cell in order to achieve any given efficacy of cell sterilisation (eg 1 log kill). The Humm-Charlton analysis was originally applied to the case of Auger emitters substitutionally incorporated in the DNA of cells proliferating with a rapid cell cycle time (48h) which would be typical for cells in exponential growth phase in culture. It is instructive to extend the model to include a range of doubling times[21]. It is then seen that doubling time influences cell kill achieved by an initially targeted bolus of radioactive atoms by two mechanisms - the dilution of the targeted radionuclide by successive cell divisions (if these can occur during the irradiation), and the rate of repopulation of surviving cells at any stage of the therapy. This is important for long-lived radionuclides such as ^{125}I for which large numbers of atoms have to be given initially, since only a fraction of these atoms will experience radioactive decay during the cell cycle time scale.

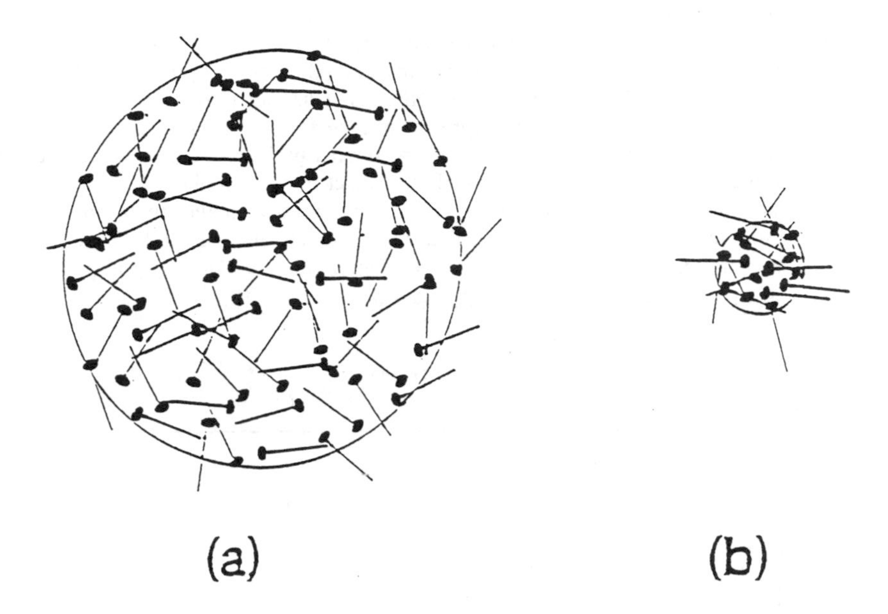

Figure 1 : *This shows the dependence of the number of atoms/ genome of ^{123}I and ^{125}I to achieve 1 log kill (10^{-1} surviving fraction) as a function of cell population doubling time.*

Figure 1 shows the number of initially targeted atoms (per DNA duplex in each cell) which have to be given to achieve 1 log kill, for the case of ^{125}I (half life ~ 60 days) and of ^{123}I (half life ~ 15 h). This shows that ^{123}I is little affected by doubling times down to 1 day, whilst the required number of ^{125} I atoms is strongly dependent on doubling time. For most of the range, ^{123}I is superior to ^{125} I (in terms of numbers of atoms required) despite the radiobiological advantage of ^{125}I over ^{123}I (1.1 DSB's / disintegration compared to 0.7 DSB's / disintegration). For clinical therapy, the number of atoms initially targeted should ideally be kept as small as possible, to avoid possible saturation of receptors or binding sites, but also to avoid the presence in the body of a large excess of radioactive atoms not destined decay whilst incorporated in tumour cell DNA.

4.1 Tumour heterogeneity and Auger electron therapy

The analysis so far has assumed that all tumour cells have the same amount of DNA-bound radionuclide. In reality, heterogeneity of radionuclide uptake is a characteristic feature of all forms of targeted radionuclide therapy and there are no reasons to expect Auger therapy will be any better. Heterogeneity of uptake translates into heterogeneity of absorbed energy (or dose) depending on the emission characteristics of the radionuclide. For high LET Auger therapy there is no cross-fire - a worse-case scenario for the effects of heterogeneity. Consider a population of tumour cells irradiated by DNA-bound ^{125}I. It is assumed that there is no molecular dissociation (eg substitutional IUDR-targeting) that the logarithm of the number of DNA-bound atoms is described by a normal distribution with mean 4 (corresponding to 10^4 atoms) and a variable standard deviation ranging from 0 (i.e. uniform targeting) to 0.5, expressed as a fraction of the mean. Figure 2 shows the log of the surviving fraction of the whole population as a function of heterogeneity (standard deviation) As the heterogeneity of radionuclide uptake increases the overall cell kill reduces dramatically. This is caused by the survival of tumour cells from the low uptake tail of the distribution. In contrast tumour cells from the high uptake tail are "overkilled". The therapeutic outcome will be dominated by any sub-population of tumour cells which experience relatively reduced biological effects. This result has an important implication for therapy : the overall log kill achievable by Auger electron therapy will usually be much less than predicted from a uniform targeting model and will be limited by heterogeneity of radionuclide distribution. In practice, it would be unreasonable to expect more than about a 2 log kill (1% tumour cell survival) which suggests that Auger electron therapy should be regarded as an adjunct to existing therapies rather than a 'stand-alone' new treatment. - a similar conclusion to that reached for beta emitters, although for somewhat different reasons. The analysis summarised in this section has been presented in more detail elsewhere[21].

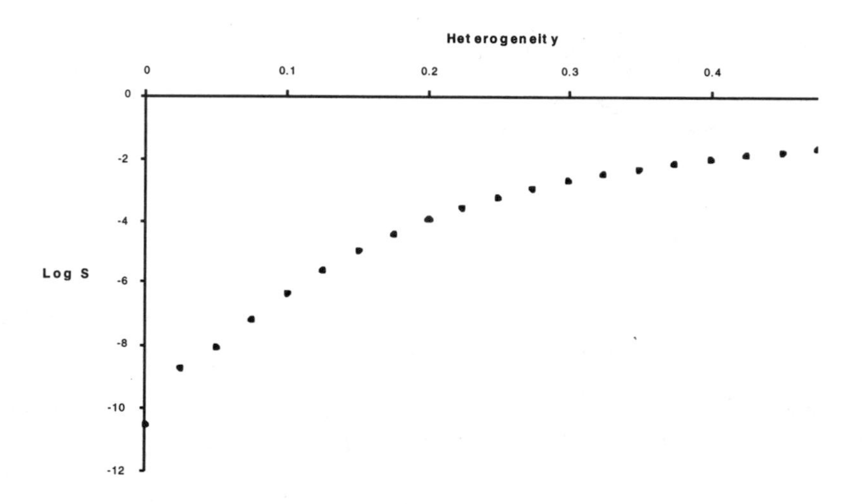

Fig 2 : *Effect of heterogeneity on Auger emitter cell kill. Tumour cells, doubling time 20 days contain an average of 10^4 atoms of ^{125}I. The logarithm of the number of bound atoms is assumed distributed normally with standard deviation corresponding to the heterogeneity parameter (x-axis).*

4.2 Alternative mechanisms for cell kill by Auger electron emitters

Recently, some workers have suggested that programmed cell death (apoptosis) could be triggered in susceptible cells by activation of a membrane-associated pathway (the sphingomyelin-ceramide pathway)[22,23] which might be stimulated by irradiation of the cell membrane rather than the nuclear DNA[24]. Cell lines are however available which differ only in their apoptosis capacity[25] ; experimental studies with such lines, using DNA-targeting agents (IUDR) and membrane-targeting agents (concanavalin A) should provide a test of this hypothesis. However, the conclusions reached above would not be affected by the existence of such a pathway.

References

1. V. Hird, , M. B. Verhoeyen, R. A. Badley, D. Price, D. Snook, C. Kosmas, C. Gooden, A. Bamias, C. Meares, J. P. Lavender and A. A. Epenetos. Br. J. Cancer, 1991, **64**, 911.
2. R. H. J. Begent, M. J. Verhar, K. A. Chester, J. L. Casey, A. J. Green, M. P. Napier, L. D. Hope-Stone, N. Cushen, P. A. Keep, C. J. Johnson, R. E. Hawkins, A. J. W. Hilson and L. Robson. Nature Medicine, 1996, **2,** 979.
3. V. Hird, D. Snook, C. Cosmas and A. A. Epenetos. In Advances in the application of monoclonal antibodies in clinical oncology (A. A. Epenetos, Ed), London, Chapman & Hall, 1991, p267.
4. L.S. Lashford, I. J. Lewis, S. L. Fielding et al. J. Clin. Oncol., 1992, **10,** 1889.
5. J. de Kraker, C.A. Hoefnagel , H. Caron et al . Eur. J. Cancer, 1995, **31A,** 600.
6. E. M. Link and R. N. Carpenter, Cancer Res., 1992, **52,** 4385.
7. S. H. Bigner, P.C. Burger, A. Wong, M. H. Werner, S. R. Hamilton, L. H. Mulbaier, B. Vogelstein and D.D. Bigner. J. Neuropath. Exp. Neurol., 1988, **47**, 191.
8. B. W. Ozanne, C. S. Richards, F. J. Hendler, D. Burns and B. Gustersen. J. Pathol., 1986, **149**,
.
9. J. Capala, and J. Carlsson. Int. J. Radiat. Biol, 1991, **60**, 497.
10. A. Andersson, J. Capala and J. Carlsson. J. Neur. Oncol., 1992, **14,** 213.
11. R. J. Mairs, W. Angerson, M. N. Gaze, T. Murray, J. W. Babich, R. Reid and C. McSharry . Br. J. Cancer,1991, **63**, 404.
12. M. Riordan and J. C. Martin. Nature, 1991, **350**, 442.
13. J. A. O'Donoghue, M. Bardies and T. E. Wheldon, J. Nucl. Med.,1995, **36**, 1902 .
14. A. E. Amin., T. E. Wheldon, J.A. O'Donoghue, M. N. Gaze and A. Barrett. Int. J. Radiat. Oncol. Bio. Phys.,1995, **32**, 713 .
15. M. N. Gaze, T. E. Wheldon, J.A. O'Donoghue, T.E. Hilditch, S. G. McNee, E. Simpson. and A. Barrett Europ. J. Cancer , 1995, **31A**, 252.
16. M. N. Gaze and T. E. Wheldon. Europ. J. Cancer ,1996, **32A,** 93.
17. T. E. Wheldon. Int . J. Radiat. Biol., 1994, **65**, 109.
18. K. F. Baverstock and D. E. Charlton (Eds). DNA damage by Auger emitters. London, Taylor and Francis, 1988.
19. R. F. Martin and W. A. Haseltine. Science, 1981, **213**, 896.
20. J. L. Humm and D. E. Charlton. Int. J. Radiat. Oncol., Bio., Phys., 1989, **17**, 351.
21. J. A. O'Donoghue and T. E. Wheldon. Phys. Med. Biol., 1996, **41**, 1973.
22. Obeid, L. M., Linardic, C. M., Karolak, L. A. and Hannun, Y. A., Science, 1994, **259**, 1769.
23. W. D.Jarvis, R. N. Kolesnick, F. A. Fornari, R. S. Traylor, D. A. Gewirtz and S. Grant. Proc. Nat. Acad. Sci. (USA), 1994, **91,** 73.
24. A. Haimovitz-Friedman, C. C. Kan, D. Ehleiter, R. S. Persaud, M. McLoughlin, Z. Fuks and R. N. Kolesnick. J. Exp. Med., 1994, **180**, 525.
25. J. Russell, T. E. Wheldon and P. Stanton. Cancer Res., 1995, **55**, 4915.

RELATIVE BIOLOGICAL EFFECTIVENESS OF Re-188 BETA PARTICLES: IMPLICATIONS FOR INTRAVASCULAR BRACHYTHERAPY

H.I. Amols, R. Miller, J. Weinberger, and E.J. Hall

Departments of Radiation Oncology, Medicine,
and Radiological Research Center
Columbia University
New York, N.Y., 10032 U.S.A.

1 INTRODUCTION

Coronary artery disease is a leading cause of death in the U.S., commonly treated via dilitation balloon catheterization or angioplasty. Proliferation of smooth muscle cells in the arterial walls following angioplasty is believed to be a major factor in restenosis, or treatment failure, which occurs in approximately 40% of all patients. Brachytherapy with beta emitters has been proposed for delivering radiation during or immediately after angioplasty to inhibit cell regrowth and hence restenosis.

Coronary arteries accessible to angioplasty have 1-2 mm radii and wall thickness ≤ 1 mm. Beta sources proposed for treatment, such as Re-188 have mean beta energies on the order of 0.8 MeV and mean beta ranges ≤ 4 mm. Such low energy beta particles may have RBE $\neq$ 1. Using an in vitro cell model and Gafchromic film dosimetry we have measured the RBE for Re-188 betas, which at treatment distances of 1-3 mm is found to be equal to or slightly greater than Co-60 gamma rays. Clinical application of beta emitters in the treatment of restenosis will therefore not be significantly altered by RBE considerations.

1.1 Background.

Over 400,000 patients undergo percutaneous transluminal coronary angioplasty annually in the U.S. for the treatment of coronary artery disease. A balloon dilitation catheter is inserted through the femoral and iliac arteries, and aorta, into the stenosed coronary artery. Fluoroscopy of the diseased artery permits the cardiologist to insert a thin guide wire across the site of the stenosis where a 2-3 cm length balloon is positioned and inflated to 5-12 atmospheres pressure for several minutes to dilate the artery, decreasing the blockage and improving blood flow. Although the initial success rate of this procedure is >90%, approximately 40% of patients have evidence of restenosis by 6 months. Several processes contribute to restenosis, the most important being abnormal proliferation of injured smooth muscle cells and neointima formation in the treated artery.[1] It has recently been shown in porcine coronary models[2-5] and human femoral arteries[6] that 10-20 Gy of radiation delivered acutely (to reduce the risk of thrombosis or ischemia to the bed supplied by the treated artery) at the time of angioplasty, or long-term via implantation of radioactive stents reduces restenosis by 50-70%.

Temporary, intraluminal insertion of a high activity ($\geq$ 1 GBq) beta source with transition energy $\geq$1.7 MeV has been suggested as an ideal irradiation technique. The source could be a thin metal seed or wire, or radioactive liquid filled balloon. The advantages and dosimetric properties of various sources such as P-32, Sr-90, Y-90, and Re-188 have been discussed previously.[7-8]

All proposed beta sources have maximum transition energies <2.3 MeV (mean energies <.8 MeV) with concomitant mean ranges <4 mm. Treatment will thus entail irradiation of some critical tissues with beta particles nearing the end of their range. In Figure 1 we plot restricted electron stopping power and residual range versus energy, [9-10] which shows stopping power >34 keV/µm near the Bragg peak. Other 'low LET' radiations such as the x-ray emitters I-125 and Pd-103, which have lineal energy spectra with significant components of 'high LET radiation'[11-12] have reported RBE's as high as 1.5-2.0.[13-14] We report here the results of an investigation designed to determine whether or not beta sources proposed for intravascular brachytherapy have similarly high RBE's.

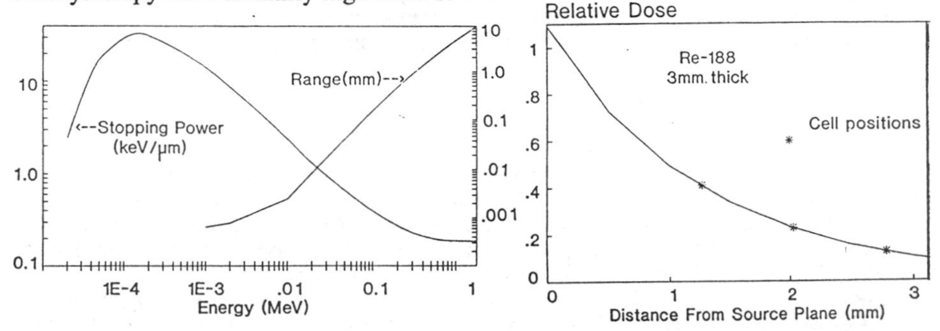

Figure 1: Restricted electron stopping stopping power (left) and range (right) vs energy. References 9 for E>.01 MeV, and 10 for E<.01.

Figure 2: Calculated dose rate vs distance from Re plane, normalized to 100% at .128 mm (18.3cGy/hr/MBq/ml), the closest distance at which measurements were made.

2 METHODS

2.1 Dosimetry.

Re-188 is a beta emitter with a transition energy 2.1 MeV, average beta energy of 0.8 MeV. and half life 17 hours. It can be obtained from a W-188 generator (half life 69 days).[15] Elution of the generator with 15 ml normal saline yields approximately 9 GBq of Re-188 solution, the exact activity being determined by calibration against an NIST calibrated W-Re solution in a standard well type ionization detector. 15 ml of Re-188 solution was deposited in a 7x7 cm^2 plastic dish with a 10 µm thick bottom window, forming a 'pool' of radioactive solution 3 mm thick. Monolayer V-79 cells were plated onto T-25 tissue culture flasks (see section 2.2), turned upside down, and placed in contact with the bottom of the Rhenium dish resulting in a minimum 'treatment distance' of 1.28 mm which can be varied by inserting sheets of Tissue-Equivalent plastic between the T-25 flasks and the Rhenium dish.

Dose rate vs. distance from the lower surface of the Rhenium dish was calculated analytically using dose kernels [16] which give dose per decay as a function of distance from a point source. Kernels were numerically integrated over the volume of the Rhenium (i.e., 7x7x0.3 cm^2volume) as described previously.[7-8] Results are shown in Figure 2 which plots dose versus distance, normalized to 100% at a distance of 0.127 mm; the closest point at which measurements (see below) could be made. Cell exposures were made at distances of 1.28, 2.03, and 2.79 mm.

The calculated absolute dose rate at the surface of the Rhenium is 20 cGy/hr/MBq/ml. For the V-79 exposures, with an eluted activity of approximately 9 GBq per 15 ml solution, the dose rate varied from 46 Gy/hr at a treatment distance of 1.28 mm to 14 Gy/hr at 2.79 mm. Because the radioactive source had a larger area than the culture flasks, the dose to the cells (which were in a plane parallel to the plane of the Rhenium) was perfectly uniform.

Dose calculations were compared to experimental measurements using Gaf-

Chromic film which is now a standard dosimetric tool.[17-18] The film contains a 20 µm thick emulsion embedded in a 0.250 mm thick polymer sheet that is approximately tissue equivalent. Optical Density (OD) vs. dose is nearly linear and no post-irradiation processing is required. Its relative insensitivity and large dynamic range (saturation dose >50 Gray) makes it particularly useful for brachytherapy measurements where dose rates and gradients are large. Film response is also nearly independent of beam energy or particle type. Films were calibrated via exposure to known doses of Cobalt-60 and 6 MeV electrons. Measured and calculated data, shown in Fig 2, agree to $\pm 7\%$ both in absolute value and depth dose distribution. The dose rate was measured experimentally prior to each cell irradiation experiment by exposing Gaf-Chromic film, determining its OD, and converting to dose.

2.2 Radiobiology

Exponentially growing V-79 cells were plated in T-25 tissue culture flasks on the morning of each experiment. Cell concentrations were determined via Coulter counter measurements. Based on survival data from previous gamma exposures cell dilutions were made to obtain approximately 100 colonies per flask after irradiation of 0, 3, 6, 9, or 12 Gy respectively. Plating efficiency was > 50%.

Flasks have a 1.28 mm thick plastic growing surface 5x5 cm^2 in area. After plating, flasks were filled 'brim-full' with medium. One hour prior to exposure flasks were cooled to room temperature to reduce cell cycle progression and growth. Exposure times ranged from several minutes up to 80 minutes per flask, which were exposed one at a time to Re-188 beta particles or Co-60 gamma rays. After exposure all flasks were drained of medium, fresh medium was replaced, and cells were incubated at 37°C for 1 week at which time they were fixed, stained, and assayed for colony formation.

Re-188 cell irradiations were made at distances of 1.28, 2.08, and 2.79 mm distance. Total irradiation time for all flasks was approximately 9 hours. To test for cell cycle progression, duplicity, or other time variant factors Co-60 exposures from 3-12 Gy were made at the beginning and end of the 9 hour irradiation period.

3 RESULTS AND CONCLUSIONS

Two independent experiments were performed two weeks apart. Fig. 3 shows the combined cell survival data vs. dose for Co-60 and beta irradiated cells. Each data point represents 2-4 flasks for beta exposures, or 4-8 flasks for Co-60. Statistical errors in survival levels are $\pm 5\%$, but dosimetric uncertainties for Re-188 may well

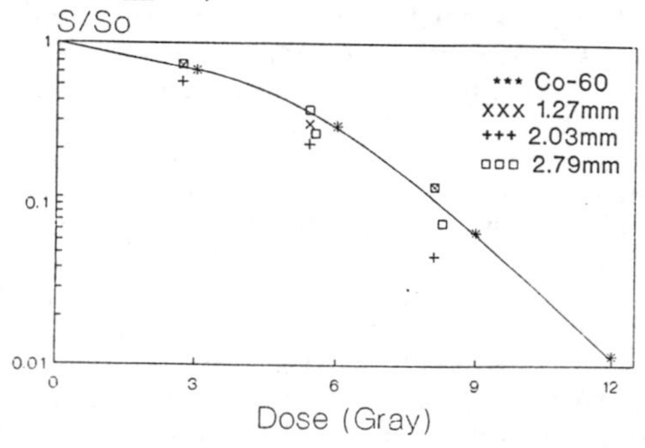

Figure 3: Cell survival vs. dose for Co-60 and beta irradiated cells. Each data point represents from 2-4 flasks for beta exposures, and 4-8 flasks for Co-60. The solid curve is drawn only as a guide to the eye, and has no statistical significance.

be larger, as discussed below. Although cell survival values for Co-60 irradiation are systematically higher than for Re-188, the 4 survival curves are not statistically different. For clarity, error bars have not been plotted in Figure 3.

The survival curves depicted in Figure 3 indicate that the RBE for Re-188 beta particles, relative to Cobalt-60 gamma rays is approximately one, or perhaps slightly higher. For beta exposures at 2.79 mm distance an RBE higher than 1.0 would not be unexpected as approximately 7% of the incident electrons have E< 100 keV, and stopping power >400 ev/um. Low energy x-ray emitters (e.g. I-125 and Pd-103), which also produce low energy secondary electrons have reported RBE's as high as 1.5-2.0 (at 5 -50 cGy/hr).[13-14]

The RBE values reported here however can be no more accurate than the dosimetry, which for low energy beta emitters presents significant experimental difficulties. As recently as 1991 for example, it was reported that the calibration of several hundred Sr-Y eye plaque irradiators (commonly used for the treatment of pterygium) were in error by as much as 35%. [19]

In our experiments dosimetric uncertainties are at least ±7%, stemming from errors in film dosimetry and source calibration. Because the dose gradient is large (>50% per mm) errors in film positioning and uncertainty in the exact location of the emulsion can lead to large dosimetry errors. Calibration of beta emitters using a well detector is also critically dependent on sample volume and geometry.

Thus, one may consider the RBE value of unity reported here to be a method of 'biological dosimetry' which confirm, at the ±7% level, the accuracy of our dosimetry. More accurate dosimetry may yield a slightly higher RBE, which could be clinically important, although imprecise knowledge of target tissues and required doses for treating restenosis may mask small RBE effects. Gillette et al however have reported that doses >30 Gy to the aorta in dogs increase the risk of aneurysms and large thrombi 4-5 years after irradiation.[20] Prevention of restenosis may require doses>20 Gy. Considering the consequences of treatment overdose, dosimetric errors, and/or RBE uncertainties cannot be ignored.

The beta irradiator was constructed by Mr. G. Johnson of Columbia University, and the W-Re generator was supplied by Dr. R. Knapp of Oak Ridge National Lab.

References

1. C. Landau, R.A. Lange, and L.D. Hillis, New England J. Med., 1994, 330, 981
2. J. Wiederman, C. Marobe, H. Amols, et. al., J. Amer. Col. Card., 1994, 23, 1491
3. J. Wiederman, C. Marobe, H. Amols, et. al., J. Amer. Col. Card., 1995, 25, 1451
4. R. Waksman, K. Robinson, I. Crocker et. al., Circulation, 1995, 91, 1533
5. C. Hehrlein, J. Zimmermann, J. Metz, et al., Circulation, 1993, 88, 1
6. H. Bottcher, B. Schopohl, D. Liermann, et. al., IJROBP, 1994, 29, 183
7. H. Amols, M. Zaider, J. Weinberger, et. al., IJROBP, 1996 in press
8. H. Amols, L Reinstein, J. Weinberger, Med Phy, 1996, in press
9. M. Berger and S. Seltzer, 'Stopping powers and ranges of electrons and positrons (2nd Ed.).' U.S. Dept. Commerce Publication, 1983, NBSIR 82-2550-A
10. R. Ritchie, H. Hamm, J. Turner, and H. Wright, In: Sixth Symposium on Microdosimetry (New York: Harwood Academic Pub.), 1978, 345
11. C. Wuu, P. Kliauga, M. Zaider, H. Amols, IJROBP, 1996, in Press
12. D. Zellmer, J. Shadley, M. and Gillin, Radiat. Prot. Dosim., 1994, 52, 395
13. M. Freeman, P. Goldhagen, E. Sierra, E., E. Hall, IJROBP, 1982, 8, 1355
14. C. Ling, W. Li, L. and Anderson, IJROBP, 1995, 32, 373
15. F. Knapp, A. Callahan, A. Beets, et. al., Appl. Radiat. Isot., 1994, 45, 1123
16. D. Simpkin and T. Mackie, Med Phys, 1990, 17, 179
17. S. Chiu-Tsao, A. de la Zerda, J. Lin, J. and Kim, Med. Phys., 1994, 21, 651
18. W. McLaughlin, Y. Chen, G. Soares, et. al., NIM, Phys. Res. A, 1991, 302, 165
19. S. Goetsch, K. Sunderland, Med Phys, 1991, 18, 161
20. E. Gillette, S. McChesney, R. Park, and S. Withrow, IJROBP, 1989, 17, 1247

BIOPHYSICAL MEASUREMENTS AT THE COSY PROTON BEAM

R. Becker[1], P. Bilski[4], M. Budzanowski[4], W. Eyrich[3], D. Filges[2], M. Fritsch[3], J. Hauffe[3], H. Kobus[2], M. Moosburger[3], P. Olko[4], H. Paganetti[1], H. P. Peterson[1], Th. Schmitz[1], F. Stinzing[3]

[1]Insitute for Medicine, Research Center Jülich, 52425 Jülich, Germany
[2]Institute for Nuclear Physics, Research Center Jülich, 52425 Jülich, Germany
[3]Physics Institute, University of Erlangen, 91051 Erlangen, Germany
[4]Institute of Nuclear Physics, 31-342 Kraków, Poland

1 PROGRAM

Conformal tumor therapy with protons requires high accuracy in treatment planning. Inaccurate calculations of the dose distribution or the maximum depth of penetration may lead to either an underdosage in parts of the tumor or to an overdosage of the surrounding healthy tissue. The change in ionization density as the proton slows down leads to a higher biological effectiveness[1]. The aim of the investigations was to compare experimental and simulated results in order to give estimations for the dose profile and the biological effectiveness of a proton beam. The advantage of using a monoenergetic proton beam is the concentration on narrow energy spectra. This allows a better isolation of effects.

The proton synchroton COSY in Jülich is operating at proton energies between 100 MeV and 2.5 GeV. The experiments were performed at a beam energy of 178 ± 1.5 MeV and an energy spread of 0.3 %. Absolute dose and microdosimetric energy deposition distributions were measured in a water phantom as a function of depth. The beam energy was determined by time-of-flight measurements. Initial biological experiments concentrate on the determination of cell killing of V79 hamster cells as a function of position on the Bragg curve for different dose levels. As reference radiation for the evaluation of the relative biological effectiveness (RBE) ^{60}Co is used.

The Monte Carlo code GEANT[2] was used to simulate the experiments. As input parameters, the measured beam characteristics were applied. No information was available for the beam divergence and the beam diameter. These parameters are thus fitted using the dose profile measurements at two different depths. The biophysical simulations use the results of microdosimetric measurements and Monte Carlo calculations to predict the RBE of protons. The microdosimetric approach[3] was applied which uses weighting functions folded with microdosimetric energy deposition spectra to obtain information on the RBE of the radiation field.

2 EXPERIMENTS

Experiments were performed in a water phantom, which is a Plexiglas tank with a volume of $50 \times 50 \times 40\,\mathrm{cm}^3$. The wall thickness is 1.1 cm. A 3-axis-scanner unit is used

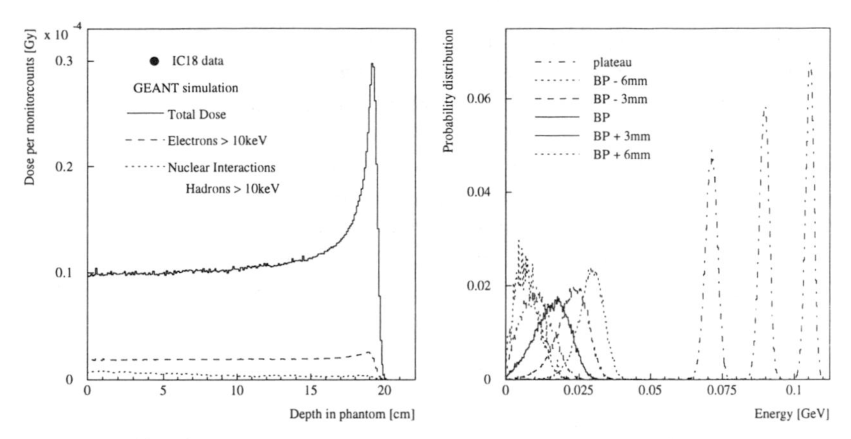

Figure 1: *(left) Depth dose in a water phantom as measured by an ionization chamber and simulated by the Monte Carlo program GEANT. In addition, the contribution of hadronic dose and δ-electrons is shown*

Figure 2: *(right) Simulated energy spectra at the position of the cell surfaces*

to position the detectors within the phantom with an accuracy better than 0.1 mm.

2.1 Physics

For beam monitoring a diode ring detector was placed at the end of the beamline. Four silicium diodes were plugged on a radius of 7 cm. As a backup beam monitor a thin scintillator was placed downstream in the beam penumbra. The proton momentum was 604.6 MeV/c with a momentum spread of $0.5 \cdot 10^{-4}$. The extraction pulse duration was about 8 s at a pulse repetition time of 16 s with 10^9 protons per extracted spill. The beam diameters on the surface of the phantom were 2.5 cm horizontally and 1.5 cm vertically and at the Bragg peak position it was 3.0 cm horizontally and 2.0 cm vertically. The dose profile was measured with a small-volume (0.1 cm^3) TE ionization chamber. Figure 1 shows the depth dose in the water phantom normalized to monitor counts. The results are given for dose to tissue. The dose calculation by GEANT is shown in the same figure. The comparison shows good agreement. In addition, the composition of dose is shown for the the contribution due to high energy delta electrons and nuclear interactions. The fraction of nuclear interactions in the Bragg peak is about 0.8 %, in the plateau region it reaches 7 %. A simulation of the proton energy spectra at various positions in the Bragg curve was done (Fig. 2).

In addition, microdosimetric dose distributions were measured at the positions of the cell carriers. In Figure 4 the spectra are plottet against the logarithmic increment of lineal energy and are normalised to unit dose. In this presentation, equal areas of the distribution represent equal fractions to absorbed dose. The shift of the spectra to higher lineal energies reflects the increase in proton stopping power as they are slowed down. Within the Bragg peak and the distal edge, larger fractions of the dose are deposited at lineal energies above 20 keV/μm.

2.2 Biology

The survial of V79 hamster cells was measured as a function of depth in the Bragg curve for different survival levels. For these experiments a box was designed, where cell carriers can be fixed over a distance of 12 cm with a minimum spacing of 3 mm. The box was mounted on the 3 axis scanner and exactly positioned in the water phantom so that the cell carriers cover the complete region of the Bragg curve. On plastic slides a circular deepening of 1 cm in diameter was made to plate the cells. The box was closed watertight and filled with growth medium. The water in the phantom was heated to 37 °C. For online dose control some carrieres were equipped with TLD dosemeters instead of cells. The TLDs were specially prepared for high dose proton measurements by the Institute of Nuclear Physics in Krakow.

3 RESULTS

Survival curves were obtained for every slide position. The data can be described by a linear quadratic fit (Figure 3). The survival curve of the ^{60}Co reference radiation is also given in Figure 3. RBE values for a survival level of 30 % and 10 % with respect to ^{60}Co are shown in Table 1.

The RBE increases from the plateau to the distal edge of the beam. The RBE in the plateau region is 1.05 at 10 % survival level and has a value of 1.2 at the Bragg peak, increasing up to 1.4 in the distal edge. The uncertainties in every step of the experiment is considered and enhance the total error. The uncertainties for every pipetting process were estimated by pre-experimental tests. The statistical error of the cell density measurement and of the number of cell colonies were considered as well as the error of the plating efficiency. Moreover the error is influenced by the small number of data points in the fit, especially at the distal edge.

The values of RBE_α are shown in Table 1. RBE_α was calculated according to the measured $f(y)$ spectra and the microdosimetric response function of Morstin[3] for V79 cells in late S and G_1/S phase. They should give bounds on the survival of non-synchronized V79 cells. Actually, they seem to overestimate the effectiveness of protons at low energies (i.e. high lineal energies). The error of data is large, because measurements were performed at doses more than 2 Gy, this leads to a high error for extrapolations down to low doses. Anyways, the calculated RBE of 2 seems incredible, because the response functions base on an investigation of Bird et al.[4]. These

Table 1: *RBE values from the cell survival measurements for various survival levels and the calculated RBE_α from the microdosimetric approach*

Position	$RBE_{10\%}$	$RBE_{30\%}$	RBE_α	RBE_α calc.
Plateau	1.06 ± 24%	1.05 ± 23%	0.98 ± 53%	1.04–1.08
BP−6 mm	1.06 ± 24%	1.06 ± 23%	1.10 ± 61%	1.05–1.12
BP−3 mm	1.06 ± 19%	1.07 ± 17%	1.14 ± 49%	1.13–1.26
BP	1.20 ± 11%	1.25 ± 9%	1.58 ± 35%	1.38–1.57
BP+3 mm	1.26 ± 17%	1.29 ± 15%	1.46 ± 55%	1.74–1.97
BP+6 mm	1.39 ± 74%	1.44 ± 80%	1.85 ± 172%	2.15–2.39

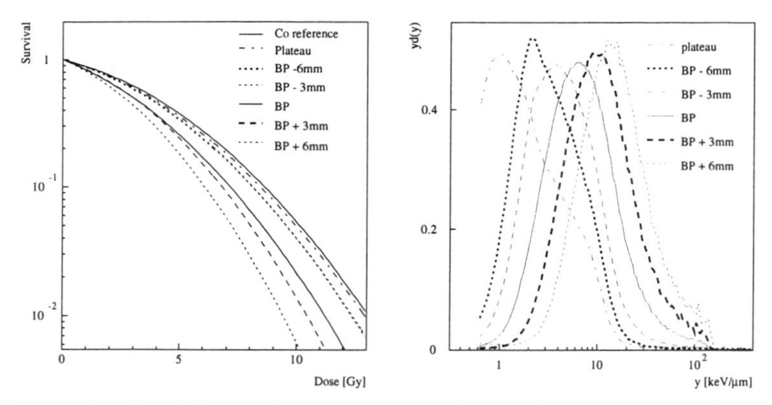

Figure 3: *(left) Survival curves obtained by a linear quadratic fit in the data. BP denotes the Bragg peak position*

Figure 4: *(right) Microdosimetric spectra measured at the position of the cell carriers*

measurements give an intervall of RBE_α between 1.4 and 1.8 for protons of the LET 10 keV/µm. Our measurements at the position of 6 mm behind the Bragg peak should be directly comparable because they correspond to a value of $\overline{LET} \approx \bar{y}_F = 10.1$. Indeed the results of Bird et al. support our measurements, not the predictions by the microdosimetric approach. The deviation of the calculations may be due to the convolution of triangular $f(y)$ distribution from LET for the determination of the response function by Morstin instead of measured spectra, which does not take into account higher energy depositions by nuclear interactions or stopping particles. This leads to an overestimation of high lineal energies.

As an alternative, the track structure model of Katz and co-workers[5] can be applied. The simulated proton energy spectra can be used to estimate the RBE at the various cell positions. This work is in progress.

4 REFERENCES

1. H. Paganetti, T. Schmitz, 'The Influence of the Beam Modulation Technique on Dose and RBE in Proton Radiation Therapy', *Phys. Med. Biol.*, 1996, **41**,1649.
2. GEANT Version 3.21, 'Detector Description and Simulation Tool', *CERN Program Library*, 1994.
3. K. Morstin, V. P. Bond, J. W. Baum, 'Probabilistic approach to obtain hit-size effectiveness functions which relate microdosimetry and radiobiology', *Radiat. Res.*, 1989, **120**, 383.
4. R. P. Bird, N. Rohrig, R. D. Colvett, C. R. Geard, S. A. Marino, 'Inactivation of Syncronized Chinese Hamster V79 Cells with Charged-Particle Track Segments', *Radiat. Res.*, 1980, **82**, 277.
5. R. Katz, S. C. Sharma, 'Heavy Particles in therapy: An Application of Track Theory', *Phys. Med. Biol.*, 1974, **19**, 413.

MICRODOSIMETRIC CHARACTERIZATION OF CLINICAL PROTON BEAMS

F. Verhaegen and H. Palmans

Dept. of Biomedical Physics, University of Gent,
Proeftuinstraat 86, B-9000 Gent, Belgium

1 INTRODUCTION

Since the mid 1980s the number of clinical proton radiation facilities in the world and of patients who received proton therapy in these facilities has increased enormously.[1-3] The advantages of tumor treatment with protons as opposed to with electrons or photons are well understood. The scattering characteristics of protons typically used for clinical treatment (50 - 250 MeV) allow administering high doses to the target region while sparing the adjacent tissues. RBE values ranging between 1 and 1.2 compared to ^{60}Co gamma radiation for clinical proton beams as well as for space originated protons has been reported,[1-5] depending on the cell type, organism and biological endpoint. Near the end of the Bragg peak the ionisation density and the lineal energy of the protons increase strongly. In the spatial region immediately beyond the Bragg peak, where there is a large variation in energy deposition, this increase is counteracted by a strong decrease in dose. Nevertheless, in this region the variation of the energy deposition distribution may result in specific biological lesions. In the usual clinical practice a homogeneous dose over the region of interest is aimed at by spreading out the Bragg peak with a modulator wheel. This however does not ensure a homogeneous lineal energy distribution over the isodose plateau.

In the present paper microdosimetric characteristics for protons in the energy range 50 - 200 MeV are calculated by the Monte Carlo method, both for unmodulated and modulated proton depth dose profiles. The influence of a small Gaussian energy spread on the incident proton energy is noted and variation of lineal energy with radial distance from the primary proton track is investigated.

2 METHOD

Microdosimetric lineal energy distributions and related frequency and dose mean lineal energies $\langle y_F \rangle$ and $\langle y_D \rangle$ are obtained in two steps. First, proton fluence spectra at the macroscopic geometric scale are calculated with the condensed history Monte Carlo code PTRAN by Berger,[6] which was extended by us for proton transport in non-homogeneous geometries.[7] The extended code allows simulation of modulator wheels. Protons are transported down to an energy of 140 keV. In its present form, the code does not include generation of secondary electrons or of secondary heavy charged particles created in nonelastic nuclear reactions in water. The effect of these nuclear reactions on the depth dose curves is however incorporated through an absorption factor. For the calculations of lineal energy, spectra for secondary protons, deuterons and alpha particles were taken from Seltzer[8]. These particles were not transported in the water. Secondary particles with higher charges were not included in this study. The effect of these omitted particles is not negligible for some of the results to be presented. In the second step, linear energy distributions for protons, deuterons and alpha particles were calculated in homogeneous gaseous water with the microdosimetric Monte Carlo code TRION,[9] using the "fluctuation detector method" in nanometer and micrometer spherical volumes. This part of the simulation includes calculation of secondary electron fluences down to 13 eV prior to the actual proton transport.

3 RESULTS AND DISCUSSION

In figure 1 $\langle y_D \rangle$ values for monoenergetic protons of 50 - 200 MeV are shown. The lineal energy varies only slightly with depth except near the Bragg peak where a strong increase is seen that becomes more prominent with increasing diameter of the simulated volume. A similar behaviour is found for $\langle y_F \rangle$. Also shown in the figure is the influence of inclusion of secondary particles (protons, deuterons and alpha particles) from nonelastic nuclear reactions compared to calculations with only primary protons.

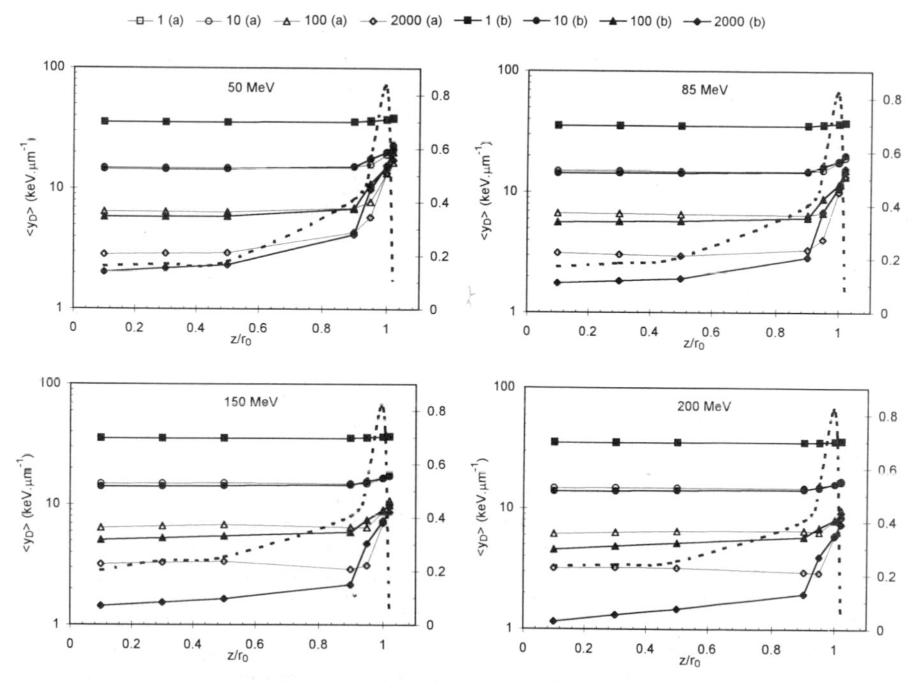

Figure 1 *$\langle y_D \rangle$ for unmodulated protons of 50 - 200 MeV in spherical volumes (1-2000 nm) versus z/r_0. z is the depth and r_0 indicates the CSDA range in water of the protons. Set (a): primary protons and secondary charged particles from nonelastic nuclear reactions, set (b): primary protons only. Dashed line: percentage depth dose curve (right scale).*

In spherical volumes with diameters in the nanometer region no differences between both sets of calculations are apparent. For the larger volumes, however, the secondary heavy particles are responsible for a marked increase in lineal energy at points before the Bragg peak. Moreover, it has to be noted that inclusion in the simulation of still heavier charged particles from the nonelastic nuclear reactions would cause a further increase. At, the Bragg peak and beyond, the secondaries are of no importance.

In figure 2 dose mean lineal energies for monoenergetic proton beams (50-200 MeV) are compared with protons with a Gaussian energy spread (σ=2%). At points before z/r_0=0.99 $<y_F>$ and $<y_D>$ values do not differ significantly for any of the volumes simulated (r_0 indicates the CSDA range of the protons).

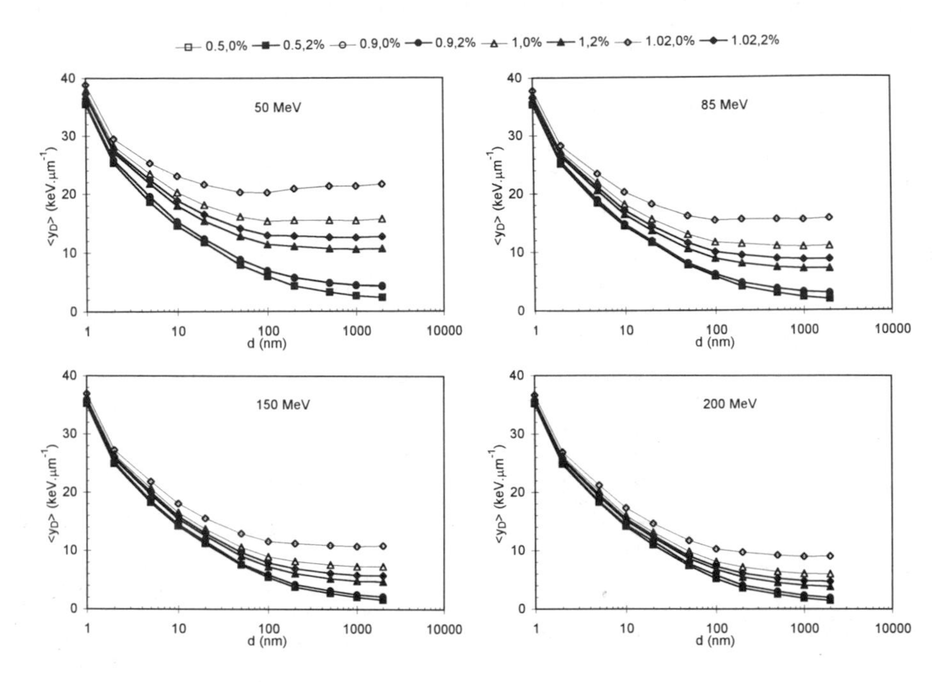

Figure 2 *Effect of 2% Gaussian energy spread for protons (50 - 200 MeV) on $<y_D>$ in spheres (1 - 2000 nm). The legend indicates z/r_0 and the energy spread.*

Starting from z/r_0=0.99 large differences are noted, due to scattering and variation in the energy deposition distribution at the end of the proton tracks. The effect is more important for the protons with the lowest energies. The differences increase with diameter of the spherical volume and reach more or less constant values for d >100 nm. Only the primary protons were taken into account here because secondary charged particles from nonelastic nuclear reactions are only important before the Bragg peak.

In the radial direction away from the beam central axis, the primary proton fluence (and therefore also the secondary charged particle fluences from nuclear reactions) does not change much. This is reflected in almost constant $<y_D>$ and $<y_F>$ values for the primary protons (50 - 200 MeV). In the vicinity of the Bragg peak differences amount to a few percent at the most for the largest simulated spheres (2 µm diameter). No biological penumbra effects are therefore to be expected.

From a practical point of view spread out Bragg peaks (SOBP) obtained by a modulation wheel resulting in flat depth dose profiles are the most important. A polystyrene modulator wheel was simulated during the PTRAN transport calculations, yielding an isodose plateau extending from 0.3 z/r_0 to 0.95 z/r_0 for 85 and 200 MeV monoenergetic protons. Figure 3 shows the variation of $<y_D>$ in spherical volumes of 10 nm - 2 µm diameter along the depth dose curves for protons with and without secondary heavy particles. A similar behaviour is found as for the simple depth dose curves (fig. 1) with slightly higher values for the modulated profiles.

Gueulette et al[4] measured <y_D> values along a 40 mm SOBP for 85 MeV protons in a 2 μm sphere. They obtained values of 5.7 keV/μm at the beginning and up to 8.4 keV/μm at the end of the isodose plateau.

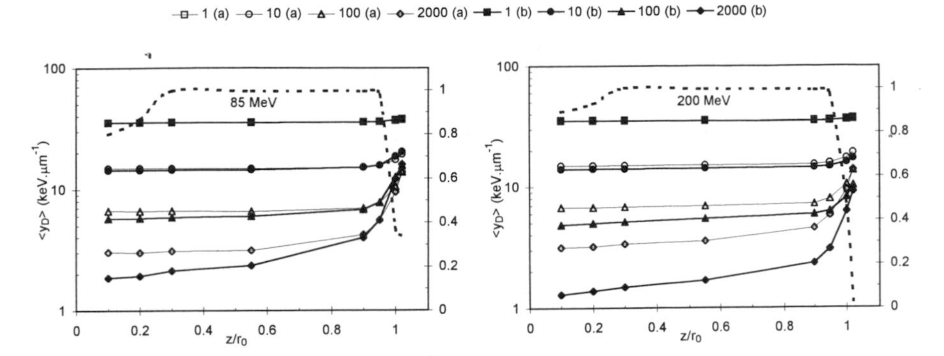

Figure 3 <y_D> *in spheres (1 - 2000 nm) for modulated proton beams of 85 and 200 MeV. Set (a): primary protons and secondary charged particles from nonelastic nuclear reactions, set (b): primary protons only. Dashed line: percentage depth dose curve (right scale).*

While the discrepancy with our results before the end of the plateau can be ascribed to the omission of the secondary heavy charged particles with Z > 2, care should be taken in comparing these experimental values with our results in view of the energy spread on a realistic proton beam as mentioned before.

4 CONCLUSIONS

Microdosimetric characteristics of clinical proton beams can be calculated with a combination of macro- and microdosimetric Monte Carlo particle transport simulations. In clinical practice, measurements of the microdosimetric spectra are advisable in view of the importance of the secondaries from nonelastic reactions and the large effect of a small energy spread of the proton beam at the position of the Bragg peak. Our results further indicate that with standard modulator wheels isodose plateaus should also be isoeffect plateaus except near the end.

References

1. M. Raju, *Radiat. Res.*, 1996, **145**, 391.
2. M. Raju, *Int. J. Radiat. Biol.*, 1995, **67/3**, 237.
3. D. Bonnet, *Phys. Med. Biol.*, 1993, **38**, 1371.
4. J. Gueulette, V. Grégoire, M. Octave-Prignot, A. Wambersie, *Radiat. Res.*, 1996, **145**, 70.
5. G. Dalrymple, I. Lindsay, J. Mitchell, K. Hardy, *Radiat. Res.*, 1991, **126**, 117.
6. M. Berger, National Institute for Standards and Technology report NISTIR 5113, Gaitersburg, USA, 1993.
7. H. Palmans and F. Verhaegen *Calculated depth dose profiles for modulated and unmodulated proton beams in some low-Z materials.* Submitted to Phys. Med. Biol.
8. S. Seltzer. National Institute for Standards and Technology report NISTIR 5221, Gaitersburg, USA, 1993.
9. A. Lappa, E. Bigildeev, D. Burmistrov, O. Vasilyev, *Radiat. Environ. Biophys.*, 1993, **31**, 1.

MEASUREMENTS OF THE LET IN A PROTON BEAM OF 62 MEV USING THE HTR-METHOD

M. Noll[1], W. Schöner[2], N. Vana[1], M. Fugger[2], E. Egger[3]

[1] Institute for Space Dosimetry, Hohe Warte 19, 1190 Vienna, Austria
[2] Atomic Institute of the Austrian Universities, Schüttelstraße 115, 1020 Vienna, Austria
[3] Department for Radiation Therapy, Paul-Scherrer-Institute, CH-5232 Villigen PSI, Switzerland

1 INTRODUCTION

In radiation protection thermoluminescence dosemeters are frequently used to obtain the absorbed dose of radiation. Using LiF:Mg,Ti information about the linear energy transfer or on microdosimetric scale about the distribution of energy deposition can also be obtained in addition to the absorbed dose. The method for evaluation of the average LET of radiation uses the fact that the high temperature peaks of LiF dosemeters increase with increasing LET if compared to peak 5. This high temperature ratio (HTR)-method was calibrated and used to determine the absorbed dose and the average LET of the complex mixed radiation fields in space. Measurements were performed during several periods on space station MIR, on satellites and during space shuttle missions[1-2]. Furthermore the method is applied for measurements of the average LET in aircrafts[3]. Investigations are under progress to understand the increase of the high temperature region by means of microdosimetry [4].

Because TL-dosemeters are very small and easy to handle investigations were performed to apply the HTR-method also in radiation therapy. Irradiations were carried out with 62 MeV protons in a therapeutic irradiation facility at the Paul Scherrer Institute, Villigen, Switzerland. Stacks of TLDs were exposed under different beam conditions, in an unmodulated and modulated proton beam. Using the HTR-method the absorbed dose as well as the LET were measured as a function of the absorber thickness. The results were compared with biological investigations performed with V79 cells.

2 MATERIAL AND METHOD

The HTR-method was developed to measure not only the absorbed dose but also the average LET in mixed radiation fields in space. The method utilizes the thermoluminescence dosemeters LiF:Mg,Ti. In standard dosimetry the integral over the glowcurve up to about 220°C (peak 5) is used to determine the absorbed dose. Information about the LET of the radiation can be obtained by analysing the high temperature peaks above 200°C. The intensity of these peaks increases with increasing LET or, on microscopic scale, depends on the distribution of energy deposition. Figure 1 shows the glowcurves after irradiation of a stack of dosemeters in the unmodulated proton beam. The increase of the LET is caused by the different positions of the dosemeters in the stack. For determination of the LET the integral of the high temperature region (220-300°C) is used. For this purpose the high temperature region of the glowcurves after high-LET irradiation is compared with this observed after Co-60 irradiation, both glowcurves being normalized on peak 5 maximum[1].

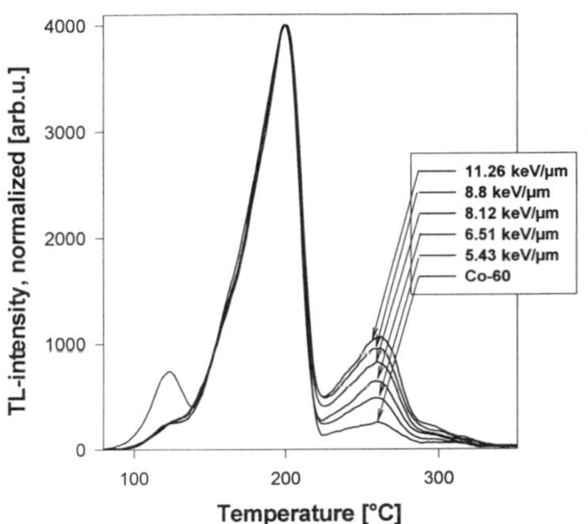

Figure 1: *Glowcurves of TLD-700 irradiated as a stack of dosemeters with 62 MeV protons compared with a Co-60 irradiation, glowcurves are normalized on peak 5 maximum*

Figure 2: *depth distribution of the absorbed dose and the HTR irradiated in the unmodulated proton beam measured with TLD 700*

3 CALIBRATION OF THE DOSEMETERS

TLDs were irradiated in the unmodulated proton beam for calibration purpose, i.e. the LET-dependence of the HTR was recorded. Stacks of thin dosemeters (down to 0.2 mm thickness) were exposed to 62 MeV protons and the depth distribution of the absorbed dose and the HTR were measured. Figure 2 shows the dependence of dose and HTR on the depth.

In order to correlate the HTR with the LET, the dose average LET, $\bar{L}_D$ of each TLD was calculated. For this purpose the energy deposited by the protons traversing the TLDs was calculated. The program used neglects scattering of the protons whereby deviations of the actual $\bar{L}_D$ are expected in the Bragg peak. To improve the calibration Monte Carlo calculations are under progress. Figure 3 shows the HTR as a function of calculated $\bar{L}_D$. The irradiation of the stack of dosemeters was performed with two different particle fluences and therefore different absorbed doses for a given depth. It can be seen that this calibration is dose dependent, i.e. that the rise of the HTR is not only a function of LET but also of the dose dependent distribution of energy deposition. Therefore the LET which is dose independent is not sufficient for the description of the differences in the HTR. Instead of the LET microdosimetric parameters describing the distribution of energy deposition should be used.

It is commonly known that the peak 5 sensitivity decreases with increasing LET[6]. Therefore the absorbed dose obtained by analysis of peak 5 is underestimated for irradiations with high-LET particles. The HTR-method allows the evaluation of the LET and therefore this decrease in sensitivity can be corrected. For this purpose the depth dose distribution obtained with TLDs (Figure 2) was compared with the depth dose distribution measured with an ionization chamber performed during irradiation. The correlation of the increasing HTR with the decrease of peak 5 sensitivity is shown in figure 4. This function allows the determination of the actual absorbed dose.

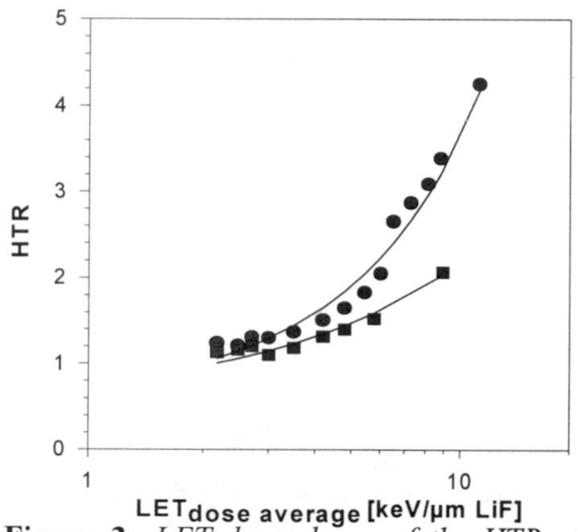

Figure 3: *LET-dependence of the HTR of TLD-700 in the unmodulated proton beam, ● 30 mGy, ■ 1 Gy in 5 mm depth (first dosmeter of the stack)*

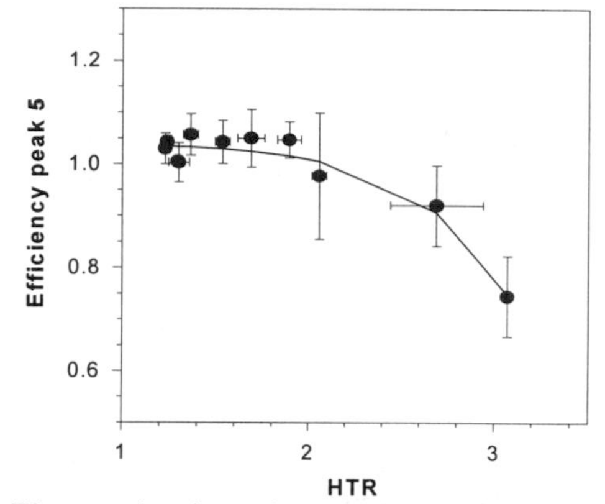

Figure 4: *dependence of the peak 5 efficiency on the HTR, measured with TLD-700*

4 MEASUREMENTS IN A MODULATED PROTON BEAM

In addition to the calibration of the TLDs irradiations were performed in a proton beam modulated with a degreader. This beam is used in radiation therapy because a constant dose over the spread Bragg curve can be obtained. The biological effectiveness over the whole proton range was assumed to be constant under these conditions. Nevertheless measured RBE increases with absorber thickness which leads to an increase of the biologically weighted dose at the end of the proton range in contradiction to the constant absorbed dose[7]. In order to verify these results and to obtain the LET distribution of the modulated proton beam, a stack of dosemeters was exposed in the modulated beam. For irradiation of the dosemeters the same absorbed dose was applied as for calibration. The dose-distribution corrected for the decreasing efficiency of peak 5 as well as the HTR-distribution measured are shown in figure 5.

It can be seen that the dose in the modulated beam is constant over the whole proton range. Nevertheless the HTR-distribution shows an increase with depth. The HTR-dependence is quite similar for the modulated and the unmodulated beam (figure 2 and 5). Based on the LET-calibration an increase of the LET from 1.78 keV/μm tissue (2 mm depth) to 4.83 keV/µm tissue (28 mm depth, last dosemeter of the stack) can be obtained. This corresponds to a quality factor of 1 respectively of 1.4 after ICRP 26.

This depth dependence of the HTR respectively LET was compared with biological investigations [7]. Chinese hamster cells V79 were exposed in the modulated proton beam with 2 Gy absorbed dose and the survival propability was determined. The decrease of survival cells is shown in figure 6 compared to the HTR-distribution.

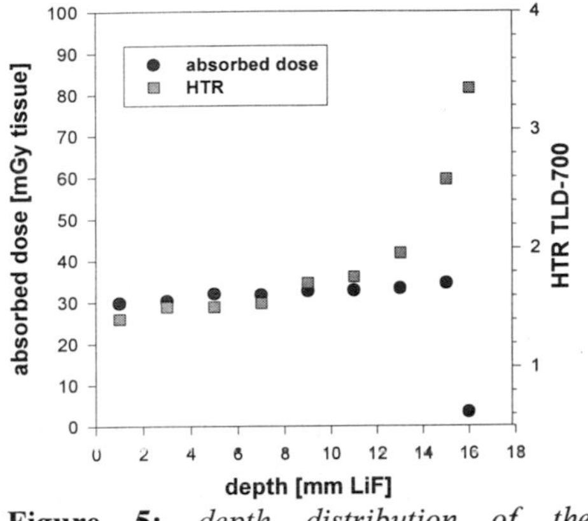

Figure 5: *depth distribution of the absorbed dose and the HTR in the modulated proton beam measured with TLD-700*

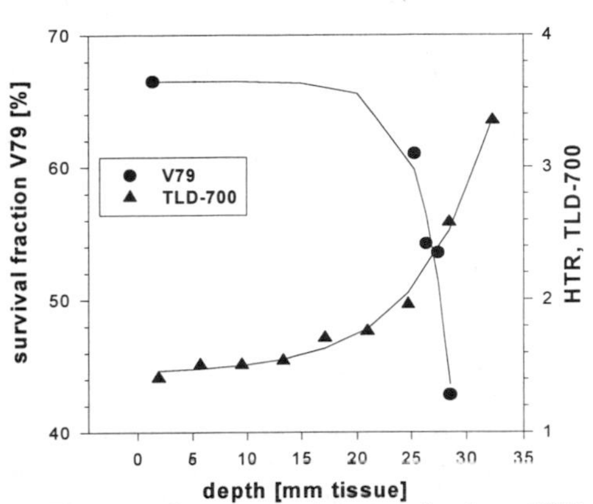

Figure 6: *comparision of the HTR measured with TLD-700 in dependence of the depth and the survival probability of V79[7] irradiated in a modulated 62 MeV proton beam*

6 CONCLUSION

The HTR-method developed to measure the average LET in mixed radiation fields was applied in a 62 MeV proton beam used for radiation therapy. A LET-calibration of the HTR was performed in the Bragg peak and used to measure the increase of LET in the modulated proton beam. Nevertheless the LET is an insufficient parameter to describe the HTR and the biological effectiveness because both are not only LET- but also dose-dependent. Therefore the HTR was correlated directly with the RBE as the comparision with biological investigations of the survival probability of V79 shows. The presented data show the possibility to use the HTR in LiF for description of the dose- and LET-dependent behaviour of the biological effectiveness. In order to describe the LET dependent processes in LiF first approaches of microdosimetric models are under progress[4].

References

(1) N. Vana et al., International Space Year Conference, Munich 1992, 193
(2) N. Vana et al., *Rad. Prot. Dosim.*, 1996, **66**, 145
(3) M. Noll et al., *Rad.Prot.Dosim.*, 1996, **66**, 119
(4) W. Schöner et al., Intern. Congress of Rad. Prot. IRPA9, Vienna 1996, Vol. 4, 4-231
(5) A.J. Mills et al, Rad. Res. 1996, **155**, 575
(6) Y.S. Horowitz, 'Thermoluminescence and Thermoluminescent Dosimetry', CRC Press, 1984, Vol.2, 152
(7) E. Egger, private communication

MICRODOSIMETRIC INVESTIGATIONS IN THE FAST NEUTRON THERAPY BEAM AT FERMI NATIONAL ACCELERATOR LABORATORY -WORK IN PROGRESS-

K. Langen*†, A. J. Lennox†‡, T. K. Kroc†, P. M. DeLuca, Jr.*

* Department of Medical Physics, University of Wisconsin-Madison, 1300 University Av., Madison, WI 53706, USA.
† Neutron Therapy Facility at Fermilab, Batavia, IL 60510, USA.
‡ Saint Joseph Hospital, 77 North Airlite St., Elgin, IL 60123, USA.

1 INTRODUCTION

Neutron interactions in tissue generate a variety of secondary charged particles with a broad range of energies. As these particles slow down they transfer energy to tissue through ionization and excitation. While it is possible to determine the macroscopic density of energy deposited in tissue, e.g. the absorbed dose, using several methods, microdosimetric measurements also provide information about how the energy is deposited microscopically.

Secondary particles generated in tissue have a wide range of linear energy transfer coefficients (LET), or ionization densities, and the distribution of LETs may be correlated with the biological effectiveness of the beam. Experimental microdosimetry using low pressure tissue equivalent proportional counters is a physical method to gain biological information by determining the spatial distribution of energy deposition on a micrometer scale. This tool is especially valuable in neutron therapy due to the large range of LETs exhibited in this beam.

Beyond microdosimetric measurements using tissue-mimicking materials, employing counters constructed of elemental materials such as carbon, metal and metal oxide paired devices and tissue equivalent material loaded with boron serve to provide specific microdosimetric information that allows a greater understanding of clinical neutron beam interactions.

1.1 Neutron Therapy

Neutron therapy is generally referred to as high LET irradiation due to the relatively large LETs of the secondary particles. High LET irradiation is advantageous in the treatment of radiation resistant tumors which do not respond to conventional photon therapy. These include rare tumors such as salivary gland tumors and sarcomas as well as adenocarcinoma of the prostate.

There are currently three operating neutron therapy facilities in the United Sates. The Fermilab facility uses 66 MeV protons from a linear accelerator (linac) to generate neutrons. The linac is part of a series of accelerators and is used to inject protons into a booster synchrotron. Since protons are only injected into the booster for a fraction of time they can be used for therapy the rest of the time. Once protons are extracted from the linac they impinge onto a beryllium target in which they lose 49 MeV. The resulting neutron field is then collimated to the desired field size. The typical dose rate, neutron and gamma, during patient treatment is 0.4 Gy/min at 100 mm depth at isocenter (1.90 m).

1.2 Microdosimetric Measurements

Microdosimetric measurements are performed with low pressure proportional counters. Counters are spherical and are 12.7 mm in inner diameter. The wall can be made out of several materials of interest and is typically 1.27 mm thick. Counters are filled with tissue equivalent gas to a low pressure to simulate a 2 μm diameter sphere of unit density tissue. Neutrons interact in the wall material and the generated secondaries enter the gas cavity where they produce ionizations in the gas. Secondary electrons travel toward the biased anode where gas multiplication occurs. Data are acquired in the usual manner with individual events expressed in terms of lineal energy,

$$y = e/\bar{l}\ [\mathrm{keV}/\mu\mathrm{m}] \tag{1}$$

where e is the imparted energy in the volume and $\bar{l}$ is the mean cord length in that volume. An internal alpha-particle source of known energy serves to calibrate energy deposition. Additionally, the location of the proton and alpha-particle Bragg edges (maxima in stopping power) were used to confirm the alpha-particle calibration. By plotting y D(y) [y^2 f(y)] versus log y, equal areas correspond to equal doses and the large dynamic range of the data are easily displayed.

2 INVESTIGATIONS

2.1 Relative Carbon and Oxygen Kerma Factors

The quantity of interest in radiation therapy is the absorbed dose in tissue delivered during treatment. A determination of this quantity is, however, not trivial since it can not be directly measured. Commonly, an ion chamber with tissue mimicking A-150 plastic walls is used to measure the absorbed dose in a neutron beam. However, the elemental composition of A-150 plastic differs from ICRU tissue in the carbon and oxygen content. Neutron cross sections can vary significantly from one element to the next and a determination of the absorbed dose in tissue requires a correction of the measured absorbed dose in A-150 plastic.

The calculation of the conversion coefficient that relates absorbed dose in A-150 plastic to absorbed dose in tissue, requires knowledge of the carbon to A-150 plastic and oxygen to A-150 plastic kerma coefficient ratios at all neutron energies as well as knowledge of the neutron energy spectrum. However, kerma coefficient information at higher energies (> 20 MeV) is sparse and the neutron energy spectrum is often not known.

Even though there are efforts to obtain kerma coefficients and kerma coefficient ratios of interest at higher energies, [1-5] we propose to determine the integral kerma coefficient ratios, e.g. the carbon to A-150 and oxygen to A-150 kerma coefficient ratios convoluted by the neutron energy spectrum, directly in the beam. This method eliminates the need to know the neutron energy spectrum as well as kerma coefficient ratios at each neutron energy.

To this end we simultaneously irradiated a carbon and A-150 plastic counter in the neutron beam to determine the integral carbon to A-150 kerma coefficient ratio. For the determination of the integral oxygen to A-150 kerma coefficient ratio we irradiated an A-150 counter simultaneously with a zirconium and then with a zirconium oxide counter. The kerma coefficient ratio can then be deduced by the subtraction technique.[6] Data are currently under analysis.

2.2 Beam Characterization

The RBE of a neutron beam is influenced by the distribution of the ionization densities of the secondary charged particles. The type and energy of secondaries produced is determined by the neutron energy spectrum. Thus, changes in the neutron energy spectrum result in RBE changes. Several parameters affect the neutron energy spectra in a given facility and we are investigating the influence of two of these parameters for the Fermilab facility. As the neutron beam penetrates tissue the lower energy neutrons are preferentially absorbed, resulting in beam hardening at depth. Scattering of the beam in the collimator assembly is responsible for changes in the neutron spectrum with field size.

Using an A-150 plastic counter we measured microdosimetric spectra at different depths and field sizes. Changes in the microdosimetric spectra indicate potential RBE changes. Figure 1 shows different spectra measured at different water depths in a 100 mm by 100 mm field each normalized to unit dose. As an indication of how this information might be used, we convolved the D(y) spectra with a weighting factor related to lineal energy, $y_{sat}(y)$, to estimate a relative biological scoring parameter, y*. That is

$$y^* = \int y_{sat} D(y)\, dy. \qquad (2)$$

y_{sat} corrects for the saturation effect at larger event sizes.[7] Preliminary results for the 100 mm by 100 mm field indicate a reduction of y* of 2.9 % at 150 mm and an increase of y* of 3.7 % at 50 mm depth relative to the y* at 100 mm depth. These results agree well with those obtained earlier for similar irradiation conditions at the neutron therapy facility at Faure in South Africa.[8] Irradiation of Chinese hamster lung cell lines in the Fermilab beam yielded a decrease of 12 % in effectiveness going from 30 to 240 mm depth in tissue which is also in agreement with the current results.[9]

2.3 Boron Neutron Capture

The general thrust of Boron Neutron Capture Therapy (BNCT) is to conform the absorbed dose to the tumor tissue by preferentially loading the tumor with a B-10 compound and then irradiating the tumor with thermal neutrons. B-10 has a high cross

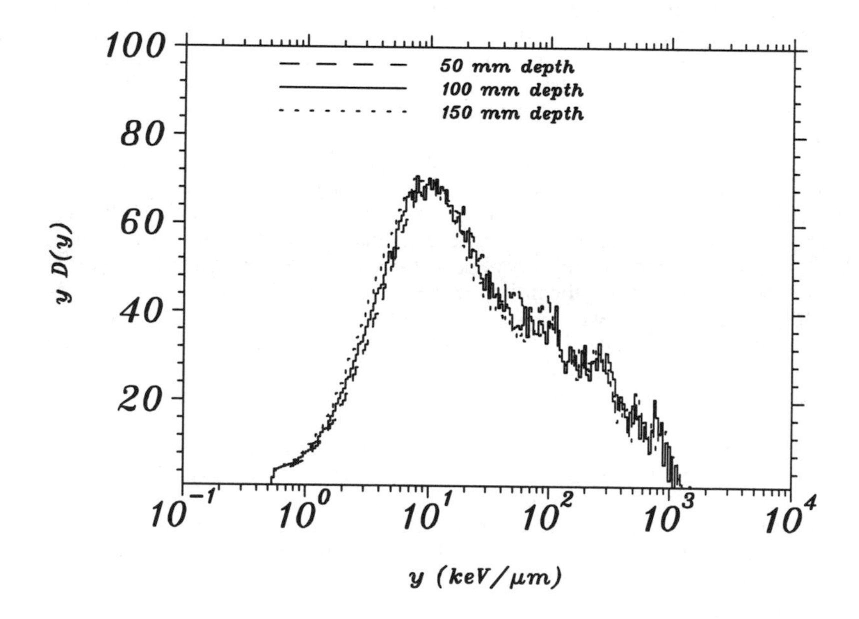

Figure 1: *A-150 tissue equivalent plastic spectra measured at 50, 100 and 150 mm water depth in the neutron beam for a 100 mm by 100 mm field size.*

section for thermal neutron capture and the reaction results in the release of a Li-7 and an α particle that share on average a kinetic energy of 2.33 MeV. Due to their short ranges in tissue (< 10 μm) their energy is deposited in the tumor tissue.

The objective of using BNCT at the fast neutron therapy facility at Fermilab will not be to treat patients with BNCT only but rather to use BNCT to boost the tumor dose. In this case the low energy component contained in the fast neutron beam is utilised.[10]

The clinical rational is based on the poor prognosis of certain brain tumor patients. The expected survival time for patients diagnosed with glioblastoma multiforme is around eight months following diagnosis and can be extended by another four months with radiation therapy. There is no difference in survival between fast neutron and photon therapy patients but the cause of death has been found to vary with treatment modality. A regrowth of the tumor is the primary cause of death in photon patients whereas with fast neutron therapy one achieves tumor control but patients sustain a high degree of normal tissue damage that causes death.[11] These clinical results indicate a small therapeutic ratio of glioma to normal brain tissue. In order to prolong patient survival the therapeutic ratio needs to be increased. BNCT in addition to fast neutron irradiation has the potential to accomplish this by increasing the tumor to normal tissue absorbed dose ratio. Hence, larger tumor doses could be used while keeping normal tissue dose at tolerable level.

Microdosimetry offers some unique capabilities for this application. To mimic tumor boron loading we constructed a tissue equivalent device loaded with 200 ppm equivalent of B-10. While anticipated tumor-to-normal tissue B-10 loading is 50 ppm, the extra B-10 in the counter affords a more easily detected response.

This counter will be simultaneously irradiated with a regular A-150 plastic counter that is identical to the boronated counter in every aspect except the boron additive to the wall. A difference in counter responses is then due to the boron additive in the one counter wall. Primarily, we are studying the optimization of the thermal neutron component as a function of moderator material and thickness. However, by using microdosimetric techniques, we anticipate the possibility of estimating the increased biological effectiveness in a more quantitative manner.

REFERENCES

1 D.J. Brenner, *Phys. Med. Biol.*, 1983, **29**, 437.
2 P.J. Dimbylow, *Phys. Med. Biol.*, 1982, **27**, 989.
3 C.L. Hartmann, P.M. DeLuca, Jr., D.W. Pearson, *Radiat. Prot. Dosim.*, 1992, **44**, 25.
4 U.J. Schrewe, H.J. Brede, S. Gerdung, R. Nolte, P. Pihet, P. Schmelzbach, H. Schuhmacher, *Radiat. Prot. Dosim.*, 1992, **44**, 21.
5 U.J. Schrewe, W.D. Newhauser, H.J. Brede, V. Dangendorf, P.M. DeLuca, Jr., S. Gerdung, R. Nolte, P. Schmelzbach, H. Schuhmacher, T. Lim *Radiat. Prot. Dosim.*, 1995, **61**, 275.
6 P. M. DeLuca, Jr., H. H. Barschall, Y. Sun, R. C. Haight, *Radiat. Prot. Dosim.*, 1988, **23**, 27.
7 ICRU, Report 36, Microdosimetry, International Commission on Radiation Units and Measurements, Bethesda, 1983.
8 J.P. Slabbert, P.J. Binns, H.L. Jones, J.H. Hough, *Brit. J. Radiol.*, 1989, **62**, 989
9 C.H. Hill, R. K. Ten Haken, M. Awschalom, *Int. J. Radiation Oncology Biol. Phys.*, 1991, **20**, 1341
10 F. M. Waterman, F. T. Kuchnir, L. S. Skaggs, D. K. Bewley, B. C. Page, F. H. Attix, *Phys. Med. Biol.*, 1978, **23**, 592.
11 K. R. Saroja, J. Mansell, F. R. Hendrickson, L. Cohen, A. J. Lennox, *Int. J. Radiation Oncology Biol. Phys., 1989, 17, 1295.*

SIMULATION OF A MICRODOSIMETRY PROBLEM: BEHAVIOUR OF A PSEUDORANDOM SERIES AT A LOW PROBABILITY

P. Meyer[1], J.E. Groetz[1], R. Katz[2], M. Fromm[1] and A. Chambaudet[1]

[1]LMN, Université de Franche-Comté, 16 route de Gray, 25030 Besançon France
[2]Department of Physics and Astronomy, University of Nebraska, Lincoln NE 68588-0111 USA

1 INTRODUCTION

The carcinogenic effects for low dose irradiations are not very well known. Estimations usually are made based on the effects observed at high doses that are then extrapolated to low doses. To estimate low dose effects, the ICRP (International Commission on Radiological Protection) uses a linear extrapolation matched with a dose-rate reduction factor equal to two. This proportionality of the effect and dose, even for the lowest doses and dose-rates, leads to two assumptions which must be questioned[1]:

- the efficiency of DNA repair in cells does not vary with the dose and the dose-rate,
- when one single particle crosses one single cell, a carcinogenic transformation may occur.

Low doses are frequently generated by fast electrons at low fluence.

We must consider the irradiated medium as an assembly of targets and the cross section as a representation of the probable interaction between the incident particle and the target[2,3,4]. Biologically, cells contain internal structures which are the sensitive elements. Physically, a hit is interpreted as a registered event caused by a particle passing through the sensitive site.

The Poisson law describes the statistic behaviour of this event:

$$P(x) = \frac{e^{-m}}{x!} m^{x}, \qquad (1)$$

where $P(x)$ is the realisation probability of the event x and m represents the average hits per target (ratio of the number of hits per number of targets). For example, consider that a flux of exactly 10^6 particles/cm^2 reaches a cell population whose sensitive elements have a geometric cross section of 100 μm^2. The average number of particles per cell will be one, but according to the Poisson statistics, about 37% of the cells will survive (0 hits), about 37% of the cells will be hit only once and 26% will be hit twice or more[5].

This problem can be extended to the response of many radiation detectors - one-hit or multi-hits detectors[6].

2 CALCULATIONS

Microphysical processes have a random character that are ruled by continuous or discrete probability laws. By using numerical simulations with Monte Carlo methods, random experiences are imitated by a sampling of pseudorandom numbers which are generated by algorithms such as the linear congruential generators (LCG) that use the integer recursion[7]:

$$X_{i+1} = (a\, X_i + c) \bmod M, \tag{2}$$

where the integers a, c and M are constant. They generate a sequence X_1, X_2,... of random integers between 0 and M-1 (into $[1,M\text{-}1]$ if c=0). Each X_i is then put into the interval [0,1).

We used a class of random empirical test methods (chi-square test, Kolmogorov-Smirnov test, gap test, run-up run-down test, serial test) that seem to give valid results[8]. Valid as they may be, these tests are not sufficient; therefore new tests created for specific applications are often advisable. We tested two LCG generators, the subroutine RAND from the Matlab software[9] and the subroutine RND from the Fortran 77 software[10]. As a systematic test, we compared the spectrum of hits obtained with the given pseudorandom numbers generators with the one predicted by Poisson's distribution law.

Calculations were performed at a HP 9000 workstation and probabilities were determined for only one series. Each pseudorandom number generated between [0,1) (original number) is put in the interval corresponding to the number of digits. As an example, for 3 digits each original number is multiplied by 10^3 and we extract the integer part (integer x); and for 4 digits each original number is multiplied by 10^4. Experimental probability Π for integer x (independently of number of digits) to be hit is defined as:

$$\Pi(x) = \frac{n_x}{N}, \tag{3}$$

where n_x is the number of drawings for integer x and N the total number of drawings.

The discrepancy $\xi(x)$ between experimental and theoretical (1) probabilities is calculated from the expression (if $P(x)$=0 and $\Pi(x)\neq0$, we put arbitrarily $\xi(x)$=-10):

$$\xi(x) = \frac{|P(x) - \Pi(x)|}{P(x)} \times 100, \tag{4}$$

We are interested in the probability of hitting a particular number, the number of hits ranges from 0 to 10 in the case of the presented application. The number m, considered as the average of hits per integer is defined as:

$$m = \frac{N}{card\{interval\}}, \tag{5}$$

where *card{interval}* is the number of integers in this interval.

Experimental probabilities for a same value of m (in our study m ranges from 0.1 to 10) were calculated as a function of the number of digits. Let us consider the case where 3 digit numbers are used:

- m=1, if we calculate the hit spectrum of integers in the interval $[0,10^3)$ for 10^3 integers generated in this interval,

- m=2, if we calculate the hit spectrum of integers in the interval $[0,10^3)$ for 2.10^3 integers generated in this interval.

3 RESULTS

We have represented the descrepancies (4) between calculations and theory for the Fortran generator on figure 2 to 5. We have obtained similar results for the Matlab generator. The Poisson law (1) is represented as a function of m and x on figure 1. Two areas can be observed: a first one corresponding to non null probabilities (relief zone) and a second one corresponding to probabilities which tend towards zero (flat zone).

Discrepancies $\xi(x)$ are localized into two types of areas; a first one which does not move with increasing number of digits and corresponding to n_x (number of hits) $\in [0,2]$ and m $\in [5,10]$, a second zone, much more important, which seems to be localized at the border of high Poisson probabilities (figure 1) and moves toward high number of hits with increasing digit numbers. As a matter of fact, we observe that an increase of the number of digits results in a decrease of the total descrepancy, in the studied ranges of n_x and m.

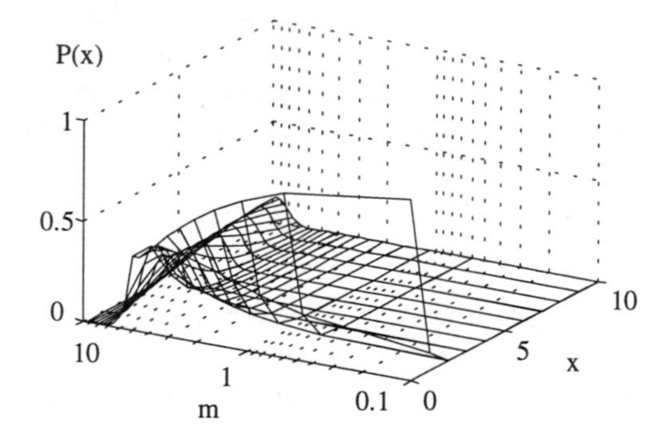

Figure 1. 3-D representation of the Poisson distribution

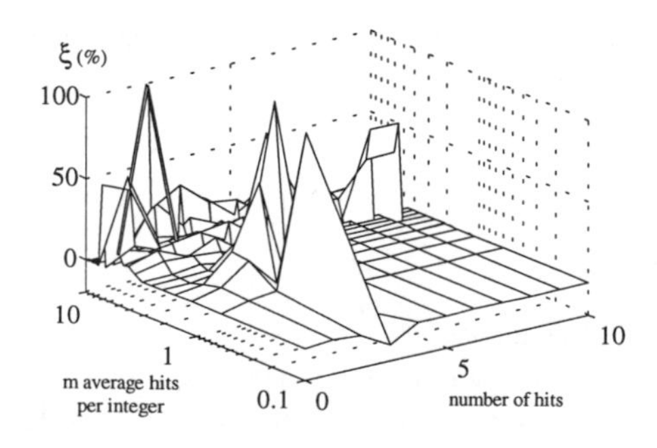

Figure 2. ξ(x) for 3 digits numbers

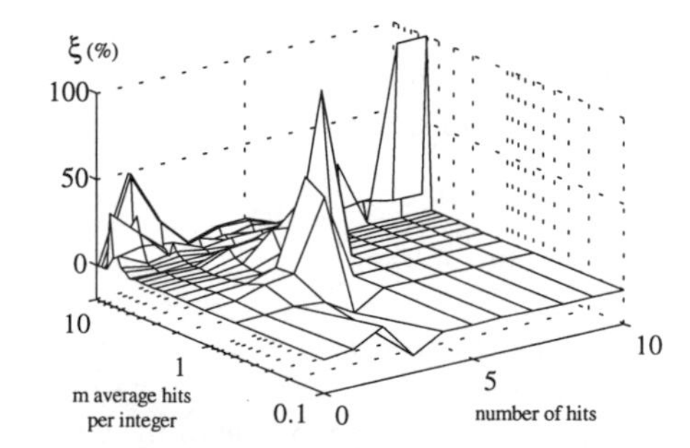

Figure 3. ξ(x) for 4 digits numbers

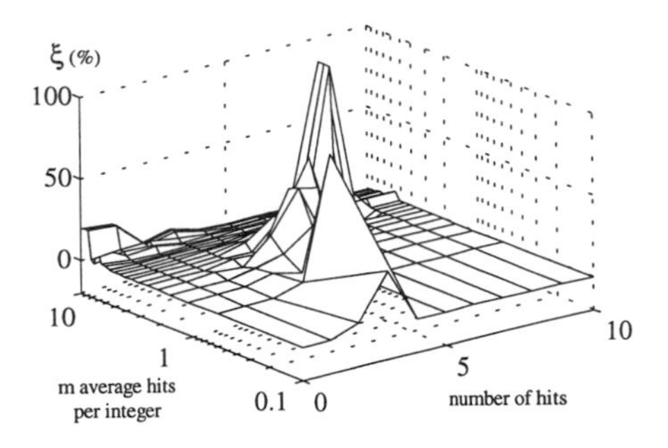

Figure 4. ξ(x) for 5 digits numbers

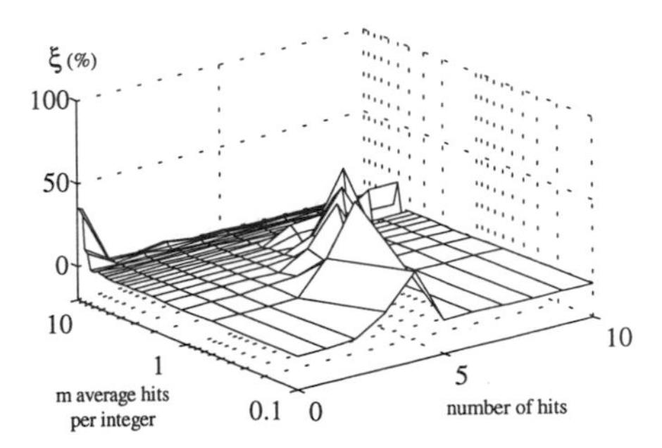

Figure 5. ξ(x) for 6 digits numbers

The results presented in the 3D graphs (figures 2 to 5) correspond to only one history. In general, many histories must be used to obtain accurate statistics. We made the calculations (average and standard deviation for $\Pi(x)$) on the largest discrepancies, and even though a large number of histories were used, an important standard deviation remains. For such calculations, we find that the general behaviour is a stabilization of the obtained average values of $\Pi(x)$ around the needed value of $P(x)$ when the number of histories increases.

4 CONCLUSION

We examined the properties of two pseudorandom generators using a test which mimics the properties of the application in which the generators will be used. In this case, these applications are ruled by the Poisson law. We have shown that the two generators fail in the described cases. It also appears that an increase of digits improves the properties of the tested generators about the Poisson law. It should be noticed that in the case of our test, the use of a number of digits greater than 6 was not possible due to the needed memory capacity and time of computation. Such a test seems necessary if a pseudorandom generator can be expected to provide reliable results, we strongly recommend this test if a number of digits lower than 7 is used.

REFERENCES

1. M. Tubiana, *Radioprotection*, 1996, **31**, 2, 155-191
2. R. Katz, *Bull. Am. Phys. Soc.*, 1966, **11**, 113
3. R. Katz, *Proc. 2nd Symposium on Microdosimetry*, 1970, 283-301
4. R. Katz, R. Zachariah, F.A. Cucinotta and C. Zhang, *Radiat. Res.*, 1994, **140**, 356-365
5. M. Durante, G.F. Grossi, M. Pugliese and G. Gialanella, *Radiat. Meas.*, 1996, **26**, 2, 179-186
6. R. Katz and F.E. Pinkerton, *Nucl. Instr. and Meth.*, 1975, **130**, 105-119
7. I. Vattulainen, "New tests of random numbers for simulations in physical system" PhD., 1994, University of Helsinki
8. P. Meyer, "Propriétés du CR 39 en dosimétrie neutron faible fluence" PhD., 1996, University of Franche-Comté (n°525)
9. Matlab Reference Guide, 1992, The MathWorks Inc., 400-401
10. HP Fortran/9000 9.16 Programmer's Guide, 1992, Hewlett Packard Company

Microdistribution in tissues

THE DISTRIBUTION OF HITS WHEN LUNG CELLS ARE IRRADIATED BY ALPHA PARTICLES

J.A. Simmons and S.R. Richards,
Univesity of Westminster,
115 New Cavendish Street,
London, W1M 8JS.,
U.K.

ABSTRACT

It is now becoming generally agreed that, for the irradiation of cells by heavy charged particles, use of the concept of "absorbed dose" has severe limitations, especially below about 3 Gy. Instead distributions of specific energy are more relevant, and examples for the case of an inhaled α-emitter lodged in the parenchymal region of the lung were shown at a previous Microdosimetry Symposium. Since many authors now present results in terms of hits to cells and nuclei, our earlier data have been re-analysed in these terms and examples will be shown of the distributions of hits as a function of specific energy for rat, beagle and human lung cells.

INTRODUCTION

The problem of the potential hazard arising from an alpha-emitting radionuclide in the lung has been under active investigation for about 20 years. Originally this hazard was considered in terms of the dose to the lung as a whole or some particular section of it. In more recent years the lines of research have been focussed on "absorbed dose" to cells derived from lung tissue or the microdosimetric distributions of specific energy within the nuclei of these cells. The newest development has been the replacement of all dosimetric measurements by measurement of numbers of hits to the nuclei, and we therefore felt it appropriate to re-work some of our earlier data in these terms.

MATERIALS AND METHODS

Measurements were carried out on material obtained from rat, beagle and human lung. Sections were cut at about 2μm thickness and stained so that the air in the alveolar sacs appeared light against a dark background of issue when viewed on the screen of a "Magiscan" image analyser. Further differential staining allowed the nuclei to be distinguished from the rest of the cell.

The source was taken to be a particulate of plutonium dioxide. The range of activities of these particulates corresponded to the range of sizes which were most likely to be trapped in the alveolar sacs. It was also assumed that the period of exposure was 100 days. Full details of measurements and the methods by which the relevant distributions were calculated have been given elsewhere(1).

RESULTS

Figure 1 shows a plot of the number of cells and nuclei hit per increment of alpha-particle track as a function of the specific energy deposited. This is for a sample site from rat lung and is typical of the results for the three different activities considered. It can be seen that, as the source activity increases, the numbers of cells and nuclei hit also increase but not in a direct linear relationship. It can also be seen that at the lowest activity, the distributions for the cells have their peaks at approximately an order of magnitude lower specific energy than the peaks for the nuclear distributions, and that the former move steadily to higher values while the latter remain substantially unchanged. Only when all of the accessible nuclei have received at least one hit do the curves coincide.

An alternative way of looking at the results is to consider the numbers of cells which are accessible, the numbers which are actually hit, and the numbers of nuclei which are hit for the various source activities. The tables show sample data for three different sites taken from rat and beagle lung respectively. From these, it can be seen that, even at the highest source activity considered here, not all of the accessible nuclei are hit and, in some cases, not even all of the much larger cells are traversed. Much higher source activities would have had to be considered for all of the nuclei to have been hit, a condition which would have arisen when the mean dose to the tissue (in traditional terms) was about 2 Gy.

DISCUSSION

Doses of 2 Gy to lung tissue are very large and therefore unlikely to arise except in extreme circumstances. For most purposes, therefore, one is

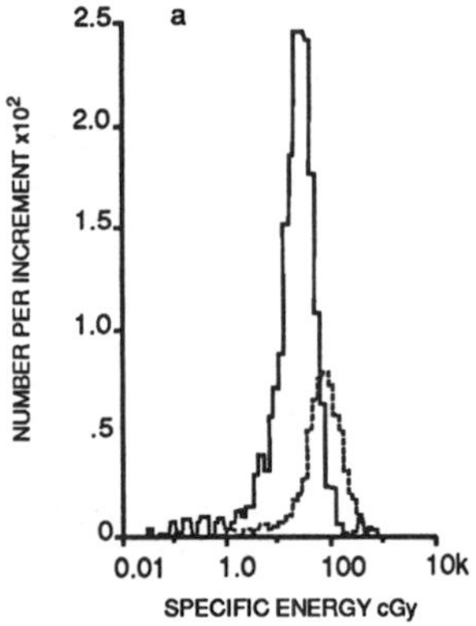

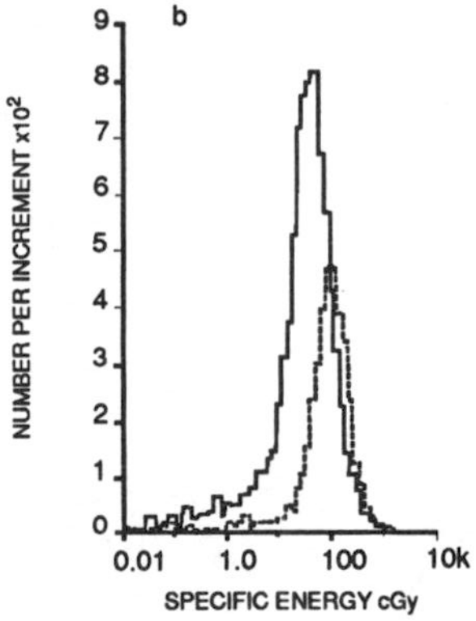

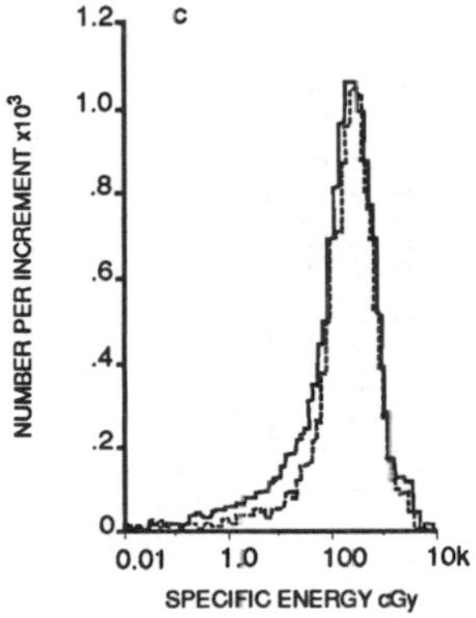

Fig 1 RA7

Distribution of specific energies to cells & nuclei for a period of 100 days

Particle activities :-

a: 10^{-4} Bq

b: 10^{-3} Bq

c: 10^{-2} Bq

— Number of Cells

--- Number of Nuclei

TABLES

Comparison of the numbers of cells and nuclei accessible and hit for 3 different source activities.

a) Sample rat sites.
b) Sample beagle sites.

Table (a)

SITE	SOURCE ACTIVITY Bq	NO. OF CELLS ACCESSIBLE	NO. OF CELLS HIT	NO. OF NUCLEI HIT
RAI	10^{-4}	9027	1509	548
"	10^{-3}	8948	6017	2899
"	10^{-2}	8933	8864	7833
RA7	10^{-4}	13501	1937	626
"	10^{-3}	13495	8331	3839
"	10^{-2}	13573	13368	11374
RA13	10^{-4}	21233	1984	616
"	10^{-3}	21134	10265	4101
"	10^{-2}	21191	20370	15427

Table (b)

SITE	SOURCE ACTIVITY Bq	NO. OF CELLS ACCESSIBLE	NO. OF CELLS HIT	NO. OF NUCLEI HIT
BE3	10^{-4}	5433	1361	361
"	10^{-3}	5435	4108	1965
"	10^{-2}	5252	5223	4481
BE9	10^{-4}	31271	2055	454
"	10^{-3}	31347	11065	3355
"	10^{-2}	31223	27629	15184
BEII	10^{-4}	35360	2268	474
"	10^{-3}	35453	13967	3650
"	10^{-2}	35407	31987	18138

considering cases where not all of the accessible nuclei have been hit and therefore (as has been argued elsewhere [2]) the concept of "dose" is meaningless. One should, instead, be considering the number of nuclei which are likely to be traversed and, as a result, what damage is likely to be caused. As an example, it has recently been shown [3] that about 31 traversals of a nucleus are required to cause one chromosome aberration in V79 cells whereas only about 12 traversals are required to cause an aberration in human lung cells. Whether such differences imply that the human lung is more susceptible to long-term ill-effects such as tumour induction is a matter that urgently needs further investigation, and it is clear that the question needs to be considered in terms of "hits" rather than of "dose".

REFERENCES

1. J.A. Simmons and S.R. Richards, Microdosimetry of alpha-irradiated parenchymal lung, Low Dose Radiation, London, Taylor and Francis, 1989, 312-324.

2. J.A. Simmons, Absorbed dose - an irrelevant concept for irradiation with heavy charged particles?, J. Radiol. Prot., 1992, 12, 173-179.

3. J.A. Simmons, P. Cohn and T. Min, Survival and yields of chromosome aberrations in hamster and human lung cells irradiated by alpha particles, Radiat. Res. 1996, 145, 174-180.

MICRODOSIMETRIC MODELLING OF DAMAGE TO HAEMOPOIETIC STEM CELLS FROM RADON DECAY IN FAT

T. D. Utteridge[1,2], D. E. Charlton[3], M. S. Turner[3], A. H. Beddoe[4], A. S.-Y. Leong[5], J. Milios[5], N. L. Fazzalari[5] and L. B. To[6]

1. Department of Medical Physics, Royal Adelaide Hospital, Adelaide, Australia 5000
2. School of Applied Physics, University of South Australia, Pooraka, Australia 5095
3. Physics Department, Concordia University, Montréal, Québec, Canada H3G 1M8
4. Division of Radiotherapy Physics, Queen Elizabeth Med. Centre, Birmingham B15 2TH
5. Division of Tissue Pathology, Institute of Medical and Veterinary Science, Adelaide 5000
6. Leukaemia Research Unit, Institute of Medical and Veterinary Science, Adelaide 5000

1 INTRODUCTION

Monte Carlo modelling of the alpha particle radiation dose to human haemopoietic stem cells from the decay of radon and its short-lived products in marrow fat cells was undertaken following Richardson et als[1] proposition that such exposure could induce leukaemia, and epidemiological observations that underground miners have not developed an excess of leukaemia[2].

2 MATERIALS AND METHODS

Fat cells were placed randomly around a haemopoietic stem cell at the centre of a box, decays randomly placed, direction cosines assigned, "hits"' determined and dose and passages calculated using a modified version of Charlton et al[3]. The modification included incorporation of measured size distribution data for human haemopoietic stem cells and measured spatial distribution in normal human marrow of stem cells relative to fat cells.

Normal human haemopoietic stem cells ($CD34^{+}CD38^{-}$ mononuclear cells[4], sorted using Fluorescent Activated Cell Sorting) were suspended in isotonic saline. Wet preparations were made, photographed and image analysis performed to determine stem cell diameter. The spatial distribution of the stem cells relative to fat cells was measured using image analysis on autopsy specimens of normal human marrow sections (stained using immunohistochemistry).

3. RESULTS

The distribution of haemopoietic stem cell diameters was found to be trimodal, similar to that found in mice[5]. The mean diameters for the three groups were 5.71, 11.63 and 14.8 μm diameter. The proportions of haemopoietic stem cells of each diameter specified were

7.3:1.7:1.0 as this was the ratio found in the measured data. The measured 2D distribution of distances between stem cells and fat cells agreed well with that predicted for a random distribution of stem cells.

Modelling of the doses to stem cells in marrows of various fat fractions exposed to the U.K. mean indoor radon concentration (20 Bq m^{-3}) was performed (see Table 1). The proportion of stem cells "hit" increases approximately linearly with fat fraction, as does the proportion of "hit" cells which survive. The mean number of passages per "hit" stem cell is unchanged (1.02) across the range of fat fractions examined (23 - 77%). The mean dose to "hit" cells similarly does not change, ie 0.86, 0.20, and 0.12 Gy for cells of 5.71, 11.63 and 14.79 μm diameter respectively. The doses received in underground miners studies were also modelled (see Table 2). The proportion of "hit" cells increases linearly with radon concentration but the mean number of passages does not vary significantly from single passages.

Survival was calculated using the mean free path method of Charlton and Turner[6]. A survival curve was generated, with a D_0 of 0.6 Gy at 120 keV/μm (which is Lorimore et als[7] D_0 value for murine 12 day CFU-S cells), as the equivalent human stem cell data has not yet been published. This gave mean free paths between lethal events of 0.97, 8.2 and 8.2 μm respectively for the three sizes of stem cells. The probability of surviving a passage was calculated to be 0.057, 0.412 and 0.329 for the three sizes respectively.

4 CONCLUSIONS

Monte Carlo modelling of the dose to marrow with different fat fractions showed a positive approximately linear relationship between the fat fraction and both the proportion of haemopoietic stem cells "hit" and the proportion of "hit" survivors. Modelling of the dose to haemopoietic stem cells in the marrow of underground miners showed a positive linear relationship between the radon concentration in external air and both the proportion of haemopoietic stem cells "hit" and the proportion of "hit" survivors. In all of the modelling the mean number of passages per "hit" cell was approximately 1.02, with doses of 0.86, 0.20 and 0.12 Gy for stem cells of diameter 5.71, 11.63 and 14.79 μm respectively. The finding that "hit" haemopoietic stem cells experience predominantly single passages across the range of radon concentrations encountered in practice in indoor and underground exposures suggests that an inverse dose-rate (dose protraction) effect such as that observed for lung cancer at high radon exposures and bone sarcomas in ^{224}Ra would not be expected to occur.

References

1. R. B. Richardson et al, *Brit. J. Radiol.*, 1991, **64**, 608.
2. S. C. Darby et al, *J. Nat. Cancer Instit.*, 1995, **87**, 378.
3. D. E. Charlton et al, *Int. J. Radiat. Biol.* 1996, **69**, 585.
4. L. W. M. M. Terstappen et al, *Blood*, 1991, **77**, 1218.
5. J. Visser et al, *Exp. Hemat. Today*, 1977, 21.
6. D. E. Charlton and M. S. Turner, *Int. J. Radiat. Biol.* 1996, **69**, 213.
7. S. A. Lorimore et al, *Int. J. Radiat. Biol.* 1993, **63**, 655

Table 1. Monte Carlo Modelling Results of Alpha Particle Radiation Dose to Marrows with different Fat Fractions (for UK national mean indoor radon concentration of 20 Bq m^{-3})

Fat Fraction in marrow	Proportion of haemopoietic stem cells "hit"	No. of passages per "hit" haemopoietic stem cell	Average dose (Gy) per "hit" haemopoietic stem cell of diameter			Proportion of haemopoietic stem cells which are "hit" and survive
			5.71 μm	11.63 μm	14.79 μm	
23%	0.0007%	1.02	0.87	0.20	0.12	0.0002%
31%	0.0009%	1.02	0.86	0.20	0.12	0.0003%
40%	0.0010%	1.02	0.86	0.20	0.12	0.0004%
45%	0.0011%	1.02	0.86	0.20	0.12	0.0004%
55%	0.0013%	1.02	0.86	0.20	0.12	0.0005%
67%	0.0016%	1.02	0.87	0.20	0.12	0.0006%
74%	0.0018%	1.02	0.86	0.20	0.12	0.0006%
77%	0.0019%	1.02	0.86	0.20	0.12	0.0007%

Table 2. Monte Carlo Modelling Results of Alpha Particle Radiation Dose to Underground Miners Exposed to Radon

Mine	Reference	Mean cumulative radon exp. on leaving mine (WLM)	O/E (Observed/expected incidence of leukaemia mortality) (95% confidence interval)	Proportion of haemopoietic stem cells "hit"	Proportion of haemopoietic stem cells which are "hit" and survive
West Bohemia, Czech Republic	Tomasek et al 1995	219	0.91 (0.44-1.67) (all leuk.) 1.03 (0.21-3.00) (myel. only)	2.7%	0.9%
Malmberget, Northern Sweden	Darby et al 1995	89	1.05 (0.39-2.30)	1.1%	0.4%
Cornwall, U.K.	Hodgson & Jones 1990	~100	1.75	1.2%	0.4%
Newfoundland	Morrison et al 1988	383	1.51 (0.30-4.42)	4.7%	1.6%
Yunnan Province, China	Xiang-Zhen et al 1993	275.4 (Male) 66.4 (Female)	not available	3.4% 0.8%	1.2% 0.3%
France	Tirmarche et al 1993	70	1.44 (0.39-3.67)	0.8%	0.3%
Collab. analysis. of 11 studies	Darby et al 1995	155	1.28 (0.51-2.64) (< 10 yrs) 1.08 (0.72-1.55) (> 10 yrs)	1.9%	0.7%

No. of passages per "hit" haemopoietic stem cell = 1.03 (Sweden, Cornwall, Yunnan (F), France); 1.04 (Collab. analysis); 1.05 (Czech, Yunnan (M); 1.06 (Newfoundland). Average dose per "hit" haemopoietic stem cell = 0.86 Gy (5.71 μm diameter haemopoietic stem cell); 0.20 Gy (11.63 μm diameter haemopoietic stem cell); and 0.12 Gy (14.79 μm diameter haemopioetic stem cell) (all studies).

MICRODOSIMETRY OF TRITIATED PARTICULATES IN ALVEOLAR SACS

R.B. Richardson and A. Hong

AECL
Chalk River Laboratories
Ontario KOJ 1JO

1 INTRODUCTION

Tritiated particulates may be released during the operation of fission and fusion facilities. Inhalation is expected to be the predominant mode of exposure from tritiated aerosols. Metal tritides such as titanium and zirconium have been examined for use in the long-term storage of tritium[1] and as neutron generator targets. Cheng et al (1995)[2] have considered the dosimetry of tritiated particulates. Preliminary results following intrathecal instillation of titanium tritide into rats indicate that the dissolution rate is moderate, similar to that observed *in vitro*.

The tritium beta particle has a maximum energy of 18.6 keV and a maximum range R_{max}, in water or tissue of 6 μm. The International Commission on Radiological Protection (ICRP) assessed the absorbed fraction of energy deposited by tritiated particulates in different regions of the lung[3]. The average absorbed fraction, AF(T←S), in target T per beta emission in source S is calculated as

$$AF(T \leftarrow S) = \frac{E(T \leftarrow S)}{E_{av}} \qquad (1)$$

where E(T←S) is the average energy absorbed by the target region T from radiation emitted by source S, and E_{av} is the average energy of the radiation.

Tritium activity deposited in the upper regions of the lung have absorbed fractions in target tissue, susceptible to radiation-induced lung cancer, of $< 3 \times 10^{-5}$, due to the depth of target tissues beneath the surface of the airways. Self-absorption by the particulates will only significantly affect the absorbed fraction in the target cells of the alveolar-interstitial (AI) region from sources in the AI region. The target cells of radionuclides deposited in the AI region are the secretory (Clara) cells of the respiratory bronchioli and the type II epithelial cells covering the alveolar surface. The ICRP assigns an absorbed fraction of unity for tritium deposited in the AI region since the particulate sources are considered to have no self attenuation and the air attenuation is assumed to be negligible. In this work the absorbed fractions are reassessed.

2 METHOD

Energy deposition calculations were carried out using the Monte Carlo ACCEPTP code of the Integrated Tiger Series (ITS) version 3.0 for the IBM PC[4]. The energy spectrum for tritium beta particles was obtained from Cross et al[5]. The trajectory and energy deposition of the beta and secondary radiation were followed in three dimensions until their energies dropped to 1 keV for all media, at which point their energy was assumed to be deposited locally. The distribution of tritium is considered to be homogeneous throughout the particle volume. The maximum concentration of tritium within the metal hydride particulates was assumed. The H/Ti ratio varies up to a maximum of 2[1]. The density was assumed to be 3.9 $g.cm^{-3}$ for TiT_2. Interaction cross-sections used in the code for hydrogen were assumed for tritium. Representative particulates are modeled as spheres. Isotropic point sources, from 30 to 3600, depending on the particle size, were randomly placed within a sphere. The number of electron histories run in each simulation was chosen so as to achieve a variation in the absorbed fraction calculated of less than 2%. For the purpose of obtaining a dosimetric model, the air-filled alveolar sacs of an adult are assumed to be spherical with an average internal diameter[3] of 200 μm and surrounded by a septum defined to have a thickness equal to R_{max}. A particle is positioned such that its surface is in point contact with the surface of the alveolar septum. The target tissue for the AI region is assumed to be the complete septum. Energy deposited by beta radiation emitted by the tritiated particles were summed in the septum in shells of 0.25 μm thickness in incremental depths to R_{max}. The method does not calculate the higher doses experienced by the septum cells near the particulate.

Table 1. *Absorbed fractions for AI tissue, varying with alveolar sac and particle size.*

Alveolar sac diameter (μm)	Particulate diameter (μm)				
	0.01	0.1	1	5	10
71	0.96	0.87	0.40	0.093	0.043
200	0.92	0.83	0.38	0.084	0.046

3 RESULTS

Absorbed fractions were calculated for the dust particle itself, the air in the sac, and the septal tissue which contains the target cells (Figure 1). The sum of these three fractions equals unity by energy conservation. Energy deposition was investigated for titanium tritide particles of 26 different diameters, ranging between 0.01 and 10 μm. Particulates with diameters >10 μm were not studied as their fractional deposition in the AI region is negligible. For TiT_2, the absorbed dose fraction in alveoli tissue was 0.38 for 1 μm diameter particles and 0.09 for 5 μm diameter particles. The fraction of energy absorbed by air is 8% of the total energy deposited from particles 0.01 μm in diameter, and decreases for particles of increasing size.

The average thickness of the interalveolar septum is ~2 μm for an adult, although large areas are less than 1 μm thick[3]. The energy deposited from tritiated particulates was tallied for shells of tissue in the alveolar sac wall, where the target cells are situated. About 80% of the total energy is deposited within 1 μm of tissue and over 95% of the energy is deposited within 2 μm. Therefore, the energy deposited beyond a primary septum of average wall thickness and into a neighboring alveolus is less than about 5%, for the particle size range studied. The thickness of the interalveolar septum decreases from birth to adulthood. The arithmetic mean thickness of the septum was measured up to 5.6 μm in a newborn. Therefore, our model better reflects the thicker septa of children than those of adults. The air cavity of the alveolar sac increases in size from birth to adulthood. The ICRP[3] proposes the airway dimensions vary to the one third power of the functional residual capacity of the lung. The mean alveolar diameter of a 3 month old child is estimated as 71 μm, when scaled assuming the alveoli of an adult has a mean diameter of 200 μm. The calculation of absorbed fraction for different-size particles and alveolar sacs shows that the energy deposition in AI tissue changes less than ±11% with age (Table 1).

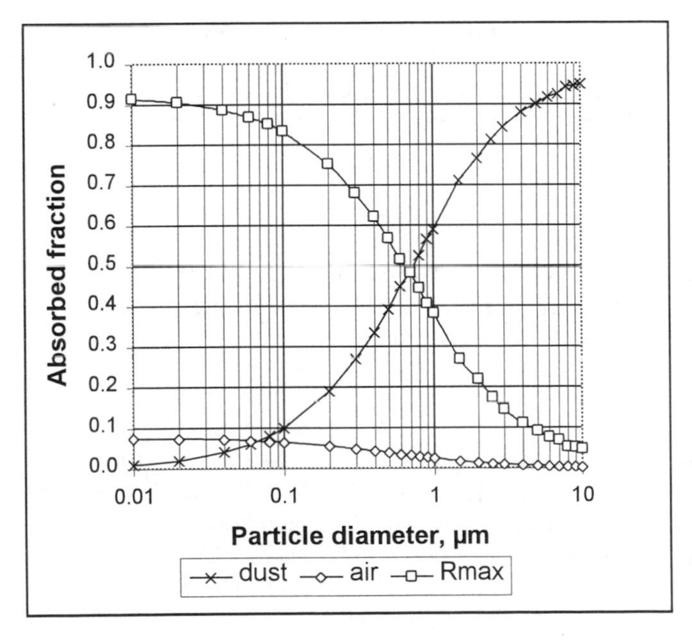

Figure 1. *Absorbed fractions for TiT_2 particulates, air and AI tissue (R_{max}) versus particle size.*

4 DISCUSSION

An effort was made to ascertain that the ITS code was giving accurate values for the absorbed fractions of small tritiated spherical particles. For the purposes of comparison, the absorbed fractions were calculated considering the energy deposition of tritium by a point kernel method[6]. By definition, the point kernel gives the variation in dose rate with distance from an isotopic point source of unit activity in an infinite medium of water. The point kernel for tritium in water was modified for titanium tritide, effective atomic number 19.7, using an empirical formula given by Cross et al (1992) [7]. The dose distribution within the sphere was computed using the method described by Howell et al (1989)[8]. Absorbed fractions for titanium tritide were calculated by the point kernel method and compared with the ITS Monte Carlo code for particles ranging from 0.01 to 10 μm in diameter and were found to be within 10% for all diameters considered.

An important clearance mechanism in the AI region is the alveolar macrophage (AM) which removes foreign substances by phagocytosis. Since the AM are between 15 and 50 μm diameter[9] most of the engulfed tritium radiation will be attenuated by the macrophage cell[10]. The extent to which alveolar macrophages are involved in the clearance of TiT_2 particulates is unclear, as the species and size of the dust greatly influence the action of the AM and the rate of phagocytosis.

In summary, this work showed, using Monte Carlo analysis and an alveolar-sac model, that there was only a small age-dependent variation in the fraction of energy absorbed by AI tissue due to anatomical changes in the alveolar sac dimensions. It was also shown that there is a significant reduction in the absorbed dose to AI tissue, when accounting for self-absorption in relatively large TiT_2 particulates. The dose to AI tissue from tritium particulates is over estimated by the ICRP 66 lung model, as an absorbed fraction of unity is assumed. Future work will consider particle size distribution, the diminution of size of the particulates during dissolution, and their effect on the energy deposited in lung tissue.

Acknowledgements

We thank Drs. W. Cross, J. Nunes, A. Trivedi of AECL and Dr. Y.-S. Cheng of the Inhalation Toxicology Research Institute, Albuquerque for their helpful advice. This work is supported by the CANDU Owners Group.

References

1. W.J. Holtslander and J.M. Miller. AECL, Canada, AECL-7757, 1982.
2. Y.-S. Cheng, M.B. Snipes, R.F. Kropf, and H.N. Jow, *Health Phys.* 1995, **68**, S53.
3. ICRP Publication 66. Annals 24 Nos. 1-4. Pergamon Press, Oxford, 1994.
4. J.A. Halbleib, R.P. Kensek, T.A. Melhorn, G.D. Valdez, S.M. Seltzer, M.J. Berger and T. Jordan, Radiation Shielding Information Center, Oak Ridge, 1994.
5. W.G. Cross, H. Ing and N.O. Freedman, *Phys. Med. Biol.*, 1983, **28**, 1251.
6. P.K. Leichner, W.G. Hawkins, and N.-C. Yang, *Antib. Immunoconj. Radiopharm.*, 1989, **2**, 125.
7. W.G. Cross, N.O. Freedman, and P.Y. Wong. AECL, Canada, AECL-10521, 1992.
8. R.W. Howell, D.V. Rao and K.S.R. Sastry, *Med. Phys.*, 1989, **16**, 66.
9. S.W. Clarke, and D. Pavia, 'Aerosols and the Lung', Butterworth and Co. London, 1984.
10. G.T. McConville and C.M. Woods, *Fusion Tech.*, 1995, **28**, 905.

ELEVATED LEVELS OF Po-210 IN HUMAN FETAL TISSUES FROM MOTHERS LIVING NEAR THE SEVERN ESTUARY

Denis L Henshaw, Janet E Allen, Paul A Keitch and Julie J Close
H H Wills Physics Laboratory
University of Bristol
Tyndall Avenue
Bristol, BS8 1TL, UK

1 INTRODUCTION

Fetal life is known to be the most sensitive period for leukaemia induction by radiation. There is current interest in the study of transplacental transfer of radionuclides and the consequences for fetal radiation dosimetry. At Bristol University, transfer of natural alpha-radionuclides from mother to fetus is being studied in autopsy samples using alpha-particle sensitive plastic track detectors. Most activity is in the fetal skeleton.[1]

Following the observation of Alexander et al. of an increased incidence of leukaemia in children living near to estuaries in England and Wales,[2] analysis of 9 autopsy fetal cases showed evidence of an association of alpha-activity concentration in fetal vertebrae with proximity of the mother's residence to the Severn Estuary.[3] Tissue samples from a total of 66 autopsy fetal cases from the area have since been obtained. We present an analysis of activity in fetal vertebrae with respect to three possible sources of environmental ^{210}Po: the Severn Estuary itself, the M5 motorway and a lead smelter at Avonmouth.

2 MATERIALS AND METHODS

Autopsy fetal vertebra samples from 18 weeks gestation to full term were obtained where the mothers lived in the county of Avon (Figure 1). Each sample, typically < 5g each, was analysed for total alpha-activity concentration by autoradiography using TASTRAK plastic detectors as described in detail elsewhere.[4] Most activity present is due to ^{210}Pb-supported ^{210}Po. The address of the mother was used to calculate the distance to the Severn Estuary, the M5 motorway and the lead smelter at Avonmouth. In the case of the Estuary both the nearest distance and the distance along the line of prevailing wind (250°) was measured.

3 RESULTS

Correlation coefficients for the various sources are given in Table 1. The results are presented for cases 1 - 26, 27 - 46, 47 - 66 and for all 66 cases. For the measurements downwind of the Estuary, 3 upwind cases from South Wales and one case from Wells over 50 km downwind were omitted from the analysis. Figures 2 and 3 show respectively the association for all data points for nearest and downwind distance from the Estuary. No correlation with either the M5 motorway or the lead smelter was found, but the total data show a significant correlation with the Estuary at the 5% confidence level, similar to that seen in the first 26 cases. Although the correlation for the separate sets of cases,

Table 1 *Correlation of ^{210}Po alpha-activity concentration in autopsy fetal vertebrae with proximity of mother's residence to known pollution sources (ns = not significant).*

Data Set	Severn Estuary [1]	Severn Estuary [2]	M5 Motorway	Lead Smelter
cases 1 - 26	0.50 ($p<0.05$)	0.43 ($p<0.05$)	ns	ns
cases 27 - 46	ns	ns	ns	ns
cases 47 - 66	ns	0.44 ($p \sim 0.05$)	ns	ns
All cases	0.26 ($p<0.05$)	0.26 ($p<0.05$)	ns	ns

[1]Shortest distance to the Estuary; [2]Distance along direction of prevailing wind.

27 - 46 and 47 - 66, generally fails at the 5% level the p-value never drops below 0.1: the positive slope of activity concentration in fetal vertebrae with respect to the estuary is retained.

4 DISCUSSION

Two features emerge from these observations. The absolute activity concentrations, up to 0.7 Bq kg^{-1} could be considered high in comparison with the average level of total alpha-activity found in autopsy adult bone from Avon[4] of 1.46 ± 0.13 Bq kg^{-1}. This may be indicative of the high level of fetal transfer of ^{210}Pb-supported ^{210}Po, especially to the skeleton. The ability of both ^{210}Pb and ^{210}Po to transfer to the fetus has also been noted by Bradley and Prosser.[5]

The correlation of alpha-activity concentration with the Severn Estuary in cases 1 - 26 is not improved when all cases are considered. However, even for the separate sets of cases, 27 - 46 and 47 - 66 where the correlation falls below significance, it is of interest that the trend of increased activity with respect to the Estuary is maintained. The lack of a correlation with either the M5 motorway or the lead smelter at Avonmouth may simply reflect the fact that these represent far smaller sources compared with the Estuary.

The Severn Estuary is known to be a source of both ^{210}Pb and ^{210}Po from general pollution and levels have increased substantially this century.[6,7] Local monitoring of airborne alpha-activity using plastic track detectors also shows elevated levels near to the Estuary.[8] If the observations in fetal tissues are indicative of a causal relationship then alpha-activity concentration is increased by about 40% in cases where the mother lives close to the Estuary. Such an increase could come about by increased inhalation of ^{210}Po followed by rapid transfer to the fetus.

Evidence for a two-fold increase in ^{210}Po activity concentration in the teeth of children living close to motorways in the UK has recently been reported.[9,10] Thus in both the teeth and fetal cases we need to address the feasibility of a substantial increase in uptake of ^{210}Po from outdoor exposure via the inhalation pathway.

Unpublished Bristol data shows that alpha-activity recorded on plastic track detectors held outdoors is often greater than that for indoor exposure to radon and its decay products. Spectroscopic analysis of the detectors held outdoors reveals that the airborne activity

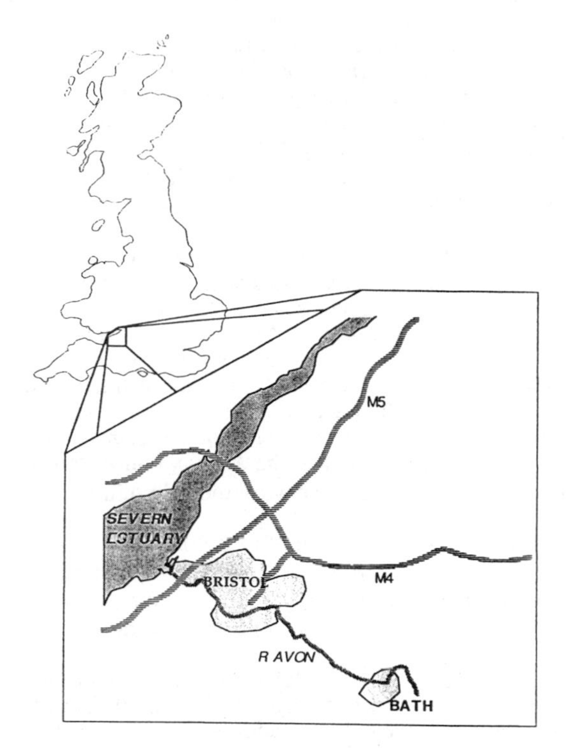

Figure 1 *Geographical area of autopsy fetal collection in Avon*

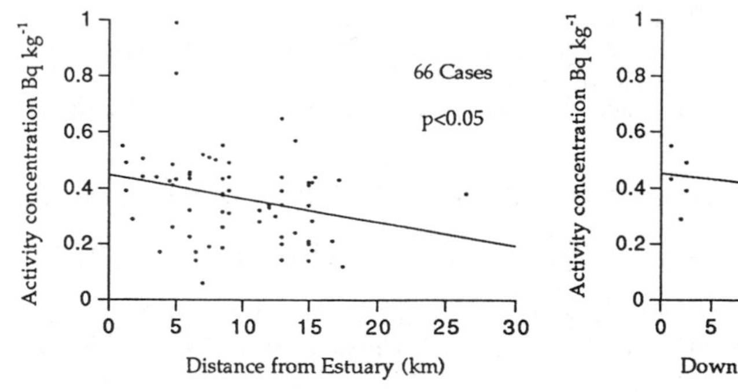

Figure 2
Alpha-activity concentration in fetal vertebrae vs distance of the mother's residence from the Severn Estuary

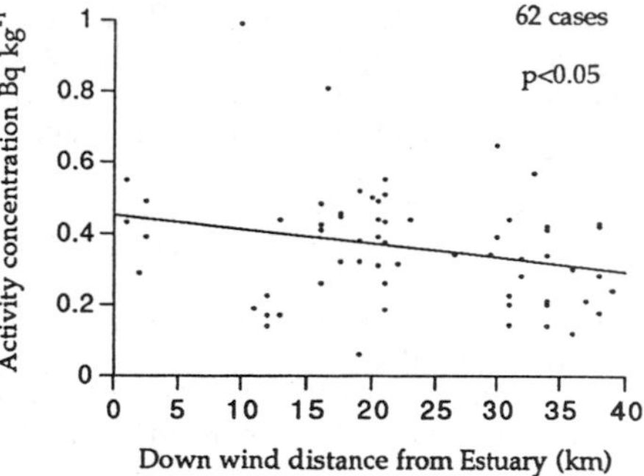

Figure 3
Alpha-activity concentration in fetal vertebrae vs downwind distance of the mother's residence from the Severn Estuary

concentration is indeed low and consistent with published data. However, it is the component of activity plated out on the plastic surface that is seen to be in excess of the equivalent indoor exposure to airborne alpha-radioactivity from radon decay products. The latter are attached to aerosols which may be conveniently divided into the so-called ultrafine aerosols, < 50 nm and attached aerosols up to 0.6 μm in size. The former are highly mobile and are the ones which preferentially plate out on the plastic detector. In indoor air about 10% of radon decay products take the form of ultrafine aerosols. Outdoors, ultrafine aerosols can be as abundant as the larger aerosols[11] and it is likely that this accounts for the higher plateout on detectors held outdoors.

This conclusion has immediate consequences for the uptake, retention and radiation dose from outdoor exposure to alpha radio-aerosols such as radon decay products and ^{210}Po. Application of the ICRP lung dose model suggests that although only 10% of radon decay product aerosols indoors are in ultrafine form, when inhaled they are nevertheless responsible for 50% of the resultant lung dose, because of their 100% retention. This dose fraction would be substantially increased if the ultrafine component had equal abundance with the larger attached aerosols. This in turn would imply a significant contribution to overall lung retention from even short exposure periods outdoors. It could also account for the apparent increase in alpha-activity concentration levels in both teeth and fetal tissues in relation to exposure to environmental pollution sources of ^{210}Po.

The importance of outdoor exposure to ultrafine aerosols has already been recognised in relation to respiratory disease and motor vehicle exhaust pollution.[12] We suggest that this knowledge be used to calculate the contribution of outdoor exposure to uptake and retention of radon decay products, ^{210}Po and other natural radio-aerosols.

Finally, if a causal relationship between alpha-activity concentration in the fetus and the Severn Estuary exists there are reasons why a highly significant correlation may not be seen. In the present data we have only one address of the mother this may not represent the residence for the major part of her pregnancy. More generally if ^{210}Po from the Estuary is predominantly in ultrafine aerosols, then these are known to migrate substantial distances from their origin and so obscure the effects of local sources.

REFERENCES

1. D. L. Henshaw, J. E. Allen, P. A. Keitch, P. L. Salmon and C. Oyedepo In: 'Health Effects of Internally Deposited Radionuclides: Emphasis on Radium and Thorium' (Eds: G van Kaick, A Karaoglou & A M Kellerer), World Scientific, 1995.
2. F. E. Alexander, P. A. McKinney and R. A. Cartwright, *J. Public Health Medicine* 1991, **12**, 109-117.
3. R B Richardson, PhD Thesis, University of Bristol, 1991.
4. D. L. Henshaw, U. Hatzialekou and P. H. Randle. *Rad. Prot. Dosim*, 1988, **22**, 231-242.
5. E. J. Bradley and S. L. Prosser, *Radiological Protection Bulletin* 1993 No. 148, 28-31, NRPB, Chilton, UK.
6. E. I. Hamilton and R. J. Clifton, *Estuarine Coastal Mar. Sci.* 1979, **8**, 271-278.
7. E. I. Hamilton, P. G. Watson, J. J. Cleary and R. J. Clifton, *Mar. Geol.* 1979, **31**, 139-182.
8. Bristol University unpublished data.
9. D. L. Henshaw, J. E. Allen, P. A. Keitch and P. H. Randle, *Int. J. Radiat. Biol.*, 1994, **66**, 815-826.
10. D. L. Henshaw, P. A. Keitch and P. R. James, *The Lancet* 1995, **345,** 324-325.
11. H. Horvath, M. Kasahara and P. Pesava, *J. Aerosol. Sci.* 1996, **27**, 417-435.
12. A. Seaton, W. MacNee, K. Donaldson, D. Godden, *The Lancet* 1995, **345**, 176-78.

MEAN SKELETAL DOSE FACTORS FOR ^{226}Ra AND ^{239}Pu

S. L. Brooke and A. H. Beddoe

Medical Physics Services
Queen Elizabeth Hospital
Birmingham

1 INTRODUCTION

There has been a revival in interest in the leukaemogenic potential of alpha emitters resulting from epidemiological studies reporting an apparent relationship between leukaemia incidence and the levels of radon through out the world.[1] While reports of the 'radium dial painters' show that skeletally absorbed radium does not appear to lead to leukaemia (as opposed to osteosarcoma)[2] there remains the possibility that occupational exposures could exhibit a different radiobiological outcome to that observed from environmental exposure. Nevertheless, the eventual explanation of the discordance, apparent or otherwise, must be based on a sound physical understanding of the relative doses to the tissues at risk.

This paper addresses the problem of the range of dose rates from given skeletal uptakes for the nominal volume- and surface-seeking alpha emitters, ^{226}Ra and ^{239}Pu. Mean skeletal dose factors have been estimated from Monte Carlo calculations of dose factors on individual bones from three male subjects aged 1.7, 9 and 44 years. These calculations are based on path-length measurements through the trabecular and cortical structure in these bones.[3]

2 METHODS

Monte Carlo calculations of energy deposition along alpha particle tracks have been performed on post-mortem bone specimens for which path length distributions were first determined using a technique developed by Darley.[4] Dose factors were calculated for (i) marrow cavities (containing active bone marrow and other mesenchymal tissues) and (ii) a 10 μm thick endosteal zone extending out from trabecular and cortical bone surfaces (including Haversian and Volkmann canals), for both the nominal volume seeker ^{226}Ra [5] and the surface seeker ^{239}Pu [6]. The corresponding dose factors were then averaged using appropriate weighting factors based on either relative volume of active marrow or relative surface area.

2.1 Calculation of Dose Factors for Each Bone

Four dose factors were defined as follows:

${}_vD_m/D_{skel}$ -the mean dose rate to the haemopoietic marrow from a volume seeking radionuclide divided by the dose rate to the skeleton.

${}_vD_e/D_{skel}$ -the mean dose rate to the endosteal tissue from a volume seeking radionuclide divided by the dose rate to the skeleton.

${}_sD_m/D_{skel}$ -the mean dose rate to the haemopoietic marrow from a surface seeking radionuclide divided by the dose rate to the skeleton.

${}_sD_e/D_{skel}$ - the mean dose rate to the endosteal tissue from a surface seeking radionuclide divided by the dose rate to the skeleton.

D_{skel} is defined as the dose rate to the skeleton calculated from the total retained radionuclide divided by the mass of the mineral skeleton[7] (5 kg [8]). Dose factors for groups of bones for adult man are shown in Table 1.

2.2 Calculation of Weighting Factors

The weighting factors, f_m, for calculating the mean skeletal dose to the active bone marrow are the ratio of mass of active marrow in each of the bone groups to the total mass of active marrow in the skeleton. Values of f_m are derived from the total marrow data and cellularity data (the mass-ratio of active marrow/total marrow) determined for different ages by Cristy.[9]

The weighting factors, f_s, for calculating the mean skeletal dose to the endosteal surface are the ratio of total trabecular area in each of the bone groups to the total trabecular area in the skeleton. Values of f_s are derived from the normalised values of the relative surface area in a given bone, A_{rel}, where A_{rel} is calculated using the equation below.[10]

$$A_{rel} = \frac{f_m}{p} \times \frac{V_b}{V_m} \times \frac{S}{V_b}$$

p - the cellularity fraction Vb/Vm - the ratio of trabecular bone to volume of marrow
S/Vb - the ratio of surface area to trabecular bone volume for a group of bones

Weighting factors for active marrow and endosteal surface are shown in Table 2.

Table 1 *Dose Factors for Groups of Bones for Adult Man*

Groups of bones	^{239}Pu		^{226}Ra	
	${}_sD_m/D_{skel}$	${}_sD_e/D_{skel}$	${}_vD_m/D_{skel}$	sDe/Dskel
cranium	1.557	8.16	0.0832	0.461
mandible, ribs, sternum	0.391	8.26	0.191	0.391
cervical vertebrae	0.766	9.09	0.0355	0.415
thoracic vertebrae	0.676	9.13	0.0309	0.409
lumbar vertebrae and sacrum	0.586	9.17	0.263	0.402
clavicles, hip, scapulae	0.667	8.76	0.0320	0.404
long bones and extremities (trabeculation)	0.518	8.76	0.0241	0.391
long bones (shafts)	-	-	-	-

Table 2 *Weighting Factors for Calculating Mean Skeletal Dose Factors*

Groups of bones	Bone representing group	1.7 years f_m	1.7 years f_s	9 years f_m	9 years f_s	44 years f_m	44 years f_s
Cranium	PB	0.228	0.440	0.122	0.265	0.076	0.257
Mandible, ribs, sternum	Rib 7	0.117*	0.149*	0.136	0.074	0.200	0.081
Cervical vertebrae	CV 5	0.207 (Cervical, Thoracic, Lumbar combined)	0.135 (Cervical, Thoracic, Lumbar combined)	0.026	0.015	0.039	0.034
Thoracic vertebrae	0.5CV5 + 0.5LV3			0.104	0.062	0.161	0.124
Lumbar vertebrae and sacrum	LV 3			0.144	0.088	0.222	0.149
Clavicles, hip scapulae	0.6 IC+0.4 LV 3	0.156	0.113	0.187	0.125	0.211	0.213
Long bones and extremities (trabeculation)	F head, neck, and prox. third of shaft	0.248†	0.140†	0.281	0.371**	0.090	0.143**
Long bones (shafts)		0.045	0.024‡	0	0	0	0

* Includes extremities in this case. † Excludes extremities in this case.
‡ Calculated on the assumption that 19% of marrow in the long bones is in the shafts, i.e. in regions where there is little trabecular bone; the percentage is an approximate estimate from radiographs. ** Includes small amount of trabecular surfaces in shafts.

2.3 Calculation of Mean Skeletal Dose Factors

The contribution of each group of bones to the mean skeletal dose factor is calculated by weighting the dose factors for groups of bones by f_m or f_s. Hence, the mean skeletal dose factor for the radionuclide ^{226}Ra for the marrow cavities is $\Sigma f_{m\cdot v} D_m/D_{skel}$ and for the endosteal surface is $\Sigma f_{s\cdot v} D_e/D_{skel}$ summed over all the bone groups. Similarly, the mean skeletal dose factor for the radionuclide ^{239}Pu for the marrow cavities is $\Sigma f_{m\cdot s} D_m/D_{skel}$ and for the endosteal surface is $\Sigma f_{s\cdot s} D_e/D_{skel}$ also summed over all the bone groups.

3 RESULTS

The mean skeletal dose factors are shown in Figure 1 plotted against age for the three age groups.

4 DISCUSSION

From Figure 1 it can be seen that for both radionuclides D_e/D_{skel} is an order of magnitude greater than the corresponding D_m/D_{skel} and that the mean skeletal dose factors for ^{239}Pu

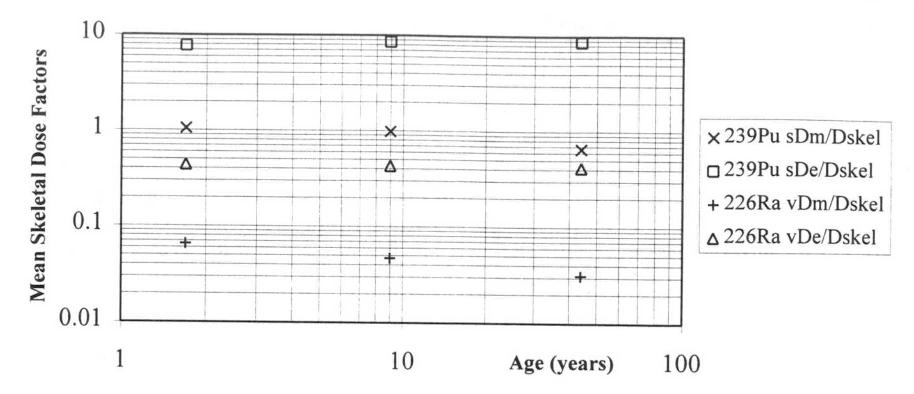

Figure 1 *Mean Skeletal Dose Factors as a Function of Age*

are an order of magnitude greater than the corresponding values for ^{226}Ra. Mean dose rates can be calculated from these dose factors for known levels of skeletal burdens on the assumption that radium and plutonium behave as true volume and surface seekers respectively (^{226}Ra may have a surface concentration in addition to the distribution within the volume[11]).

Further work needs to be done to estimate the effects of bone apposition and the consequent burial and/or redistribution of surface layers of plutonium and other so called surface seekers. Presumably remodelling will gradually produce situations where $_{s}$D approaches $_{v}$D.

5 ACKNOWLEDGEMENTS

We wish to acknowledge the key leadership role played by Professor Bill Spiers (now deceased) in the genesis of this project, as well as the major contributions made in the early stages by Drs Phil Darley and Joan Whitwell.

References

1. D.L. Henshaw, J.P. Eatough and R.B. Richardson, *Lancet*, 1990, **335**, 1008
2. F.W. Spiers and J. Vaughan, *Leukaemia Res*, 1989, **13**, 347
3. A.H. Beddoe, P.J Darley and F.W. Spiers, *Phys. Med. Biol*, 1976, **21**, 589
4. P.J. Darley, In: Proceedings of the Symposium on Microdosimetry, EAEC Publication EUR-3747, 1968, 509
5. J.R. Whitwell and F.W. Spiers, *Phys. Med. Biol,* 1976, **21**,16
6. F.W. Spiers, J.R. Whitwell and A.H. Beddoe, *Phys. Med. Biol,* 1978, **23**, 481
7. F.W. Spiers, A.H. Beddoe and J.R. Whitwell, *Brit. J. Radiol*, 1981, **54**, 500
8. ICRP Publication No. 26, Pergamon Press, Oxford, 1976
9. M. Cristy, *Phys. Med. Biol*, 1981, **26**, 389
10. F.W. Spiers, A.H. Beddoe and J.R. Whitwell, *Brit. J. Radiol*, 1978, **51**, 622
11. R.A. Schlenker and B.G. Oltman, *Rad. Prot. Dos,* 1986, **16**, 195

COMPARISON OF EXPERIMENTAL AND CALCULATED DOSE ENHANCEMENT FACTORS IN TISSUE ADJACENT TO HIGH-Z IMPLANTS FOR DIAGNOSTIC X-RAYS.

H. Průchová[1,3], D. Regulla[1], L.A.R. da Rosa[1,2], R. Seidlitz[1]

[1] GSF - National Research Center for Environment and Health,
Institut of Radiation Protection,
85764 Neuherberg, Germany
[2] on leave from Comissao Nacional de Energia Nuclear, Rio de Janeiro, Brazil
[3] on leave from Czech Technical University, Prague, Czech Republic

1 INTRODUCTION

Implants with high atomic number (Z) in the human body lead to dose discontinuities during radiation exposures as is known from megavoltage therapy [1]. Significant dose enhancement has been found, for diagnostic x-rays, in the tissue close to a gold layer by using thermostimulated exoelectron emission technology with a spatial resolution in the micrometer range [2]. For low energy photons ($E \leq 100$ keV) the Compton effect is the dominant interaction in tissue, but in a high-Z medium the photoelectric effect becomes the most important interaction mode due to the dependence of this effect on the fourth power of the atomic number.

The enhanced energy deposition gives rise to a so called "dose enhancement factor". This factor is defined as the ratio of dose in soft tissue in the vicinity of the high-Z implant to the dose in soft tissue, under equilibrium conditions in the absence of the implant. The objective of the present study was to verify the experimental results by calculations. The calculations of deposited energy described here take into account not only scattering of photons within the high Z material, but also the production and transport of secondary electrons into the adjacent tissue region.

2 METHOD

The experimental method used for dose assessment near plane interfaces was based on the thermally stimulated exoelectron emission technique (TSEE). The surface character of TSEE dosimeters and their low atomic number permit to measure the energy depositions in tissue of micrometer thickness. In the present work the TSEE detectors used were BeO thin film detectors, manufactured by evaporation of a beryllium layer onto graphite discs and subsequent oxidation and sensitization in wet nitrogen. The sensitive layer of BeO has a thickness $\approx 1\ \mu m$. A multineedle counter is then used to evaluate the detector readings.

The experimental setup consists of two PMMA slabs ($100 \times 100 \times 4\ mm^3$) with the TSEE detector and 150 μm gold foil in between. All experiments and computations were performed for backscatter conditions.

The irradiation experiments were performed with heavily filtered X-ray spectra of ISO A-quality : A-40, A-60, A-100, A-120 of mean energies 33 keV, 48 keV, 85 keV, 100 keV.

The photon-electron transport code PARTRAC [3] was used to calculate deposited energy in the sensitive layer of TSEE detectors adjacent to the gold plate, using monoenergetic primary photons of 33 keV, 48 keV, 85 keV and 100 keV. The electron track simulation covers energies from 10 eV to 100 keV. Electron cross sections for water vapor with appropriate density scaling were taken to simulate the BeO thin film detector.

In the case of gold the ionization and excitation cross sections for electrons were calculated using Seltzer method [4] with input values of the mean kinetic energies of orbital electrons U_j, the expectation values of the electron orbital radius $\langle r_j \rangle$ and the binding energies B_j taken from Hartree - Fock calculations recently made by P. Indelicato [5]. These cross sections are shown in figure 1 for ionization and figure 2 for excitation. Elastic cross section data were taken from Fink et al. [6] and Riley et al. [7].

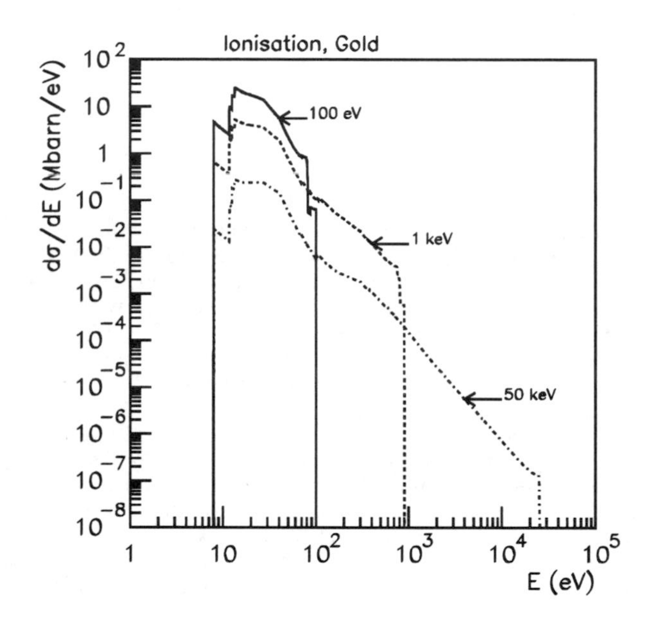

Figure 1: *Electron differential cross sections for ionisation as a function of energy transfer E for electron primary energy T (100 eV, 1 keV and 50 keV).*

3 RESULTS

The table 1 shows the experimental and calculated results of dose enhancement factors. It is obvious that there is significantly increased energy deposition in tissue that is adjacent to the gold surface. Maximum dose enhancement is found for 48 keV X-rays reaching values of 113 in experiment and 122 in calculation. These results are in a good agreement within the limits of uncertainty. The remaining difference may be due to response properties of the TSEE detector which were not included in Monte Carlo modelling. The degree of agreement decreases for energies 33 and 100 keV. The calculated values in Table 1 are however systematically higher than the measured values. This discrepancy could be explained by existence of an air gap between the BeO and gold layers causing the different secondary electron spectra and therefore lower experimental enhancement factor. Moreover there is some uncertainty in the determination of the thickness of the BeO detector sensitive layer. Additional source of uncertainty could arise from the fact that cross sections for water vapor

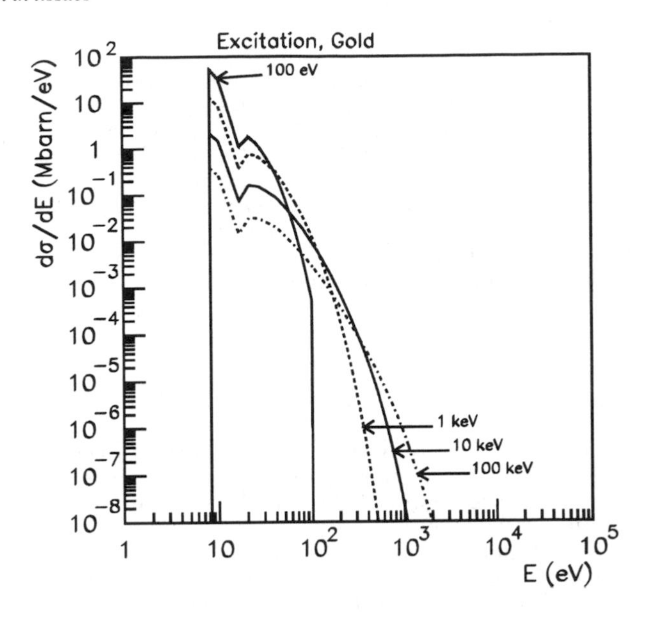

Figure 2: *Electron differential cross sections for excitation as a function of energy transfer E for electron primary energy T (100 eV, 1 keV, 10 keV and 100 keV).*

(appropriately scaled) were used instead of those for BeO.
The biological significance of the dose enhancement was verified by *invitro* experiments, in which monolayers of CH310T1/2 mouse cells were irradiated close to the gold surface, with a 2 μm PET foil for separation [2]. Those experiments clearly demonstrate an enhanced cell killing of the same order of magnitude as in our physical experiments and calculations. A significantly reduced cell survival is shown in [2] for macroscopic doses already above 50 mGy. It also becomes evident from that reference, that the range of secondaries is between some tens to hundred μm for the energies of X-rays considered.

Table 1 *The experimental and calculated results of dose enhancement factors in 1 μm of tissue for various energies of X-rays.*

Energies of X-rays	*TSEE measurements*	*Calculated values*
33 keV	98.3±12	115.5±20
48 keV	113.6±11	122.4±22
85 keV	72.6±6	86.5±16
100 keV	55.1±2	74±16

4 CONCLUSION

The Monte Carlo code PARTRAC was used to simulate the experiment using beryllium oxide thin-layer detectors of sensitive thickness $\leq 1\ \mu m$. Significantly enhanced energy deposition by more than two orders of magnitude was found in the thin slice of tissue of 1 μm thickness adjacent to a gold foil of 150 μm. For the enhancement factor there is a good agreement between the experimental and the calculated results, within the limits of uncertainty. The results could be of interest particularly for patients in medical X-ray diagnostics. These preliminary results confirm that the PARTRAC code offers a large potential to analyze enhanced energy deposition in tissue at the interface to materials with high atomic number.

References

1. M. Farahani, F.C. Eichmiller, W.L. McLaughlin, Phys. Med. Biol., 1990, **35**, No. 3, 369.
2. D. F. Regulla, L. B. Hieber, M. Seidenbusch, Physikalische und biologische Wirkung von Sekundärstrahlung in grenzschichtnahem Gewebe. In: Tagungsbericht, Jahrestagung Fachverband für Strahlenschutz, Hannover, 23.-25. Oktober, 1996, 299.
3. H. G. Paretzke, Simulation von Elektronenspuren im Energiebereich 0,01-10 keV in Wasserdampf. GSF-Bericht 24, 1988
4. S. M. Seltzer, Cross sections for Bremsstrahlung Production and Electron-Impact Ionization. In: Monte Carlo Transport of Electrons and Photons. Plenum Press, New York, 1988
5. P. Indelicato, Private communication. CERN 1994
6. M. Fink, A.C. Yates: Theoretical electron scattering amplitudes and spin polarizations.Part I. Atomic Data, 1970, **1**, 385.
7. M. E. Riley, T. MacCallum, E. Biggs: Electron-Atom Elastic Scattering. Atomic Data and Nuclear Data Tables, 1975, **15**, No.5, 444.

Single particle effects

MICRODOSIMETRY AND MICROBEAM IRRADIATION

L. A. Braby

Department of Nuclear Engineering
Texas A&M University
College Station, Texas 77843

1 INTRODUCTION

Microdosimetric studies of energy deposition in cell sized objects by high LET radiations have made it clear that the biological effects of environmental doses are driven by the stochastic nature of the radiation. Excluding medical exposures, the majority of the equivalent dose received by most individuals is delivered by alpha particles emitted by radon and its progeny in the lungs. Other significant exposures to high LET radiation occur during commercial aircraft flights, particularly when high altitude polar routes are utilized, and through occupational exposures at power plants, high energy accelerators, and other facilities which produce neutrons. For alpha particle irradiation of 8 μm diameter spherical objects, typical of the size of cell nuclei, the mean specific energy is approximately 0.4 Gy. For typical neutron spectra the specific energy in this size site ranges from around 0.1 to 0.4 Gy. These values are much larger than the average annual dose received by most individuals exposed to these radiations. Consequently, individual cells within an irradiated tissue seldom experience a direct interaction with a heavy charged particle. Furthermore, cells are known to rejoin within a few hours, many of the DNA strand breaks which are produced by a high LET event. Thus it is very unlikely that the effects of two or more high LET particles can interact in a single cell. We are forced to conclude that the health risks of environmental exposures to high LET radiations originate with the effects of single charged particle events in individual cells, and the biological processes that those events initiate.

Even though it is very unlikely that a single cell will experience more than one high LET event at the doses of concern, the long term health effects of irradiation may be influenced by various types of interactions initiated by two or more independent charged particle tracks. For example, when populations of cells are irradiated with alpha particles from a radioactive source, the average number of cell nuclei traversed by one or more alpha particles can be calculated. In two independent series of experiments, [1,2] when populations of cells were exposed to low doses of alpha particles, the fraction of cells with elevated sister chromatid exchange rates was significantly higher than the fraction of the cell nuclei which had been traversed by the charged particles. A similar experiment, conducted to evaluate the effect of irradiation on P53 levels in cells, showed that levels were elevated in more cells than could have been traversed by alpha particles.[3] These experiments and others suggest that the response of a tissue to irradiation may be more complex than the sum of the effects in the individual cells. Damaged cells apparently can communicate the occurrence of damage to neighboring (and possibly more distant) cells, and cells receiving the message may alter the way that they respond to radiation damage.

Thus we have two distinct, but related, types of problems in radiation biology. One is to adequately address the problems of low doses, overcoming the limitations of conventional experiments and providing data which can be used in connection with detailed models of the distribution of charged particle tracks in the cells of a tissue. The second is to determine the effective targets for initiating various types of biological response, provide information on the mechanisms which are relevant to the extrapolation from damage at the cellular level to health effects in the whole animal and to identify the differences between species which must be understood in order to extrapolate from animal systems to man. These two problems are interrelated because, to be meaningful, the experiments to study mechanisms must be conducted at low doses, with the same kind of precision that is needed for the study of low dose effects in individual cells.

2 CHARACTERISTICS OF AN IDEAL IRRADIATION

The experiments to answer these fundamental radiation biology problems can often not be conducted successfully using conventional radiation biology techniques. With any conventional irradiation procedure, the occurrence of a charged particle track in a specific location is a random event, independent of the occurrence of a track in any other volume, or previous tracks in the volume of interest. Thus the number of tracks through a defined structure is a Poisson random variable. Furthermore, the average numbers through adjacent volumes will be equal. If the dose is made low enough, the average number of tracks per target can be as small as desired and, if the biological process of concern is a linear function of the number of tracks, the effect for a specified average number of tracks will be the same as that for exactly the specified number. However, most biological effects are not linear functions of the initiating insult, and the apparent linearity of dose response curves for high LET radiations does not assure that the underlying effect is a linear function of the number of tracks. In fact, if the effect is a linear/quadratic function of the number of tracks, the observed effect for a Poisson distributed number of events will be more nearly linear than the effect observed for specific numbers of tracks. Furthermore, the random positions of the tracks relative to the biological targets makes it impossible to investigate the role of the position of the track, relative to a specific structure, on the probability of producing a specific response.

In order to efficiently conduct the needed experiments, a different approach to irradiation is needed. The irradiation needs to be controlled so that the occurrence of charged particle tracks in space and time is not independent of other interactions, but is precisely controlled. Specifically, we would like to be able to specify

- the position of each track relative to specified biological structures
- the number of tracks at each point
- the energy deposited at each point (the stopping power of the track)
- the timing between tracks

One approach to achieving this type of control is to select cells with the desired exposures from a randomly irradiated population. Utilizing conventional track segment irradiation techniques and sophisticated track recording methods, combined with careful measurement of the position of the biological targets at the time of irradiation, all of the required parameters can be measured. This approach can be very efficient for some types of experiments, particularly when data for all possible combinations of the number and position of the tracks will be utilized. Irradiations can be quick and the required experimental techniques, possibly based on active or passive position sensitive detectors and image processing techniques may be relatively inexpensive. However, if the experiment does not require data for all possible track positions, or if the desired distribution of numbers of cases does not match that produced by random irradiation, this approach can be very inefficient, requiring measurement of the response of many cells which do not contribute to the success of the measurement.

The alternative, which will be discussed in detail here, is to control each charged particle interaction, so that irradiation is not a Poisson random process. In order to evaluate alternative methods, we begin by considering the specifics of each of the four criteria listed above. Limiting charged particle interactions to a specific position relative to a biological structure requires measurement and control of several variables. First, one must be able to reliably recognize the structure of interest and accurately measure its position. Since the objective is to understand the response of the biological system to irradiation, it is important that imaging the target should not disrupt it unnecessarily. Clearly this precludes fixing the cells and the use of electron microscopy. Optical microscopy and use of specific stains which can label structures of interest is feasible. The resolution of optical microscopy depends of the wavelength of the light and numerical aperture of the lens used, but is typically limited to a fraction of a micrometer. The use of very short wavelength light presents a problem since ultra violet light creates specific biological damage. The use of molecular probes to help identify the structure to be targeted may have similar drawbacks. Depending on the endpoint to be investigated, the molecular probe may interfere with normal function of the target or with the normal biological response to the radiation event. It is difficult to establish that these probes or stains are not altering the response, but is probably impossible to do some of the desired experiments without them. Optical microscopy of unstained live cells is difficult and it is often impossible to distinguish regions suspected of being significant in determining the biological response even though they are clearly defined by other techniques.

The other variable that must be controlled to limit the irradiation to the specified biological structure is the spatial distribution of the charged particles. If the resolution of the biological target is limited to a few tenths of a micrometer, it will usually not be beneficial to control the variation in charged particle track positions to significantly less than that. One type of experiment, study of the effect of distance between tracks on the biological effectiveness of the energy deposited by two or more tracks, could benefit from tighter control of the track position, but this type of experiment, using random track spacings produced by breakup of molecular ions, has already been done.[4] Thus, for the foreseeable future, the goal for spatial resolution of charged particle irradiation is likely to be a few tenths of a micrometer. Since many of the biological questions of current concern deal with discriminating between the cells being irradiated, or between irradiation of the nucleus and some other portion of the biological system, and since the nuclei of attached cells tend to be on the order of 10 micrometers in diameter or larger, beam sizes as large as about 8 micrometers are useful for many experiments.

Control of the number of charged particle tracks through the individual target is critical to understanding the effects of low doses. Partial cell irradiation has been utilized before,[5] and was instrumental in demonstrating that the nucleus is the critical target for many biological effects. However, the early experiments did not allow for the detection of the individual charged particle interactions, and thus were limited to relatively large doses, where the variation in the number of tracks per target was acceptable. For the mechanisms to be studied in the future, the effects of a single charged particle may be critical. As a consequence, particle interactions should be detected with high reliability and the response time for terminating the exposure should be much less than the mean time between events. The goal would be to have 100% of irradiated cells with exactly the number of charged particle tracks specified. This is unrealistic for any real measurement system, but can be approached very closely.

Controlling the energy deposited per track involves both physical and biological variables. Use of a charged particle beam from an accelerator provides monoenergetic (energy spread a small fraction of a percent) particles at the specific energy needed to provide the desired stopping power in the target, eliminating the spread in energies that results from attenuating particles from a radioactive source. Assuming monoenergetic particles at the point where they enter the cell, there will be a small variation in energy deposited due to the fact that the energy is transferred through individual ionization and excitation events. The thickness and composition of the biological target clearly affects the energy deposition by each charged particle track. When cells have attached to a substrate, they tend to flatten, and the thickness of the nucleus, in the direction perpendicular to the substrate, is relatively uniform. However, the variability between cells in a given culture, or between cells on different dishes, is not easily determined. The limitation of the vertical resolution of scanning confocal microscopes is typically 0.4 micrometers, possibly 10% of the path length through the nucleus. However, these uncertainties may be small compared to those which are encountered when more structured biological systems such as cell aggregates from normal tissue, or self organized structures grown on synthetic extra cellular matrix[6] are utilized. The description of the energy deposition presents additional complications. As particles lose energy their stopping power increases, so the rate of energy deposition may vary significantly from the side of the cell attached to the substrate to the opposite side. This effect is most significant in the Bragg Peak, near the ends of charged particle tracks, but the Bragg Peak is a major fraction of the total dose deposited by alpha particles from natural sources.

Finally, the timing of successive charged particle tracks may be significant. This is not an issue with single particle irradiation, but if a cell receives two or more tracks, the time between them may influence the expression of some biological effects. Furthermore, if damage in one cell produces a signal which modifies the response of other cells, this signal will almost certainly be time dependent. Most processes which modify DNA strand breaks and modify cell survival have half times on the order of a few minutes to a few hours. However the time course of other processes such as cell communication are only beginning to be studies. It would seem reasonable to require irradiation systems timing accuracy of a few seconds or less.

3 OPTIONS FOR CONTROLLED PARTICLE IRRADIATION

It seems evident that the use of particle beams from a charged particle accelerator is the best approach to achieve the objectives of controlled irradiation experiments. Particles are inherently monoenergetic, beams are essentially unidirectional resulting in large fluences and short exposure times. However, there are many options for the control and detection of such a beam. Three elements of the overall system are critical and determine the applicability for different biological experiments. These are the method of defining the beam, the method for detecting individual particle interactions, and the method for identifying the structures to be targeted.

3.1 Beam Definition

In principal, the irradiation could be limited to a specific region of a cell by either collimating the beam or by using a reduced image of a small aperture. A scanned focused beam approach has been used quite successfully for imaging a wide variety of materials, and the focusing technology has been highly refined. Since the beam spot is created as a reduced image of a larger aperture, the effects of edge scatter are nearly eliminated and a 0.1 μm diameter irradiation volume could be easily achieved. However, there is a fundamental problem with this approach for irradiating living cells. The exact position of the beam is not easily determined, and it is difficult to specify where to position a target to achieve the desired irradiation. This is not a problem in scanning microbeam applications because only the relative position of the beam is required to form an image. Some form of position sensitive detector with spatial resolution better than 0.1 μm could be used to determine the beam position, but exceptional mechanical and electronic stability would be required to assure that the beam remained in the specified position while a number of subcellular targets were being irradiated.

The alternative, collimating the beam to the desired spot size, has the advantage that the final portion of the collimator can be placed very near the cells to be irradiated, and can serve as a visual reference for the position of the beam. However, the physics of scattering at a knife edge, and current manufacturing technique limit the size of the beam which can be achieved with a collimator. Scattering occurs when particles pass near any defining edge, and the fraction of the beam which has been scattered increases in proportion to the ratio of the perimeter to area. Consequently, edge scattering becomes a significant problem with apertures around 10 μm in diameter and becomes worse at smaller sizes. The scattered particles will strike the target plane outside the projected area of the collimator, and will have degraded energy, two undesirable characteristics for microbeam irradiation.

The conventional method for overcoming slit edge scatter is to utilize two apertures in succession. The first (upstream) aperture is slightly smaller than the second aperture and defines the size of the beam. The second aperture intercepts the particles scattered by the first, but because the flux density in the penumbra of the beam is relatively low, few particles are scattered from its edge. The problem with this approach is that the two apertures must be aligned very accurately, and they are so small that it can be very time consuming to scan one relative to the other until the beam is aligned. Two approaches have been used by groups developing microbeam irradiation systems. One is to make an assembly with two thin aperture foils and a spacer, and laser drill the two holes in a single operation. This assures that the holes are aligned, but requires that they be relatively close together. The second approach is to use a fixed, laser drilled hole, for the second aperture and construct the first aperture using a set of adjustable knife edges. In this way the beam defining aperture can be constructed to give the desired spot size and to be aligned with the final aperture. However, this approach requires a relatively large distance between the first and second aperture, making it very dependent of the stability of the beam focusing and steering components of the accelerator.

Collimator approaches which rely on fabricated holes generally are limited to relatively low energy beams. The techniques for drilling holes are generally limited to a maximum hole depth about five times the hole diameter. Since the aperture foils must be thick enough to stop all of the particles being used, the minimum collimator diameter is related to the range of the particles in the collimator material. This limits collimators to about 2 μm diameter for 8 MeV alpha particles, but smaller apertures could be used with lower energy beams.

An entirely different approach to beam collimation has also been tried.[7] Instead of apertures separated by a drift space, this approach uses a long narrow capillary to eliminate particles that were scattered near the entrance to the aperture. Although there are many technical challenges associated with this type of collimator, including the difficulty in keeping the channel straight, and aligning it with the path of the particles, it appears to be a very promising approach. It may be possible to make much smaller beam diameters using capillaries than can be achieved using laser drilled holes, in part because there is no immediate limitation on the length of the capillary and therefore the thickness of the aperture plate.

3.2 Particle Detection

Detection of the individual particles as they interact with the biological sample is critical for the study of effects at low doses. In some cases it is satisfactory to detect the particles after they have passed through the biological sample. This approach has the advantage that the detection process does not contribute to the divergence of the beam. Several different types of detectors could be used after the cells, and since the particle can stop in the detector, the signal can be quite large, eliminating the signal to noise ratio problems which effect most other detection systems. However, the requirements that the particle have significant energy remaining after passing through the target, and that the medium be removed from the cells so that the particles can reach the detector, prevent this approach from being used for some biological experiments, particularly those that require study of the full range of stopping power.

The alternative is to place the particle detector between the collimator and the biological target. This eliminates the need to remove tissue culture media in order to make a measurement, and allows use of all particle energies, down to those which stop in the biological target. The resulting problem is that the detector must be thin, in both mass per unit area and actual path length, in order to minimize effects on the beam size. The limitation on actual distance comes about because there will be some small angle scatter of the beam as it enters almost any type of detector, and these small angles translate into significant displacements from the intended target position if there is a significant distance after the scattering event. The ideal detector would be a thin film that produced a large signal (large signal charge per keV of energy deposited) and had the mechanical strength to prevent damage from petri dishes placed on it. This ideal detector has not been found. Thin plastic scintillators are rugged enough, but produce a very small signal and require a very efficient photomultiplier and light gathering system to produce a useful signal. Improved detectors will probably be developed as time progresses.

3.3 Target Imaging and Positioning

The efficient recognition and positioning of biological targets for irradiation is probably the critical step in determining the number of cells that can be irradiated. The difficulty comes primarily in the recognition of the targets. Automated recognition is essential if more than a few cells are to be irradiated per minute. However, automated recognition generally requires a high contrast between the structures of interest and the remainder of the biological system. To date, this high contrast has been achieved only by use of dyes which label a molecule that is specific to the target to be irradiated. However, image processing techniques may be devised to identify specific structures if they have unique shapes or other properties which can be recognized through the use of special illumination techniques. Unfortunately, the need for the beam collimation system immediately below the biological targets limits the illumination techniques that can be used to those that can be implemented with all components on the same side of the sample as the microscope objective. Acquisition of a high contrast image is further inhibited by the optical background which may be dictated by the structure of the charged particle detector, and by the presence of numerous changes in index of refraction in the light path, between air, tissue culture medium, polyester film and sample.

Many optical techniques for imaging remain to be explored, and new hardware and software for image processing is becoming available frequently. The image recognition component of microbeam irradiation systems is probably going to evolve rapidly over the next few years.

4 CHARACTERISTICS OF ONE SYSTEM

The single particle microbeam initiated at Pacific Northwest National Laboratory, and now being relocated to Texas A&M University, was designed to provide irradiation of targets without utilizing stains to allow automated recognition for rapid target positioning. In this system, the collimator consists of an adjustable primary aperture followed by a fixed laser drilled hole. Particle detection is accomplished with a thin plastic scintillator and an array of photomultiplier tubes which gives the high light gathering efficiency needed for acceptable signal to noise ratio. Targets to be irradiated are imaged in white light using epi illumination and a low numerical aperture, long working distance objective. Both phase contrast and oblique illumination have been used, with oblique illumination proving to be more reliable. The image is acquired by a personal computer based image processor, and the positions of targets is marked by the operator using a computer mouse. A computer controlled, microstepping, three axis positioning system is used to lift the cell dish and position it for the irradiation.

It has proved to be surprisingly difficult to develop reliable quantitative measures of the accuracy of the irradiations produced by this system. Track etch techniques in plastic films can be used to give a measure of the number of tracks delivered to specified target areas. However, it is difficult to count overlapping tracks or to measure their positions precisely enough to specify the probability density of tracks as a function of radial distance from a specified target. When the system is properly adjusted in terms of beam current and focus, we believe that if one track is specified, approximately 90% of the targets receive exactly one track, about 5% receive no track (due to the electronic system triggering on noise), and 5% receive 2 tracks due to one track not being detected or a second track occurring before the shutter closes. We believe that about 90% of the tracks fall within the 5 μm diameter circle centered at the designated spot, but have not yet been able to characterize the spatial distribution of those that fall outside the spot. The count rate through the collimator is generally set at 100 per second, so actual irradiation times are very short, and intervals between pairs of events can be set quite precisely. However, the time needed to find and position a target for irradiation depends on the density of targets in the sample, and averages around 20 seconds, limiting experiments to about 180 cells per hour.

This system has been used for several experiments[8] utilizing nuclear irradiation. It has also been used for preliminary experiments to study effects of cell communication by delivering specified numbers of tracks at specified intervals without regard to the positions of the cells or other targets on the dish. The most intriguing result to date comes from a study by Metting (this volume) which utilizes a new technique to visualize the location of DNA strand ends at various times after irradiation. The microbeam irradiated cells show that there is a major change in the staining pattern when a cell nucleus has been traversed by alpha particle tracks. The effects can also be recognized in cell populations which have been exposed to random tracks from a radioactive source. However, without the evidence of the microbeam irradiation, it is doubtful that the pattern produced by random irradiation would have been recognized.

References

1. A. Deshpande, E. H. Goodwin, S. M. Bailey, G. L. Marrone, and B. E. Lehnert, *Radiat. Res.* 1996, 145, 260-267.
2. H. Nagasawa, and J.B. Little, *Cancer Res.* 1992, 52, 6394-6396.
3. A. W. Hickman, A.W., R. J. Jaramillo, J.F. Lechner, and N.F. Johnson, *Cancer es.* 1994, 54, 5797-5800.
4. H. H. Rossi, *Radiat. Res.*, 1979, 78, 185-191.
5. R. E. Zirkle, 'Advances in Biological and Medical Physics' 1957.
6. M. J. Bissell, H. G. Hall, and G. Parry, *J. Theor. Biol.* 1982, 99, 31-68.
7. K. J. Hollis Ph.D. Thesis, Brunel University, 1995.
8. J. M. Nelson, A. L. Brooks, N. F. Metting, M. A. Khan, R. L. Buschbom, A. Duncan, R. Miick and L. A. Braby, *Radiat. Res.*, 1996, 145 568-574.

This research funded by the Office of Health and Environmental Research, US Department of Energy and by Texas Engineering Experiment Station.

TARGETING CELLS INDIVIDUALLY USING A CHARGED-PARTICLE MICROBEAM: THE BIOLOGICAL EFFECTS OF SINGLE OR MULTIPLE TRAVERSALS OF PROTONS AND $^{3}He^{2+}$ IONS.

M. Folkard, K. M. Prise, B. Vojnovic, A. G. Bowey, C. Pullar, G. Schettino and B. D. Michael.

Gray Laboratory Cancer Research Trust
PO Box 100
Mount Vernon Hospital
Northwood HA6 2JR

1 INTRODUCTION

In recent years, there has been a resurgence of interest in the use of microbeams in radiation biology[1,2,3]. Charged-particle microbeams provide a unique opportunity to control precisely, the dose to individual cells and the localisation of dose within the cell. By this means, it is possible to study a number of important radiobiological processes in ways that cannot be achieved using conventional 'broad-field' exposures. The Gray Laboratory is now routinely operating a charged-particle microbeam capable of delivering targeted and counted particles (either protons, or $^{3}He^{2+}$ ions) to individual cells. This report describes the rationale, development and current status of this facility. Preliminary data for cell survival and micronuclei formation after exposure to targeted and counted 3-3.5 MeV protons are presented.

2 PROPOSED EXPERIMENTS

The microbeam approach is now recognised as an important technique for investigating the cellular basis of hazards associated with occupational and environmental exposure to low doses of charged particles, where exposed cells are unlikely to receive more than one particle traversal. Using the charged-particle microbeam, we are developing an *in vitro* experimental model that reproduces the levels of exposure that occur *in vivo.* In essence, this utilises the unique ability of the microbeam to deliver precisely one particle to each cell. An appropriately sensitive assay of biological radiation damage will then be used to assess radiobiological effect at this level of dose.

We will also use the microbeam to ascertain the microdistribution of sensitivity across the cell nucleus. A study by Raju and colleagues[4] demonstrated that α-particles which completely traverse the nucleus are more effective per unit average dose, than those terminating within it. Interestingly, this finding differs from the earlier work of Cole[5] who showed that the DNA close to the nuclear membrane is the most easily damaged by ionizing radiation.

Finally, we will investigate the role of radiation-induced cell-to-cell and sub-cellular signal-transduction pathways, with regard to p53 expression, apoptosis, genomic instability and induced radioresistance. The existence of radiation effects transmitted from irradiated cells to neighbouring un-irradiated cells has been indicated in a number of recent studies[6,7].

3 METHODS

3.1 Microbeam Irradiation Facility

The Gray Laboratory microbeam collimation, alignment, detection and imaging system are depicted in figure 1. The microbeam has been designed to individually irradiate mammalian cells (in the first instance, V79-379A Chinese hamster cells) attached to a 4 μm thick polypropylene membrane that forms the base of a cell dish. The dish is supported on a 3-axis, precision motorised microscope stage. The cells can be viewed *in situ* using an epi-fluorescent microscope (Olympus, type BH2), either through the eyepiece, or more usually, with a charge-coupled device (CCD) imaging system (Xillix).

Each cell dish is divided into 4 x 5 mm square regions (two control, and two exposed regions). Cells are seeded to form an attached, sparse monolayer with roughly 250 cells per region and are stained with 1 μM Hoechst 33258. An automated cell-finding routine is used to locate and record the position of all the cells in each region prior to the irradiation phase.

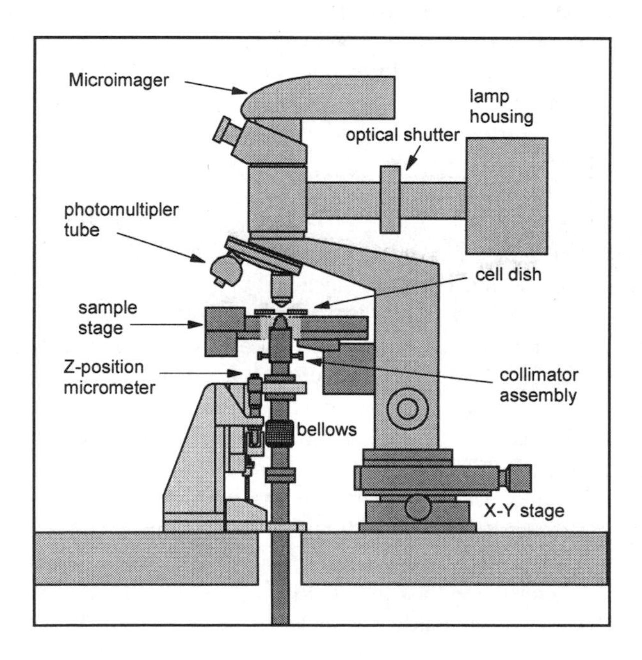

Figure 1 *The microbeam collimation, alignment, detection and imaging system.*

A 250 μm diameter by 1 mm long fine-bore glass capillary, with a ≥1 μm diameter bore is used to collimate monoenergetic protons, or $^3He^{2+}$ ions (produced using a 4 MV Van de Graaff accelerator). Irradiations are performed by precisely positioning each cell, in turn, above the collimator and exposing the cell through the supporting membrane. The number of particles passing through each cell is controlled by detecting photons generated in a thin scintillator (18 μm thick NE102A) located between the collimator and the base of the cell dish. The photons are detected using a miniature photomultiplier (PM) tube sited close to, and above the cell dish. During the irradiation sequence, the experimental area is in darkness, and the PM tube replaces the microscope objective used previously to locate the cells. As a consequence, cells are not normally viewed during this phase and the alignment of each cell (which is automated) utilises the stored co-ordinates for the collimator and cells. The exposure of each cell is terminated by a fast electrostatic shutter, triggered by a pre-set number of counted pulses from the PM tube. Currently, cells are irradiated at a rate of just less than one cell per second. Overall, its takes about 5 minutes to locate and log the positions of ~250 cells in each region, a further 5 minutes to perform a focus and alignment check, then 5 minutes to complete the irradiation sequence. Therefore to process the whole dish (assuming 4 regions and 1000 cells) takes about 1 hour.

3.2 Assays for Radiation Damage

Two assays of biological radiation damage are currently being used; a single-cell clonogenic assay to measure cell survival and the micronucleus assay to measure chromosome damage. To ascertain clonogenic survival, cell dishes are incubated for three days following irradiation and viable colonies (those with more than 50 cells) are scored. To do this, the stored co-ordinates are used to re-visit and determine the status of each irradiated cell. The micronucleus assay requires that cells are treated with cytochalasin-B (3 $\mu g\ ml^{-1}$) for 24 hours to allow nuclear, but not cytoplasmic division. Cells are then fixed in methanol/acetic acid (3:1) and stained with 0.5% (w/v) acridine orange. Micronuclei are scored using the protocol of Almassay *et al.*[8], as micronucleus frequency per binucleate cell. All data are corrected with regard to an accompanying control region which receives the same treatment as the exposed region (except that it is not irradiated).

4 RESULTS AND DISCUSSION

4.1 Collimator and Detector Performance

The collimator and detector performance have been evaluated using CR39 track-etch plastic. Currently, the probability of delivering the required number of particles to within 4 μm of the selected position is 96%, and to within 2 μm, the probability is 90%. The track-etch method has also been used to evaluate the accuracy of the particle counting system. The detected pulses from 18 μm thick NE102A scintillator are sufficiently well separated from the background 'noise' (i.e. due to stray light and to 'dark current') such that it is apparently possible to detect particles (and to reject spurious pulses) with 100% efficiency. However, approximately 1-2% of exposures indicate the passage of one unwanted extra particle which we attribute to a particle arriving after the pre-set number of particles has been reached, but before shutter has closed (which takes several microseconds).

4.2 Radiation Damage using Counted Particles

Table 1 *The surviving fraction of V79-379A cells following exposure to 3 MeV protons (errors are standard deviations).*

Number of Particles through the Nucleus	Surviving Fraction
5	0.93 ± 0.025
15	0.89 ± 0.050
30	0.82 ± 0.025
60	0.47 ± 0.034

Preliminary data for the survival of V79-379A Chinese hamster cells following the targeted exposure of the cell nucleus to pre-set numbers of near-monoenergetic 3 MeV protons have been obtained. The mean surviving fraction after exposure to 5, 15, 30 and 60 protons is indicated in table 1 and indicates a shouldered response consistent with the LET of this radiation. The same data has been compared to 250 kV X-rays by expressing the number of particle traversals in terms of the dose to the cell nucleus. Compared in this way, the RBE for cell survival (at 50% surviving fraction) is 1.80.

We have also obtained preliminary data for the induction of micronuclei from exposure to counted 3.2 MeV protons in the range 30 particles, down to just a single particle traversal. The number of cells with micronuclei (per binucleate cell) increases linearly with number of proton traversals at a rate of 1% cells per traversal. Expressed as dose, the RBE compared to 250 kV X-rays is 1.6.

References

1. B. D. Michael, M. Folkard, and K. M. Prise, *Int. J. Radiat. Biol.,* 1994, **65**, 503.
2. M. Folkard, K.M. Prise and B. D. Michael, *Radiat. Prot. Dosim*, 1994, **65**, 215.
3. J. M. Nelson, A. L. Brooks, N. F. Metting, M. A. Khan, R. L. Buschbom, A. Duncan, R. Miick, and L. A. Braby, *Radiat. Res.,* 1996, **145,** 568.
4. M. R. Raju, Y. Eisen, S. Carpenter, K. Jarrett and W. F. Harvey, *Radiat. Res.,* 1993, **133,** 289.
5. A. Cole, R. E. Meyn, R. Chen, P. M. Corry, and W. Hittleman, 'Radiation Biology in Cancer Research' Eds: R.E. Meyn and H. R. Withers, Raven Press, New York, 1980.
6. A. Deshpande, E. H. Goodwin, S. M. Bailey, B. L. Marrone and B. E. Lehnert, *Radiat. Res.,* 1996, **145,** 260.
7. H. Nagasawa and J.B. Little, *Cancer Res.,* 1992, **52,** 6394.
8. Z. Almassay, A. B. Krepinsky, A. Bianco and C. J. Koteles, *Int. J. Radiat. Appl. Instr.,* 1987, **38,** 241.

MICROBEAM MEDIATED CELLULAR EFFECTS: SINGLE α PARTICLE INDUCED CHROMOSOMAL DAMAGE, CELL CYCLE DELAY, MUTATION AND ONCOGENIC TRANSFORMATION

C.R. Geard, G. Randers Pehrson, T.K. Hei, G.J. Jenkins, R.C. Miller, L.J. Wu, D.J. Brenner and E.J. Hall

Center for Radiological Research
Columbia University College of
Physicians and Surgeons
New York 10032 USA

1. INTRODUCTION

The use of micro-irradiation techniques in radiation biology dates back to the 1950's to the work of Zirkle and Bloom[1]; Zirkle[2]; Davies and Smith[3] and Smith[4]. However, we are now able to take advantage of recent developments in particle delivery, focusing and detection, image processing and recognition and computer control, coupled with the benefits of new assays of individual cellular response.

The biological interest in the microbeam stems from the potential to define the ionizing energy absorbed by a cell, in terms of space, time, and number:

1.1 The microbeam allows irradiation of many cells each in a highly localized *spatial* region, such as part of the nucleus, or the cytoplasm, or through the immediate neighbor cells of a given cell. This capability allows questions regarding cell-to-cell communication, functionality of sub-components of the cell, and intra-cellular communication, to be directly addressed. The microbeam also allows particles to be passed through a cell with a known *temporal* separation, to investigate, for example, the dynamics of cellular repair.

1.2 At the low doses of relevance to environmental radiation exposure, individual cells only rarely experience traversals by a densely-ionizing particle, the time intervals between the tracks typically being months or years. The biological effects of exactly one particle are unknown because, due to the random (Poisson) distribution of tracks, this cannot practically be simulated in the laboratory using conventional broad-field exposures. Microbeam techniques can overcome this limitation by delivering exactly one particle per cell site.

2. METHODS

Precise charged particle delivery has been achieved with the construction of a microbeam at the Nevis Laboratories Radiological Research Accelerator Facility (RARAF).

The objective was to construct a system in which individual nuclei of cells (with diameters of about 10 μm) could be hit with single α-particles. At RARAF the microbeam facility was constructed in a dedicated laboratory on the second floor. The laboratory was equipped with all necessary cell handling equipment (incubators, cell handling bench, etc.) such that minimal time (seconds) would elapse between handling and irradiation. A dedicated beam line from the Van de Graaff accelerator was constructed to direct the beam vertically into thc laboratory. Physical details are presented elsewhere. The system can be reliably turned on each day, or as needed, to produce a parallel, vertical beam of adequate intensity. A He-Ne laser is positioned below a 70 degree up bend and defining aperture which establishes collimation through to the microscope. An XYZ motion translation stage is installed above the exit aperture. The stage is moved at adjustable rates under computer control, and is integrated with the image analysis system to sequentially visit locations corresponding to the centroids of cells, found in the video image. A 4X objective is used to observe cell cultures and locate the centroids of each cell, and cells are positioned using a 40X objective. Cells are irradiated in specially constructed dishes attached to a polypropylene film. A silicon solid state detector was constructed to mount as if an objective in the nosepiece of the microscope, while a pulsed ionization chamber is mounted on the 40X objective to count particles while the cells are being observed.

Preliminary results with asynchronous HeLa (human cervical carcinoma), human hamster hybrid A_L (containing human chromosome 11) and C3H 10T1/2 (mouse fibroblast) as well as synchronised G1 phase AG1522 human fibroblasts, have been obtained. Endpoints considered have been frequencies of micronuclei and/or chromatin bridges - per cell pair, surviving fraction, the progression of synchronized G_1 phase cells through the cell cycle after irradiation monitored by BrdU uptake and immuno-fluorescence, mutation at the S1 locus (ALcells) and oncogenic transformation (C3H 10T 1/2 cells). These studies have established the utility of the system constructed, to evaluate the consequences of single particle irradiations. This system provides a precise tool for evaluating number, time and spatially dependent effects.

3. RESULTS

All microbeam studies undertaken to date have involved delivery of a precise number of α-particles per cell nucleus. The cell lines chosen for study encompass a range of endpoints and of nuclear cross sectional areas. Here we will concentrate on the effects of one α-particle, recognising that these results reflect work in progress. Table 1 shows nuclear cross sectional areas, and dose per particle. Micronuclei/chromatin bridge frequencies per post mitotic cell pair were recorded after irradiating individual cells with exactly one α-particle at 90 keV/μm. When converted to induced effect per cell pair per Gray, the asynchronous A_L, HeLa and C3H 10T 1/2 cells produce similar responses, while the synchronised G1 phase normal human fibroblasts show a lesser response.

In these latter cells (p53, wild type) it was determined (by BrdU uptake into cells undergoing DNA synthesis and fluorescent antibody detection) that 1 α-particle slows entry from G1 phase into Sphase. Relative to controls 43% and 23% of cells remained in G1 at 24 and 48 hrs. post single particle irradiation respectively.

Cell survival was assessed by a colony forming assay, with indications of linear dose responses comparable with "track-segment"irradiation. One α-particle traversal (90 keV/μm) kills ~15 - 30% of cells with D_{37} 's of ~0.4 - 0.7 Gy at 3-7 α-particles per cell nucleus.

When mutation at the S1 locus of human chromosome 11 in A_L cells was assessed, it was determined that 1 α-particle induces mutation at a frequency of ~1 x 10^{-3} per surviving cell.

When oncogenic transformation in mouse C3H 10T 1/2 cells was assessed it was found that 1 α-particle (90 keV/μm) induces oncogenic change at a frequency of ~5 x 10^{-5} per surviving cell. As was found in the cell survival assessments, these results from microbeam single cell irradiations are similar to, and consistent with, those found after "track segment" α-particle irradiation. Ongoing studies will establish, with a greater degree of certainty, the results and implication from these preliminary findings.

4. CONCLUSION

A microbeam provides the means to define energy absorbed by a cell in terms of space, time and frequency. We have shown the ability of exactly one α-particle delivered by microbeam to elicit biologically deleterious effects. This approach has the potential to impact significantly on assessments of the consequences of low level environmental exposures (the radon problem) and to aid in defining radiobiological mechanisms. Exact numbers of particles can be placed in nuclei of cells, in the cytoplasm of cells, or in neighboring cells in a controlled manner.

That is intranuclear, intracellular and intercellular biological responses can be assessed with a microbeam which has a spatial precision lacking from all other sources of ionising radiation.

ACKNOWLEDGMENT

This work was supported by Program Project CA 49062 from the National Institute of Health.

REFERENCES

1. R.E. Zirkle and W. Bloom, Science, **117**: 487 1953.
2. R.E. Zirkle. Partial cell irradiation. In: Advances in Biological and Medical Physics.
3. Eds., H.J. Curtis, L.H. Gray and B. Thorell, Academic Press, N.Y. 1957
4. M. Davies and C.L. Smith. Exptl. Cell Res. **12**: 15, 1957.
5. C.L. Smith. Proc. Royal Soc. B. **154**: 557, 1961.

Table 1. *Microbeam single α-particle irradiation of individual cell nuclei (90 keV/μm)*

Cell type	*Mean nuclear cross-sectional area (μm²)*	*Dose per particle Gy*	*Micronuclei/bridge induced by 1 particle per cell pair*	*Micronuclei/bridges per cell pair per Gy*
A_L (human-hamster hybrid)	85	0.17	0.18	1.10
HeLa	120	0.12	0.16	1.34
AG 1522 (normal human fibroblast)	120	0.12	0.04	0.27
C3H 10T1/2 (mouse fibroblast)	196	0.074	0.09	1.22

Visualization of Charged Particle Traversals in Cells

N.F. Metting
Pacific Northwest National Laboratory
Richland, WA 99352

L.A. Braby
Texas A&M University
College Station, TX 77843

Our research addresses the early events that occur in the cell, and particularly in the cell nucleus, after passage of a charged particle. We present an assay system which locates the path of a charged particle through the cell nucleus, and speculate that this will be a valuable tool to define a start point for cell signalling of DNA repair processes, as well as signalling of cell-cycle checkpoint proteins. This study of the biological effects of low doses of high LET particles stems from the need to understand molecular mechanisms of long term health effects originating from the heavy particle component of galactic cosmic rays, a major concern in extended space missions. In the deep-space environment each target cell would be traversed only once a month, on average, by a heavy charged particle (1); therefore it was important to use very low particle fluences for subsequent analysis and understanding of resulting measurements. The Single-cell/Single-particle Irradiator at PNNL[1] was used to deliver particles from an electrostatic accelerator, and thus eliminate most of the experimental variability in the exposure of cells to high LET radiation. The number of tracks through each cell can be specified, rather than the random number obtained with conventional irradiation. Irradiation can be limited to a specified portion of the cell, and the variation in stopping power of the particles as they enter the cell can be minimized.

Methods. Cultures of the human carcinoma cell line, HeLa S3 were synchronized by mitotic shake-off, to obtain cells in late mitosis. Specially designed mylar-bottom dishes were seeded with two 10 μl drops of cell suspension at 50 cells/drop, with a distance of approximately 5 mm between the drops. They were then incubated for two hours at 37 °C in a humidified CO_2 incubator to allow for attachment, after which each dish, now containing two separate patches of attached cells, was flooded with warmed medium and incubated for one additional hour. The dishes were irradiated using the PNNL Single-cell/Single-particle Irradiator (2), which will hereafter be called the microbeam. For each dish, only one of the two patches of cells was irradiated, one cell at a time, with a predetermined number of 3.2 MeV helium ions, to simulate alpha particle irradiation. Cells in the second, control patch

[1]The Single-cell/Single-particle Irradiator has been relocated by the Department of Energy to Texas A&M University.

were scanned but not irradiated. Duration of irradiation was ~30 min. Each dish was then returned to the incubator for 2 hours. The cells were fixed by addition of ice-cold methanol, and stored in a -20°C freezer until assayed.

In a second experimental protocol, a larger number of asynchronously growing cells were seeded in mylar dishes and allowed to attach and incubate for 24 hours. These dishes were then exposed to a planar, collimated Pu-238 alpha particle source (3) at a fluence of 1 particle per 98 μm^2, or 0.8 particles per cell-nuclear cross-sectional area, and an energy of 3.65 MeV at the cell layer. As in the first protocol, each dish was then returned to the incubator for 2 hours, followed by methanol fixation, and storage at -20°C. Control dishes were sham-irradiated for comparable times.

To ascertain the status of the DNA, the cells were probed for the presence of DNA 3'-OH ends by enzymatic addition of biotinylated dNTP's, followed by streptavadin-fluorescein secondary. The results described here utilized components and instructions from the apoptosis labelling kit TACS 1, provided by Trevigen, Inc. Briefly, cells were rehydrated in PBS, pretreated with dilute proteinase K for 5 minutes at room temperature, then incubated for one hour at 37 °C with Trevigen labelling buffer containing biotinylated dNTPs and Klenow fragment. After suitable rinsing, the cells were incubated with streptavadin-FITC.

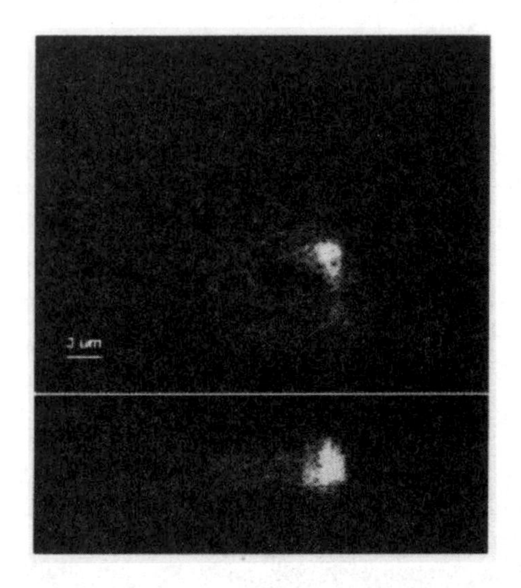

Figure 1 HeLa cell irradiated by three helium ions at PNNL microbeam.

Visualization and analysis of the fluorescence signal was performed with a Sarastro confocal scanning laser microscope (Molecular Dynamics, Inc) equipped with an argon laser light source filtered to provide 488 nm excitation wavelength. Detection of the fluorescein label utilized a 530 df30 bandpass filter. The cell contour was visualized by measuring its autofluorescence through a 600 nm longpass filter. Detector PMT gains and voltages were held constant for correlated controls and samples, as were software analysis parameters.

Results After attachment, the HeLa cells selected by mitotic shake-off presented nearly 100% as doublets on the microbeam dishes; thus cells were in early G1-phase at the time of irradiation. Figure 1 shows a single HeLa cell through which exactly three 3.2 MeV helium ions, collimated by a 5 μm diameter aperture, have passed. The upper picture shows a horizontal section parallel with the attachment surface, 0.19 μm thick, and located approximately midway in the cell nucleus. The lower picture is a vertical section 0.17 μm thick through the same cell at the point indicated in the upper view. The two views clearly show a column of concentrated fluorescein labelling within the cell nuclear volume. In the entire irradiated patch of cells at least 80% had distinct, bright columns of labelling, most often one, but occasionally there was a lesser second and perhaps third focus. In addition to the bright foci, most of the irradiated cells had a very light labelling that extended to the nuclear membrane. The remaining 20% of cells were labelled similarly to that of control cells. The tracks produced by the microbeam system are approximately randomly distributed over a 5 μm diameter circle. In most cases the maximum distance between the three tracks will be less than 5 μm. Apparently, when the tracks are within a few micrometers of each other, their effects overlap, and by the same reasoning multiple foci may occur when the tracks are at nearly the maximum separation.

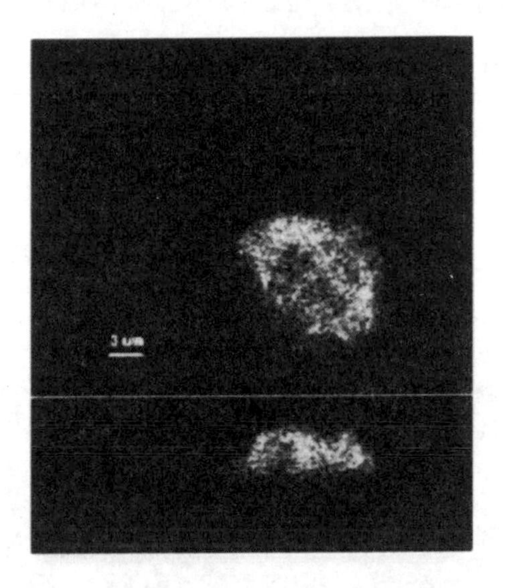

Figure 2 HeLa cell, Control. Scale bar is 3 μm.

Figure 2 shows a representative control-unirradiated cell from the same dish. The control cell is more homogeneously labelled throughout this volume, showing, however, distinct areas of lighter labelling which are consistent with thc usual appearance and number of nucleoli. Virtually 100% of the control cells showed a similar pattern of labelling, though there were slight differences in overall amount.

Figures 3 and 4 show the results of an experiment using unsynchronized HeLa cells. In figure 3, the cells were irradiated with the Pu-238 alpha particle source such that each cell nucleus received an average of about one particle. A number of cells appear to show one, two, or even three labelled foci, while others are labelled throughout the nuclear volume. The controls from this experiment, shown in figure 4, are all homogeneously labelled across the nuclear volume to some degree, and they do not show the foci.

To test for specificity of the streptavadin-FITC binding, both irradiated and unirradiated cells were processed through the entire assay with the exception of the primary enzymatic labelling step. There was no labelling in the nucleus, and a very small amount of label showing in the cytoplasm of a few cells (data not shown).

Discussion There is a clear and distinct difference in the labelling of irradiated and unirradiated cells in these experiments. The pattern of labelling in microbeam-irradiated cells (figure 1) is consistent with the expected spatial distribution of charged particle tracks, and the pattern of labelling in the ^{238}Pu-alpha-irradiated plates (figure 3) is consistent with a Poisson-distributed particle delivery. The clear implication is that we are visualizing the actual location of the initial DNA damage imparted to a cell nucleus as a result of the passage of a charged particle.

Based on our initial results of the microbeam experiments, in which the cells were pre-synchronized, one might first assume that the pattern of labelling is adequately explained in terms of cell-cycle kinetics. If the timing had been such that the unirradiated controls were well into DNA-synthetic, S-phase of the cycle, there would be an abundance of 3'-OH groups, one for each unligated Okazaki fragment, which might be available for enzymatic labelling. In addition, ongoing maintenance repair (unscheduled DNA synthesis) might supply suitable substrate. Likewise, the lack of overall labelling in the irradiated cells might be a result of a cell-cycle delay occasioned by DNA damage, and the ongoing repair of that damage would provide DNA strand break ends suitable for the enzyme assay. However, our cells were plated in late M- or early G1-phase. The G1 phase of the unperturbed HeLa cell cycle lasts an average of 11 hours, which would encompass the entire duration of the experiment, from plating to fixation.

In summary, we believe we have visualized the actual location of the initial DNA damage imparted to a cell nucleus as a result of the passage of a charged particle. We are now investigating the mechanism of this labelling, and believe we will gain new insight on repair protein responses to cell damage.

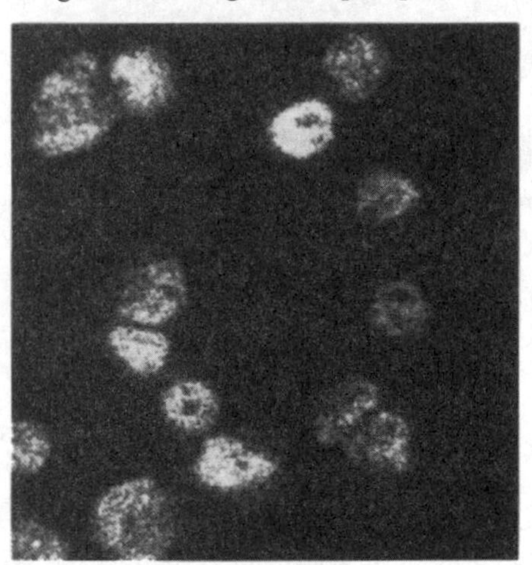

Figure 3 Unsynchronized HeLa cells exposed to Pu-238 alpha particles.

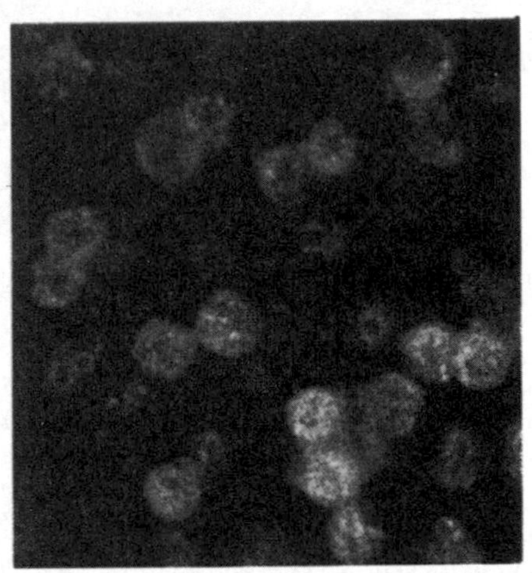

Figure 4 Control unsynchronized HeLa cells on mylar dishes.

References

1. S.B. Curtis and J.R. Letaw, *Adv. Space Res.* 1989, **9**, 293-298.
2. J.M. Nelson, A.L. Brooks, N.F. Metting, M.A. Khan, R.L. Buschbom, A. Duncan, R. Miick, and L.A. Braby, *Radiat. Res.* 1996, **145**, 568-574.
3. N.F. Metting, A.M. Koehler, H. Nagasawa, J.M. Nelson, and J.B. Little, *Health Phys.* 1995, **68**, 710-715.

Acknowledgements *Our thanks to Dr. JB Little for the generous use of his laboratory, as well as for his very helpful suggestions. Work supported by the National Aeronautics and Space Administration (NASA) 93-OLMSA-07: Space Radiation Health Program, under DOE contract DE-AC06-76RLO 1830.*

VISUALIZATION OF DAMAGE GENERATED ALONG ALPHA-PARTICLE TRACKS IN IRRADIATED RAT TRACHEAL EPITHELIAL CELLS

J. R. Ford, N. F. Metting*, S. J. Marsden, D. L. Stevens, K. M. S. Townsend, and D. T. Goodhead
Medical Research Council Radiation and Genome Stability Unit, Harwell, Didcot, Oxon, OX11 0RD, United Kingdom.
*Molecular Biosciences, Pacific Northwest National Laboratory, Richland, WA 99352 U. S. A.

1 ABSTRACT

Using reagents for labelling DNA strand breaks we have succeeded in visualizing the path of alpha-particles through cells grown on Hostaphan films. Track-like structures can be easily detected within a few minutes of irradiation of the cell. In order to verify that these were due to the traversal of alpha-particles, the cells were irradiated with collimated alpha-particles of varying incident angles. The angle of the track-like structures corresponded to the incident angle of the alpha-particles. We have examined the effects of dose and time on the appearance of these tracks and we now hypothesize that the track-like structures are due to the initial events of apoptosis which begin at the original sites of damage.

2 INTRODUCTION

One of the Authors had noted that there appeared to be track-like structures in cells after exposure to an alpha-particle microbeam when they were fluorescently stained for DNA breaks.[1] We became interested in verifying whether these actually coincided with the track of a particle or whether they just coincidentally resembled a particle track through the cell nucleus. In addition we wanted to extend the original observations to other cell types, in particular primary epithelial cells in culture and ***in situ***. If this were found to be generally applicable it could be useful as a method of dosimetry and for the examination of damage clusters within a track.

3 MATERIALS AND METHODS

3.1 Cell Culture

Cells were obtained from 12-20 week old Fischer F344 rats. The cells were harvested and cultured in a modified Ham's F-12 medium which has been described previously.[2] The cells were seeded into glass dishes with 2.5 μm thick Hostaphan film as the culture surface. For the following experiments a total of 250,000 viable cells were seeded into each 3 cm diameter dish and allowed to attach and grow for 48 hours before irradiation.

3.2 Cell Measurements and Cell Irradiation

Prior to irradiation one or more dishes were selected at random and rinsed with fresh medium. Nuclear measurements were obtained from these using a modification of the method of Townsend and Marsden.[3] In this case the cells were stained with 0.5 µg/ml 3,3'-dihexyloxacarbocyanine iodide (Sigma) in place of rhodamine 123.
Cells were irradiated by using the Plutonium-238 source which has been described previously.[4] For some exposures the cells were irradiated on a small stand which held the dish at a forty-five degree angle to the collimated beam of alpha-particles. The dose was calculated from previous measurements and verified by exposure of CR-39.

3.3 Cell Fixation and Track Visualization

Cells were either fixed using 2% paraformaldehyde within one minute of irradiation or the dishes were returned to the incubator. Those dishes were then fixed at 2 or 4 hours after irradiation depending on the dose and the number of dishes available.
After fixation all the dishes were stored at 4°C until stained. For the experiments described this was for 1 to 14 days. The cells were stained with FITC-conjugated dUTP by a terminal deoxynucleotidyl transferase reaction using a kit from Boehringer Mannheim and following the manufacturer's instructions. The cells were counter-stained with 0.5 µg/ml propidium iodide and coverslipped using an anti-fade agent (Vector). The dishes were then examined immediately after or stored at 4°C for up to one week. The tracks were examined using a Bio-Rad MRC 600 confocal laser scanning system attached to a Nikon Diaphot inverted microscope. The objective lens used was a Nikon Plan-apochromat 63XN.A.1.4. The argon ion laser beam, operating at 488nm wavelength, was attenuated with a neutral density filter.

A cell was considered apoptotic if it stained brightly and exhibited nuclear condensation with apoptotic bodies. A cell was considered to have a track if a signal was detected which was located in the nucleus and was found to extend the length of the nucleus without extending beyond it. Also to be counted as a track the signal had to exhibit a discrete columnar appearance which extended in the same direction as the incident beam of alpha-particles.

4 RESULTS AND DISCUSSION

Rat tracheal epithelial cells require 24 hours to become firmly attached to the Hostaphan surface and they were allowed a further 24 hours to spread and begin cycling. At the time of irradiation the cells had a mean nuclear cross-sectional area of 86 ± 13 µm^2.
This measurement was used for the determination of the expected number of particle traversals per nucleus.

In the first experiments we wanted to examine the response of the appearance of the tracks with dose and incubation time after irradiation. We irradiated cells from both male and female rats but found no difference in the frequency of apoptotic cells or the appearance of tracks. The cells were irradiated on a rotating wheel and the cells were oriented perpendicular to the incident alpha-particles. The results for the highest and lowest doses were from a single experiment whereas the results of the 0.25 and 0.5 Gy exposures are from that experiment and an additional experiment. We combined all of these results for Table 1.

TABLE 1 *Variation in the number of apoptotic cells and nuclei with tracks with changes in dose and incubation time.*

Treatment (Gy)	*Time (hr)*	*Total Number*	*Fraction of Cells Traversed (%)**	*Apoptotic Cells (%)*	*Tracked nuclei (%)*
+ control	all	300	not applicable	100	0
- control	all	900	not applicable	0.1	0
Sham	all	1560	0	2.2	0.1
0.125	0	115	57	1.7	0
0.25	0	107	81	3.7	0
0.5	0	317	97	3.6	2.7
0.125	2	106	57	1.9	0
0.25	2	540	81	3.1	0.3
0.5	2	685	97	2.5	0.6
1.0	2	110	99.9	5.5	0
0.25	4	338	81	4.1	0.6
0.5	4	453	97	1.5	0.4

*Calculated from a mean nuclear area of 131μm^2 and a fluence of 5.14×10^{-2} $\alpha / \mu m^2/Gy$.

The greatest number of tracks were observed when the cells were fixed immediately after receiving 0.5 or 1.0 Gy. Tracks were seen in the 0.25 Gy exposure but only at the later timepoints. In all cases there appeared to be fewer nuclei exhibiting tracks than would be expected from the particle fluence. We postulate that this may be due to some cell cycle dependence of the phenomenon as these primary cells have only a small fraction of reproductively active cells. In fact flow cytometric analysis has shown that around five percent of the cells are in G2/M at the time of irradiation.

Also it is possible that a number of cells which might have shown tracks were lost from the samples by apoptosis. The data in Table 1 for the percentage of apoptotic cells give only a snapshot of the cells which undergo apoptosis. In the later timepoints the cells which underwent apoptosis earlier are lost from the sample as they break up into small bodies which scatter about the dish and in some cases are phagocytosed by nearby cells. In fact we suspect that the tracks may be the result of nucleases that were present in the vicinity of the damage track and which had begun the process of apoptosis. The enzyme which we use to label the tracks recognizes the ends of nuclease digested DNA and would be unlikely to react with unprocessed damage. This could explain why we only see tracks very early with the 1.0 Gy exposure. The cells fragment rapidly as the amount of damage necessary for the initiation of apoptosis has been greatly exceeded. A preliminary experiment with V79 cells which do not readily undergo apoptosis seems to support this idea. They exhibit tracks at frequencies several magnitudes below that seen with rat tracheal epithelial cells.

To insure that the tracks we were seeing were indeed localized to the path of the particles and not constrained by some coincidental arrangement of the cell nuclei, we irradiated the cells with alpha-particles whose incident angle was 45 degrees to the Hostaphan surface. We used an exposure of 0.5 Gy to the midpoint of the dishes and looked at several timepoints. Figure 1 depicts a typical track within a few minutes of

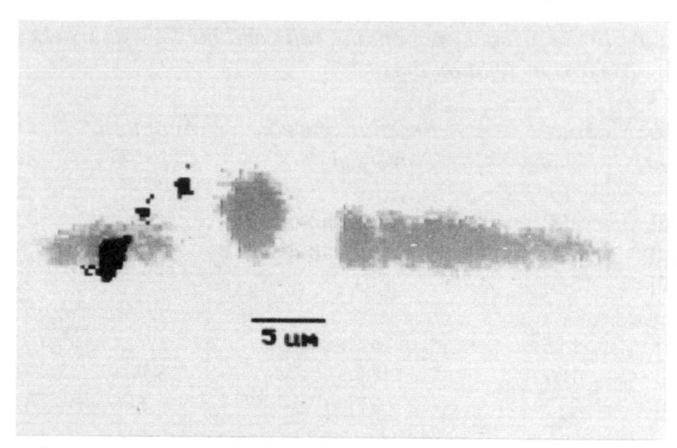

Figure 1 *Negative composite vertical confocal image of the damage track through a propidium iodide stained cell nucleus.*

irradiation. As can be seen the track through the nucleus is at the same angle as the incident particles. The track appears broken but this may be due to the difficulty we had in aligning the dishes precisely so that the confocal microscope software could compose a vertical section through the cell of interest. Conversely this may actually represent some sort of clustering of damage sites along the particle path. This reassured us that what we are seeing has some relation to the initial damage caused by a particle traversal.

We hope to demonstrate these tracks in the intact tissue in the near future.

Acknowledgements

This research is supported by Grant #PF-4181 from the American Cancer Society and partially by EC contract #F14P-CT95-0011.

References

1. N. F. Metting and L. A. Braby,
   Abstract from the Forty-fourth Annual Meeting of the Radiation Research Society, Chicago, 1996.
2. J. R. Ford and M. Terzaghi-Howe, *Rad. Res.*, 1993, **136**, 89.
3. K. M. S. Townsend and S. J. Marsden, *Int. J. Radiat. Biol.*, 1992, **61**, 549.
4. D. T. Goodhead, D. A. Bance, A. Stretch, and R. E. Wilkinson, *Int. J. Radiat. Biol.*, 1991, **59**, 195.

MEASURED PARTICLE TRACK IRRADIATION OF INDIVIDUAL CELLS

E. Heimgartner, H. W. Reist, A. Kelemen, M. Kohler, J. Stepanek and L. Hofmann

Institute for Medical Radiobiology of the
University of Zurich and the Paul Scherrer Institute
CH-5232 Villigen PSI
Switzerland

1 INTRODUCTION

At the low doses generally of concern in radiological protection the probability for exposure of cell nuclei by more than one particle is negligible, and the energy transfer to the nuclei by these particles is independent of the dose absorbed in the cell system. Single traversals of protons or α-particles through the nucleus of an individual cell are in particular relevant for understanding the biological significance of exposure to radiation (neutrons, radon) in low doses, both with respect to risk assessment and mechanistic studics. Thc biological effects of exactly one particle per cell are largely unknown because they cannot be simulated in the laboratory using conventional broad-ficld exposures due to the random distribution of tracks.

2 EXPERIMENTAL METHOD

The irradiation technique with measured track positions relies on the following prerequisites: (i) The beam divergence has to be < 1 mrad, (ii) scatters should not increase the uncertainty of the track positions in cell nuclei by more than 0.5 μm and (iii) the detection efficiency of the track detector has to be 100 %.

1.1 Single track particle beam

A special beam line was developed providing beam intensities of 10^4 to 10^6 particles per cm^2s with a beam spot 3 cm in diameter. The extracted beam intensity from Injector I at the Paul Scherrer Institute is 0.01 to 1 μA. The beam line is set up in the low energy area NE-C with a 90°-magnet that bends the beam upwards to the target cells to permit horizontal exposure to protons or α-particles at variable energies from 2.5 to 45 MeV. The position and the size of the beam are monitored with beam profile monitors using 1 mm plastic scintillator fingers coupled to photo multiplier tubes by fiber optics cables. The intensity of the beam is measured with a 1 mm thick scintillator counter in place of the target[1].

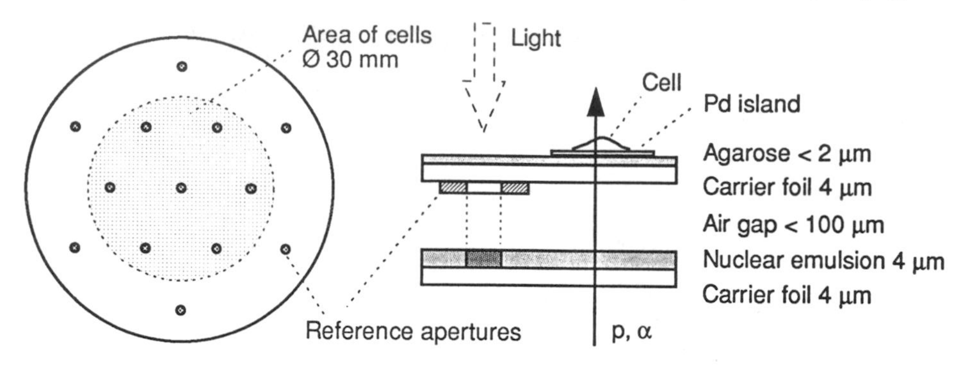

Figure 1 *Top view and cross section of the cell chamber with cells growing on palladium islands with the attached track detector. The positions of the cells and of the reference apertures are shown.*

1.2 Cell chamber assembly

The cell chamber assembly combines the possibility of irradiating a large number of cells simultaneously with an efficient colony-forming ability and the capability of localizing the particle tracks through the cell nucleus.[1,2] It consists on the one hand of a cell chamber with more than 1000 palladium islands on an agarose layer, a method by which cells can be confined to restricted areas, and on the other hand of a technique by which a particle passage through the cell nucleus can be accurately localized by measuring the track position in the attached detector (Figure 1). The track detector consists of a 4 µm thick nuclear emulsion (K5, Ilford) on a 4 µm thick polypropylene foil (Trespaphan, Hoechst). The position of the particle traversals are measured by digitally superimposing the image of the tracks on the phase contrast image of the cell. Distortions due to the processing of the detector film are corrected by measuring the relative position of circular reference apertures (100 µm in diameter) on the carrier foil of the cells and their light projected images on the detector. The accuracy of the method is assessed using additional apertures in place of cells.[4]

An inverted microscope (LSM 410, Carl Zeiss) together with a step motor driven stage (Märzhäuser, Wetzlar, Germany) is used to measure the cell and track positions. Glass scales were installed for the Y-axis giving a reproducibility equal to that of the X-axis of 0.3 µm. Systematic errors are controlled and corrected by laser interferometry (length of the axes) and by a certified angle gauge block with an electronic calliper (straightness and rectangularity). The temperature is held constant within (25.0 ± 0.3) °C. The correlation accuracy as given from the non-correctable displacements of additional apertures in place of cell positions is shown in Figure 2a. It was verified to be (0.9 ± 0.5) µm. In this measurement glass scales were not yet installed. Their use will further improve the accuracy to ~ 0.6 µm.

Using the PSI version of the code GEANT[5] the distribution of proton and electron tracks in the cell nucleus relative to the track position in the detector was simulated allowing for scatters in the vacuum window of the beam pipe and in the set-up.[6] As shown in Figure 2b the calculated uncertainty (FWHM) of the track positions inside the nucleus, relative to the tracks in the detector is 0.22 µm for 9.2 and 0.30 µm for 4.8 MeV

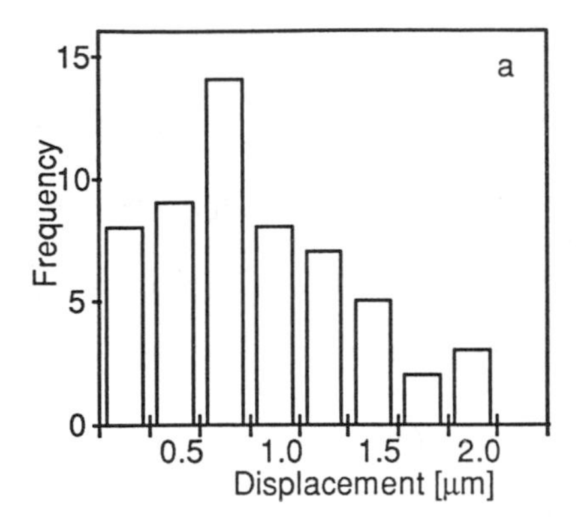

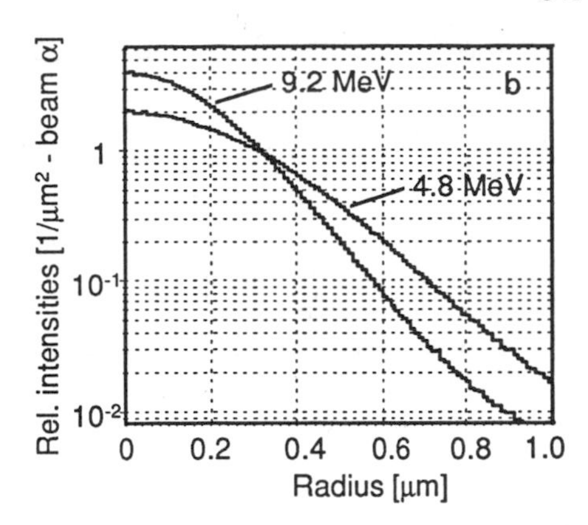

Figure 2 *(a) Frequency distribution of the remaining displacement of additional apertures in place of cell positions. (b) Radial distribution of particle tracks at the center of the nucleus relative to their positions in the detector due to scatters for 9.2 and 4.8 MeV α-particles.*

α-particles in the nucleus of V79 cells with the 8 mm gap between vacuum window and cell chamber assembly filled with air. In the meantime it is being improved by replacing the air with helium.

1.3 Irradiation procedure

Asynchronous V79 hamster fibroblast cells are seeded 16 hours before irradiation on palladium islands in a cell suspension with 2×10^3 cells/ml medium. Single cells are detected two hours before irradiation. In a first step out of 1000 palladium islands 200-400 are selected where a single cell is attached on the island. This takes 3 seconds per island. Then the positions of the reference apertures and of the cells are measured within 10 seconds per cell. This procedure minimizes the error due to a possible movement of the cells. Immediately after irradiation the apertures are light-projected on the track detector by a collimated light beam. The track detector is removed and processed. At the moment all position sensitive measurements are done at room temperature.

1.4 Determination of energy transfer

The track position relative to the nucleus (impact parameter) is determined from the superposition of the image of the tracks on the phase contrast image of the cell. Intersection length and energy transfer to the nucleus are calculated using a statistical model of the surface contour of the cell nuclei.[7]

1.5 Investigation of cell proliferation

In the present study cell survival and colony growth are investigated. Cells are monitored at regular times after irradiation and from the digital images the colony size (number of cells per cell clone) is determined. In Figure 3 the colony size is plotted against time after irradiation for a given amount of energy transfer to the nucleus.

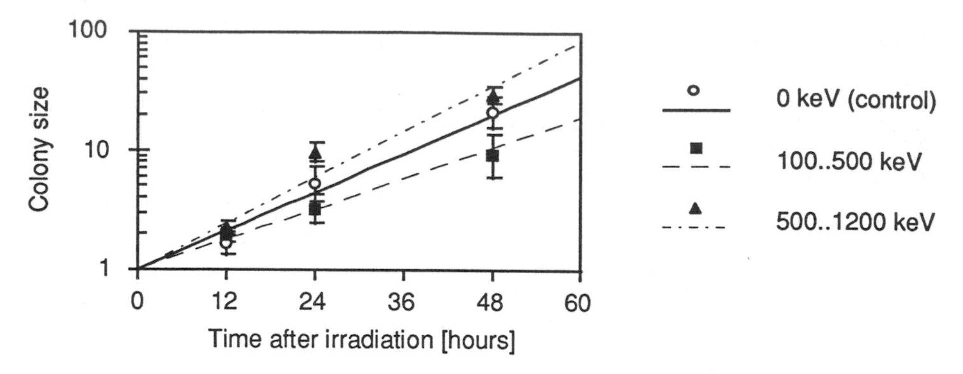

Figure 3 *Preliminary data of colony growth of V79 hamster fibroblast cells for given energy transfers. The colony size is plotted against the time after irradiation.*

3 DISCUSSION

The method developed (irradiation with measured track positions combined with the palladium island technique) and the results achieved enable to study colony growth for cells assigned to groups of successive intervals of transferred energy by charged particles and to determine cross sections for survival. Preliminary data on the colony growth show the feasibility of the method. Its capacity is now being improved in order to permit future investigations of the induction of chromosomal aberrations after single particle traversals through the nuclei of the exposed cells.

References

1. H. W. Reist, P. Doria, E. Heimgartner, D. Knespl, M. Kohler, B. Larsson B, Ch. Markovits, Ch. Michel, J. Stepanek, S. Teichmann and St. Tuor, 'Proceedings of the Swiss Society for Radiobiology and Medical Physics (SSRMP) and the Swiss Society for Medical and Biomedical Engineering (SSMBE)', H. Blattmann (Ed.), Villigen, 1993, p. 211.
2. H. W. Reist, W. Burkard, D. Calabrese, P. Doria, E. Heimgartner, D. Knespl, M. Kohler, B. Larsson, R. Leemann, Ch. Markovits, Ch. Michel, J. Stepanek, S. Teichmann and S. Tuor, *Radiat. Prot. Dosim.*, 1995, **61**, 221.
3. W. Burkard and B. Larsson, 'Proceedings of SSRMP-SSMBE', H. Blattmann (Ed.), Villigen, 1993, p. 231.
4. E. Heimgartner, H. W. Reist, A. Kelemen, M. Kohler and J. Stepanek, 'Proceedings of the V International Conference on Applications of Physics in Medicine and Biology', Trieste, Italy, 2.-6. September, *Physica Medica*, 1996.
5. D. Knespl, H. W. Reist and J. Stepanek, 'Proceedings of SSRMP-SSMBE', H. Blattmann (Ed.), Villigen, 1993, p. 241.
6. J. Stepanek J and H. W. Reist, 'Internal PSI-Report TM-29-96-1', Paul Scherrer Institute, Switzerland, 1996.
7. A. Kelemen, H.W. Reist, G. Gerig and G. Szekely, 'Visualization in Biomedical Computing. Lecture Notes in Computer Science 1131', K. H. Höhne and R. Kikinis, Springer, Heidelberg-New York, 1996, p. 193.

MICROBEAM SYSTEM FOR LOCAL IRRADIATION OF BIOLOGICAL SYSTEMS AND EFFECT OF COLLIMATED BEAMS ON INSECT EGG

Y. Kobayashi*, H. Watanabe*, M. Taguchi*, S. Yamasaki** and K. Kiguchi**

*Japan Atomic Energy Research Institute, Takasaki, Gunma 370-12, Japan
**Faculty of Textile Science and Technology, Shinshu University, Ueda, Nagano 386, Japan

1 INTRODUCTION

Recently laser microbeam irradiation has been used for cell biology and developmental biology; however ion microbeam has not been applied. Therefore we have designed and installed a microbeam apparatus at TIARA (Takasaki ion accelerators for advanced radiation application, JAERI-Takasaki), as we reported before[1], to develop a novel cell manipulation technique, which is called Cell Surgery technique. To investigate the effect of local irradiation of heavy ions on various biological system, *e.g.* fertilized eggs of insect and meristematic tissue of plant, the energy of the ions must be high enough to penetrate the region of interest in the target. Therefore the apparatus has been connected to a vertical beam line of AVF cyclotron giving heavy ion beams ranging from 12.5 MeV/u ^{4}He to 11.0 MeV/u ^{40}Ar. This particle spectrum covers an LET range between 14-1800 keV/μm. The beams have been collimated to about 10 μm in diameter using a set of apertures. In this paper an outline of microbeam system for local irradiation of biological systems and biological effect of collimated ion beams will be discussed.

2 EXPERIMENTAL SETUP AND THE FIRST RESULTS

2.1 Outline of Microbeam System for Local Irradiation

The apparatus has been installed under the vertical beam line of AVF cyclotron. This microbeam system has two vacuum chambers, connected to each other with an isolation gate valve and a conductance pipe, and two sets of pumps for differential vacuum pumping to pass the ions from the second chamber into the air through an aperture on the microprobe. Electrical status signals are given by a main control unit which supervises the vacuum system, positions of collimators and the status of Faraday cups. CCD-camera image of beam spot on beam monitors can be observed from a neighboring room. Almost all functions of this microbeam system can be controlled by both local control unit beside the apparatus and remote control unit from a neighboring room.

Figure 1(A) shows a schematic diagram of the biological microbeam system. Heavy ion beams from the AVF cyclotron are introduced to the first chamber through a bending magnet and a pair of beam steering magnets. The shape and the position of the beam spot can be observed with beam monitor 1. Two aluminum collimators with 5 mmϕ aperture and 0.5 mmϕ aperture, respectively, are set in the first chamber. The lower collimator with 0.5 mmϕ aperture is set on the micropositioning X-Y stage 1. The intensity of the ion beams through the collimators can be measured with the Faraday cup 1 and 2, respectively. After adjustment of the position of the beam spot on the beam monitor 2, the ion beams are extracted from the second chamber into the air with a microaperture in a tantalum disk perforated using a spark erosion method. The microprobe is attached to the tilt-adjuster and fixed on the micropositioning X-Y stage 2, which is connected to the second chamber with a flexible bellows. The alignment of the beam is achieved by the use of the beam steering magnet system upstream of the vertical beam line and two sets of micropositioning X-Y stages in the apparatus. The intensity and the energy of ion beams on the target micropositioning stage or after penetrated the target are measured with a plastic scintillator, CR-39 track detector and solid-state detector (SSD) in the atmosphere. During irradiation, the biological target can be observed with optical microscope system under the microprobe. The X-Y stage 2 and the whole optical microscope system are held with an anti-vibration damper. Almost all functions of optical microscope system, *e.g.* focus, lighting, objective and stage positioning can be controlled from a neighboring room. Detailed view of the end of the microprobe is shown in Figure 1(B). The minimum aperture size of available microprobe is about 10 μm in diameter.

2.2 Spatial Distribution of Etched Ion Tracks on the Target

To characterize the collimated ion beams passing through the microaperture, intensity and energy of the ions were measured with a plastic scintillator. The particle fluence at the

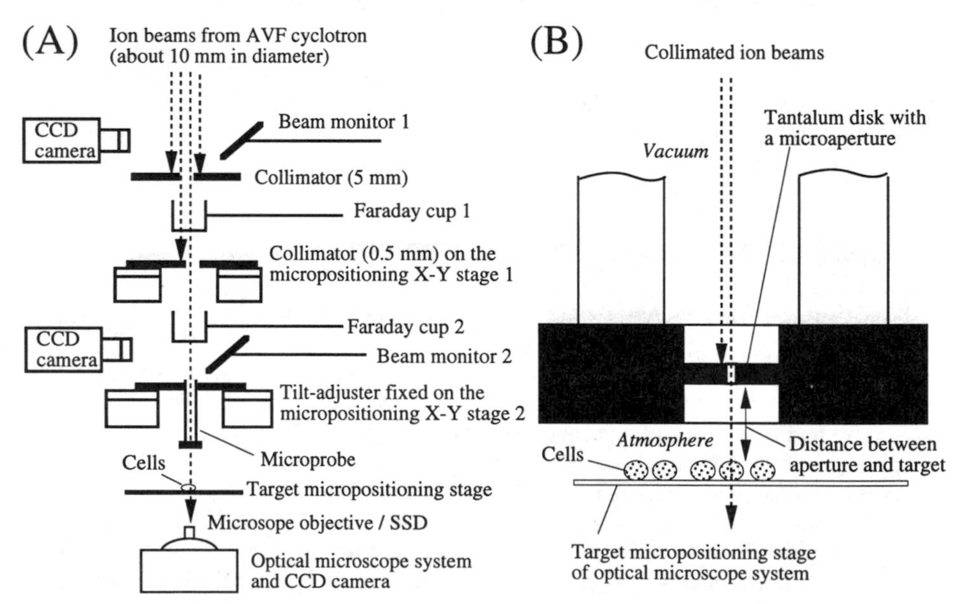

Figure 1 *Schematic diagram of microbeam system for local irradiation installed under the vertical beam line of AVF cyclotron at TIARA, Takasaki. (A) Beam collimation system of the microbeam apparatus. Heavy ion beams introduced from the AVF cyclotron are about 10 mmϕ at the Beam monitor 1. (B) Detailed view of the end of the microprobe.*

target position was also determined by electron microscopic counting of etched particle tracks on track detector CR-39 exposed to the ion beams on the target micropositioning stage. Obtained beam flux was about 1-10^4 ions/sec.

Figure 2(A) shows spatial distribution of etched ion tracks on CR-39 exposed to collimated 17.5 MeV/u ^{20}Ne ions at distances of 1 mm and 5 mm between aperture and target. Microprobes with apertures of 40, 90 and 250 μmϕ on 0.5 mm-thick tantalum disk, respectively, were used. Normalized particle fluence was flat in the "core" region around a center of beam distribution. The size of the core was equal to the size of the microaperture. On the CR-39 exposed to ions at a distance of more than 1 mm, two components of halo were observed outside of the core; "inner halo" where a ratio of particle fluence decreases with a steep slope from 1 to about 10^{-2}, and "outer halo" with a gradual distribution of widely scattered ions. At every exposure, the ratio of particle fluence in halo was decreased to 10^{-2} at about 25 μm of radial distance from the edge of the core. The width of inner halo increased as the ion path through air became longer (Figure 2B). It suggests that the size of inner halo may reflect the divergence of ion beams passed through the microaperture. Microbeams available at present for local irradiation of biological systems are listed on Table 1.

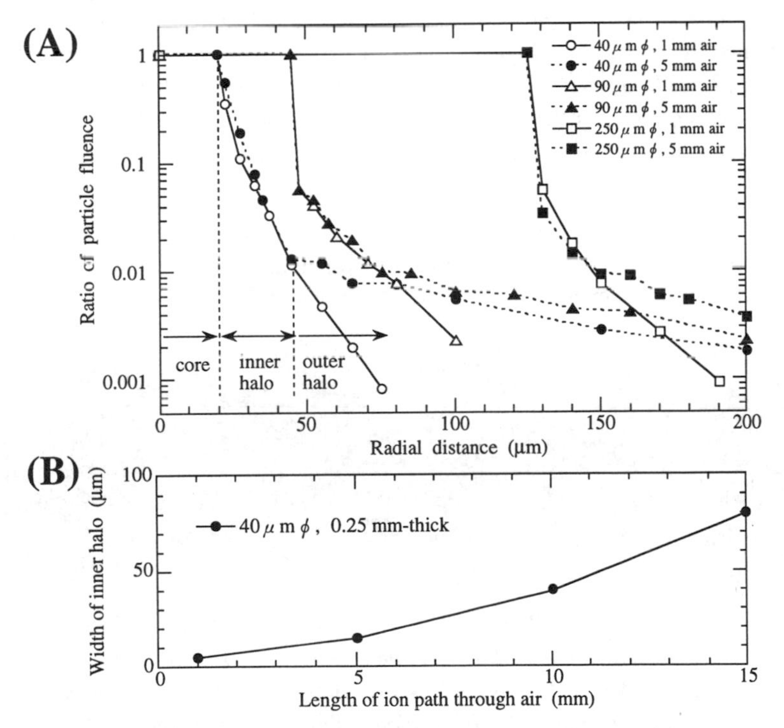

Figure 2 *Effect of diameter of microaperture and air gap (distance between aperture and target) on spatial distribution of etched 17.5 MeV/u ^{20}Ne ion tracks on CR-39. (A) Spatial distribution of etched ion tracks relative to a center of beam distribution collimated with 0.5 mm-thick tantalum microaperture. (B) Width of the inner halo around the core of beam distribution collimated with a 0.25 mm-thick tantalum microaperture as a function of the length of ion path through air (air gap).*

Table 1 *Available microbeams at present for local irradiation of biological systems*

Ions	*Specific Energy* MeV/u	*Minimum Beam Size* μmφ	*Projectile Range* mm in water
$^{12}C^{5+}$	18.3	20	1.2
$^{20}Ne^{8+}$	17.5	10	0.7
$^{40}Ar^{13+}$	11.0	10	0.24

2.3 Effect of Local Irradiation of Ions on Fertilized Eggs of Silkworm.

To ascertain the biological effect of collimated ion beams, the collimated beams were applied to fertilized eggs (1.20 mm long, 0.95 mm wide, 0.63 mm thick) of silkworm *Bombyx mori*. The eggs were irradiated at cellular blastoderm stage by 18.3 MeV/u ^{12}C ions at various doses. At the position of 50% VD (ventrodorsal diameter) and 50% EL (egg length), 5.4×10^9 p/cm^2 (1 kGy) of local irradiation less than 200 μmφ did not decrease hatchability while 8.6×10^7 p/cm^2 (16 Gy) of whole-egg irradiation resulted in 0% of hatching rate. The damaged cells inside the local irradiation area seem to be complemented by not-irradiated cells outside. However, 250 μmφ local irradiation with 2.7×10^9 p/cm^2 (500 Gy) ions inactivated the eggs, suggesting that the size of local irradiation site is critical for the complementation of damaged cells. The position of the local irradiation site is also critical for the inhibition of embryogenesis because 2.7×10^9 p/cm^2 of local irradiation in 180 μmφ was lethal at 25% VD and 50% EL position. Furthermore, induction of cuticle defects was observed after local irradiation of 2.7×10^9 p/cm^2 in 200 μmφ at the 25/50 site (Figure 3). Detailed analysis of the site-specificity of the inhibition of embryogenesis and anomalies of larva after local irradiation is in progress.

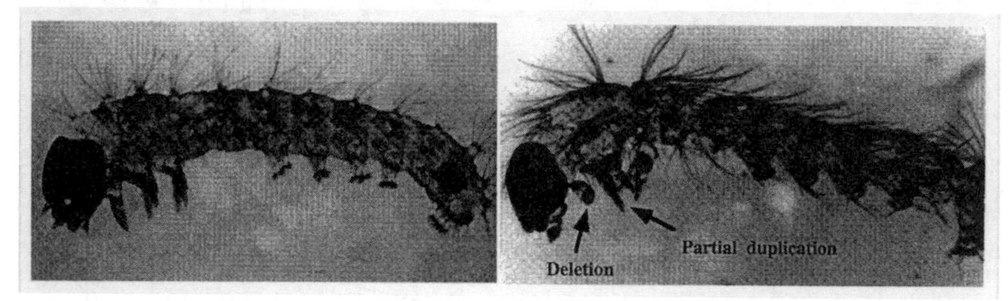

Figure 3 *Cuticle patterns of Bombyx mori larvae. A normal larva from non-irradiated control (left). A defective unhatched larva induced by local irradiation of 2.7×10^9 p/cm^2 (500 Gy) of 18.3 MeV/u ^{12}C ions at 25% VD and 50% EL position at the cellular blastoderm stage: deletion and partial duplication of thoracic appendages (right).*

References

1. H. Watanabe, Y. Kobayashi, T. Kamiya and K. Arakawa, Abstracts of 4th L. H. Gray Workshop 8th-10th July 1993 "Microbeam Probes of Cellular Radiation Response", Gray Laboratory, Northwood, UK, 1993, 4.4.1-4.4.4.

THE SOFT X-RAY MICROPROBE: A FINE SUB-CELLULAR PROBE FOR INVESTIGATING THE SPATIAL ASPECTS OF THE INTERACTION OF IONIZING RADIATIONS WITH TISSUE.

G. Schettino, M. Folkard, K. M. Prise, B. Vojnovic, T. English*, A. G. Michette*, S. J. Pfauntsch*, M. Forsberg and B. D. Michael.

Gray Laboratory Cancer Research Trust
PO Box 100
Mount Vernon Hospital
Northwood HA6 2JR

*King's College London
Strand
London WC2R 2LS

1. INTRODUCTION

A new *in vitro* technique for the micro-irradiation of cells is being developed at the Gray Laboratory. The soft X-ray microprobe facility is designed to generate a very fine beam of carbon-K (278 eV) X-rays to irradiate sub-cellular targets and assess their sensitivities. The microprobe uses ultrasoft X-rays generated by a beam of 10-20 keV electrons, focused on a carbon target, and employs a circular grating (zone plate) to focus the radiation to a fine spot of about 100 nm. A mirror situated between the carbon target and the zone plate eliminates the bremsstrahlung component from the beam. Computerised imaging, cell recognition and micro-positioning techniques will be used to identify and align the cells with the probe.

2. RATIONALE

The microprobe will be used to address a number of key questions relevant to spatial aspects of the interaction of ionizing radiation with tissue:

(i) To study how the clustering of ionizations influences the biological response to radiations. The much higher biological effectiveness of densely ionizing radiations for inactivating and mutating cells is believed to be related to increased clustering of damaged sites on DNA[1]. The extent to which ionizations are clustered can be precisely controlled using the microprobe.

(ii) To ascertain how sensitivity to ionizing radiation is distributed across the cell nucleus. Raju *et al.*[2] showed that α-particles which completely traverse the nucleus are more effective per unit average dose than those that only partly traverse the cell. This contrasts with earlier work of Cole[3] who showed that the DNA close to the nuclear membrane is the most sensitive.

(iii) To seek pathways to cell death by ionizing radiations other than through DNA damage.

(iv) To investigate the possibility that radiation effects may be transmitted from irradiated cells to neighbouring unirradiated cells. This has been observed by Nagasawa *et al.*[4] who find an unexpectedly high frequency of sister chromatid exchanges following exposure to doses of α-particles low enough that only a fraction of the cells are hit. A similar finding has been reported recently by Deshpande *et al.*[5].

3. THE MICROPROBE SOURCE

Carbon-K X-rays are generated by an electron beam incident on a 4.4 mm thick graphite target. The electrons are emitted by a heated tungsten filament, accelerated to 10-20 kV and focused to a 3 μm diameter spot on to the target by water-cooled focusing magnets. This produces unwanted bremsstrahlung in addition to the characteristic-K X-rays. The soft bremsstrahlung X-rays are absorbed by the target while the hard bremsstrahlung component is removed by a silica mirror situated between the carbon target and the zone plate. Using the phenomenon of total reflection that occurs at shallow incidence angles and low energies, it is possible to use a mirror to 'filter' the X-ray spectra. A ray tracing program, optimised for soft X-rays ('Shadow') has been used to investigate reflectivity for different mirror materials and different incident angles. A low atomic number mirror, such as silica (SiO_2), demonstrates high reflectivity for low energies (up to 70 %) and low reflectivity for high energies such that it should be possible to obtain a near-monochromatic X-ray beam. The optimum angle for simultaneously maximising characteristic X-ray output and minimising the bremsstrahlung component is 3 degrees. By increasing the incident angle it is possible to obtain a less contaminated beam at the expense of low reflectivity.

A zone plate is used to focus the radiation to a spot of about 100 nm. A zone plate is a circular diffraction grating about 100 μm in diameter with alternate opaque and transparent zones, arranged such that the density of zones increases with the radius.

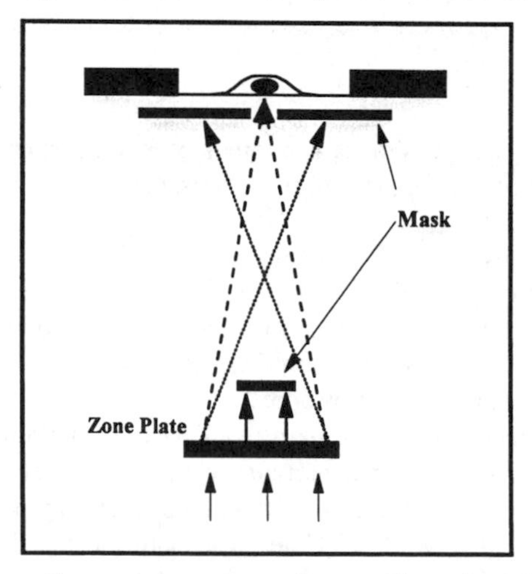

Figure 1. *Schematic representation of zone plate focusing action. Zero order (solid arrows) and third order radiation (dotted arrows) are stopped by masks and only the first order (dashed arrows) is focused on he cells.*

Since a zone plate is a form of diffraction grating, it will focus the radiation in more than one positive and negative (virtual) foci (m = ±1, ±3, ±5, ...). There is also some undiffracted radiation corresponding to the order m = 0 that will not be focused. An appropriate set of masks, positioned between the zone plate and the cells, stops all but the first order radiation, as shown in figure 1. Computerised simulations performed using 'Shadow' indicate that a zone plate with a diameter of 90 μm and 225 zones will focus X-rays in a 110 nm radius spot with an efficiency of 10 % and a depth of focus of ±6 μm. The focal length (that will also be the distance from the zone plate to the cells) is proportional to the square of the radius of the zone plate and is inversely proportional to the number of zones and to the wavelength of the radiation. For the above zone plate, the focal length is 2.7 mm.

4. DEVELOPMENT OF A SINGLE-CELL CLONOGENIC ASSAY

Computerised imaging and micro-positioning techniques are required to recognise biological targets and to align them with the probe. These techniques are already in use in connection with the Gray Laboratory charged particle microbeam facility[6]. A comparable system is currently being implemented on the microprobe. The arrangement uses fluorescent dye to stain biological targets (i.e. the cell nucleus). A UV microscope and computerised image analysis system are used to identify the cells and record their position. To irradiate the cells, a 3-axis, computer-controlled micro-positioning stage precisely aligns each pre-selected target with the probe. The established microbeam system has been used to develop a single-cell clonogenic assay. V79-379A mammalian cells are plated at low density, stained with Hoechst 33258 and positions of each cell logged before exposing them to a broad field of 250 kVp X-rays. Cells are then individually revisited to check the damage after three days; this has been found to be a suitable time to distinguish between surviving and damaged colonies (surviving colonies are defined as those with more than 50 cells). Using this assay, statistical fluctuations due to uncertainties in cell plating and colony scoring are avoided. Measurements

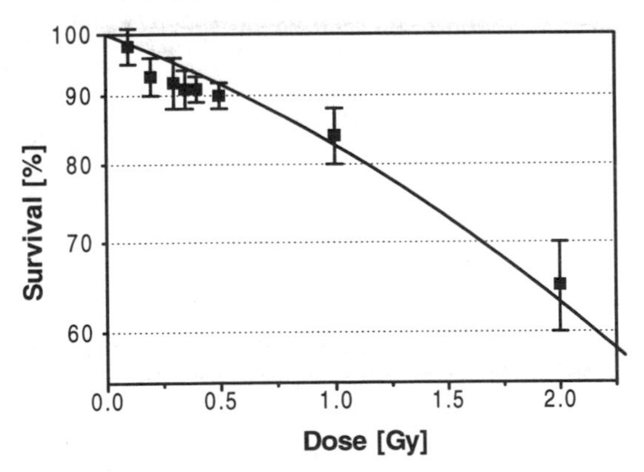

Figure 2. *Single-cell clonogenic assay data after 250 kVp X-ray exposure. The solid line has been obtained from data using a conventional clonogenic assay.*

indicate a hypersensitivity at doses below 0.5 Gy (figure 2) in agreement with other published data[7], confirming the high sensitivity of this technique at low doses.

5. CONCLUSION

The microprobe source has been successfully installed at Gray Laboratory and it is now running routinely. Measurements of the X-ray output using a thin window gas flow proportional counter are in good agreement with the theoretical data and indicate that 10 kV electrons can generate carbon-K X-rays at a rate of about 800 photons s^{-1} through a 100 nm thick silicon nitride window. Under the same conditions the bremsstrahlung count rate is approximately 12 10^3 counts s^{-1}. The silica mirror has also been tested confirming that it removes the bremsstrahlung component, while reflecting a significant fraction of the characteristic X-ray beam. Finally, a single-cell clonogenic assay has been developed such that we will be able to begin X-ray micro-irradiation experiments once the zone plate focusing arrangement is fully operational, which is anticipated in the near future.

References

1 K. M. Prise, M. Folkard, H. C. Newman and B. D. Michael, *Int. J. Radiat. Biol.,* 1994, **66,** 537
2 M. R. Raju, Y. Eisen, S. Carpenter, K. Jarrett and W. F. Harvey, *Radiat. Res.,* 1993, **133,** 289.
3 Cole, R. E. Meyn, R. Chen, P. M. Corry, and W. Hittleman, 'Radiation Biology in Cancer Research' Eds: R.E. Meyn and H.R. Withers, Raven Press, New York, 1980.
4 Nagasawa and J. B. Little, *Cancer Res.,* 1992, **52**, 6394.
5 A. Deshpande, E. H. Goodwin, S. M. Bailey, B. L. Marrone and B. E. Lehnert, *Radiat. Res.,* 1996, **145,** 260.
6 M. Folkard, K. M. Prise and B. D. Michael, *Radiat. Prot. Dosim*, 1994, **65**, 215.
7 Marples and M. C. Joiner, R*adiation Research*, 1993, **133**, 41.

Proportional counter microdosimetry

A PORTABLE DEVICE FOR MICRODOSIMETRIC MEASUREMENTS

I. Almasi[(a)], E. Anachkova[(b)], T. Bartha[(a)], K.Erdelyi[(a)], A.M. Kellerer[(b,c)] and H. Roos[(b)]

[(a)] Microvacuum Ltd. H-1147 Kerégyártó u.10, Budapest, Hungary
[(b)] Strahlenbiologisches Institut der Universität München, Schillerstrasse 42, D-80336, München, Germany
[(c)] Institut für Strahlenbiologie, GSF, Forschungszentrum für Umwelt und Gesundheit, Postfach 1129, D-85758 Oberschleißheim, Germany

1 ABSTRACT

A portable device has been developed for determination of the dose average lineal energy, y_D, and the dose equivalent, in terms of the variance-covariance method. The device includes battery operated high voltage supply, two low noise and low offset electrometers, two 20-bit resolution ADCs, and a microcontroller. The range of applications is broad, including measurements at very high dose rate, pulsed fields of different intensity, and fields with low dose rate. The device has been tested with two different TEPCs.

2 INTRODUCTION

Various measurement systems based on tissue equivalent proportional counters (TEPC) are used in radiation protection practice[1,2], two of these being commercially available[3,4]. Most of the systems measure pulses from single events, proportional to the energy imparted in the counter gas, convert the pulse height distribution to a distribution of dose in lineal energy, and evaluate dose rate, or dose equivalent rate by different approximations of the dependence of the quality factor on lineal energy.

An alternative technique for determination of the microdosimetric parameters is the variance method. In this case the electric charge proportional to the energy imparted over a specified time interval (including multiple events) is measured by an electrometer connected to the TEPC. The fluctuations of the energy imparted in the counter are used to determine the dose average lineal energy or the dose average specific energy.

While for the single event measurements the conventional and widely developed pulse height technique is applicable, the variance method requires high precision current measurements. For the variance-covariance method[5] simultaneous measurements in two detectors and two independent channels of signal processing are needed. They must contain low noise electronics with high resolution to allow the precise determination of the fluctuations of the energy deposition. Furthermore, in the measurements at high dose rate high sampling frequencies are desirable. These requirements are not readily met and - for lack of a generally applicable multi-purpose design - considerable work is usually invested in the signal processing electronics, when the variance or the variance-covariance method is applied. In spite of the efforts that are involved, the resulting instrumentations tend to be bulky and to lack versatility, a notable exception being a portable system for performing variance and variance-covariance measurements (the *Sievert* instrument) that has recently been designed by Lindborg et al[6] and was applied for measurements of the ambient dose equivalent and the average quality factor on board aircraft.

In view of the potential of the variance-covariance technique it was felt desirable to create a suitable multi-purpose device for microdosimetric measurements in terms of the variance-covariance method. The resulting design that is here reported is portable and fully battery operated. It provides simple communication with computers via standard RS 232 interface, and it equally provides the possibility to connect different TEPCs or ionisation chambers. The large range of sampling frequencies permits broad applicability, including measurements at very high dose rate, pulsed fields of different intensities, or fields with low dose rate. The evaluation program can be readily modified and adapted for special applications.

3 INSTRUMENT DESIGN

The operational amplifier AD549 is applied as low-noise switched electrometer with guarded input lines. The offset voltage of each channel can be adjusted by a potentiometer. Each electrometer converts the input current to an output voltage by integration over a high precision (270 pF ±1%) integration capacitor. Two fully symmetrical measuring electrometer units are built into a separately guarded box.

Each electrometer output is connected directly to an analog-to-digital converter DDC101 (Burr Brown). The DDC101 units are programmed in CDS (Correlated Double Sampling Mode) which allows to compensate for internal errors related to factors such as steady state, charge injections, and thermal noise. The DDC101s operate in integrating unipolar mode with 20-bit resolution: the sampling time is programmed by the control unit. The resolution of the voltage measurements is 5.96 μV.

The controlling unit includes 80C537 Microcontroller, 32 kbyte EPROM, 32 kbyte RAM, 8 kbyte EEPROM (Electrically Erasable Programmable Read - Only Memory), an 8 character LCD display and four control buttons. The evaluation algorithm is put on the EPROM. The operating parameters (sampling frequency, high voltage, number of samples) and the detector parameters (gas multiplication factor for each detector, mean chord length, and air-kerma calibration factor for photons) are put on the EEPROM. The configuration can be readily read or can be promptly changed via the RS 232 interface. The electrometer readings are stored on the RAM (maximum 2000 in each channel) and can be read out via the RS 232 port. The operating parameters need to be set through the control buttons before each measurement.

The following operation parameters can be chosen through the control unit :

- sampling frequency: 2, 10, 100, 1000, 4000 Hz. The additional frequencies 23, 57, 230, and 568 Hz are implemented in a second variant of the instrument.
- high voltage: from 200 V to 1300 V in steps of 50 V
- number of sampling intervals: 500, 1000, and 2000.

The second variant provides the possibility of repeated measurement, the number of cycles (1 to 63) being chosen from the control panel. This allows better statistics in pulsed radiation fields.

The quality factor, Q, is approximated by the relation [6]

$$Q = a + b\, y_D \quad \text{with } a = 0.8 \text{ and } b = 0.17\ \mu\text{m/keV}.$$

This relation holds below 150 keV/μm, beyond this value it provides an overestimate of Q on the 'safe' side.

At the end of each measurement the following information can be read off the display:

- dose average lineal energy, y_D (in keV/μm), obtained as the mean from the two detectors
- dose equivalent rate (in μSv/h)
- the mean value of the signal in each channel (in relative units)
- the relative variance of the signals in each channel
- the relative covariance of the signals from the two detectors
- the number of sampling pairs actually used in the calculation

The last four readings facilitate the optimisation of the operation parameters. There is also a possibility to check the electrometer readings for each sampling interval.

4 TEST MEASUREMENTS

The instrument has been tested with cylindrical tissue equivalent proportional twin-counter[7] and a ^{137}Cs-source. Although the geometrical shape and dimensions of the two single detectors were identical, the gas multiplication was seen to differ, at the highest possible voltage, up to 30 %; but this difference did remain constant during the measurements. The gas multiplication was determined with the ^{37}Ar calibration technique[8] and with a built-in Am α-source[7]. The tests of the new device were mainly concerned with its function under different operating conditions and with its range of applicability. The dose average lineal energy, y_D, was determined in measurement series with different numbers of sampling pairs, N, and at different sampling frequencies between 2 Hz and 4000 Hz. Figure 1 illustrates the measurements at a dose rate of about 200 μGy/h (simulated dose rate : about 6 Gy/s).

Figure 1 *Examples of detector signals, U, (left panel), signal fluctuations at different sampling frequencies, i.e. the same values rescaled to their mean and on a log-scale (middle panel), and detector noise, i.e. signals with the radiation field switched off (right panel).*

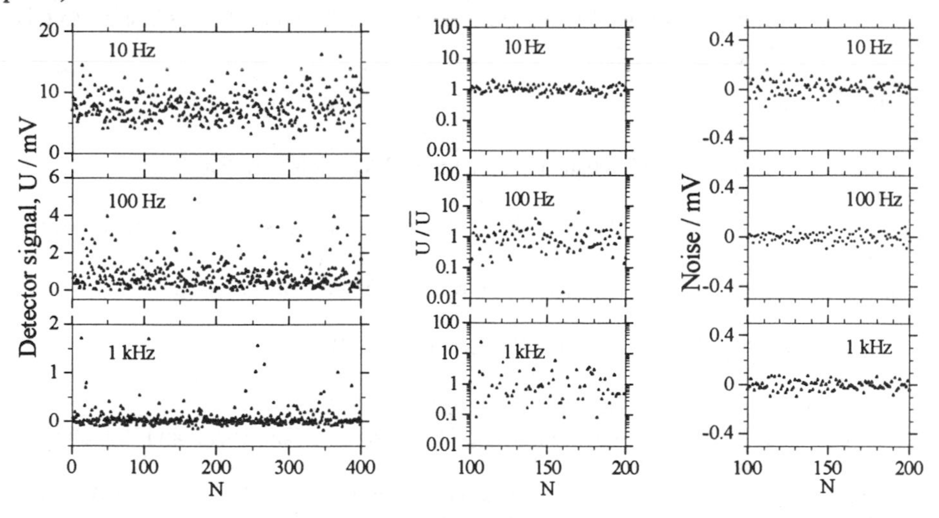

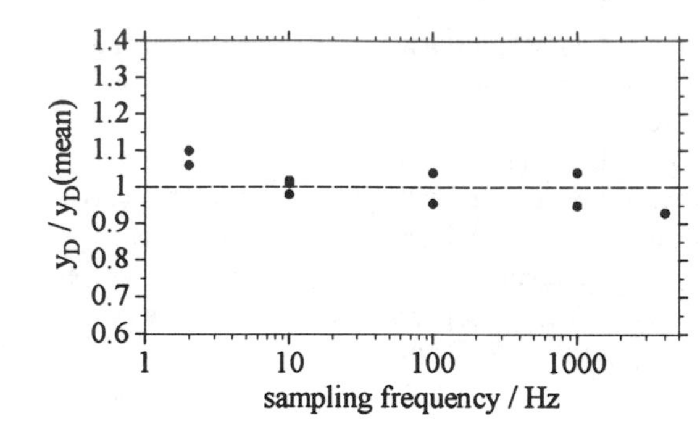

Figure 2 *A set of measured values, y_D, of the dose average lineal energy at different sampling frequencies. The number of sampling intervals, N, was 2000 in all cases.*

Measured values of y_D are represented in Figure 2 as ratio of the y_D determined at specified frequency to the average value y_D *(mean)* from all measurements shown in the picture. At 4 kHz it was necessary to compute y_D with higher numerical precision. In all other cases there was full agreement between the calculations in the portable device and a more precise analysis.

The new algorithm, proposed recently by Kellerer[9] for the case of changing dose-rate ratio in two detectors, was also applied. The results were generally identical, as is to be expected in a constant radiation field. But even under these simple conditions the new algorithm was found to provide more stable results in the presence of strong electronic noise. The implementation of the new algorithm and an increase of accuracy in the numerical calculation will be the next steps in the development of the device.

ACKNOWLEDGEMENTS

This research was supported by EURATOM contract FI3P-CT 920039 and by PHARE TDQM contract HU-9305-02/10008; it is part of the cooperation within EURADOS Working Group 10. We are grateful to Dr. L. Lindborg for his interest to this work.

References

1. L. Lindborg, D. Bartlett, H. Klein, Th. Schmitz and M. Tichy, Radiat. Prot. Dosim.,1995, **61**, 89.
2. H. G. Menzel, L. Lindborg, Th. Schmitz, H. Schumacher and A. Waker, Radiat. Prot. Dosim., 1989, **29**(1-2), 55.
3. A. W. Kunz, P. Pihet, E. Arendt and H. G. Menzel, Nucl. Instrum. Methods, 1990, A**299**, 696.
4. A. Aroua, M. Höfert and A. V. Sannikov, Radiat. Prot. Dosim., 1995. Radiat. Prot. Dosim., 1995, **59**, 49.
5. A. M. Kellerer and H. H. Rossi, Radiat. Res., 1984, **97**, 237.
6. L. Lindborg, J. E. Grindborg, O. Gulleberg, U. Nilsson, G. Samuelson and P. Uotila , Radiat. Prot. Dosim., 1995, **61**(1-3), 119.
7. J. Chen, J. Breckow, H. Roos and A.M. Kellerer, Radiat. Prot. Dosim.,1990, **31**, 171.
8. E. Anachkova, A. M. Kellerer and H. Roos, Radiat. Environ. Biophys.,1994, **33**, 353.
9. A. M. Kellerer, Radiat. Environ. Biophys., 1996, **35**, 117.

USE OF MICRODOSIMETRY FOR MIXED-FIELD RADIATION MEASUREMENTS AT THE AFRRI TRIGA REACTOR

H. M. Gerstenberg, R. C. Bhatt, B. A. Torres, and K. D. Bolds

Armed Forces Radiobiology Research Institute
8901 Wisconsin Avenue
Bethesda MD 20889-5603 USA

1 INTRODUCTION

Microdosimetry has been used at the Armed Forces Radiobiology Research Institute (AFRRI) for comparing radiation fields between different radiation facilities[1,2] as well as a tool to characterize, in terms of energy deposition, both the neutron and photon components of mixed radiation fields available for conducting radiobiology experiments at the AFRRI TRIGA reactor. The reactor core resides in, and is movable through, a water pool such that different thicknesses of water may be interspersed between the core and the biological samples located in an adjoining exposure room. Changes of core position in the pool and experimental conditions in the exposure room were used to vary the proportion of neutron and gamma radiation doses received by the target. These field characteristics have been measured for both standard (normally used) and nonstandard positions. The results have been compared with those using paired-ionization chamber methods[3] for mixed-field neutron/gamma radiation dosimetry.

2 METHODS

In this work, four mixed neutron/gamma free-in-air radiation fields were selected from among a variety obtainable by varying reactor and measurement parameters. These fields had been used to investigate the $LD_{50/30}$-based relative biological effectiveness (RBE) values for AFRRI studies with mice. Historically, the fields were initially labeled 10:1, 1:1, 1:3, and 1:20 to indicate their "expected" free-in-air neutron/gamma dose ratios. Table 1 gives the shieldings, as well as the experimental distances from the reactor core center, which were used to approximate the stated nominal fields.

2.1 Paired Ionization Chamber Measurements

From measured and calculated spectral measurements[4], NIST (National Institute of Science and Technology) compiled an AFRRI Neutron Spectrum Directory[5]. The complete neutron and gamma spectral information available from this directory for the nominal field 10:1 was used in the paired-chamber method for the evaluation[3] of neutron dose (D_n) and gamma dose (D_γ) components of the total dose (D_T). However, such spectral information was lacking for the other three nominal fields. Therefore, as a first approximation, it was assumed that the nominal fields 1:1, 1:3, and 1:20 had neutron and gamma radiation spectra similar to that of field 10:1. The divergence from this approximation was examined by comparing the neutron- to total-dose ratios from the paired-chamber method with those obtained independently by the microdosimetric method employing a tissue-equivalent proportional counter (TEPC). This comparison is important because, unlike the paired-chamber method, the TEPC method does not require advance information on neutron and gamma radiation spectra in the measurement of neutron-to total-dose ratios. In fact, the two methods play complementary roles in mixed-field dosimetry. The total dose-rate can be accurately and conveniently measured by the paired chamber method in the range of 1 to 10^3 cGy·min^{-1}. This is possible because accuracy of measured total dose-rate is not a strong function of the accuracy of the spectral data. On the other hand, the TEPC method — valid only for single-event energy depositions and therefore limited in dose rate (< 1. cGy·min^{-1} in this experiment) — can accurately estimate neutron-to gamma-dose ratios without *a-priory* spectral information and thus complements the paired-chamber method.

Table 1 *Shieldings and Distances Used for the Experiment*

Nominal Field	*Water Shield* cm	*Lead Shield* cm	*Distance from Core Center*, cm
10:1	2.5	15	100
1:1	2.5	15	285
1:3	11.4	15	285
1:20	16.5	15	285

2.2 Microdosimetry Measurements

For the four nominal mixed fields the microdosimetric distributions were measured using a TEPC chamber at the paired-chamber locations in the reactor exposure room with shielding and experiment conditions enumerated in Table 1. Figures 1 to 4 give differential dose distributions, y·d(y), in lineal energy, y, for the four nominal fields. The data point shown in each figure in the region of 0.1 keV·μm^{-1}, is at the limit of the experimental threshold.

The shape of the distribution between 0.1-10. keV·μm^{-1} is thought to be caused by both low-energy photons from neutron-capture gamma-rays in the exposure room walls as well as by higher-energy photons from the reactor core. In the distribution, the low-energy photons contribute a lineal energy between 1-10. keV·μm^{-1}, while the high-energy-photons contribute in the region between 0.1-1.0 keV·μm^{-1}. This is consistent with recent AFRRI y spectra measurements of 250-kVp x-rays[6] and ^{60}Co γ-rays[7], which have a dose-mean lineal energy, $\bar{y}_D$, of 4. and 1.94 keV·μm^{-1} respectively. Reference 7 also reports, for the mixed reactor radiation field, a $\bar{y}_D$, of 63 keV·μm^{-1} for a neutron-dose fractions, D_n/D_T, of 0.96.

Although more rigorous methods are available[8] to estimate the neutron and gamma radiation dose fractions of the total dose , we took advantage of the symmetrical shapes of the dose distribution curve in Figures 2 to 4. For all the curves, as a first approximation, areas under the curves above and below y = 15 keV·μm^{-1} respectively were assigned to the neutron- and gamma-dose fractions.

3 RESULTS

The D_n/D_T values obtained by using TEPC and paired chambers are compared in Table 2.

Table 2 *Comparison of Measured Values of D_n/D_T by TEPC and Paired Chambers*

Nominal Field	Paired Chamber	TEPC
10:1	0.90	0.89
1:1	0.67	0.56
1:3	0.46	0.40
1:20	0.35	0.31

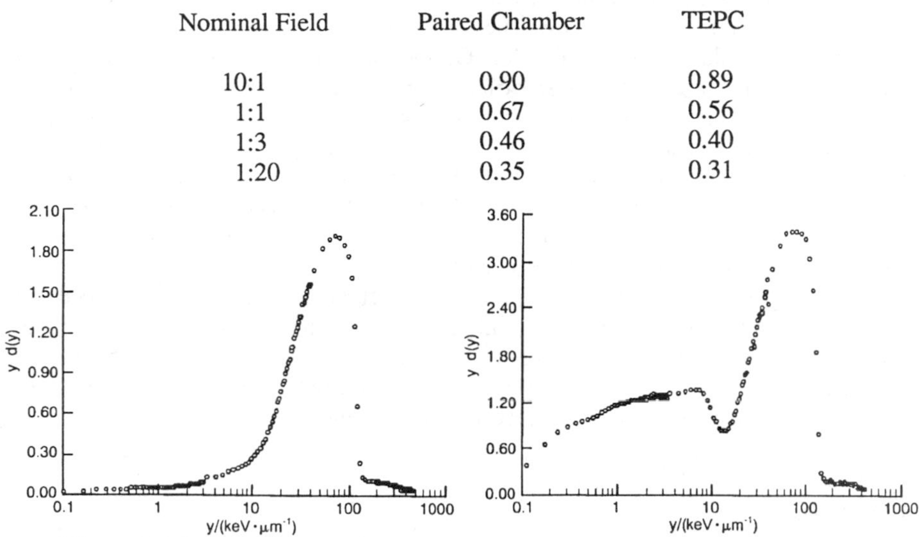

***Figure 1** Measured microdosimetric dose distribution for the 10:1 nominal field with conditions listed in Table 1*

***Figure 2** Same as Figure 1 but for the 1:1 nominal field*

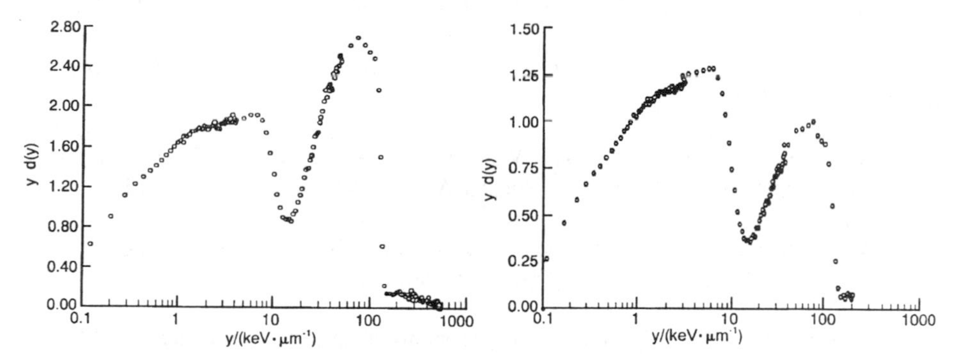

***Figure 3** Same as Figure 1 but for the 1:3 nominal field*

***Figure 4** Same as Figure 1 but for the 1:20 nominal field*

Note that whereas the dose distribution with respect to type and energy of radiation in the mixed field was measured by TEPC without making any assumptions about the spectra, evaluation of neutron and gamma dose fractions using paired chambers required assumptions for all but the 10:1 field for which the spectral data were available from the earlier work. Table 2 shows a good match of D_n/D_T values obtained by the two methods for the nominal field 10:1. However, a wide disparity is present in the values for the other three nominal fields.

4 DISCUSSION

The paired-chamber D_n/D_T values depend on accurate neutron and gamma radiation spectra, and clearly the simple assumption made in Section 2.1 about the unknown spectral data are not valid. Since the AFRRI Neutron Spectrum Directory does not include any spectral measurement made at 285 cm from core center, it is clear that the paired-chamber method cannot be used to furnish accurate estimates of the neutron and gamma dose components for the nominal fields 1:1, 1:3 and 1:20. On the other hand, the TEPC chambers are specifically designed to generate the spectrum information in terms of y distributions. As a consequence, the method of microdosimetry is obviously better suited in making more reliable estimates of the neutron- and gamma-dose fractions, D_n/D_T and D_γ/D_T respectively, in a mixed field. This useful characteristic of TEPC chambers can supplement the accurate information on dose rates obtained using paired chambers. The paired chambers will continue to be useful in the measurement of wide range of total dose mainly because they measure higher-dose-rates than the TEPC detector.

If meaningful comparisons of radiobiology measurements are to be made between different reactor facilities, then detailed information must be known about the radiation field at the point of measurement[1]. It is not enough to quote average values of the neutron and photon spectra. Extracting spectroscopic information from microdosimetric measurements is an area of work that deserves more attention[9].

References

1. H. M. Gerstenberg, 'Proc. 9th ICRR', J. D. Chapman et al, Ed, Academic Press, London, 1991, Vol. 1, p. 110.

2. H. M. Gerstenberg, 'Abst. 11th Symp. On Microd., Gatlinburg, Tennessee, 1992, p. 37.

3. 'Ionizing Radiation: Protection and Dosimetry', Guy Paić, Ed, CRC Press, Boca Raton, Florida, 1988.

4. V. V. Verbinski, C. G. Cassapakis, W. K. Hagan, K. Ferlic, and E. Daxon, Report: DNA 5793F-1 and 2, Defense Nuclear Agency, Washington, DC, 1981.

5. C. M. Eisenhauer, 'AFRRI Neutron Spectrum Directory', National Institute of Standards and Technology, Gaithersburg, Maryland, 1991.

6. P. G. S. Prasanna, C. J. Kolanko, H. M. Gerstenberg, and W. F. Blakely, *Health Phys.*,**1997** (in press).

7. A. A. Stankus, M. A.Xapsos, C. J. Kolanko, H. M. Gerstenberg, and W.F. Blakely, *Int. J. Radiat. Biol.*, 1995, **68**, 1.

8. H. G. Menzel and H. Schuhmacher, 'Ion Chambers for Neutron Dosimetry', J. J. Broerse, Ed, EUR 6782, Brussels and Luxembourg, CEC, 1980, p. 337.

9. A. J. Waker, *Radiat. Prot. Dos.*,1995, **61,** 297.

PAIRED TEPCS FOR VARIANCE MEASUREMENTS

J. E. Kyllönen, L. Lindborg and G. Samuelson

Swedish Radiation Protection Institute
S-171 16 Stockholm, Sweden

1 INTRODUCTION

In this paper a portable instrument for radiation protection dose measurements in combined low- and high-LET fields is presented. The instrument called the Sievert-instrument[1] has been modified to fit the need for measurements of cosmic radiation on board aircraft. The new instrument is based on two tissue equivalent proportional counters and the variance technique.[2] Basic features of the instrument as well as results from measurements in a high energy stray neutron field at CERN and from a flight between Geneva and Stockholm in 1996, are described.

2 DESCRIPTION OF THE INSTRUMENT

In radiation protection the quantities of interest are the absorbed dose D and the quality factor Q.[3] The dose is proportional to the electric charge and the quality factor may approximately be determined from the dose mean value, $\overline{y}_D$, of the lineal energy.[4] In the variance method $\overline{y}_D$ is determined from the variance and the expectation value of energy imparted in a microscopic site.[5] When variations occur in the radiation field, the beam variation is accounted for by correcting for the covariance between two detector signals according to the variance-covariance method.[2] Further, an algorithm for a generalised variance method can be applied in situations when the variations in the radiation source affect the two detectors differently.[6] The instrument can present results according to any of the three methods.

At common flight altitudes the dose equivalent rates are typically a few μSv/h.[7] The new instrument had to be smaller than the existing spherical detector due to the limited space on board aircraft, but should still be able to measure the low rates at pressures simulating diameters of 2 – 3 μm. By choosing a cylindrical shape the detector volume was increased by 50 % as compared to a sphere of the same diameter. The height of the cylinder was chosen equal to the diameter since the variance from the chord length distribution was then minimised. The variance in the chord length distribution is though generally small compared to the variance due to straggling.[8] A more influential variance source can be a non-uniform electric field in the detector. This effect is reduced by field adjusting tubes at a potential corresponding to the potential at that distance from the detector wall.[9]

Table 1 *Technical data*

Sensitive volume diameter	115.4	mm
Sensitive volume height	115.4	mm
Wall thickness	5	mm
Central wire diameter	100	µm
Field tube diameter	24.6	mm
Field tube length	12.3	mm
Pressure	2.38	kPa
Simulated mean chord length	1.92	µm
Vacuum container diameter	150	mm
Vacuum container height	221	mm
Weight of one detector	2.4	kg
Weight of complete equipment	10.1	kg

The two detectors are made of the tissue equivalent plastic A-150 and contained in vacuum containers of 2 mm aluminium. Experience shows that for polarising voltages of 1000 – 1500 V and gas pressures of about 2 kPa, an electrical break-down safety distance of 10 mm is necessary between the detector wall and the vacuum container. The containers are filled with a methane based tissue equivalent gas. Before filling them up the detectors were evacuated at 60 – 80 °C for 10 days. The long term stability of the instrument is under investigation.

The central electrodes are connected to low-leakage current-to-frequency converters and the resulting pulses are counted under the control of a microprocessor. The converters are contained in boxes of size 101 mm × 50 mm × 46 mm, attached to the vacuum containers. A new electronics design allows integration times down to 0.1 s.

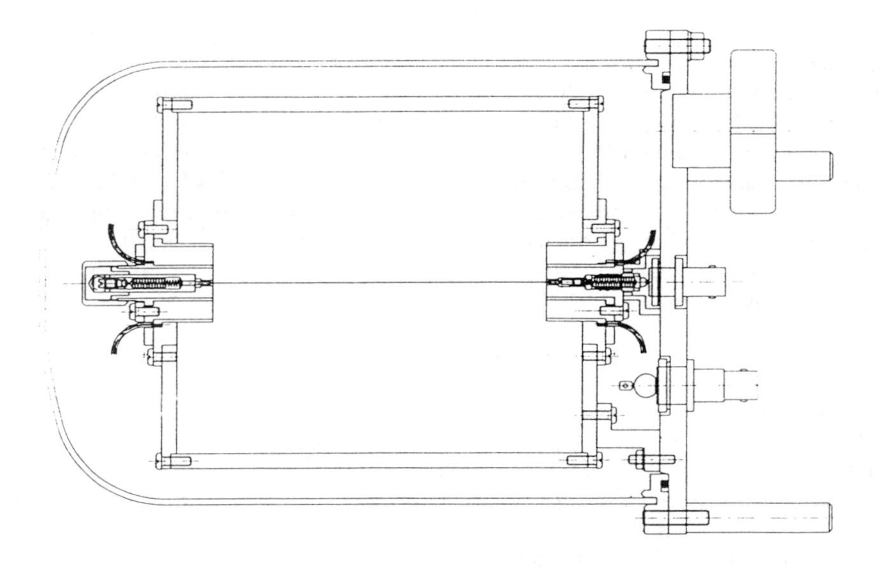

Figure 1 *Cross-section of the new detector construction.*

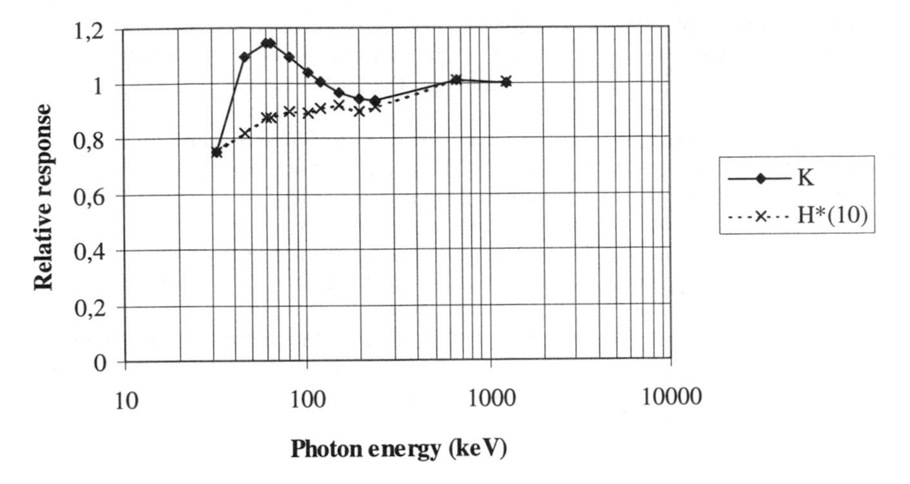

Figure 2 *Relative kerma K and ambient dose equivalent H*(10) response.*

A complete measurement equipment includes two detectors with electrometers, a GM-counter, two portable battery powered high voltage supplies and a counting unit. It is also possible to connect a computer for extended data analysis. Basic technical data are presented in Table 1 and a cross-section of the detector construction is shown in Figure 1. For a more detailed discussion concerning the proportional counter construction, see reference 10.

3 EXPERIMENTAL RESULTS

The photon energy dependence of the instrument was studied in Am-241, Cs-137, Co-60 and X-ray fields. From Figure 2 it can be concluded that the relative kerma and ambient dose equivalent responses are within ± 20 % for photon energies between 40 keV and 1.3 MeV. All data are normalised to the Co-60 result.

Both the cylindrical and the spherical instrument versions were used for measurements in a stray neutron field at CERN. The neutron energy distribution of this field is known to be similar to that found on board aircraft at normal cruising altitudes.[11] The results, with $\bar{y}_D$ calculated according to all three methods, are presented in Table 2.

Table 2 $\bar{y}_D$ *calculated according to the variance (V), the variance-covariance (VC) and the generalised variance-covariance (GVC) method. Values are in keV/μm and the uncertainty corresponds to a 95 % confidence interval.*

	instrument	$\bar{y}_D$ (V)	$\bar{y}_D$ (VC)	$\bar{y}_D$ (GVC)
CERN	cyl	34 ± 6	17 ± 4	24 ± 18
	sph (4.4 μm)	38 ± 4	16 ± 3	17 ± 14
	(1.9 μm)	39 ± 4	14 ± 3	28 ± 15
Geneva-Stockholm	cyl (10 km)	8.4 ± 2.8	8.4 ± 2.8	8.6 ± 3.3
	sph (11 km)	7.8 ± 1.0	-	-

It is concluded that a covariance correction was necessary in the pulsed accelerator beam and that the variance-covariance results for the two detector versions do not differ significantly. Considering the greater uncertainties, the results calculated according to the generalised variance method also agree.

Results from measurements on a flight from Geneva (46° N) to Stockholm (59° N) on July 31st 1996, are also presented in Table 2. These are compared with results from measurements performed three years earlier along the same flight route with the spheres.[1] Both investigations indicate that the covariance correction has no effect on these altitudes and latitudes. It is also seen that the generalised variance method, correcting for radiation fields varying differently across the two detectors, has no effect. The greater uncertainties with this method observed at CERN are not seen here.

One safety aspect in aircraft measurements that needs further consideration is how to safely secure the instrument. It will also be investigated if the levels of electro-magnetic radiation emitted from the instrument are within the safety requirements.[12]

4 CONCLUSIONS

The results from CERN and the flight measurements are in agreement for the spherical and the new cylindrical detector versions. From the aircraft measurements it is noted that the new instrument is rigid enough to handle the mechanical stress on board and that the instrument only occupies one passenger seat when operated manually. The three methods (the variance, the variance-covariance and the generalised variance method) gave very similar $\overline{y}_D$ values and uncertainties for the aircraft measurements. At CERN the generalised variance method yielded much larger uncertainties than the variance-covariance method, which is not completely understood.

References

1. L. Lindborg, J. E. Grindborg, O. Gullberg, U. Nilsson, G. Samuelson and P. Uotila, *Radiation Protection Dosimetry,* 1995, **61**, no. 1-3, 119.
2. A. M. Kellerer and H. H. Rossi, *Radiation Research*, 1984, **97**, 237.
3. ICRP Publication 60, Pergamon Press, Oxford, 1990, **21,** no. 1-3.
4. ICRU Report 36, Bethesda, Maryland, 1983.
5. L. G. Bengtsson, in 'Proc. of 2nd Symp. on Microdosimetry', Report No. EUR 4452, H.G. Ebert, Ed., 1970, Commission of the European Communities, Brussels.
6. A. M. Kellerer, *Radiat. Environ. Biophys.*, 1996, **35**, 111.
7. G. Reitz, *Radiation Protection Dosimetry*, 1993, **48**, no1, 5.
8. A. M. Kellerer, 'The Dosimetry of Ionising Radiation', K. R. Kase, B. E. Bjärngard and F. H. Attix, Academic Press, 1985, Vol. 1, Chapter 2, p. 77.
9. A. L. Cockroft and S. C. Curran, *Review of Scientific Instruments,* 1951, **22**, 37.
10. J. E. Kyllönen, M.Sc.Thesis, Dep. of Technology at Uppsala University, 1994.
11. S. Roessler and G. R. Stevensson, CAN/TIS-RP/IR/93-47, 1993.
12. EUROCAE ED-14C/RTCA DO160C

COMPARATIVE STUDY OF TE-GASES AND DME IN A PROPORTIONAL COUNTER

I. Krajcar Bronić [1,2] and B. Grosswendt [2]

[1] Rudjer Bošković Institute, P.O.Box 1016, HR-10001 Zagreb, CROATIA
[2] Physikalisch-Technische Bundesanstalt, Bundesallee 100, D-38116 Braunschweig, GERMANY

1 INTRODUCTION

Detection of ionising radiation by a proportional counter consists of two independent stochastic processes: (i) primary ionisation and (ii) electron multiplication. Interaction of ionising radiation with the counter gas (primary ionisation) is dependent on the kind and energy of the incident particle and on the type of the gas. The distribution of the number of primary ion pairs can be characterised by its mean value and the variance. Usually, the mean energy required to form an ion pair (**W**) and the Fano factor (**F**) are used, instead.[1]

Electron multiplication does not depend on the incident particles (unless high dose rates are applied), but depends on the type of the gas and its pressure, as well as on the counter geometry and the applied high voltage. The principal physical quantity that determines electron multiplication is the ionisation coefficient. Electron multiplication can be characterised by the mean gas amplification factor (**M**), and by the so-called single electron (SE) distribution. The SE spectrum represents the distribution of the sizes of avalanches triggered by a single primary electron. It is usually described by its relative variance (**f**).

The energy resolution of a proportional counter depends on gas gain and is a function of all above mentioned quantities. Its theoretical lower limit is given by $\mathbf{R}^2 \sim \mathbf{W}(\mathbf{F}+\mathbf{f})/\mathbf{T}_0$ ($\mathbf{T}_0$ is the initial particle energy). In order to completely understand the signal formation and the line shape measured by a proportional counter, it is therefore necessary to know accurately the **W** and **F** values for a given gas or gas mixture, and to know the gas gain **M** and the statistical fluctuations **f** of the gas amplification.

We performed a comparative study of propane, propane-based TE gas (p-TE; 39.6% CO_2, 55% C_3H_8, 5.4% N_2)[2], isobutane, isobutane-based TE gas (ib-TE; 42.3% CO_2, 51.4% i-C_4H_{10}, 6.3% N_2)[3] and DME (dimethyl ether) in a proportional counter. (It has recently been proposed that DME mixed with a small amount of N_2 may be used as a TE gas.[4]) The mean value **M** and the statistical fluctuations **f** of the gas amplification, the energy resolution **R** and the **W** value were measured. From the measured gas gain data, the first Townsend ionisation coefficient was determined. More detailed analysis of the mean gas gain **M** and the results concerning the ionisation coefficient are presented in ref. 5.

2 EXPERIMENT

A cylindrical stainless steel proportional counter of cathode radius **b** = 25 mm, specially designed for low-energy X-ray measurements, was used.[6-9] The anode radius (**a**) was 12.5 μm in the first part of measurements, and 15 μm in the second part. The gas pressure **p** was adjusted so that several values of the product **p**×**a** were the same in both series. Propane, isobutane and DME of purity >99.95% were used. The gases, including p-TE and ib-TE mixtures, were used without special purification.

3 RESULTS

3.1 Gas Amplification

For each gas and each pressure the mean gas amplification factor **M** was measured as a function of the applied high voltage **V**. The gas gain in all gases is an exponential function of the applied voltage over the whole investigated gain range ($10 - 2.5\times10^4$). No over-exponential increase at high voltages is observed, as opposed to argon-isobutane mixtures with low isobutane content.[8] When a group of gases with the same value of **p**×**a** is concerned, the highest gas gain for a fixed voltage is obtained with DME, followed by isobutane, ib-TE, propane, and p-TE. Curves representing the reduced gas gain, $\ln\mathbf{M}/(\mathbf{paS_a})$, as a function of the reduced electric field strength at the anode surface, $\mathbf{S_a} = \mathbf{V}/[\mathbf{p}\ \mathbf{a}\ \ln(\mathbf{b}/\mathbf{a})]$ for different pressures for each gas should form a single line if the electrons attain the equilibrium with the electric field. However, the reduced gas gain is lower for lower gas pressure at the same $\mathbf{S_a}$,[5] indicating that the equilibrium is not reached.[10] Deviations from the equilibrium conditions are observed for $\mathbf{S_a} > 3\times10^5\ \mathrm{Vm^{-1}\ kPa^{-1}}$, and increase with the decrease of the gas pressure.

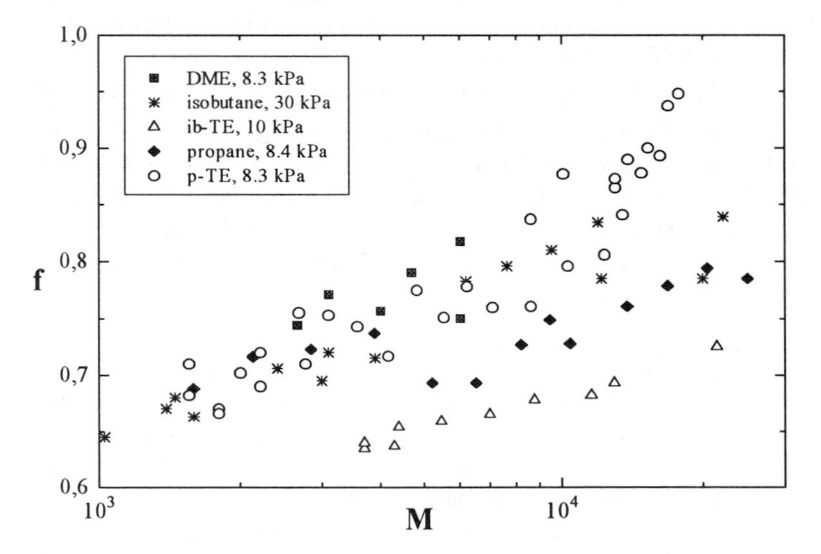

Figure 1 *The relative variance of SE spectra as a function of the mean gas amplification factor*

3.2 Single-Electron Spectra

The SE spectra were fitted by a Polya distribution of the form $\mathbf{P(x) = A\, x^B} \exp(-\mathbf{Cx})$, where $\mathbf{x}$ is a channel number, and **A**, **B**, and **C** are constants obtained by the fitting procedure. The relative variance of the SE spectra is then calculated as $\mathbf{f}=(1+\mathbf{B})^{-1}$ and is shown in Figure 1 as a function of the gas gain for each of the investigated gases at a single pressure. The most obvious feature of **f** in all gases is its increase with **M**. However, even at the highest gains (2.5×10^4), the value of **f** is always below 1, meaning that the SE spectra remain of the peaked shape. In contrast, in argon-isobutane mixtures[8] with relatively low isobutane partial pressure, **f** reached values ≥1 (exponential or hyperbolic shape of the SE distribution) due to insufficient quenching and avalanche chain formation. The present **f** values may be compared with the single **f** values at low gain (**M**~2000) in propane (**f** = 0.64), p-TE (0.69) and butane (0.66).[6,11] The **f** value is found to be very sensitive to the contamination of the counting gas due to the outgassing. This effect was specially conspicuous when DME was used as the counting gas.

3.3 W Value, Energy Resolution, Fano Factor

The mean energy required to form an ion pair (**W**) was determined for each pair of SE and Fe spectra taken at the same gas gain as $\mathbf{W}=\mathbf{T_0}(\mathbf{x_{SE}}\,\mathbf{G_{Fe}})/(\mathbf{x_{Fe}}\,\mathbf{G_{SE}})$, where $\mathbf{x_{SE}}$ is the mean value of the SE spectrum calculated as $\mathbf{x_{SE}}=(1+\mathbf{B})/\mathbf{C}$, $\mathbf{x_{Fe}}$ is the mean value of the Fe spectrum, and **G** are the corresponding electronic gains. The value reported here (Table 1) is obtained as the mean value of several measurements performed at different gas gains (below ~8000) and at different gas pressures. At higher gas gains, the influence of the space charge was observed as an increase in the calculated **W** value, because the gas amplification for a high number of primary ion pairs (Fe) was lower than for a single electron.[9] DME is recently often used in proportional counters and multistrip gas chambers,[12] however, the only known **W** value is that of Leblanc and Herman[13] measured for β particles.

The energy resolution of the proportional counter at 5.89 keV was also measured. The best resolution was obtained in all gases at low gas gains of several hundreds.[5] The resolution degraded with the increase of the gas gain, and for the same gain it degraded with the increase in gas pressure. The instrumental contribution to the resolution was determined by measuring the resolution of the ^{244}Cm α line[16] and subtracted from the measured resolution in order to obtain the theoretical resolution.[14] Since **W** and **f** were measured, it was possible to determine also the Fano factor in these gases (Table 1).

4 CONCLUSION

The measured gas gain in various gases in a cylindrical proportional counter shows that electrons do not reach the equilibrium with the electric field at rather moderated electric field strengths. Single electron spectra in pure DME, isobutane and propane, and in TE mixtures remain of the peaked shape described by the Polya distribution function having the relative variance **f** < 1 over the whole gas gain range. The gain dependence of the SE

Table 1 *W values and the Fano factors **F** in the studied gases for 5.89 keV photons*

Gas	W (eV)	other data	F	other data
propane	25.1	24.0[1], 25.8[7]	0.22-0.26	0.25[14]
propane-based TE	27.2	27.0[14,15]	0.24-0.26	
isobutane	23.4	23.4[1,14]	0.26-0.27	0.26[14]
isobutane-based TE	25.1*		0.25*	
DME	24.6	23.9[13]		

* **W** and **F** values, measured in this experiment, have been preliminary published [16]

spectral shape, however, does not enable the use of a single **f** value for all gas gains. At relatively low gas gains (below several thousands) the value of **f** may be taken as 0.70 ± 0.05 in all gases presently studied.

W value for 5.89 keV photons in propane is higher than that recommended earlier for high energy electrons,[1] and is closer to the W value obtained by extrapolation of data of Combecher[17] and Srdoč[7] to high energy (25.6 eV).[15] The difference between the measured **W** value for p-TE, and that calculated by the appropriate mixing formula[18] is 1.5%, what is within the present experimental uncertainties (±2%). The agreement of **W** in isobutane and p-TE gas with the previously measured data is very good, and the **W** for photons in DME is presented here for the first time. In pure DME the count rate was much lower and the energy resolution worse than in other gases. DME was also much more sensitive to small contamination due to outgassing than other gases.

Work supported by the Alexander-von-Humboldt Foundation and by the Ministry of Science and Technology of the Republic of Croatia.

References

1. D. Srdoč, M. Inokuti and I. Krajcar Bronić, "Atomic and Molecular Data for Radiotherapy and Radiation Research", Ed. M. Inokuti, IAEA TECDOC 799, Vienna, 1995, Chapter 8, p.547.
2. D. Srdoč, *Radiat. Res.*, 1970, **43**, 302.
3. U.J. Schrewe, H.J. Brede and G. Dietze, *Radiat. Prot. Dosim.*, 1989, **29**, 41.
4. F. Sauli, *Radiat. Prot. Dosim.*, 1995, **61**, 29.
5. I. Krajcar Bronić and B. Grosswendt, 3rd Symposium of the Croatian Radiation Protection Association (HDZZ), Zagreb, Croatia, 1996, Proceedings, pp. xx (1996).
6. D. Srdoč, The Response of the Proportional Counter to Unit Charge, Annual Report on Research Project COO-4733-2 (Columbia Univ., New York), pp. 65-71 (1979).
7. D. Srdoč, *Nucl. Instrum. Methods,* 1973, **108**, 327.
8. I. Krajcar Bronić, *Radiat. Prot. Dosim.*, 1995, **61**, 263.
9. I. Krajcar Bronić and B. Grosswendt, *Nucl. Instrum. Methods in Phys. Res.*, 1996, **B 117**, 5.
10. P. Segur, P. Olko and P. Colautti, *Radiat. Prot. Dosim.*, 1995, **61**, 323.
11. D. Srdoč and I. Krajcar Bronić, Symposium of the Physics of Ionized Gases (SPIG), Šibenik, Croatia, 1986, Contributed Papers, pp.131 (1986).
12. F. Hartjes, *Nucl. Instrum. Methods in Phys. Res.*, 1995, **A 368**, 249.
13. R.M. Leblanc and J.A. Herman, *J. Chim. Phys.*, 1966, **63**, 1055.
14. D. Srdoč, B. Obelić and I. Krajcar Bronić, *J. Phys. B: At. Mol. Phys.*, 1987, **20**, 4473.
15. I. Krajcar Bronić, *Radiat. Prot. Dosim.*, 1996, **xx**, in press.
16. I. Krajcar Bronić, *Nucl. Instrum. Methods in Phys. Res.*, 1994, **B 84**, 300.
17. D. Combecher, *Radiat. Res.*, 1980, **84**, 189.
18. I. Krajcar Bronić and D. Srdoč, *Radiat. Res.*, 1994, **137**, 18.

MONTE CARLO SIMULATIONS OF ELECTRON MOTION IN GAS COUNTERS USED IN MICRODOSIMETRY. STRONG DEPENDENCE OF THE RESULTS ON THE ELECTRON-MOLECULE CROSS-SECTIONS.

H. Průchová[a,1] and B. Franěk[b]

[a] GSF - Forschungszentrum fur Umwelt und Gesundheit,
85758 Oberschleissheim, Germany
[b] Rutherford Appleton Laboratory, Chilton, Didcot, OX11 0QX, UK
[1] on leave from Czech Technical University, Prague, Czech Republic

1 INTRODUCTION

In our earlier publication [1] we described a Monte Carlo program that we developed to simulate, in microscopic details, electron motion and electron interactions with gas molecules in gas counters operating under any operating conditions. This is especially important for counters used in microdosimetry where the region close to the anode is characterised by extremely high values of electric field and field gradients. The program allows the study of any aspect of processes in these counters (such as an electron avalanche development) in any level of detail.

Important inputs into the program are tables of cross-sections for various electron-molecule processes for a given gas. The quality of results of simulations therefore depends on the quality of the 'cross-section set' used. In this paper, we shall be concerned only with data for CH_4. As one of the checks of our program, we have compared the published measurements of the electron drift velocity with the results of our Monte Carlo simulations and also studied the changes in the Monte Carlo results due to changes in the input cross-sections. The reason why we used the measurements of the drift velocity rather than Gas Gain, for example, is because the drift data is very reliable. We found a strong dependence of our results on the input cross-sections.

In order to make sure, that in our future simulations we use the latest (and hopefully the best quality) available cross section data, we have made an extensive search through literature and collected a 'cross-section set' (referred to as *Set II*), which we believe is the most up to date set of cross-sections for electron-molecule interactions for methane in the energy region $0.003\ eV - 700\ eV$. Our selection of the data is described in the following section. Also the results of the Monte Carlo simulations of the drift velocity using this 'cross-section set' are compared with those using an earlier set collected by J.Groh et al [2] (in the following referred to as *Set I*).

2 RESULTS

The cross-section data collected are summarized in Figure 1.

Elastic cross-section - In the region $0\ eV$ to $0.1\ eV$ we took the measurements of $\Delta\sigma_3$ by B.Schmidt et al[3]. The $\Delta\sigma_l$ is defined as $\sigma_0 - \sigma_l$, where σ_0 is the integrated elastic cross-section and

$$\sigma_l = 2\pi \int_{-1}^{1} \frac{d\sigma}{d\Omega} P_l(cos\theta) dcos\theta. \quad (1)$$

From this definition it is clear that $\Delta\sigma_l$ converges to σ_0 with increasing l. In the region 0 to $0.1\ eV$ the $\Delta\sigma_2$ is quite close to $\Delta\sigma_3$ (see B.Schmidt et all [3]) suggesting that

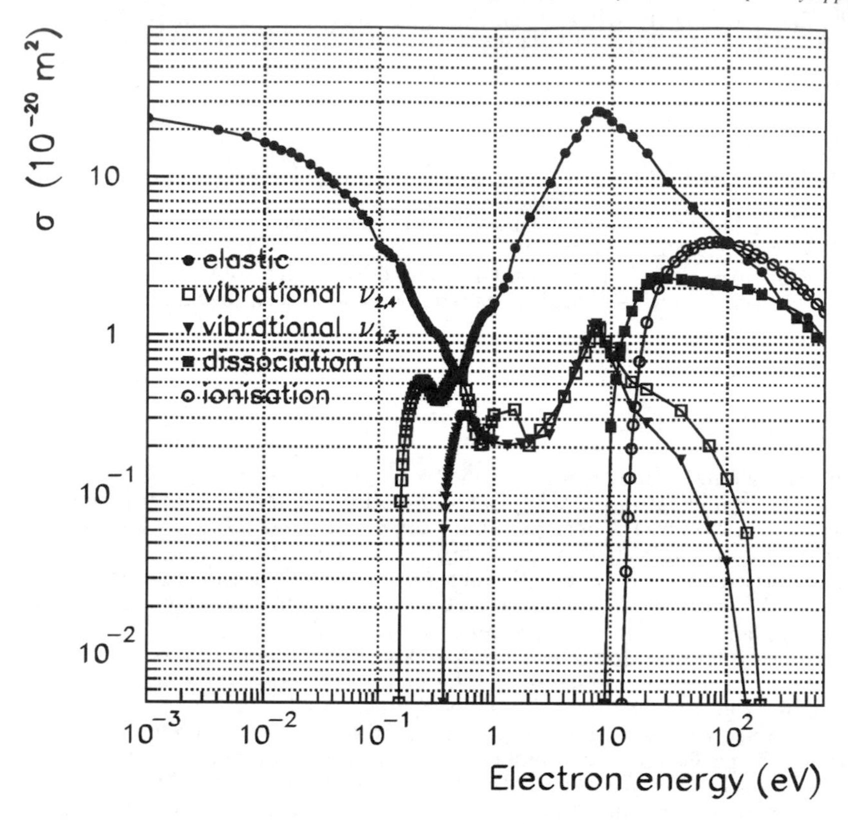

Figure 1: *Electron molecule cross-sections for methane.*

the later is a good approximation of σ_0. This however is not the case in the energy region 0.1 to 1.3 eV where there is the Ramsauer minimum. We therefore proceeded as follows: As the only inelastic processes in this region are the vibrational collisions (we are neglecting the rotational collisions whose cross-section is very small), we took the measurement of the total cross section by Lohmann et al [4] and subtracted from it the measurements of vibrational cross sections by B. Schmidt et al [5]. From 1.5 eV to 50 eV we continued with the measurements of Tanaka at al [6]. At 100 eV we took the average between the measurements by Tanaka et al and Sakae at al [7] and then continued with the measurements by Sakae at al.

Vibrational cross section - For the two vibrational modes $\nu_{2,4}$ and $\nu_{1,3}$, following measurements were taken: In the region from their respective thresholds up to 2 eV the measurements by B. Schmidt et al [5] were used. Even though there have been some recent measurements above 2 eV (e.g. [8]), the most consistent set of values still seems to be that assembled by D.K. Davies et al [9] and we used his values above 2 eV.

Total ionization cross section - This includes the direct ionization of the molecular gas (i.e. $e + CH_4 \rightarrow CH_4^+ + 2e$) and also the dissociative ionization for the production of atomic or molecular fragment ions (i.e. CH_3^+, CH_2^+, CH^+ and C^+). We considered measurements by three groups [10, 11, 12]. The measurements by Rapp and those of Nishimura are in good agreement in the overlap region. In the energy region from 17.5 eV to 700 eV, we therefore took the more recent measurements by Nishimura. From threshold (assumed to be 12.6 [9]) to 16.0 eV, measurements by Rapp were taken. In our calculation only this total ionization cross section has been used and the onset of the various above processes has been taken into account by varying the energy losses with energy.

Neutral dissociation cross-section - There are no known stable, electronically excited states of CH_4 [9]. All electronic excitations lead to dissociation where two or more fragments are produced. The number of possible neutral dissociation processes is very large. Also a large fraction of dissociation products are in their ground electronic states which are therefore difficult to detect. Thus, consideration of the cross sections for all possible channels is impractical. The total cross section for neutral dissociation, i.e. dissociation of CH_4 into neutral fragments was estimated by Davies by subtracting the total dissociative ionization cross section of Chatham et al [13] from the total dissociation cross section of Winters [14]. This procedure was subsequently improved by Biagi [15] and we took his values. To approximate the continuous energy loss spectrum, we generate a random energy loss between 9 eV (assumed to be the threshold for the neutral dissociation) and the energy e_{max} which is taken to be either the electron's energy or 18 eV whichever is smaller.

We used this new 'cross-section set' (*Set II*) in our Monte Carlo to simulate the electron drift velocity as a function of E/N. The results are shown in Figure 2 together with the measured values of the drift velocity. In the same figure also shown are the simulation results when we used the earlier set (*Set I*). As can be seen, the results of the Monte Carlo simulations as quite different for the two sets. Also, the agreement with the measured data for *Set II* is much better than for *Set I*. With the exception of the region $0 - 4.0$ Td, the agreement is, in fact, quite satisfactory. This fact suggests that the new 'cross-section set' is indeed nearer to the truth.

3 CONCLUSION

We have assembled a new 'cross-section set' for electron-molecule interactions for methane gas in the energy range 0.003 eV - 700 eV. We have simulated with our Monte Carlo program the electron drift velocity and compared the results with the measured data. We have also simulated the drift velocity using an earlier 'cross-section set' and found the differences between the two results quite large. This demonstrates the importance of availability of good measurements of these cross-sections if reliable quantitative results from Monte Carlo simulations are to be obtained. The fact, that the Monte Carlo results are very sensitive to the input cross-sections could be used in future for making judgements about the 'correctness' of cross-section data. We intend to extend these studies to making also comparisons of the measurements of transversal and longitudinal diffusion of electrons with results of the Monte Carlo.

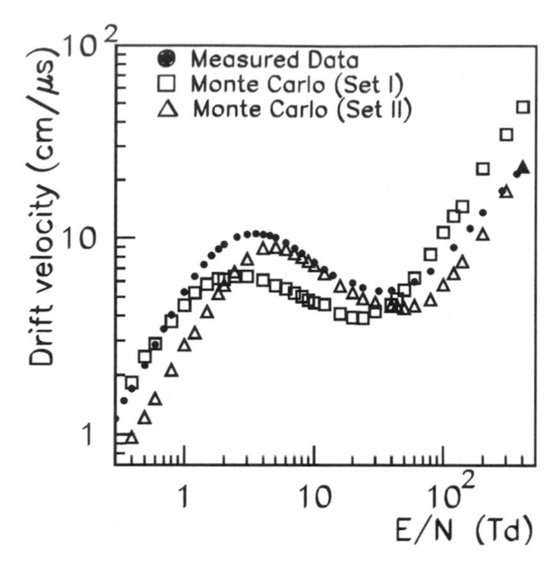

Figure 2: *Comparison of measured drift velocity with that from Monte Carlo calculations*

References

1. H. Pruchova and B. Franek, *NIM-A* , 1995, **366**, 385.
2. J. Groh, E. Schenuit and H. Spitzer, *NIM-A* , 1989, **283**, 730.
3. B. Schmidt, *J.Phys.B*, 1991, **24**, 4809.
4. B.Lohmann and S.J.Buckman, *J.Phys.B*, 1986, **19**, 2565.
5. B. Schmidt, HD-HE Swarm group, Heidelberg, June 1995; private communication of tabulated cross sections.
6. H. Tanaka and L. Boesten, *J.Phys.B*, 1991, **24**, 821.
7. T. Sakae, S. Sumiyoshi, E. Murakami, Y. Matsumoto, K. Ishibashi and A. Katase, *J.Phys.B*, 1989, **22**, 1385.
8. B. Mapstone and W.R. Newell, *J.Phys.B*, 1994, **27**, 5761.
9. D.K. Davies, L.E. Kline and W.E. Bies, *J.Appl.Phys.*, 1989, **65**, 3311.
10. H. Nishimura and H. Tawara, *J.Phys.B*, 1994, **27**, 2063.
11. O. J. Orient, S. K. Srivastava, *J.Phys.B*, 1987, **20**, 3923.
12. D. Rapp and P. Englander-Golden, *J.Chem.Phys.*, 1965, **43**(5), 1464.
13. H. Chatham, D. Hills, R. Robertson and A. Gallagher, *J.Chem.Phys.*, 1984, **81**, 1770.
14. H.F. Winters, *J.Chem.Phys.*, 1975, **63**, 3462.
15. S.F. Biagi, *NIM-A*, 1989, **283**, 716; and private communication (1995).

Nanodosimetric devices and other detectors

A NANODOSIMETER BASED ON SINGLE ION COUNTING

S. Shchemelinin, A. Breskin, R. Chechik and A. Pansky

Department of Particle Physics,
Weizmann Institute of Science,
Rehovot 76100, Israel

P. Colautti

INFN Laboratori Nazionali di Legnaro,
Via Romea, 4, I-35020 Legnaro, Italy

1 INTRODUCTION

According to microdosimetry, a biologically significant radiation damage distribution should be related to a living cell size or to that of its subsystems: $\sim 10\ nm$ for the nucleosomes and $\sim 2\ nm$ for the double-helix of the DNA. Some biological models[1] indicate that the maximum radiobiological damage is related to the coincidence of two ionization clusters, each having a size of a few nanometers and being a few tens of nanometers apart. Therefore, it is desirable to develop experimental methods enabling measurement of ionization statistics within volumes of a nanometer across as well as measurement of spatial correlations between such two separated damage sites.

There is an approach according to which a small tissue-equivalent sensitive volume is simulated by a low-pressure (a few $Torr$) tissue-equivalent gas (TEG); the number of ionizations in this volume is measured by counting the number of deposited electrons.[2] Electron diffusion limits the resolution of this method to about 10 nanometers.

We proposed a new approach[3] based on counting single deposited ions, which permits to reach significantly better spatial resolution. The development of such ion counting nanodosimeter requires the knowledge of ion diffusion parameters in TEG, which are not available. In this work a method for measuring such parameters was developed and applied to propane. The data enable us to calculate the expected performance of a nanodosimeter based on the proposed approach.

2 MEASUREMENT OF ION TRANSPORT PARAMETERS

The experimental setup built for our measurements (Figure 1 a) consists of two main sections. The first is filled with the gas under study (typically a few tenths of a $Torr$) and contains a movable ion source and a drift column with a narrow slit ($0.1 \times 10\ mm$) centered at its bottom end. This slit connects the first section to another one containing a vacuum operated microsphere plate electron multiplier (MSP)[4] used as a single ion detector. It requires ion acceleration up to $\sim 4\ keV$, which necessitates the second section to be maintained at a few $10^{-5}\ Torr$.

In the ion source, α-particles from an ^{241}Am source pass 20 mm of gas in the direction perpendicular to the figure plan, form a "linear" ion cloud (trail), and hit a PIN diode. This diode provides the "$start$" signal for measuring the ion drift time. Under

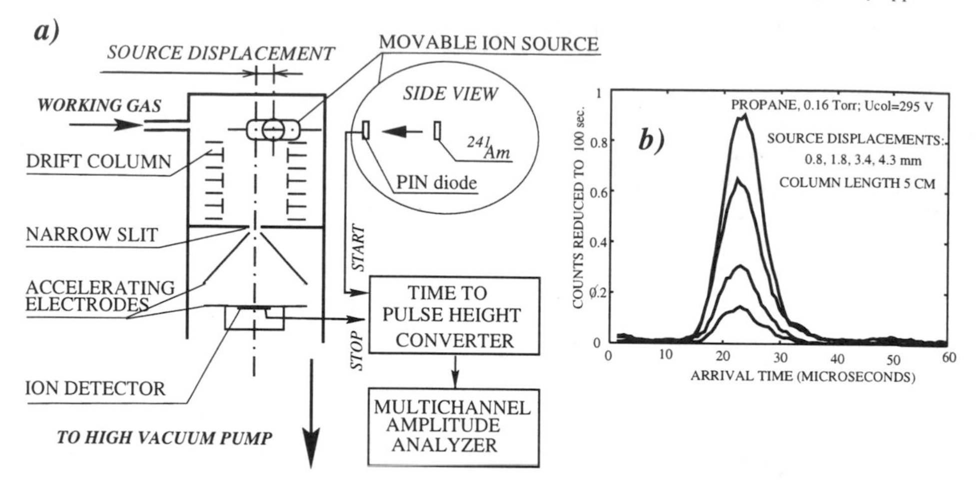

Figure 1 *Measurements of arrival time spectra and transverse diffusion*

an electric field the ion cloud drifts towards the slit, spreading in three dimensions. Some of the ions, that pass the narrow slit, are counted by the ion detector. The whole source unit can be moved perpendicular to the drift column axis. Examples of arrival time spectra corresponding to various ion source displacements are shown in Figure 1*b*. The distribution of the transverse displacement of the ions during their drift is obtained by integration of such spectra over the time.

It should be noted that in multiatomic gases (of interest for nanodosimetry), a multitude of ion species can be produced, each having a different diffusion coefficient and mobility. Moreover, various ion-molecule reactions are possible, which lead to subsequent transformations of the ions during the diffusion process. Therefore, the data required for our nanodosimeter development should be averaged over the spectrum of the initially produced ion species and over various "transport channels", connected to various transformations during the diffusion. In our experiment, where no selection of ions exists, such averaging naturally occurs. However the same feature of our technique hinders the comparison of our data with data of other experiments. For testing our method we measured mobilities of the ions induced by α-particles in argon and drifting in the same gas, as some data for an indirect comparison (Ar^+ and Ar_2^+ ions in argon gas) are available in literature[6]. Such comparison shows reasonable agreement, which

Table 1 *Comparison of ion RMS transverse displacements (from present measurements) and those of electrons (computer simulations) in propane. Centimeters are used for gas and nanometers – for simulated tissue.*

	Pressure	Electric field	Drift length		RMS displacement	
	$Torr$	V/cm	cm	nm	cm	nm
Ions induced	0.15	50	1.5	5.4	0.13	0.5
by α-particles	0.15	5	1.5	5.4	0.17	0.6
Electrons (initial	0.15	50	1.5	5.4	0.55	2.1
energy of 10 eV)	0.15	5	1.5	5.4	0.95	3.4

gives confidence in the technique.

Series of measurements were made in propane, in a reduced electric field range of 10 to 300 $Vcm^{-1}Torr^{-1}$. In Table 1 examples of measured transverse diffusion-induced RMS displacement of the ions in propane are compared with corresponding values for electrons in the same gas. The latter are obtained by computer simulations according to our previous work,[5] for an initial energy of 10 eV, which is typical for electrons induced in ionization tracks. One can see that ion transverse displacement is about five times lower than that of the electrons. This is connected to the fact that the mean free path of the electrons induced in an ionization track in living tissue is more than 1 nm, and many such lengths are required to reach equilibrium conditions. Consequently, in a gas, quasi-ballistic electron transport dominates at corresponding simulated distances and causes high diffusion. This sets limits on the space resolution of methods based on electron counting. The induced ions have low initial energies and effectively exchange it with the gas molecules. Furthermore, moderate drift velocities

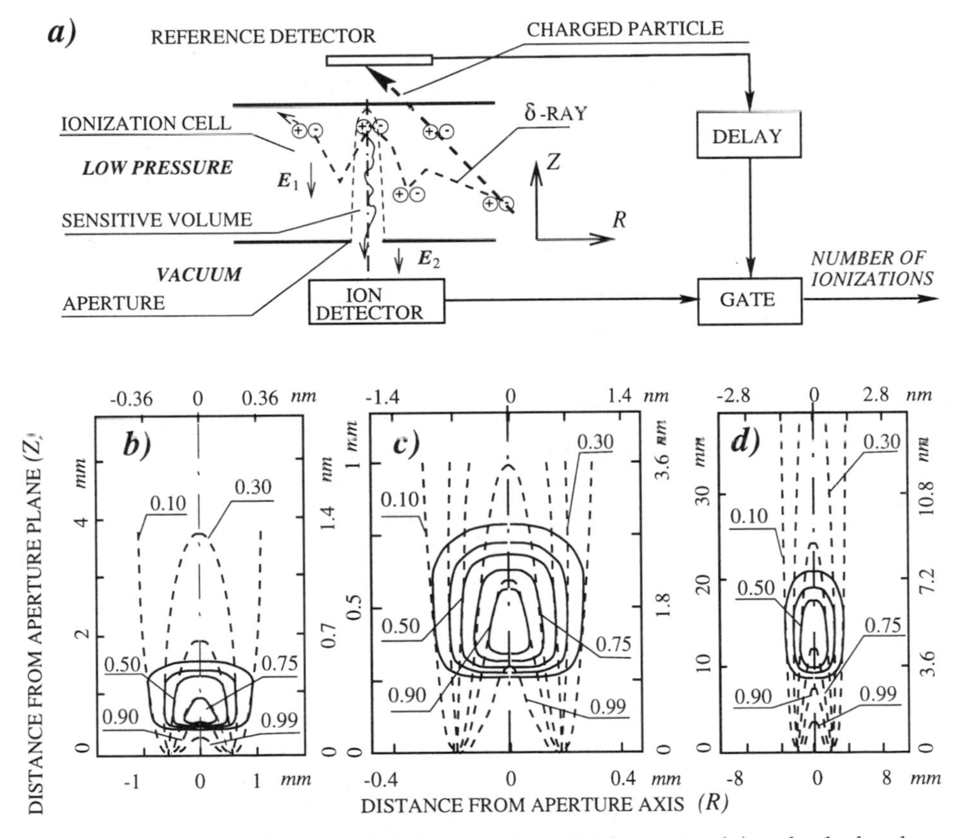

Figure 2 *Conceptual diagram of an ion counting nanodosimeter (a) and calculated sensitive volume configurations for propane (b – d). The contours correspond to the marked efficiency levels for collecting ions with time gating (solid) and without it (dashed). b) 0.15 Torr, E_1 = 50 V/cm, 1.12 mm aperture diameter; c) 1.5 Torr, E_1 = 500 V/cm, 0.35 mm aperture diameter; d) 0.15 Torr, E_1 = 50 V/cm, 4.0 mm aperture diameter.*

of ions (in comparison to those of electrons) enable additional simple selection of the sensitive volume size using delayed time gating.

3 AN ION COUNTING NANODOSIMETER

A concept of the proposed device in shown in Figure 2*a*. An energetic charged particle traverses a low-pressure gas ionization cell. It induces ionizations directly and through the mediation of δ-rays. Radiation-induced ions drift under an electric field E_1. Those ions which start from a certain sensitive volume, schematically shown in Figure 2*a*, pass through a small aperture and are counted by an ion detector. The electric field configuration, the size and geometry of the aperture, and ion diffusion, define the sensitive volume. An additional selection of the sensitive volume along E_1 is provided by time gating, triggered by pulses from the reference detector. Ions are accelerated (under E_2) to an ion counter operating in high vacuum, which requires an appropriate differential pumping system (not shown).

Based on results of our measurements and a known diffusion model[6] we calculated the sensitive volume configurations for a number of cases (Figure 2 *b* – *d*). The vacuum conditions in cases *b* and *c* are achievable with the technique used in our present measurements. One can see that sensitive volumes of about 0.2 *nm* and 1 *nm* across, respectively, are already possible. Case *d* indicates another way of affecting the sensitive volume size and configuration, by enlarging the aperture size.

To conclude, the proposed ion counting nanodosimeter should enable subnanometer level spatial resolution. This permits good definition of *completely closed wall-less* sensitive volumes. Due to the absence of electron avalanche (like in electron counting devices), the choice of working gas is not restricted. This permits the use of various gases for simulation of the chemical composition of living cell fine subsystems.

This work was supported by the Foundation Mordoh Mijan de Salonique and the Basic Research Foundation of the Israel Academy of Sciences and Humanities. S. S. is grateful to the Israel Ministry of Absorption for his partial support. A. B. is the W. P. Reuther Professor of Research in the peaceful use of Atomic Energy.

References

1. D. T. Goodhead and D. J. Brenner, Phys. Med. Biol. 28 (1983) 485.
2. A. Breskin, R. Chechik, P. Colautti, V. Conte, A. Pansky, S. Shchemelinin, G. Talpo and G. Tornielli, Radiat. Prot. Dosim. 61 (1995) 199 and references therein.
3. S. Shchemelinin, A. Breskin, R. Chechik, A. Pansky, P. Colautti, V. Conte, L. De Nardo and G. Tornielli, *Nucl. Instrum. and Meth.*, A 368 (1996) 859.
4. A product of *El-Mul Technologies*, Soreq, POB 571, Yavne 81104, Israel.
5. S. Shchemelinin, A. Breskin, R. Chechik, A. Pansky. The ionization density distribution in a low pressure gas sample measured by single electron counting. A computer simulation. WIS, Detector Physics Group internal report, December 5, 1994.
6. E. W. McDaniel, E. A. Mason, *The mobility and diffusion of ions in gases* (Wiley, New York, 1973).

THEORETICAL BASIS FOR SOLID-STATE MICRODOSIMETRY USING PHOTOCHROMIC ALTERATIONS

D. Emfietzoglou and M. Moscovitch

Departments of Radiation Medicine and Physiology & Biophysics
Georgetown University Medical Center
3970 Reservoir Rd., NW, Washington DC, 20007, USA.

1 INTRODUCTION

Of special importance for assessing RBE is the study of single HCP (heavy charged particle) tracks in biomatter, because for this type of radiation fields and at low dose levels critical microscopic sites of the body (e.g. cell nucleus, DNA) will experience single particle traversals. However, currently, no experimental technique can provide such information in a TE condensed-state medium, important for simulating the cellular environment. Therefore, a theoretical study was carried out investigating the feasibility of using the radiation-induced color decay of photochromic molecules doped in a polymer matrix as a probe for solid-state microdosimetry of HCPs.

As implied by the name, if photo-excited, a photochromic molecule is capable of changing into a chromophore. Spirobenzopyran (SP) molecules embedded in a polymethylmethacrylate (PMMA) matrix is the model system employed in the present study, due to its tissue equivalence and extensive study as a computer optical memory device.[1] Originally in the thermodynamically more stable -colorless- state, called spiropyran, the SP molecules may be selectively excited by the simultaneous absorption of two photons (total excitation energy in the UV spectrum) yielding the colored merocyanine isomer. The prompt optical radiation emitted by the latter as a result of, again, a two-photon absorption process (total excitation energy in the visible spectrum) has a longer wavelength as compared to the absorption band of either isomers. Therefore, by avoiding self-absorption, and using appropriate photomultipliers that are capable of single photon detection, extremely high sensitivity can be achieved.

The proposed method is based on photo-exciting the SP molecules prior to irradiation turning them into chromophores, the color-decay rate of which is increased (above background levels) within the HCP track. In principle, by measuring at the appropriate time after the formation of the track the reduction of the optical radiation emitted from the matrix, information relevant to the microscopic distribution of energy deposition in the HCP track may be obtained.

The objective of the present study was to calculate the color-decay (or bit-flip) probability of the SP molecules as a function of the radial distance from the HCP track axis. The theoretical model has been developed for single tracks of protons, α-particles, carbon and oxygen nuclei of 1-10 MeV/amu.

In the following section a brief summary of the theoretical analysis is presented. More details can be found in reference 2.

2 THEORETICAL MODEL

2.1 Track Structure

Along the lines of reference 3 an analytical formula for the radial dose distribution, D(r), within the HCP track was developed for the PMMA matrix. The principal assumptions underlying the model are: (i) energy deposition by HCP-induced primary ionizations and excitations is neglected. In other words, only energy deposition by the first and subsequent generations of electrons is considered (it is a δ-ray model where Auger electrons are not considered), (ii) atomic electrons are considered to be unbound and at rest, and (iii) δ-rays are assumed to travel in straight paths of length equal to their ranges, where the range is assumed to be the depth of penetration in the direction of ejection. It can be shown that:

$$D(r) = \frac{1}{2\pi r \rho} \int_{W_{0,1}}^{W_{0,2}} \left(\frac{dn}{dW_0} \right) \left[\frac{dW(r, W_0)_{res}}{dr} \right] dW_0 \quad (1)$$

where dn is the number of δ-rays produced with initial kinetic energy between W_0 and W_0+dW_0 per unit path length of incident HCP, $W(r,W_0)_{res}$ is the residual kinetic energy of a δ-ray at radial distance r having initial kinetic energy W_0 , and the limits of integration are such that δ-rays with an initial kinetic energy within these limits are capable of reaching radial distance r (accounting also for angular distribution).

2.2 Temperature Diffusion

The spatial and temporal distribution of temperature in the track was calculated by solving the differential heat diffusion equation for a cylindrical symmetric system (as dictated by our non-stochastic treatment) based on the Fourier-Bessel series. Then, according to the initial and boundary conditions of the problem the following equation holds:

$$\Delta T(r, t) = \frac{2}{r_{max}^2} \sum_n \frac{J_0(\lambda_n r)}{J_1^2(\lambda_n r_{max})} e^{-\lambda_n^2 a t} \int_0^{r_{max}} R(r') J_0(\lambda_n r') r' dr' \quad (2)$$

where a is the thermal diffusivity of the material, $J_0(\lambda_n r)$ and $J_1(\lambda_n r)$ are Bessel functions, λ_n refers to the positive roots of $J_0(\lambda r_{max})=0$ where r_{max} is the penumbra radius, and R(r) is the initial distribution of temperature in the track as calculated iteratively by the temperature-dependent heat capacity of the matrix and Eq. (1).

2.3 Matrix Degradation

Ionizing radiation induces only main-chain scission (degradation) on polymer molecules in PMMA, producing a reduction in the average molecular weight, in the absence of almost any cross-linking effects which would result in a respective increase. Radiation-induced matrix-degradation increases the local free-volume around the photochromic molecules according to the following equation:

$$\Delta f(r) = 2\Theta\rho \frac{G_s}{100} D(r) \quad (3)$$

where Θ is excess free-volume per molecular chain end and G_s is the radiation chemical yield for main-chain scission. In turn, it can be shown that the above increase enhances the bit-flip rate, due to the explicit dependence of the color decay-rate constants on the available free-volume in the vicinity of the chromophores. Also, Eq. (3) divided by the appropriate thermal expansion coefficient, was used for deriving the "new" glass transition temperatures, T_g, after irradiation.

2.4 Kinetics

The spatial and temporal distribution of bit-flip probability in the track was calculated according to experimental data for the color-decay at room temperature,[1] as well as theoretical considerations regarding higher temperatures and different conditions of the matrix. The following general equation was used:

$$P(r,t) = 1 - \sum_i [\alpha_i(r) \exp(-\int_0^t k_i(r,t)\,dt)] - \alpha(r) \qquad (4)$$

For $T<T_g$ the k_i values were calculated based on an Arrhenius-type function where the pre-exponential term was taken to be inversely proportional to the viscosity. The dependence of the latter on temperature was calculated through the free-volume theory. While, for $T>T_g$ a Williams-Landel-Ferry type equation was used as derived from the free-volume theory.

3 RESULTS

The results presented in the following Figures (1 to 3) show the different effects expected in the tracks of the HCPs studied giving grounds for the possible application of the technique to experimental solid-state microdosimetry. However, the experimental observation of radiation-induced bit-flips, alone, is of limited importance to microdosimetry (or even dosimetry) if a correlation with a relevant quantity does not exist. The latter is mainly a theoretical enterprise. The theoretical model developed, although non-stochastic in nature, does attempt to provide such a correlation by enabling to link the microscopic distribution of energy deposition in the "average" track to the rate of bit-flip in it.

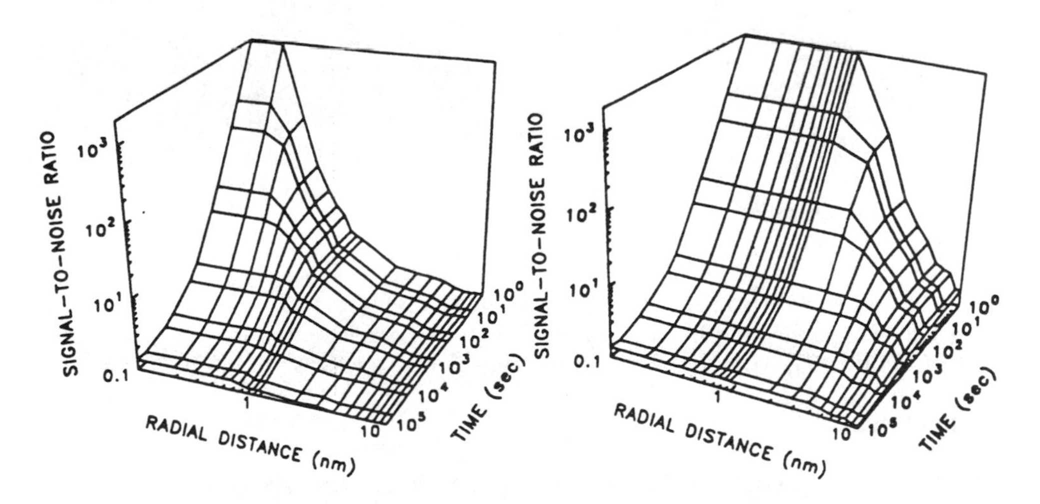

Figure 1 Signal-to-noise ratio in 1 MeV/amu (left) proton and (right) oxygen nucleus tracks.

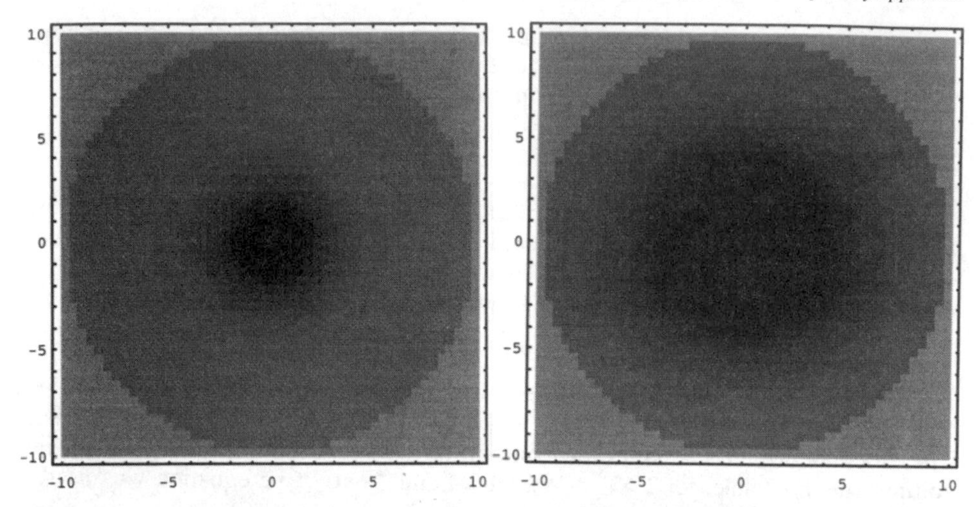

Figure 2 Decoloration in 1 MeV/amu (left) α-particle and (right) carbon nucleus tracks at $t=10^3$ sec (track axis perpendicular to the paper surface); X-Y axis in nm.

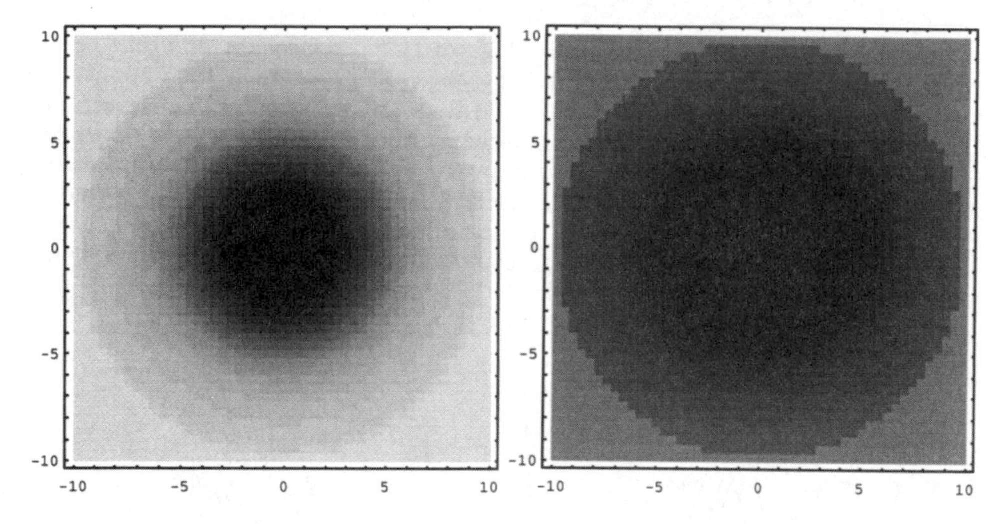

Figure 3 Decoloration in 1 MeV/amu oxygen nucleus tracks at (left) $t=10^2$ sec and (right) $t=10^3$ sec.

ACKNOWLEDGMENTS: Work supported by NASA (Grant No. 9307-0347) and by the US DOE under the Health Physics Faculty Research Award (Contact No. DE-AC05-76OR0003) with Oak Ridge Associated Universities.

References

1. D. A. Parthenopoulos and P. M. Rentzepis, *Science*, 1989, **245**, 843.
2. M. Moscovitch and D. Emfietzoglou, *J. Appl. Phys.*, 1996 (in press).
3. J. Kiefer and H. Straaten, *Phys. Med. Biol.*, 1986, **31**, 1201.

EXPERIMENTAL MICRODOSIMETRY WITH MICROSTRIP GAS COUNTERS

J. Dubeau [a], M.S. Dixit [b], E.W. Somerville [b], A.J. Waker [c],
R.A. Surette [c], F.G. Oakham [b], D. Karlen [a]

[a] Department of Physics and [b] CRPP, Carleton University, Ottawa, Ontario, K1S 5B6, Canada [c] AECL, Chalk River Laboratories, Chalk River, Ontario, K0J 1J0, Canada

1 INTRODUCTION

New gas detector developments have been taking place in the particle physics community and the potential of these devices for microdosimetry was discussed at a recent workshop on advances in radiation measurement [1]. One such device is the Microstrip Gas Counter (MSGC) [2]. Microstrip gas counters were principally designed as high count rate and high spatial resolution devices. As such they are normally operated under different conditions than those required for applications in microdosimetry. This paper describes a microstrip counter developed at the Centre for Research in Particle Physics (CRPP) and at the Department of Physics at Carleton University, Ottawa and initial work carried out to investigate the basic counting properties of MSGC when operated under conditions more appropriate to applications in microdosimetry.

2 THE CARLETON UNIVERSITY MSGC DESIGN

MSGCs are similar to multiwire proportional chambers except that the wires have been replaced by alternating anode and cathode strips formed on a supporting substrate using photolithographic techniques. A schematic profile of a generic MSGC is shown in Figure 1. The MSGC itself acts as one wall of the gas volume while the location of a drift electrode defines the effective thickness. The electron signal at the anodes is read out with anodes at ground, while cathodes and drift electrode are held at negative voltages. Over most of the gas volume field lines are parallel but converge onto the anodes were the high field values yield gas gains of up to 3000 in argon based mixtures.

Our two preferred substrate materials are 100 μm ion-implanted or carbon coated polyimide plastic (Upilex, manufactured by UBE, Japan) [3] and 1-2 mm electronic conductive (Schott) glass. An effective surface resistivity in the range of 10^{13} Ω/square to 10^{15} Ω/square is required to prevent surface charging effects. These devices display very good energy resolution as shown on Figure 2 where a 15% FWHM is obtained using the 5.9 keV x-ray of ^{55}Fe. The gas used was Ar-isobutane (90%-10%) and the Ar escape peak is well separated from the 5.9 keV (channel 1700) full energy peak.

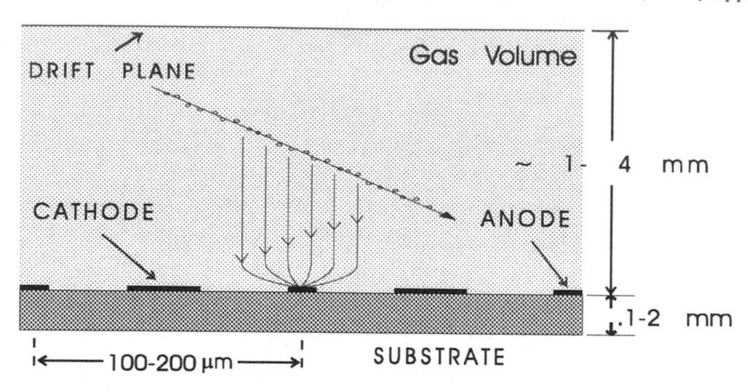

Figure 1 *Cross section and critical dimensions of an MSGC print and gas volume.*

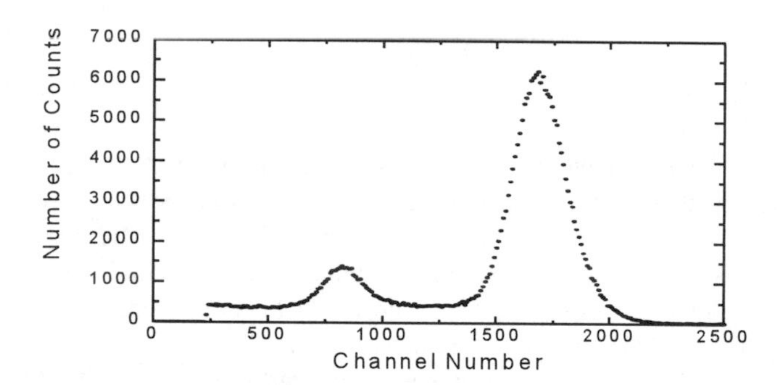

Figure 2 *Pulse height spectrum of 5.9 keV x-rays from ^{55}Fe.*

The operating characteristics of a 200 µm MSGC on a Upilex substrate were studied using 5.9 keV photons for various mixtures of tissue equivalent (TE) gas (55% propane, 45% CO_2, 5% N_2) and argon at a pressure of 1 atmosphere. The detector performed well but a small concentration of Ar was always required to maintain a reasonable gain and a good energy resolution. Gain curves as functions of the cathode bias are shown in Figure 3. In these measurements all anodes were coupled to a single Ortec 142PC charge sensitive preamplifier. Similar results were obtained for a 400 µm detector on Schott glass at 0.33, 0.66 and 1 atm.

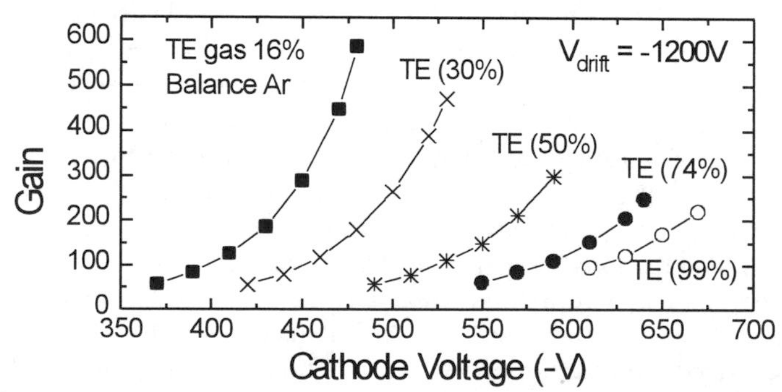

Figure 3 *Gain as a function of the V_{cath} for a 200 µm MSGC on Upilex for a TE mix.*

3 EXPERIMENTAL MICRODOSIMETRY

Apart from the high resolution and high count-rate capability the principal feature of the MSGC of value for applications in experimental microdosimetry is its spatial resolution. This feature may be useful to obtain more information on the actual path length of charged particles in the counter and would enable better discrimination between radiation types in a mixed-field. Having established that the MSGC can operate with dosimetric gases and at pressures typical of microdosimetric requirements the next operating characteristic investigated was the dynamic range of the signal response of the counter in grouped and single anode configurations. It should be again noted at this juncture that we were testing the basic counting characteristics of the MSGC and not optimizing the design with regard to the tissue equivalence of construction materials and the overall dosimetric response of the device.

Figure 4 shows the event-size spectrum recorded for ^{60}Co and for ^{252}Cf neutrons free in air from a 200 μm MSGC on Schott glass. The spectrum has been recorded from a group of 8 anodes coupled to a single charge sensitive preamplifier. The signal processing, pulse height and data analysis was carried out using the same conventional experimental microdosimetry system described previously by Waker [4]. The active volume described by this arrangement is a rectangular prism of cross section 4.35 mm × 1.6 mm and length 5 cm. The μ-randomness [5] geometric mean chord length for this configuration has been calculated to be 2.33 mm with a most probable chord length of 1.61 mm. The simulated mean chord length at 1 atmosphere of the propane TE(76%)/Ar gas mixture used is 4.1 μm. In order to accommodate the entire range of pulse heights expected the operating gain of the MSGC was set at around 50 with a cathode voltage of -508 volts and a drift plane voltage of -1080 volts. This is at the lower end of the cathode voltage-gain characteristic shown in Figure 3. At the cathode voltage used noise levels restrict the

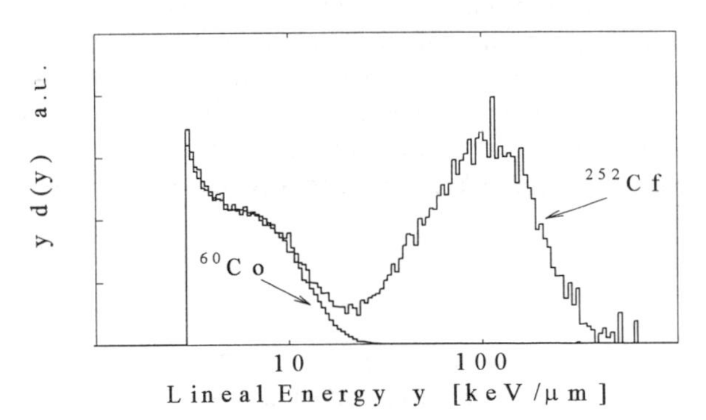

Figure 4 *Microdosimetric event-size spectra for ^{60}Co and ^{252}Cf recorded with a 200μm MSGC, 1 atmosphere TE(76%)/Ar gas filling.*

lower level of the event-sizes that can be recorded such that only the upper 'shoulder' region of the photon spectrum can be seen. This is sufficient to discriminate between photon and neutron induced events and to make a photon event subtraction. Above the photon component of the event-size spectrum a proton recoil peak is seen, however it does not have a distinct 'proton edge' that is observed in conventional spherical tissue equivalent counters [6]. The position of the valley between photon and neutron events at

20 keV/μm and of the proton peak at 100 keV/μm is higher on the lineal energy scale than expected, by about 40%. This difference could be due to a calibration error or to differences in the actual charged particle path length distribution compared to what had been calculated. Calibration of the MSGC was carried out by measuring the position of the full energy peak of an ^{55}Fe source similar to that shown in Figure 2. However this could only be done at a higher gain with a cathode voltage of more than 540 volts. To obtain a calibration point at the lower operating cathode voltage of 508 volts the ^{55}Fe peak position was determined at 10 volt intervals between 600 volts and 540 volts and extrapolated to 508 volts. This $log(gain)\text{-}V_{cathode}$ characteristic gives a linear regression coefficient of correlation of 0.9998 and the uncertainty of the extrapolation is much less than the 40% observed shift in the event-size spectrum.

An even more marked shift to higher lineal energies in the event-size spectrum for a single anode arrangement has been observed, where a (μ-randomness) mean chord length of 390 μm "simulates" a 0.7 μm tissue size at 1 atmosphere. Note that a smaller site size of 0.23 μm is achievable for a MSGC with TE gas at 0.33 atmosphere. Our conclusion is that an accurate evaluation of the actual chord length distribution for this rather unusual geometry is required and, to this end, we are commencing a series of Monte Carlo simulations.

CONCLUSIONS

Our work to date indicates that MSGC can be operated under conditions suitable for experimental microdosimetry and that different anode counting arrangements with different sensitive volume geometries can be used. From a general instrumentation point of view the tasks ahead will involve the optimization of the MSGC performance with regard to noise, gas gain and anode configurations. Concerning neutron dosimetry in particular, optimization of the dosimetric response will also have to be carried out by investigating theoretically and experimentally the effect of changing the tissue equivalence (or hydrogen content) of the counting gas, drift plane and substrate materials. Energy deposition modeling will also be required to enable interpretation of event-size spectra which appear significantly different to those obtained with conventional counters.

References

1. F. Sauli, *Radiat. Prot. Dosim.*, 1995, **61**, No 1-3, 29.
2. A. Oed, *Nucl. Instrum. Methods.*, 1988, **A263**, 351.
3. M.S. Dixit, F.G. Oakham, J.C. Armitage, J. Dubeau, D. Karlen, G.C. Stuart, S. C. Taylor, I. Shipsey, E. Johnson, A. Greenwald, *Nucl. Instrum. Methods*, 1994, **A348**, 365.
4. A.J. Waker, *Radiat. Prot. Dosim.*, 1994, **52**, No 1-4, 415.
5. A.M. Kellerer, *Radiat. Res.*, 1971, **47**, 359.
6. A. J. Waker, *Radiat. Prot. Dosim.*, 1995, **61**, No 4, 297.

A COMPARISON OF MEASURED AND CALCULATED $\bar{y}_D$ VALUES IN THE NANOMETRE REGION FOR PHOTON BEAMS

J.-E. Grindborg and P. Olko*

Swedish Radiation Protection Institute
S-171 16 Stockholm, SWEDEN
* Health Physics Lab, Institute of Nuclear Physics,
Radzikowskiego 152, PL 31-342 Krakow, POLAND

1 INTRODUCTION

A comparison of measured and calculated dose mean lineal energies, $\bar{y}_D$, have been made for two photon beams, at simulated object diameters between 6 nm and 2 μm. The measurements have been made with ionisation chambers using the variance-covariance method and have been reported elsewhere.[1] The calculations have been made with the PHOEL-2[2] code to obtain the first collision electron spectrum and the code TRION[3] for simulating the electron tracks. Only ionisations were taken into account in the calculations. In both the measurements and the calculations $\bar{y}_D$ was calculated by applying a constant mean energy to produce an ion pair, *W*. In the nanometre range the relationship between energy deposition and ionisation becomes less clear as the energy is deposited in only a few ionisations.[4-6] A calculation of $\bar{y}_D$ from ionisation measurements is nevertheless made in this paper in order to get results, which are comparable with earlier results.

2 RADIATIONS

A collimated X ray beam and a collimated ^{60}Co γ ray beam have been investigated.

The X ray beam had a tube voltage of 100 kV and an added filter of 3.14 mm Al (HVL 0.141 mm Cu). The photon fluence of the X ray beam is known and was used as input for the Monte Carlo calculations.

The measurements in the ^{60}Co γ ray beam were made on a conventional therapy unit. In the calculations an ideal ^{60}Co photon spectrum was assumed, without taking into account photon transport in the ^{60}Co therapy head, the vacuum shield and the counter wall.

3 MEASUREMENTS

The detectors were two spherical ionisation chambers, both with a 2.5 mm thick wall made of C-552 air equivalent plastic. The inner diameters were 38 mm and 57 mm respectively. The two detectors were positioned in a cylindrical plastic vacuum shield made of polystyrene with a diameter of 230 mm and a wall thickness of 10 mm. This shield was filled with air at different pressures. The object diameters simulated by the ionisation chambers were in the range 6 nm to 2 μm.

From the measurements $\bar{y}_D$ is calculated using the variance -covariance method.[1] In the calculation a constant W equal to 33.97 eV was used. The mean chord length, $\bar{l}$, was calculated with a stopping power ratio of air to tissue, $s_{air,t}$, equal to 0.89 for both radiation qualities[7]. The tissue density value 1.00 g.cm^{-3} was used.

4 CALCULATIONS

The calculation of the microdosimetric distributions and their moments for the X ray beam and the ^{60}Co γ ray beam were performed using Monte Carlo simulation of electron tracks in water vapour. The calculation procedure starts from computing the first collision electron spectra, ϕ(E), produced in water by photons using the Monte Carlo code PHOEL-2.[2] No further transport of scattered photons were considered since the probability for the scattered photon to interact in the sensitive volume can be neglected due to the small volume as compared to the photon path length.

Next, a set of electron tracks in water vapour, scaled to unit density material, were generated for 27 discrete energies ranging from 1 keV to 1000 keV.[3] The tracks included delta rays and were calculated for ionisation events only. The number of tracks followed was such that typically 200,000 ionisations were created in the sphere for each electron energy E_i. These tracks were then used to score dose distributions of ionisations $d_I(j;\ E)$ in spherical volumes of diameters from 5 to 2000 nm by sampling over individual ionisation points.[8]

Finally, $d_I(j)$ distributions were calculated by weighting calculated $d_I(j;\ E)$ distributions over the first collision electron spectra.

The $\bar{y}_D$ values were calculated using a W value for water vapour of 29.6 eV.

5 RESULTS AND DISCUSSION

The measured and calculated $\bar{y}_D$ values as a function of the simulated object diameter are given in Figure 1, both for the X ray beam and the ^{60}Co γ ray beam. The uncertainty in the measurements was less than 15 % for simulated object diameters larger than 18 nm.

For the ^{60}Co γ ray beam the ratio between measured and calculated $\bar{y}_D$ value is about a factor 1.2 for simulated object diameters between 9 nm and 2 μm. For the simulated object diameter 6 nm the factor is 2. The corresponding ratio for the X ray beam is less than 1.15 for simulated object diameters above 20 nm. For smaller simulated object diameters the ratio increases rapidly and is about a factor 2 at 6 nm.

The calculations of $\bar{y}_D$ for the ^{60}Co γ ray beam were made for the primary photon energies 1.17 and 1.33 MeV only. As the scattered radiation has a higher $\bar{y}_D$ value than the primary photons this explains part of the difference between the measured and the calculated $\bar{y}_D$ values. Calculations and measurements of spectra from ^{60}Co therapy units have been reported in the literature. The kerma rate free in air from the scattered radiation is normally reported to be about 20 % of the kerma rate free in air from the primary

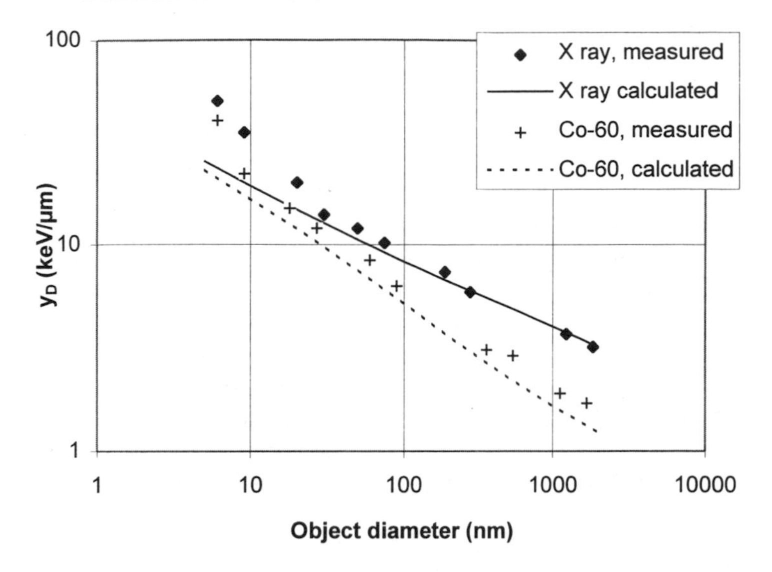

Figure 1 *A comparison of measured and calculated $\bar{y}_D$ values for different simulated object diameters in a ^{60}Co γ ray beam and an X ray beam (100 kV and HVL 0.141 mm Cu)*

photons in the central point of the beam at a normal SSD and a field size of 10 cm x 10 cm.[9,10] If $\bar{y}_D$ values are calculated for such a spectrum, $\bar{y}_D$ increases about 5 % at the largest simulated object diameters, while for the lowest the difference will be less than 1 %. The measurements in this investigation were made with a larger field size and with a polystyrene shield in the beam. The amount of scattered radiation is then probably larger than 20 %. To check the influence of a phantom a calculation of $\bar{y}_D$ was now made for ^{60}Co γ rays incident on an infinite water phantom. The $\bar{y}_D$ values increased at 2 µm with a factor 1.20 and this value is in close agreement with the measured one. This calculation probably overestimate the amount of phantom scattered radiation, however, the source scattered component was not included. It is then not unlikely that the scattered component may explain a difference of as much as 20 % between the calculated and the measured $\bar{y}_D$ values at the largest simulated object diameters.

Another possible explanation for the difference could be the wall effect, which increases $\bar{y}_D$ values measured with walled detectors. Theoretically the wall effect will increase with decreasing simulated object diameters and also increase with chamber radius. For photons the wall effect reported in the literature from calculations and measurements is less than 10 % for objects larger than 0.1 µm. The effect below that is unknown.

A comparison of the present measurements in the ^{60}Co γ ray beam with measurements published by Braby and Ellett[14] and Bengtsson and Lindborg[15] shows an agreement better than 10 % for simulated object diameters larger than 1 µm.

For the measurements in the X ray beam a comparison with measurements by Forsberg et. al.[16] shows an agreement better than 14 % for simulated object diameters larger than 75 nm. For the measurements in the X ray beam at the smallest simulated object diameters no correction for covariance was done which gives an overestimation of the $\bar{y}_D$ value.

6 CONCLUSIONS

The measurements with the variance-covariance method in the 100 kV X rays beam resulted in $\bar{y}_D$ values which well agreed with results of Monte Carlo track structure calculations for site diameters above 20 nm. For ^{60}Co γ ray beam calculated $\bar{y}_D$ values were systematically 20 % lower than those measured for simulated object diameters between 9 nm and 2 μm. The difference is supposed to result from omitting in the calculation the wall effect and the photon scattering in the ^{60}Co therapy head, the vacuum shield and the counter itself. For simulated diameters below 20 nm for the X ray beam and below 9 nm in the ^{60}Co γ ray beam the measured values exceeded the calculated up to a factor of 2. More work is needed to understand energy deposition in such small targets.

ACKNOWLEDGEMENT

We are grateful to Dr L. Lindborg for many useful discussions.

References

1. J.-E. Grindborg, G. Samuelson and L. Lindborg, *Radiat. Prot. Dosim.*, 1995, **61**, 193.
2. J. E. Turner, R. N. Hamm, H. A. Wright, J. T. Modolo and G. M. A. A. Sardi, *Health Phys.*, 1980, **39**, 49.
3. A. V. Lappa, E. A. Bigildeev, O. N. Vasilyev and D. S. Burmistriv, *Radiat. Environ. Biophys.*, 1993, **32**, 1.
4. P. Kliauga, *Radiat. Prot. Dosim.*, 1990, **31**, 119.
5. H. I. Amols, C. S. Wuu and M. Zaider, *Radiat. Prot. Dosim.*, 1990, **31**, 125.
6. E. A. Bigildeev and A. V. Lappa, *Radiat. Prot. Dosim.*, 1994, **52**, 73.
7. ICRU, 'Microdosimetry. Report 36', International Comission on Radiation Units and Measurements, Washington, DC, 1983.
8. A. M. Kellerer and D. Chmelevsky, *Radiat. Environ. Biophys.*, 1975, **12**, 321
9. D. W. O. Rogers, G. M. Ewart and A. F. Bielajew, 'Proc. of a Symposium on Dosimetry in Radiotherapy', IAEA, Vienna, 303, 1987.
10. P.-O. Löfroth, S. Westman and G. Hettinger, *Acta Radiologica Ther. Phys. Biol.*, 1973, **12**, 553.
11. A. M. Kellerer, *Radiat.Res.*, 1971, **48**, 216.
12. L. A. Braby and H. Ellett, *Radiat. Res.*, 1972, **51**, 569.
13. R. Eickel and J. Booz, *Rad. and Environm. Biophys.*, 1976, **13**, 145.
14. L. A. Braby and E. H. Ellett, 'Report AD 731709', Oregon State University, 1971.
15. L. G. Bengtsson, and L. Lindborg, 'Proc. of the 4th Symp. on Microdosimetry', EUR-5122, Luxembourg, 832, 1974.
16. B. Forsberg, M. Jensen, L. Lindborg and G. Samuelson,'Proc. of the 6th Symp. on Microdosimetry', EUR-6064, Brussels, 260, 1978.

A SLDD BASED NANODOSIMETER FOR ELECTRONS AND PHOTONS

T.M. Evans and C.K. Wang

Nuclear Engineering/Health Physics Program
Georgia Institute of Technology
Atlanta, GA 30332

1 INTRODUCTION

The assessment of biological effects of radiation for low dose and low dose rate exposures is an important component of radiation protection. The parameters required for a complete risk assessment are the distributions of energy deposition in biological targets and the biological effects of radiation action. Traditionally, the energy depositions of low dose/dose rate radiation from photons and electrons have been represented by $f(z_1)$ and $f(y)$.[1] The measurement of these quantities is performed using TE equivalent proportional counters simulating target sizes as small as 0.3 μm.[2,3] Using these quantities as the physical description of radiation action on biological systems, some authors have expressed radiological risk.[4–6]

One limitation of the above approach is the scale in which physical energy depositions may be measured. TE proportional chambers can only simulate target sizes down to ~0.3 μm. An alternative method of measuring the energy distribution of radiation tracks on a nanodosimetric scale is through the use of the superheated liquid drop detector (SLDD). The SLDD is a threshold detector which measures the distributions of cluster energy depositions. In this paper we introduce the SLDD as a viable nanodosimeter. The measured quantity, $c(\varepsilon)$, the cluster spectrum, will be briefly described.

2 THE CLUSTER SPECTRUM

Using microdosimetry track codes, much information about electron track structure has been discovered. In particular, analysis of the event-by-event spatial distributions of energy transfer points along a radiation track shows that large energy depositions tend to form in clusters of several events.[7] The distribution of these clusters will, to some extent, determine the efficacy of the radiation to cause complex biological damage.[8] A quantity which defines the distribution of clustered energy deposition events is the cluster spectrum, $c(\varepsilon)$.

The cluster spectrum is the probability distribution function for cluster formation along a charged particle track. As opposed to $f(z)$, $c(\varepsilon)$ is not explicitly dependent on a particular site size. Instead, $c(\varepsilon)$ includes all events which are spatially localized

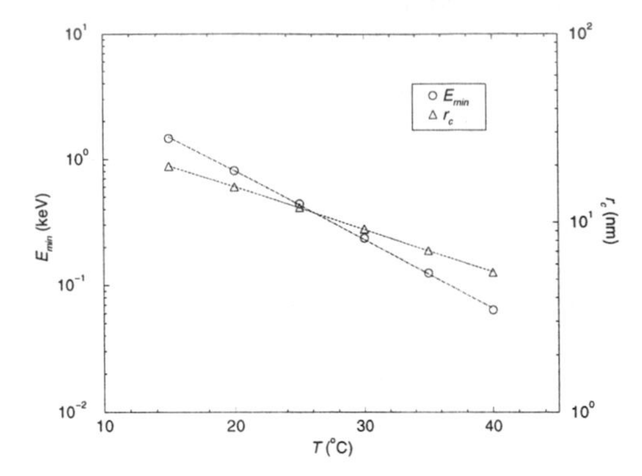

Figure 1: *The variation of E_{min} and r_c with temperature for Fr-115 SLDDs. Notice the different scales along the y-axis.*

in a single grouping. In this sense there is a particular minimum size which denotes a cluster; however, a cluster may be considerably larger than the minimum size. Preliminary calculations suggest that the minimum diameter of cluster volumes is on the order of 5 nm. Using the cluster spectrum as the physical definition of energy deposition which is responsible for biological action, models may be built to incorporate total effect provided a biological function is available.[1,4–6]

3 THE SLDD NANODOSIMETER

A SLDD records radiation induced events through the initiation of vapor bubbles in superheated liquids by energetic charged particles.[9] The SLDD is created by introducing tiny droplets of superheated liquid (Fr-115) into an inert gel host medium. Each droplet has a diameter of $\sim$10 μm and each detector has a concentration of $\sim 10^4$ droplets per ml of host gel.

When exposed to ionizing radiation, the secondary charged particles generated in the detector medium interact with the superheated droplets which act as nucleation centers. If the energy deposited in the droplets exceeds the threshold energy for bubble nucleation the droplet will vaporize. Vaporization is accompanied by a release of acoustic energy which may be measured electronically; thus, the number of bubbles generated is a measurement of the energy deposited in the system.

Bubbles are created by thermal spikes produced by the heat originating from the recombination of ions at the end of δ-rays according to Seitz's thermal spike theory.[10] Thermal spike theory postulates that the minimum energy, E_{min}, required for the formation of a bubble with critical radius, r_c, is provided by secondary charged particles generated from the passage of ionizing radiation. The variation of r_c and E_{min} with temperature and pressure allows for the selection of different site and cluster sizes for measurement. Figure 1 shows E_{min} and r_c versus temperature for Fr-115 SLDDs at a constant liquid pressure of 1 atm.

Because cluster sizes are smaller than r_c, E_{min} is equivalent to the threshold energy deposition, ε_{min}. One expects to observe a distribution of clusters inside the

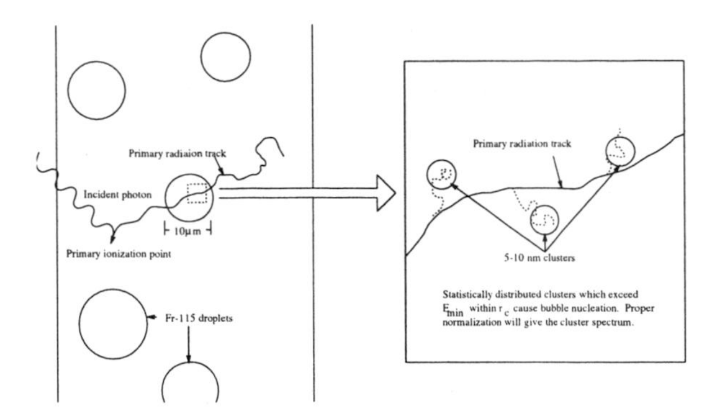

Figure 2: *Illustration showing the distribution of clusters in a Fr-115 droplet. Each cluster has the possibility of causing bubble nulceation; however, the distribution of such clusters will be statistically equivalent in each droplet.*

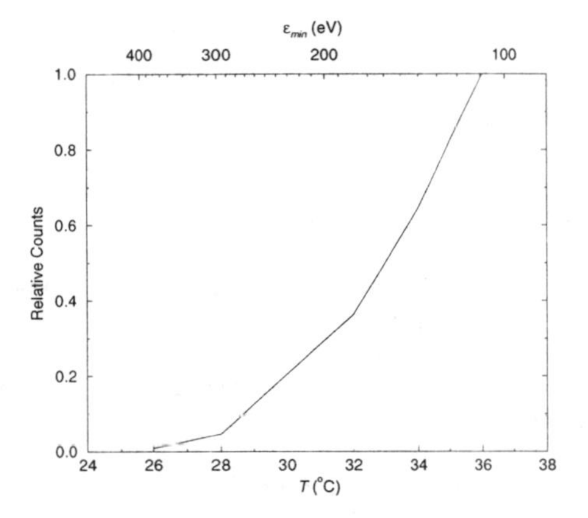

Figure 3: *The measured Fr-115 SLDD response versus temperature for* ^{60}Co. *The top scale shows the corresponding threshold energy deposition,* ε_{min}.

Fr-115 droplets which are capable of causing bubble nucleation as shown in Figure 2. With proper normalization, the frequency of cluster formation with energy may be measured in increasing temperature intervals. The cluster spectrum may be unfolded from the SLDD response with temperature shown in Figure 3. The measured cluster spectrum for ^{60}Co is shown in Figure 4.

4 CONCLUSIONS

We have introduced the SLDD as a nanodosimeter which measures small scale energy depositions. The quantity which defines these measurements is $c(\varepsilon)$. Computational studies which confirm that $c(\varepsilon)$ is measured by the SLDD have been performed and will be the subject of a future paper.

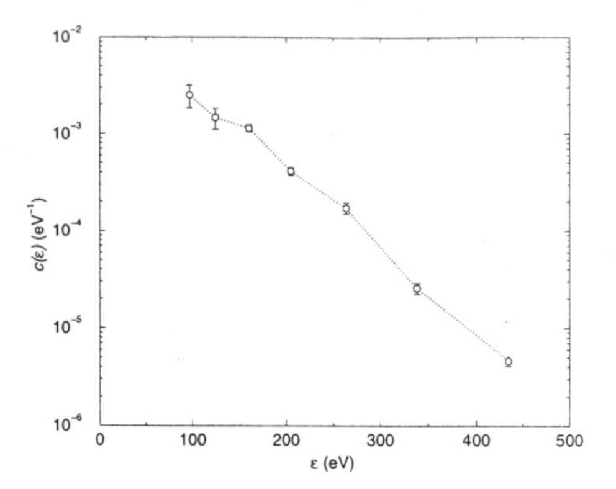

Figure 4: *The cluster spectrum for* ^{60}Co.

References

1. H. Rossi and M. Zaider, 'Microdosimetry and Its Applications', Springer-Verlag, Berlin, 1996.

2. ICRU, Microdosimetry, Technical Report 36, International Commission on Radiation Units and Measurements, 1983.

3. M. Zaider and H. Rossi, 'The Dosimetry of Ionizing Radiation', Academic Press, San Diego, 1987, Vol. 2, Chapter 4, p. 171–242,

4. M. Varma, V. Bond, L. Braby, L. Feinendegen, and M. Zaider, *J. of Rad. Protection*, 1993, **13**, 243.

5. M. Varma, C. Wuu, and M. Zaider, *Rad. Protection Dos.*, 1994, **52**, 339.

6. V. Bond, M. Varma, L. Feinendegen, C. Wuu, and M. Zaider, *Health Phys.*, 1995, **68**, 627.

7. R. Hamm, H. Wright, J. Turner, and R. Ritchie, *Proc. Sixth Sym. Microdosimetry*, 1978, p. 178–186.

8. D. Goodhead, *Int. Journal Rad. Biol.*, 1989, **56**, 623.

9. W. Lim and C. Wang, *Nucl. Instr. Methods Phys. Res. A*, 1993, **335**, 243.

10. F. Seitz, *Phys. Fluids*, 1958, **1**, 2.

IONISATION MEASUREMENTS IN NANOMETRE SIZE SITES WITH JET COUNTER - RECENT EXPERIMENTAL RESULTS

S.Pszona

Soltan Institute for Nuclear Studies
05 - 400 Otwock/Swierk, Poland

1. INTRODUCTION

The attempts to find an experimental method which is able to characterise the interaction pattern of radiation with different quality at nanometre size sites was started in the early seventies [1-3]. This approach was based on differential pumping technique for simulating nanometre sizes (SNS) and single ion counting. It has been quickly learnt that with this technique the SNS more than 1nm can not be achieved. To overcome this barrier the pulsed flow of gas through an orifice instead of constant flow has been devised[4]. The practical implementation of this concept appeared very difficult. Just recently the differential pumping technique was revived by a group from Legnaro and Rehavot[5]. Another approach in which a pulsed expanded flow of gas, Jet Counter, JC, has been proposed by Pszona and Gajewski[6] for studying the delta electron spectra escaping from a nanometre sites when irradiated by charged particles. By this method an indirect estimate of the mean value of restricted LET for low energy electrons was obtained. In this first JET COUNTER, (presented at 11 Microdosimetry Symposium), the gas jet expanded to a large dimension of an interaction chamber. In present paper a description of a modified Jet Counter has been presented.

2. DESCRIPTION OF MODIFIED JET COUNTER

The principle of the operation of the modified JET COUNTER, JC, is explained in Figure 1. A simulated nanometre - size, SNS, is obtained by a short lasting gas jet (nitrogen in this case). This jet is created due to pulse operated valve, PZ, which injects gas from a volume, R, over a valve, through a nozzle with a 1 mm diameter orifice to an interaction chamber, IC, below a nozzle. The interaction chamber is a cylinder, dia. 10 mm, made of tissue equivalent plastic or other material when a secondary particles equilibrium spectrum of incoming radiation have to be investigated. The ions created at the specific volume of this chamber during the gas flow are removed from that volume by an electric field created by a grid, G. The ions are then guided in an electric field Eh to a counting device, CH2. The effective thickness of a SNS is controlled by a gas pressure inside chamber R as well as by the electrical parameters of valve PZ.
The scaling procedure applied for the SNS needs to have an electron gun, EG, as well as the electron counting detector, CH1, installed inside the device as shown schematically in Figure 1.

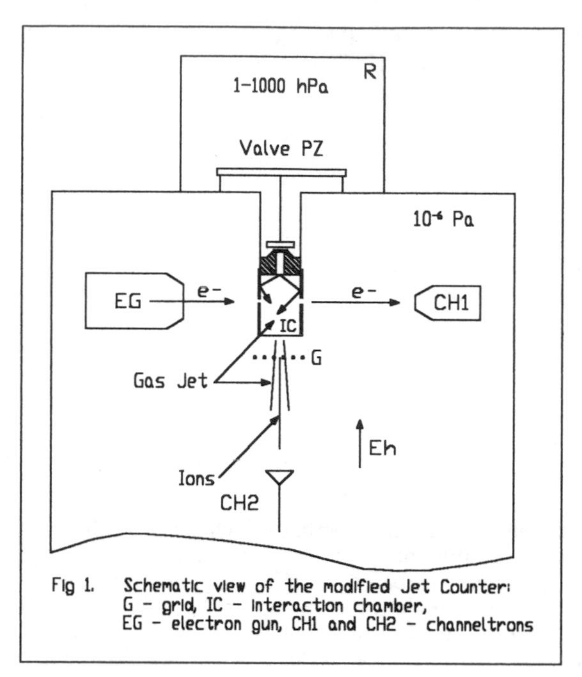

Fig 1. Schematic view of the modified Jet Counter: G - grid, IC - interaction chamber, EG - electron gun, CH1 and CH2 - channeltrons

3. RESULTS

The modified Jet Counter differs from previous one in a construction of an interaction chamber IC. In modified JC an interaction chamber has a cylindrical shape made of TE material which provides the particle equilibrium, necessary for dosimetric experiments, especially for photons and neutrons. The special attention has been paid to scaling procedure for the 2 - 10 nm dia. cylindrically shaped SNS. For this purpose a 100eV electron beam generated by EG enters through a slit to IC chamber, crosses a SNS and exits through another slit to a channeltron, CH1, where electrons are detected. The attenuation of 100 eV electron beam by nitrogen jets as seen on a multiscaler screen, MCS, are shown in Figure 2 as a function of nitrogen pressure in R chamber above PZ valve. The same effects have been obtained when for constant pressure of nitrogen in R chamber, the steering voltage to PZ valve are changed. The results are shown in Figure 3. The effective thickness of nitrogen volume have been derived from known Rao[7] transmission function and taking the Groswendt[8] data for practical range of 100eV electrons (3.2 nm in unit density scale). The 2 nm SNS is at transmtion ratio equals to 0.55, which is attained at 42 Torr of nitrogen in R chamber.
The set up shown in Figure 1. has been designed for studying the types of ions created in nanometric volumes by time of flight method, TOF,. The 5nm cylindrical volume has been irradiated by 500 eV electrons with 2 μs pulsed beam . The pulsed beam with repetition of 3Hz hits synchronously the nanometric volume at maximum gas flow. Ions created in this volume are extracted by grid G with negative potential, 0 -200 V then are guided through 150 mm distance to a channeltron CH2 with entrance voltage of 3400 V.
The time of flight spectra were registered by Multiscaler with 10 μs resolution and are shown in Figure 4.

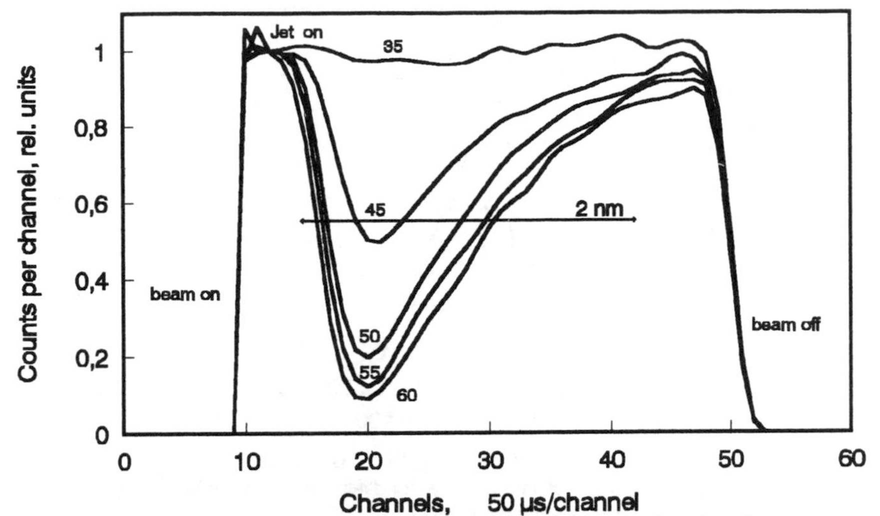

Fig.2. Attenuation of 100 eV electron beam by nitrogen jets - timing chart for different pressure (in Torr) in R chamber above a valve.

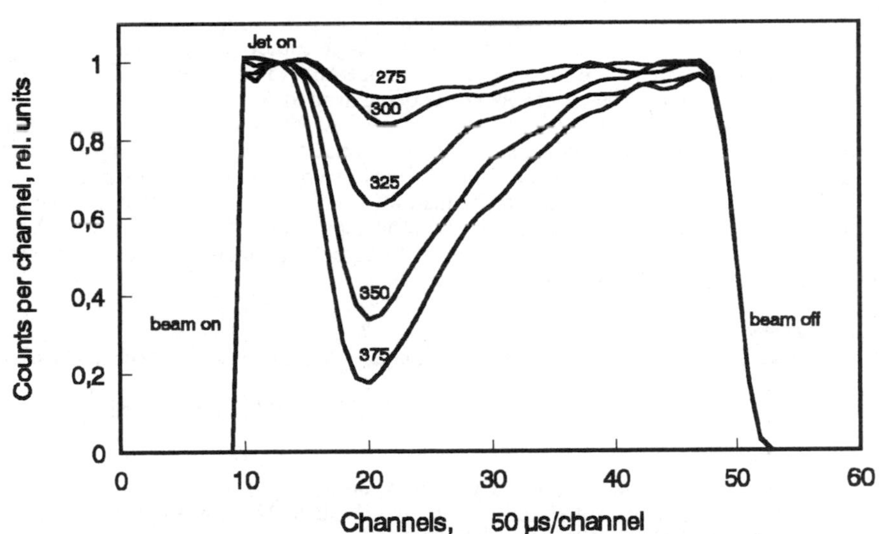

Fig.3. Attenuation of 100 eV electron beam by nitrogen jets - timing chart for different voltage on PZ valve (in volts) for constant pressure in R chamber.

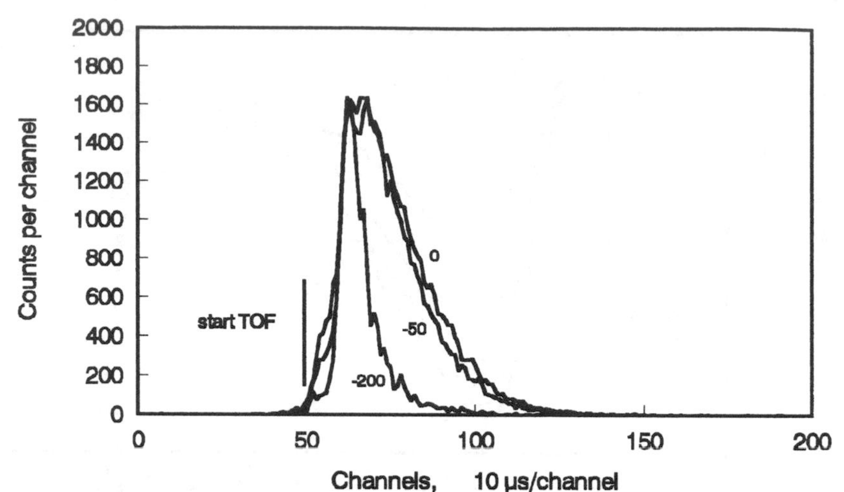

Fig.4. Time of flight spectra of nitrogen ions for different potential in Volts on grid G. Start of TOF spectra shown as vertical line.

4. DISCUSSION

As seen from Figures 2 and 3 the achievable range of SNS around 2 nm is well controlled by an electron gun with 100 eV electron beam. The accuracy of such estimates depends on the accuracy of evaluation of the practical range of 100 eV electrons in molecular nitrogen which seems to be relatively good, being around 7 - 10 %.
The results of TOF experiment with nitrogen gas, which is shown in Figure 5, indicate rather large time spread over 400 μs of nitrogen ions reaching the counting channeltron, CH2. The spectrum of TOF depends on the values of voltage applied to the grid, i.e for -200 V the TOF spectrum is much sharper than for -50V, which means that the main cause of this spread is due to interaction of ions with nitrogen molecules within IC chamber. The TOF studies were necessary for optimisation of interaction volume and for estimating the necessary time resolution of counting equipment.

Acknowledgements
Author is very indebted to S. Marjańska, J. Kula and A. Dudziñski for technical assistance during the experiments.
Support by CEC under the subcontract ERBCIPDCT 930407 is appreciated.

References

1 S. Pszona, COO 3243-2 Report, USAEC,New York, 1973 .
2. D. Chmelewski, N, Parmentier and J. Le Grand, EUR 5122 d-e-f Report, 869, 1974.
3. S. Pszona, EUR 3452d-e-f Report,1107, 1975.
4.A.M. Kellerer, COO-3243-3, USAEC, New York, 1974.
5. S. Shchemelinin et all, *Nucl. Instrum. and Methods*, 1996, **A315**, 82.
6. S. Pszona and R. Gajewski, *Radiat. Prot. Dosim*. 1994, **52,** 427.
7. B. N. Rao Subba, *Nucl. Instrum. Methods*, 1966, **44** ,155.
8. B. Groswendt and E. Waibel, *Nucl. Instrum. Methods*, 1978, **155**, 145.

MICROSTRIP GAS CHAMBER WITH TEG AS A DETECTOR FOR MICRODOSIMETRY*

B. BEDNAREK, K. JELEŃ, T. Z. KOWALSKI, E. RULIKOWSKA-ZARĘBSKA

University of Mining and Metallurgy, Faculty of Physics and Nuclear Techniques, Al. Mickiewicza 30, 30-059 Cracow, Poland

INTRODUCTION

Tool commonly used in microdosimetry to simulate the radiation interaction in small tissue volumes, such as an individual cell or a part of it, is the low pressure tissue equivalent proportional counter.

Since the late 1950s the spherical, so-called, Rossi counter with helical grid electrode has widely been applied in microdosimetric measurements [1]. However some disturbing phenomena in the Rossi counter have been observed [2]. Therefore, it was recommended to stop working with helix counters and to use cylindrical counters, for example counters simulating a square cylinder whose sensitive length is equal to its diameter [3]. The typical microdosimetric square cylinder proportional counter have sensitive volume of at least 0.25 litre and is filled with tissue equivalent gas mixture under the pressure in the range 100-200 torr. This counter could be applied as a microdosimestric detector simulating the tissue volumes with diameters of about a few μm. However, to simulate smaller tissue volumes, such as a nucleus cell, the counter dimensions have to be reduced.

Basing on our many year experience in the field of proportional counters developments and investigations the proportional counter for that purpose has been designed and checked up [4]. Further diminishing of the simulated diameters till nanometer scale could be done by reducing the counter geometric parameters and the pressure of filling gas. However, the construction of a sufficiently small proportional counter is technologicaly very difficult and lowering of the gas pressure makes the size of avalanche multiplication region comparable with that of the counter.

Microstrip Gas Chamber (MSGC) has been invented by A. Oed for neutron detection purposes [5]. F. Sauli [6] has presented the possible applications of high energy physics gas detectors in microdosimetry field. To overcome the above mentioned problems the use of one of them a MSGC seems very attractive. To verify the possibility of MSGCs application as detectors for microdosimetry Microstrip Plates elaborated by us have been investigated using different TEG mixtures.

EXPERIMENTAL

Microstrip Plates with several electrodes configurations for the purpose of this work have been designed and constructed in Warsaw at the Institute of Electronic Materials

*) This work was supported by the Project Grant KBN No. 18.220.59, of the Polish State Comittee for the Scientific Research

Technology using the elecronolitography technique [7]. Four different configurations of the strip electrodes have been investigated for the fixed parameters as follows: the anode width - 10, 15 and 20 μm ; the cathode width - 50 and 70 μm, the anode-anode pitch of 200 and 400 μm. Two types of glass plates with different surface resistivities, D263-semiconducting glass and S8900-electron conductive glass were applied, the thickness of the glass plates was 0.7 mm and the strips made in chromium. The active area of each plate was 50×50 mm^2. The plates were put into a special tight metal box with the investigated gas flowing through it (Fig.1). The cathodes were grounded and connected together each other. The measurements were carried out by means of a standard spectrometric set up.

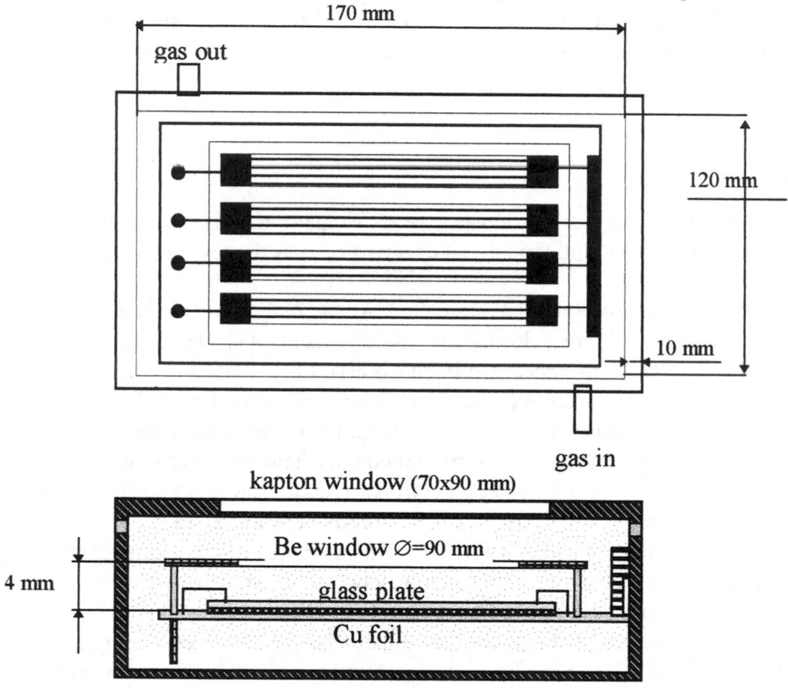

Fig.1. The cross section of MSGC detector

In order to verify the performance of the developed MSGC detector as a proportional counter the characteristic parameters, such as stability in time, gas gain curve and energy resolution, have been determined for the classical gas proportional mixture of Ar +18% CO_2. As the results were satisfactory the detector has been investigated for the TEG mixtures. The following TEG mixtures under the pressure of 765 torr and temperature 300K, were investigated:

- methane based - 64 % CH_4 + 32.5 % CO_2 + 3.5 % N_2
- propane based - 55 % C_3H_8 + 39.6 % CO_2 + 5.4 % N_2
- neon based - 40.57 % Ne + 39.6 % C_2H_4 + 16.7 % C_2H_6 + 3.13 % N_2

The results of the measurements have been compared with those obtained using the proportional counter developed earlier for microdosimetric purposes [4]. The counter sensitive volume had the shape of a cylinder, 8 mm in diameter and a length of 40 mm (0.002 litre) to simulate the small tissue volumes of at least 1 μm. Along the axis of the cathode the anode wire of 25 μm is streched. As radiation sources ^{55}Fe (Mn K) and ^{109}Cd (Ag K) were used.

RESULTS

First experiments were carried out to evaluate the stability in time of the MSGCs operating with the investigated gas mixtures. The results which are summarized in Fig 2 were observed for ^{55}Fe source at the small rate 500 mm^{-2}s^{-1}. Such small rate of radiation has been applied to avoid count rate effect. The measurement was done for the following electrodes configuration: anode width - 10 μm, cathode width - 50 μm, anode-anode pitch of 200 μm. Similiar dependences have also been observed for other electrodes configurations.

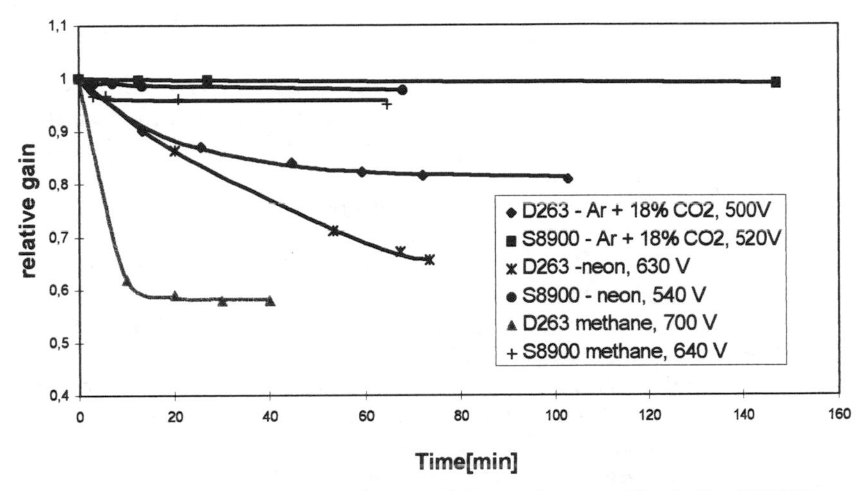

Fig.2.Time dependence of the pulse amplitude for MSGC

The gas gain curves were also determined. The maximal reached gas gain values for D263 plate compared with those for the proportional counter are following:

	MSGC		proportional counter
Ar + 18 % CO_2	$9 \cdot 10^2$	720 V	10^4
TEG (methane)	$4 \cdot 10^2$	840 V	$4 \cdot 10^2$
TEG (propane)	$2 \cdot 10^2$	800 V	-
TEG (neon)	$2 \cdot 10^3$	950 V	$2 \cdot 10^3$

For the S8900 glass plates the maximal gas gain is a little lower than this received for D263 plates.

The energy resolution and pulse amplitude dependence on the incident radiation energy were determined. As a result of the above measurements a very good linear dependence and good energy resolution have been obtained for MSGC as well as for the miniature proportional counter.

The relative detection efficiency, with respect to classical gas mixture (Ar +18% CO_2) were determined using collimated ^{55}Fe for MSGC and compared with that for the proportional counter

	MSGC	proportional counter
Ar + 18 % CO_2	1	1
TEG (methane)	0.21	0.44
TEG (propane)	0.11	-
TEG (neon)	0.15	0.21

As it is shown, the MSGC could work with TEG mixtures as well as the developed earlier proportional counter does.

CONCLUSIONS AND DISCUSSION

The results of preliminary investigations of MSGCs filled with TEG mixtures confirm their possible application as microdosimetric microstrip chambers for μm simulated size even at atmospheric pressure. The next stage of the research will include the investigations on:

a) MSGC operating conditions with TEG mixtures under low pressure - below 100 torr
b) application of TE materials as elements of the MSGC structure (window, tight box, drift electrode - at future the replacement of glass plates by plastic one)
c) design of the electrode configurations to reach higher „maximal" gas gain
d) evaluation of the chord length distribution

For microdosimetry investigations the determination of microdosimetric distribution is indispensible, which means that the local ionisation density and its variation have to be measured. The results are reported of experiments on the response and sensitivity of MSGC operating in the flow gas system with TEG. The experiments are preliminary to design of a fully microdosimetric monitor based on the MSGC filled with TEG under low pressure of about a few hPa to simulate DNA dimensions.

The application of microdosimetric microstrip gas chambers connected in a series gives the possibility of simulating the diameters in the wide range from micrometers to nanometers.

REFERENCES

[1]. H.H.Rossi and W.Rosenzweig, Radiology 64,404,1955
[2]. B. Bednarek, P. Olko and J. Booz, Nucl. Instr. and Meth. in Phys Res. A ,271, 349-358,1989
[3]. J.Booz, Rad. Environ. Bioph., 23, 155 - 170,1984
[4]. B. Bednarek and E. Rulikowska-Zarębska,
Proc. of the 10th Congress of the Polish Society of Medical Physics,
Kraków, Poland, Sept. 15-18, 1995
[5]. A. Oed,
Nucl. Instr. and Meth., A263, 351 (1988)
[6]. F. Sauli,
Rad. Prot. Dos. vol. 61, 1-3, 29-38 (1995)
[7]. L. Dobrzański, Private communication

A RADIATION QUALITY DOSEMETER BASED ON THIN ORGANIC FILMS

Carole E. Tucker, F. A. Smith and J. Oriel

Physics Department
Queen Mary and Westfield College
Mile End Road
London E1 4NS

INTRODUCTION

To date, the experimental determination of energy loss distributions in simulated sub-micron-sized volumes of tissue has been largely confined to the use of gas proportional counters. Charge amplification and collection after the initial energy deposition, followed by data reduction and correction, leads to event-size spectra after using the appropriate W-value conversion between charge released and energy deposited. Similar studies in condensed organic media have now become feasible at the nanometre level following the development of techniques for extracting electronic charge from conducting organic films. In this work the latter are formed from polymerized Langmuir-Blodgett (LB) layers which have sufficiently good in-plane conductivity to permit charge collection. The inherent advantages of a radiation dosemeter based on such a system are that:

- the thickness can be controlled to nanometre precision by varying the number of individual layers,
- the sensitive volume is condensed and organic,
- the system is suitable for modification with molecules which have biological significance.

Previous investigations into the response to uv radiation of LB multilayers of porphyrins, phthalocyanines and diacetylenes[1,2,3] have revealed the existence of anisotropic inter-molecular electron tunnelling rates in which conduction within, and across, the molecular layers takes place on picosecond and nanosecond time scales respectively. This paper reports preliminary work on the suitability of a thin layer LB molecular system as the sensitive element of a dosemeter for ionizing radiation.

EXPERIMENTAL SYSTEM

Two sets of interdigitated Al electrodes, with an electrode spacing of 140μm, were vapour deposited to a thickness of 100nm onto the upper side of a taut 7500nm thick film of kapton. Multilayer films of 10,12 pentacosadyanoic acid (also known as (12,8) poly-diacetylene) were then deposited over and between the electrodes to thicknesses of 6nm and 60nm on each of the two sets of electrodes. These monomer films were then polymerized, by exposure to uv radiation, to form the first sensitive element of the detector. Each of the kapton supports was then placed over the front surface of a silicon detector to make the second sensitive element of the combination.

Low polarization voltages on the electrodes (<10V) were used to extract charge from each film, which was then converted to a mV output using a buffer amplifier. Output from the silicon detector substrate was amplified using a charge sensitive preamplifier. Both elements of the dosemeter therefore gave outputs which were proportional to the rate of production of charge in their respective sensitive volumes.

The two combinations were mounted on a printed circuit board, each one presenting a total sensitive area to the radiation of ~100 mm^2. In order to minimize leakage between the two charge-collecting circuits on each of the two sets, a physical separation between them of 20mm was necessary in this prototype arrangement.

The two detector combinations were simultaneously exposed to two radiation sources: photons in the range 3 - 4.5 eV from a large area lamp, and 4.5 MeV α-particles from an ^{241}Am source occupying an area of 2mm diameter. Since there were large uncertainties and differences in the areas and intensities of these sources, there were correspondingly large errors in the corrections necessary for solid angle effects. The separation between incident flux and energy deposited in the detectors was therefore not made in the presentation of the results (Figures 1,2).

THEORY

A series of different film thicknesses having 1, 2, ...n layers is exposed to fluxes of radiation of different LET. If N_L and N_H represent the fluxes of low and high LET photons or particles, the number of secondary electrons, e_n , produced in the different thicknesses when the irradiation is uniform over all film areas is :

$$e_1 = y_{1H}N_H + y_{1L}N_L$$
$$e_2 = y_{2H}N_H + y_{2L}N_L$$
$$\cdots\cdots\cdots\cdots\cdots\cdots$$
$$e_n = y_{nH}N_H + y_{nL}N_L$$

where y_{nL} and y_{nH} are the secondary electron yields due to low and high LET radiation. For low LET radiation, in which energy deposition is uniform with depth into the film, the electron yields will scale with film thickness such that, e.g. $y_{2L} = 2y_{1L}$. This will not be the case for high LET radiation if the energy deposition varies sufficiently with depth over the thickness of the film. A set of n film thicknesses will therefore provide a quantitative measure of the distribution of energy deposition rates with thickness from a set of differential equations. A solution for the case of two such films will be :

$$N_L = \frac{e_2 y_{1H} - e_1 y_{2H}}{y_L(2y_{1H} - y_{2H})}$$
$$N_H = \frac{2(e_1 - e_2)}{2y_{1H} - y_{2H}}$$

The yields, which reflect the energy and particle dependence of the W-values for the material of the films, are not known with any degree of precision for the condensed organic media used in this work. Their measurement can be attempted using a combination of the techniques described in the work of Byrne et al[4].

RESULTS

The linear response of the LB film system to increasing fluxes of 3 - 4.5 eV photons is shown in Figure 1, after using a calibrated uv detector to determine absolute photon fluxes. Care had to be taken to correct for the measureable thermal response of the LB film system when using the uv lamp.

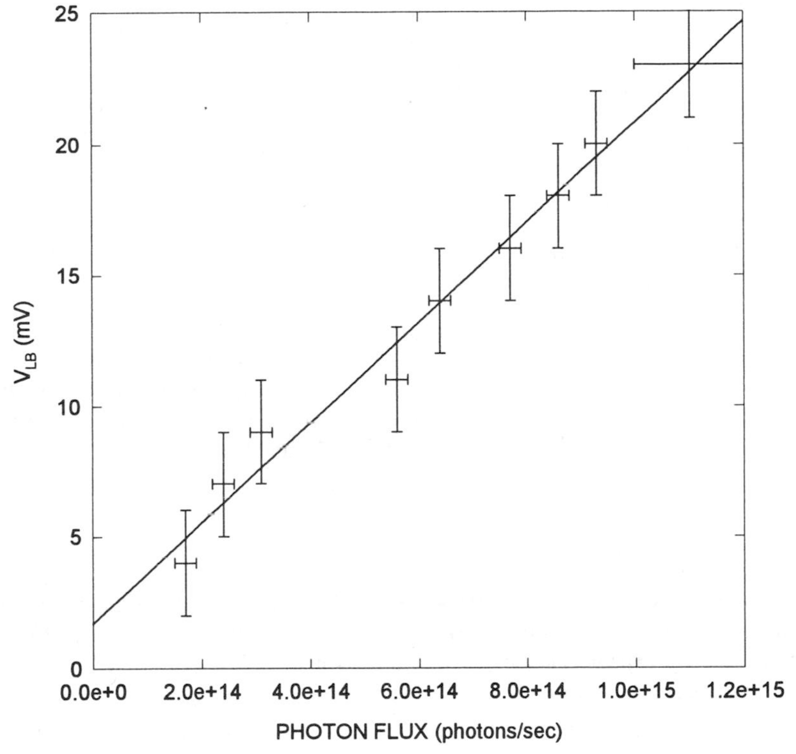

Figure 1 Response of 60nm of (12,8) polydiacetylene to 3 - 4.5 eV photons

The response to α-particles was tested using two methods of charge collection. In Figure 2a the charge deposition rate is converted to an amplified voltage with a conversion factor of 1pA ≅ 1mV. Because of the α-particle energy losses in the insulating layers between the two sensitive elements of the combined detector, the responses shown in Figure 2a do not refer to the same α-particle energy.

In Figure 2b, the charge generated within the film was stored using a modifed Baldwin Farmer electrometer, so that the reciprocal of the time needed to record a full-scale deflection was proportional to the time rate of energy deposition. Particles at the maximum energy (4.5 MeV) impinged on thicknesses of 15, 32 and 90 layers (3nm per layer) and showed a linear response, suggesting that the polarization voltages used (2.16V and 10.29V) were not high enough to overcome charge trapping effects.

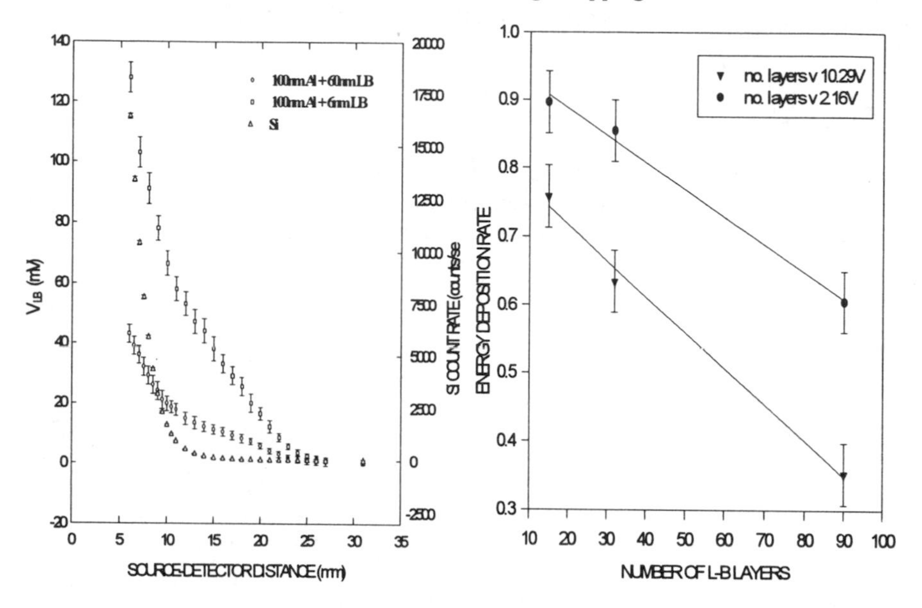

Figure 2 (a) Voltages proportional to charge deposition rate in 6nm and 60nm thick LB films and the response of the silicon detector substrate to the same, but energy-degraded, α-particle flux. (b) energy deposition rate using capacitor method of charge storage for 4.5 MeV α-particles impinging on thicknesses of 15, 32 and 90 layers.

CONCLUSIONS

The work has shown that nanometre layers of organic film will respond to ionizing radiation and may be used to form part of the sensitive element of a new type of radiation dosemeter which is capable of giving information on radiation quality. The work forms part of British Patent Application No: 9606752.5 - Queen Mary and Westfield College.

References

1. K.J.Donovan, R.V.Sudiwala and E.G.Wilson, Thin Solid Films, 1992, **210**, 271.
2. K.J.Donovan, R.Paradiso, K.Scott, R.V.Sudiwala, E.G.Wilson, R.Bonnett, R.F.Wilkins, D.A.Batzel, T.R.Clark and M.E.Kenny, Thin Solid Films, 1992, **210**, 253.
3. K.J.Donovan, K.Scott, R.V.Sudiwala, E.G.Wilson, R.Bonnett, R.F.Wilkin, R.Paradiso, T.R. Clark, D.A.Batzel, and M.E.Kenny, Thin Solid Films, 1994, **244**, 923.
4. C.Byrne, F.A.Smith, M.A.Hill, C.E.Tucker and J.Oriel, to be published (1997)

MICRODOSIMETRY OF LiF:Mg,Cu,P SOLID STATE DETECTORS - WHAT IS THE TARGET?

P. Olko and P. Bilski
Health Physics Laboratory
Institute of Nuclear Physics, ul. Radzikowskiego 152,
PL 31-342 Kraków, POLAND

1 INTRODUCTION

Phenomenological biophysical models, such as Track Structure Theory,TST[1,2] or the microdosimetric One-Hit Detector Model,[3] have been successfully applied to study dosimetric characteristics of physical detectors such as alanine, Fricke dosemeters or thermoluminescence detectors, TLDs etc. Even without any deeper understanding of the underlying physical mechanisms, e.g. of the phenomenon of thermoluminescence, it was possible to fit model parameters to measured dose- and energy response curves and to predict responses for other dose ranges and radiation modalities.

In each of these models there appears a free parameter of dimension of length, which should correlate with the size of some hypothetical sensitive site in the detector. In microdosimetric models, one of the main model parameters is the diameter of the sensitive site, *d*, which is assumed *a priori* or estimated by fitting the model to experimental data. However, in the case of solid state detectors the question arises what is the physical meaning of the evaluated diameter, since it is usually impossible to identify in the detector the structure of the sensitive site.

In this paper the microdosimetric one-hit detector model[3] was used to study the response of a new generation of highly-sensitive thermoluminescent detectors LiF:Mg,Cu,P (MCP-N) after doses of different radiations. It will be shown that the model-calculated target size correlates with the concentration of Mg ions which presumably constitute thermoluminescence trapping centres. With increasing Mg concentrations the average distance between the ions decreases, which also decreases the target diameter.

2 METHOD

2.1. Experimental data on LiF:Mg,Cu,P thermoluminescent detectors

Thermoluminescence of highly sensitive LiF:Mg,Cu,P detectors has been extensively investigated in the last years and several measurements of TL saturation, photon energy response, TL efficiency have been published. It was found that the response of this TL material strongly depends on the material (batch) and on handling procedures. For that reason only MCP-N detectors produced at the Institute of Nuclear Physics in Kraków, of standard dopant composition (0.05% Cu, 0.2% Mg and 1.25 % P molar weight), were chosen for all experiments. In a series of experiments[4,5] performed over the past few years, the relative TL efficiency, η_i, defined as the TL light output of the whole glow curve per unit dose of the investigated radiation and normalised to standard radiation (Cs-137 γ-rays), was determined for 24 different radiation modalities (X-rays, γ-rays, beta electrons, thermal neutrons and 1.5 to 5 MeV α-particles). The dose response curve for 662 keV γ-rays was measured at the Institute of Nuclear Physics over a dose range from a few mGy up to 1000 Gy. In addition, TL efficiency for 5.3 MeV stopping α-particles, relative to Cs-137 γ-rays was measured for a specially prepared, also at the INP Krakow, set of detectors with Mg concentration, C_{Mg}, equal to 0.05, 0.1, 0.15 and 0.2 % of molar weight. All this experimental data was analyzed using microdosimetric models.

2.2. The Microdosimetric One-Hit Detector Model for LiF:Mg,Cu,P (MCP-N)

The microdosimetric One-Hit Detector model assumes that the detector contains a number of radiation sensitive sites. As radiation energy is deposited within the site (a hit occurs) there is a certain probability that the effect will take place. The term "one hit" means that a single hit deposited within the target volume is sufficient to produce the full effect.

The probability of TL light emission, $P_{TL}(D)$ for the i-th radiation is an inverse probability that a sensitive site will not be hit

$$P_{TL}(D) = 1 - S(D) = 1 - \exp\left[-\frac{D}{\overline{z_F}}\int_0^\infty (1-\exp(-\alpha z)) f_1^i(z)dz\right] \tag{1}$$

For low doses i.e. when D<< $\overline{z_F}$, TL efficiency, η, is expressed:

$$\eta_i = \frac{\dfrac{1}{\overline{z_F^i}}\displaystyle\int_0^\infty (1-\exp(-\alpha z)) f_1^i(z)dz}{\dfrac{1}{\overline{z_F^{Cs137}}}\displaystyle\int_0^\infty (1-\exp(-\alpha z) f_1^{Cs137}(z)dz} \tag{2}$$

Two free model parameters are used i.e. saturation parameter, α, and target diameter, d, which influences $f_j^i(z)$ distributions.

3 RESULTS

For evaluating the model parameters three sets of data are required i.e. microdosimetric distributions $f_j^i(z)$ for all i radiations considered, measured TL efficiencies, η_i, and measured dose -response $P_{TL}(D)$ e.g. for Cs-137 γ-rays. Microdosimetric single-event distributions $f_1^i(z)$ in 1 gcm^{-3} water for photons, beta electrons, tritons and alpha particles for target sites from 5 to 500 nm were calculated using Monte Carlo track structure calculations[5] employing MOCA-8, MOCA-14, TRION and PHOEL-2 codes and scaled next to targets of density 2.5 gcm^{-3}. Next, using a minimising routine based on the least-squares method, values of model parameters α=0.091 Gy^{-1} and d=24 nm were found which best fitted Eqs.1,2 to the set of values of TL data listed in 2.1.

The agreement between measured and calculated $P_{TL}(D)$ dose response for Cs-137 γ-rays as well as for measured and calculated η_i for a broad spectrum of radiations is fairly good[5], particularly for alpha particles and for photons with energies above 20 keV.

A rapid decrease of response per unit dose with increasing y_D can be observed not only for densely ionizing α-particles but even for low-LET photons, which is typical for one-hit detectors.

The measured values of η_i, for 5.3 α-particles for LiF:Mg,Cu,P samples with different Mg concentration and absolute TL sensitivity of these samples for Cs-137 γ-rays (in arbitrary units) are presented in Fig. 1. The bold line is a fit with a function $\eta_i = a\,(C_{Mg} - b)^{-1/3}$, where a= 0.034 and b= 0.0357.

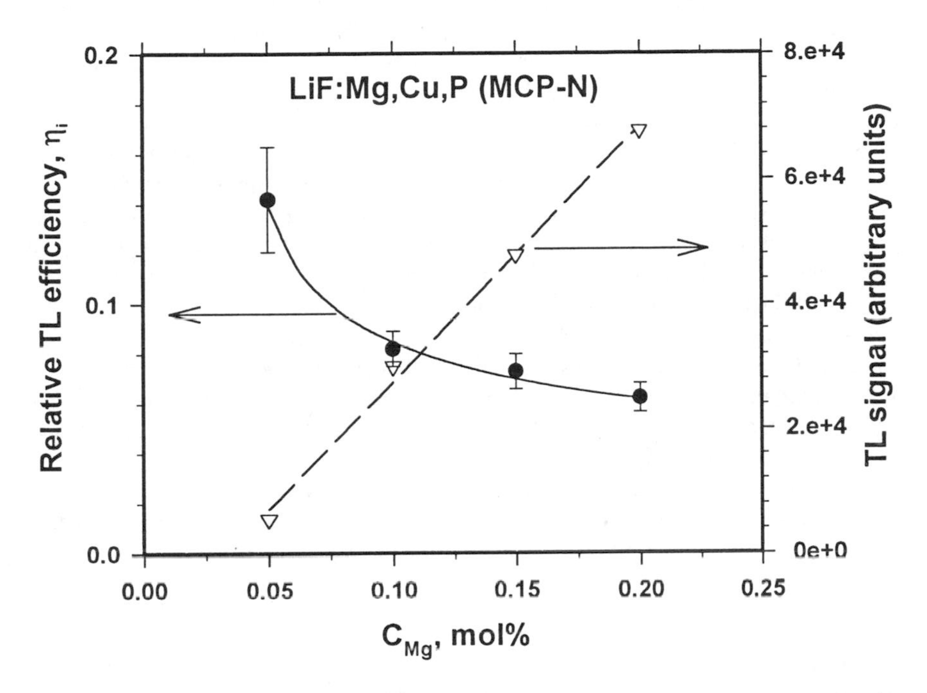

Figure 1. *Relative TL efficiency, η_i, of LiF:Mg,Cu,P detectors for 5.3 MeV stopping α-particles versus Mg concentration (left) and total TL light output of these detectors per 1 mGy of Cs-137 gamma-rays (right).*

4 DISCUSSION AND CONCLUSIONS

It was assumed in this paper that TL processes in LiF:Mg,Cu,P take place in certain distinct regions (volumes) of the detector. The first question which arises is whether the volume is associated with spatial distribution of radiation induced tracks or with any particular spatial organisation of activators or defects within the cristal. The first option was postulated in Horowitz's Track Interaction Model[6], where an effective radius r_0 parametrizes the maximum distance from the alpha particle track from which TL photons originate. A typical value of r_0 for 4 MeV α-particle is about 10 - 20 nm for different peaks of LiF:Mg,Ti. Good agreement between calculated and measured η_i values obtained in the present work for a one set of parameters for all 24 radiation modalities (d=24 nm and α=0.091 Gy^{-1}) could suggest that the sensitive site is independent of radiation type and is rather associated with the detector structure. The value of 24 nm in

LiF is of about 1.5 times higher than the radius of the sensitive volume obtained by Waligórski et al[7] (40 nm diameter in 1 g cm^{-3} in water) from the Katz model analysis of the response of LiF:Mg,Cu,P after proton and α-particle doses. For one-hit detectors, averaging of the effect in mixed radiation fields appears to lead to identical results from Katz's track structure theory and microdosimetric model calculations[8].

Thermoluminescence is a two-stage process. It was suggested by McKeever[9] that the glow peak structure of LiF:Mg,Cu,P is governed by the presence of Mg. Mg ions, which are supposed to be built into the LiF crystal structure in the form of dipols with Li vacancies, seem to be involved in the absorption stage of thermoluminescence i.e. constitute the thermoluminescence trapping centers. This hypothesis may be qualitatively analysed using results of experiments with different Mg concentrations. Over Mg concentrations ranging from 0.1 to 0.2% the overall TL sensitivity for γ-rays increases almost in proportion with Mg concentration (see Fig. 1 - broken line). It is also worth mentioning that the ratio of γ-ray sensitivity between LiF:Mg,Cu,P and LiF:Mg,Ti (25:1) is not much different from the ratio of Mg concentrations in both phosphors (10:1).

If the target size is correlated with the average distance between Mg ions (defects, complexes?) then an increase of Mg concentration should reduce, within a certain range of concentrations, the target size. The reduction of the relative TL efficiency for α-particles for higher Mg concentration, or in other words, slower increase of α-particle-induced TL signal in comparison to γ-rays, may result from faster saturation of the response of the smaller size targets after high-LET hits. As seen in Fig.1, η_i varies with Mg concentration as $C_{Mg}^{-1/3}$, i.e. is inversely proportional to the average distance between Mg ions. This may support the view that the target diameter is correlated with the average distance of Mg complexes in LiF.

However, since parameters d and α are partly correlated, more experimental data are necessary to settle this issue. In particular, dose-response saturation experiments for low energy X-rays and α-particles can prove or disprove our interpretation of target size.

Acknowledgements
The authors are indebted to Prof. M.P.R. Waligórski for valuable discussions and comments on the manuscript. One of us (PO) thanks EURADOS for partly supporting his participation at the 12th Symposium on Microdosimetry.

References

1. Katz, R., *Nucl. Track Detection,* 1978, **2**, 1
2. Waligórski, M.P.R. and Katz, R. *Nucl. Instr. Meth.,* 1980, **172**, 463
3. Zaider, M. Radiat. Res. 1990, **124**, S16
4. Bilski, P., Olko, P., Burgkhardt, B., Piesch, E. and Waligórski, M.P.R. *Radiat. Prot. Dosim*. 1994, **55**, 31
5. Olko, P., *Radiat. Prot. Dosim*. **65**, 1996, No. 1-4, 151
6. Horowitz,Y., Moscovitch, M., Dubi, A. *Phys. Med. Biol.* **27**, 1982, 1325
7. Waligórski, M.P.R., Olko, P. Bilski, P., Budzanowski, M., Niewiadomski, T. *Radiat. Prot. Dosim*. **47**, 1993, 53
8. Katz, R. *Radiat. Res.* **133**, 1993, 390
9. Mc Keever, S.W.S, *J.Phys.D: Appl.Phys.* **24**, 1991, 988

THE PEAK-HEIGHT RATIO (HTR)-METHOD FOR LET-DETERMINATION WITH TLDs AND AN ATTEMPT FOR A MICRODOSIMETRIC INTERPRETATION

W. Schöner, N. Vana, M. Fugger and E. Pohn

Atomic Institute of the Austrian Universities, A-1020 Vienna, Schüttelstraße 115

1 INTRODUCTION

For determination of the absorbed dose and the average LET in complex mixed space radiation fields, the HTR method was developed. The method utilizes the changes of peak height ratios in thermoluminescence glowcurves obtained from LiF-dosemeters. This method was used for measurements of absorbed dose and average LET in space. Measurements were carried out during several missions on space station MIR, on the surface of satellites and in space shuttles. Our results of the average LET measurements respectively the assessment of radiation quality factor obtained with the HTR-method are in good agreement with results of French-Russian measurements with a LET-spectrometer (CIRCE) on MIR station and results of TEPC-measurements in shuttles by the Space Radiation Analyses Group (NASA). In the course of further development a microdosimetric interpretation of the LET dependent effects was achieved. The applied track structure model suggests that the changes of the peak height ratios is caused by the variation of the microdosimetric energy density distribution. This opens the opportunity to use the HTR-method as a tool for microdosimetric investigations.

2 SCIENTIFIC METHODS

In standard TL-dosimetry with LiF dosemeters light emission at temperatures below about 240 °C is analysed. The intensity of the dominant peak (peak 5) in the glowcurve appearing at about 200 °C is linear over a wide range of absorbed dose. For evaluation of absorbed dose either peak amplitude or peak integral is used. After absorption of radiation with higher LET, peak ratios change significantly. In general high temperature peaks are enhanced after absorption of high LET radiation. These peaks appear in the temperature range from 240 to 350 °C. This effect is used in the HTR-method, developed at the Atomic Institute, Vienna and was calibrated with test irradiations of different LET (1). As parameter for the average LET of absorbed radiation the HTR-method uses the ratio of the intensity of TL-light in the high temperature region to the intensity of emission in the same temperature range after Co-60 irradiation.

3 EXPERIMENTAL

Measurements on space station MIR were carried out during the Austrian-Russian mission AUSTROMIR (May-Oct.91) and during the Russian-Austrian joint project RLF (Russian Long Duration Flight, Jan 94-March 95), experiment ADLET. In experiment ADLET absorbed dose and average LET were recorded over three periods. During each period dosemeters were exposed in low and heavy shielded positions, in order to measure the variations of doserate and average LET within the habitable area of the space station MIR. During the shuttle missions STS-60, 63 and 71 (1994-95) measurements with the HTR-method were performed to determine quality factor and equivalent dose. All Austrian TL-dosemeters were read out with the labor-made read-out system TL-DAT-II and evaluated by a specifically developed software.

4 RESULTS FROM MEASUREMENTS IN SPACE

All dosemeters were calibrated with Co-60 gamma radiation. Gathered data show that the measured average doserate depends very strongly on position. During period ADLET-1 the ratio of measured absorbed dose in the two positions was 1.48. During period ADLET-2 the ratio was only 1.27. Austrian and Russian results of absorbed dose measurements are in good agreement. A summary of results from space station MIR and STS-60, 63 and 71 is shown in table 1. The errors given with the results are only statistical. The total error for dose measurements is assumed to be less than 5% and the error for LET measurements less than 10%. For the evaluation of the quality factor the relation between LET and Q proposed in ICRP-26 was used because "average LET" is not meaningful using ICRP 60, which demands the assessment of the various components of the radiation field.

Table 1. Summary of Results from Dose Measurements in Space Crafts (2)

Mission	Duration (d)	Position	Absorbed dose rate (mGy/d)	average LET (keV/μm tiss)	Q	Equivalent dose rate (mSv/d)
ADLET-3	437	Pos 1	0.288±0.010	6.4±0.1	1.8	0.518±0.018
		Pos.2	0.220±0.010	6.7±0.2	1.9	0.418±0.019
STS-60	8.3	TLD-600	0.165±0.012	10.6±0.6 [1)]	2.7 [1)]	0.446±0.032 [1)]
		TLD-700	0.149±0.010	6.5±0.4	1.8	0.268±0.018
		NASA	0.149±0.004		2.55 [2)]	0.380±0.010
STS-63	8.2	TLD-600	0.316±0.021	10.2±0.4 [1)]	2.6 [1)]	0.822±0.055 [1)]
		TLD-700	0.303±0.011	8.8±0.6	2.3	0.697±0.025
STS-71	9.8	TLD-600	0.233±0.006	14.6±0.3 [1)]	3.4 [1)]	0.792±0.020 [1)]
		TLD-700	0.209±0.004	8.0±0.2	2.2	0.460±0.009

[1)] includes thermal neutron-component [2)] NASA TEPC data

In general average LET is lower at the position with weaker shielding and higher behind stronger shielding. The values obtained by TLD-600 are higher because of the evidence of thermal neutrons. This difference is more significant for the results from space shuttles. During STS-63 mission the measured absorbed doserate was higher by a factor of 2 compared to STS-60. The measured average LET, neglecting thermal neutrons (TLD-700), was also higher.

5 AN ATTEMPT FOR A MICRODOSIMETRIC INTERPRETATION

In order to improve the understanding of LET dependent mechanism of TLDs we started to investigate the microdosimetric properties of TLDs. For this purpose we applied a model of track structure theory. **It is commonly known that the LET is not** the optimal parameter to describe the biological effectiveness of radiation. For this reason in microdosimetry the LET is replaced by statistical parameters like the lineal energy or the distributions of energy density or the specific energy. Of course, from the point of view of microdosimetry, "LET"-dependent effects are a result of the microdosimetric distribution of energy density. The results of our investigations confirm, that the shape of the glowcurves respectively the HTR is determined by the distribution of the deposited energy density in the crystal.

5.1 Track Structure Model Used

Track Structure Theory was developed in order to understand the behaviour of different detectors as well as the sensitivity of biological material in connection with the absorption of radiation with high LET (3). In contradiction to the concept of LET this theory takes the radial distribution of energy density around the track of a charged particle into account. Several authors have applied this theory on thermoluminescent dosemeters in order to calculate their efficiency against HCP (4, 5). We developed a computer code for the calculation of the average grain dose distribution for various projectile ions and folding with the measured, dose-dependent beta efficiencies of TL-crystals in order to calculate the relative peak efficiencies after absorption of HCP compared to beta irradiation. With this computer code it is possible, to do this calculation for various diameters of the sensitive sites as suggested by Hansen (6). For the calculation of the radial energy density distribution track structure model introduced by Zhang (7) was used.

5.2 Experimental

Results from various types of dosemeters irradiated at the accelerator center in Dubna with high energetic heavy ions (2) were used for calculation The dose response of the dosemeters (peak 5, 6 and 7) after absorption of sparely ionizing radiation was recorded up to 40 kGy after irradiation with a Sr-90/Y-90 beta source. For testing the model data obtained after alpha irradiations (Am-241) were used.

5.3 Results

Table 2 shows the excellent agreement between measured and calculated peak efficiencies. The radius of the sensitive site a_0 was fitted to the data. Radii obtained from this experiment were used for calculations of the efficiencies of high energetic heavy ions.

Table 2: Measured and calculated peak efficiencies of TLD-600 after alpha irradiation:

Dosemeter	Peak number	Measured efficiency	Target radius a_0 (nm)	Calculated efficiency
TLD-600	5	0.125	15	0.124
TLD-600	6	0.093	15	0.094
TLD-600	7	0.070	11	0.070

Although our calculations for the alpha efficiencies of single peaks lead to very good results in agreement with measured and reported values of $\eta_{\alpha\beta}$, the model is underestimating the efficiencies after absorption of high energetic ions. Nevertheless the model describes the observed relative increase of the high temperature peaks (peak 6 and 7) with increasing LET of absorbed radiation.

6 DISCUSSION

The HTR-method was successfully applied for measurements of the absorbed dose and the average LET in spacecrafts and aircrafts (8). The method, using small TL-crystals, is also applicable to measure the LET distribution within matter. Good results were obtained in a therapeutic modulated proton beam (9). The interpretation of LET-dependence leads to microdosimetric approches. The applied track structure model confirms, that the peak ratios in glowcurves from LiF-TLDs are determined by the microdosimetric distribution of the energy density. For high energetic ions the model does not describe the absolute values of peak efficiencies sufficiently, but qualitatively the increase of peak ratios with increasing local energy density is described. The deviation of the calculated results from the experiments with high energetic ions is caused by at least two reasons:
1. The model used for the calculation of energy density in the particle tracks is neglecting the angel distribution of delta electrons.
2. The model is assuming homogeneous tracks. This is more or less true for low energy alpha tracks. For high energetic particles the statistical nature of energy deposition (treelike shape of tracks) has to be taken into account.
Improvements of the model are intended by the intoduction of statistical models for particle tracks. A better understanding of the microdosimetric properties of TLDs would open the opportunity to use TLDs as micro-dosemeters, providing important information for a better understanding of biological effects after absorption of different kinds of radiation.

References

(1) Vana, N.et al, *Radiat.Prot.Dosim.*,1995, **66,** 145-152
(2) Vana, N. et al, COSPAR 31st Scientific Ass., *Adv.in Space Resarch*, 1996, in press
(3) Katz, R. et al, *Supplement 1, Academic Press*, New York, 1972, 317-383
(4) Kalef-Ezra, J., Horowitz, Y., *Int. J. Appl. Radiat. Isot.*,1982, **11**, 1085
(5) Zimmerman, D.W., *Radiat. Eff.*, 1972, **14**, 81
(6) Hansen, J.W., Riso National Laboratory, Riso-R-507, 1984
(7) Zhang, C.X. et al., *Radiat.Prot.Dosim.*,1994, **52**, 93-96
(8) M.Noll et al, *Radiat.Prot.Dosim.*, **66**, S. 119 - 124, 1996
(9) M. Noll et al, 12th Symposium on Microdosimetry, Oxford, Proc. 1996

MICRODOSIMETRY RESEARCH: A HISTORIC OVERVIEW

H. H. Rossi

105 Larchdale Ave
Upper NYACK
N.Y. 10960

At the time of the 12th Symposium on Microdosimetry it seemed fitting to look at the records of the early ones I note in the massive tome containing the Proceedings of the first Symposium that G.J. Neary expressed the hope that there would be others - a wish that was amply fulfilled. I should add that Dr Neary's scientific presentation anticipated much of my own work.

I also note that at the 2nd Symposium I voiced cautious optimism that microdosimetry would amount to more than a collection of definitions and measurements of physical quantities. To what extent has this hope been met and what is the future of microdosimetry?

The realization that equal absorbed doses of different radiations usually entail different biological effect probabilities made it essential that radiation quality be defined quantitatively. A logical index appeared to be the rate of energy loss (LET) in the tracks of charged particles that deliver the absorbed dose. Apart from other objections[1] a principal difficulty is that the required distribution of absorbed dose in LET is a theoretical concept devoid of physical reality. The energy absorbed in the mass contained in any volume of irradiated matter is never entirely proportional to LET. Hence, the distribution can be measured only approximately.

What can be measured is the probability distribution of actual energy deposits in terms of the lineal energy, y, with a mean value which differs from the LET. The construction of an instrument to accomplish this turned out to be relatively easy[2] following the development of the tissue-equivalent ionization chamber and its tissue equivalent plastic[3]. The second or third design has been retained with minor changes for forty years in the walled version, and also for a long time in the wall-less type. The device has almost perfect sensitivity because it can register nearly all the electron avalanches produced by even single ionizations. Its main limitation is a restriction of simulated site diameters to more than several tenths of a micrometer. In recent years there has been considerable activity in producing designs intended to operate at the nanometer level. Examples are covered in an EURADOS report[4] and in the proceedings of an international workshop[5].

In radiation fields of unknown composition, microdosimetry provides what seems to be the only method of obtaining the information required in radiation protection. Here counters that are large (or contain large inside surfaces[6]) are required for adequate counting rates. This is another direction for further technical development.

In therapy, especially with radiations other than photons, microdosimetric determinations are now commonly made and a, thus far, limited correlation between lineal energy and clinical RBE has been established for neutron therapy[7]. In therapy there is the opposite problem of excessive counting rates indicating the need for miniature counters that can also be suitable for phantom studies.

Since the distribution, f(y) depends on the site diameter, it is necessary to select a diameter that represents the intracellular region where energy concentration determines biological effects. The justification of the choice of about 1 μm was a rigorous theorem by Kellerer[8]. It implies that the departure from linearity of survival curves of cells irradiated with low LET must be due to action by multiple events in sites that are at least several hundred nanometers in diameter.

For reasons, that will be given below, site diameters of about 1 μm represent the best compromise for measurements in both radiation protection and radiotherapy.

When hoping for a fundamental contribution by microdosimetry, I was primarily considering applications to radiobiology although well aware [1]of widespread scepticism that the study of the distribution of absorbed energy in irradiated tissues could have immediate significance to an understanding of radiation effects. No doubt it needs to be the first step, but only the first step in a detailed accounting of a series of complexities which, even now, some 25 years later have not been unravelled. My hope was that without struggling with the involved chain of processes one might find some general rules that, being based on physical facts, are more credible than the various hypotheses and models that assure no more than mathematical fits to dose-effect curves.

From a number of considerations and especially the dependence of RBE on absorbed dose, it appeared that many radiobiological data could be accounted for in terms of a mechanism in which two entities, termed sublesions, combine over average distances of about 1 μm to form lesions that are responsible for radiation damage. This results in the simplest relation short of simple proportionality of radiation damage and specific energy* because it makes the former proportional to the square of the latter.

The resulting "α-β model" applies to lesions and not necessarily to such effects as cell inactivation (which must at any rate be subject to lesion saturation) or to the far more complex carcinogenesis.

In the first treatment of "Dual Radiation Action"[9] some possible complications were mentioned.

1. The combination of sublesions may depend on their initial separation which makes the geometrical distribution of energy deposits in the "site" important.
2. The yield of sublesions may depend on radiation quality.
3. Biological effects may be caused not only in dual radiation action but also as a result of single injuries (possibly single "sublesions").

On the basis of the probability distribution, f(y), one may readily derive f(z) the probability distribution of specific energy, z. This function represents the fluctuation of local energy density about its mean value, which is the absorbed dose, as produced by one or more microdosimetric events with an event frequencey, Φ.

With the usual complexity of nature all three of these possibilities turned out to be significant. In dealing with the first one, a separation-dependent treatment was based on Kellerer's introduction of proximity functions[10], and the molecular ion experiment[11] provided some data indicating two classes of interaction distances. Further such experiments (especially with heavier ions) are desirable.

Compound dual radiation[12] action which deals with the second and third complications abandons the "black box" approach with the supposition that "sublesions" are damaged DNA molecular (or RNA molecules in the case of plants) and that the "lesions" can be any of the variety of chromosome aberrations that result from two DNA damages [most likely DSBs (double strand breaks) followed by chromosome breaks]. The term "sublesion" became somewhat misleading because single breaks as well as pairs of interacting ones can have major consequences.

While dual radiation action may well be subject to further modification, it seems very likely that the application of microdosimetry to radiobiological findings has established a number of facts.

1. Single events of any ionizing radiation can cause decisive deleterious biological damage. DSBs and subsequent chromosome breaks can result in cell death, mutations and carcinogenesis. At the absorbed doses involved in most radiobiology the probability that DSBs are caused by two events is negligible even for relativistic electrons. The RBE of neutrons or α particles is therefore dose independent and it is probably no more than 2[13] but larger for heavier ions. Less than one in 5000 electrons traversing the nucleus of V79 hamster cells causes lethality by this mechanism[14].
2. Pairs of DNA damage interact with distance-dependent probability over a initial separation of the order of 1μm. The probability that a single high energy (>100 keV) electron causes both the DNA damages is negligible but this is of primary significance for high LET radiation. The RBE therefore increases inversely as the square root of the absorbed dose of high LET radiation.
3. Biological Effects are generally due to a varying superposition of these two mechanisms which explains the differences in maximum RBE for biological effects.

Whilst there considerations indicate that in practical microdosimetry "micrometer" sites are more important, there remains a theoretical interest in the "nanometer" scale of radiation action. There a number of considerations are involved.

The continuous f(y) spectrum is a diffuse representation of the histogram of integral number of ionizations. However, largely due to the work of Srdoc and his collaborators these numbers can be identified[15].

In only a few collisions charged particles (and their delta rays) impart sufficient energy to cause ionizations which may be called <u>relevant energy transfers.</u> This term, rather than "ionization", which is somewhat vague in the condensed phase can also refer to those changes that can contribute to biological effects. There are some indications that the yield of relevant transfer points does not differ substantially for irradiated tissue-equivalent gases and tissues[16] and at least in a first approximation the geometric distribution of relevant transfer points is the same in the two media. More definite analyses are needed.

This suggests modification of Kellerer's proximity function of energy transfer to the proximity function of relevant transfer points which (hopefully) should be the same in the two media. In an elegant combination of experiment and theory the measurement of the dose average lineal energy as a function of site diameter, as determined with the variance covariance technique, permits calculation of the proximity function of energy transfers[17, 18]. This non-stochastic relation eliminates the need for the site concept and, in principle, permits the correlation of the average response of the biological system to an average distribution of relevant energy transfer points. Its application to radiobiology and specifically to compound dual radiation action poses considerable theoretical challenges.

At the tens-of-nanometer level transport of energy (e.g. by active radicals) becomes important and the location of a relevant transfer point may differ from that of the target of the DNA. It should be an intriguing task to exploit what seems to be a deficiency and to derive the spatial aspects of energy transport by relating the yield of DSBs to various configurations of relevant transfer points. Such studies rather than being performed with equal absorbed doses of different radiations may, more efficiently, be made with ions of different mass but equal LET. Data on the yield of DSBs are already available[19].

Microdosimetry has obvious significance in (relatively) simpler non-biological systems. There have -thus far- been rather few applications.

Dr Li Kaibo pointed out that the combination of electrons and holes in thermoluminescent materials is a perfect case of dual radiation action and a paper was published on this subject[20] which merits further investigation.

Radiation damage to microelectronic devices is another important subject[21].

The basic features of radiation chemistry cannot be understood without microdosimetry[22].

Mathematical aspects -especially those of geometric probability- have been strikingly developed by Kellerer[23].

Much has been done and more needs to be done.

References

1. H.H. Rossi, *Radiat. Res.*, 1959, **10**, 522.
2. H.H. Rossi and W. Rosenzweig, *Radiat. Res.*, 1955b, **2**, 417.
3. H.H. Rossi and G. Failla, *Nucleonics*, 1956, **14(2),** 32.
4. Design, Construction and Use of Tissue Equivalent Proportional Counters. Schmitz, T., *et al* Eds. *Rad. Prot. Des.*, **61/4** 1995.
5. Advances in Radiation Measurements: Applications and Research Needs in Health Physics and Dosimetry. Waker, A.J. *et al* Eds. *Rad. Prot. Dos.* **61/1-3** 1995.
6. P.J. Kliauga, H.H. Rossi and G. Johnson, *Health Phys.,* 1989, 57/4: 631.
7. P. Pihet, H.G. Menzel, W.G. Alberts and H. Kluge, *Rad. Prot. Dos.*, 1989, **29/1-2**, 113-118.
8. O. Hug and A.M. Kellerer, 'Stochastik der Strahlenwirkung', Springer Verlag, Berlin, 1966.
9. A.M. Kellerer and H.H. Rossi, *Radiat. Res.,* 1972, Q8: 85-158.
10. A.M. Kellerer, "Analysis of Patterns of Energy Deposition EURATOM Report EUR 4452 d-e-f", Brussels, H.G. Ebert (ed): Second Symposium on Microdosimetry: 107-134.
11. A.M. Kellerer, Y.P. Lam and H.H. Rossi, *Radiat. Res.,* 1980, **83**, 522-528.

12. H.H. Rossi and M. Zaider, *Radiat. Res.*, 1992, **132**, 178-183.
13. H.H. Rossi, *Radiat. Prot.,* 1994, *Dos.* **52**, 9-12.
14. H.H. Rossi and M. Zaider (To be published).
15. D. Srdoc, B. Obelic and I. Kajar Bronie, *J. Phys. B. Atom. Molec. Phys.*, 1987, **20**, 4473.
16. H.H. Rossi and M. Zaider, 1996, 'Microdosimetry and its Applications, Springer, p.86.
17. L. Lindborg *et al*, *Rad. Prot. Dos.*, 1985, **13/1-4,** 347-351.
18. M. Zaider, D.J. Brenner, K. Hanson and G.N. Minerbo, *Radiat. Res.*, 1982, **91**, 95-103
19. G. Kampf, K. Eichhorn, *Stud. Biophys.*, 1983, **93**, 17-26.
20. K. Li, P. Kliauga and H.H. Rossi, *Radiat. Res.*, 1984, **99,** 465-475.
21. P.J. McNulty, 'Predicting single event phonomena', In: IEEE 1990 International Nuclear and space Radiation Effects Conference, Reno, Nevada, 3/1-3/93, 1990.
22. M. Zaider and D.J. Brenner, *Radiat. Res.,* 1984, **100,** 245-256.
23. A.M. Kellerer, 'Concepts of geometrical probability relevant to microdosimetry', In: J. Booz, H.G. Ebert and H.D. Hartfiel (Eds): Seventh Symposium on Microdosimetry. EUR 7147 de-en-fr. Harwood, London: Academic Publishers, pp. 1049-62, 1981.

USE OF NEW COLLECTION SYSTEMS ASSOCIATED WITH A MULTI CELLULAR TISSUE EQUIVALENT PROPORTIONAL COUNTER FOR INDIVIDUAL NEUTRON DOSIMETRY.

C. HOFLACK*, J.M. BORDY*, Y. CHARBONNIER*, T. LAHAYE*, M. LEMONNIER**, M.S. DIXIT***, J. DUBEAU***, and P. SEGUR+

* Institut de Protection et de Sûreté Nucléaire, DPHD, SDOS, IPSN, BP 6, 92265 Fontenay aux Roses Cedex, France.
** LURE, 91405 Orsay, France
*** Physics Department, CRPP, Carleton University, Ottawa, Canada
+ CNRS, CPAT, Université P. Sabatier, 31062 Toulouse Cedex, France

1 INTRODUCTION

Currently, in the field of radiation protection, research is in progress to develop personal electronic dosemeters for mixed radiation fields (neutron-photon). The individual neutron dosemeters available do not meet the requirements of the International Standard Organisation and the International Electrotechnical Commission as well as the recommandations of the International Commission for Radiological Protection[1] (ICRP 60). This paper deals with Tissue Equivalent Proportional Counters (TEPC). These detectors are well-known for area monitoring but their application as individual dosemeters presents some difficulties. The main problems are: inadequate sensitivity because of the small size, and ageing, mostly due to chemical gas interactions with material, gas cracking and outgasing of tissue equivalent materials.

This paper describes briefly the state of the art: the previous prototype, the results achieved and the problems which remain to be solved. Then, some improvements are presented and discussed.

2 STATE OF THE ART AND SUGGESTED SOLUTIONS OF PROBLEMS

The initial studies were focussed on the crucial problem of sensitivity. They led to the design of a new (neutron, Heavy Charged Particles) parallelepipedic converter shape: a MultiCellular Converter (MCC) allowing a high neutron sensitivity[2].

The energy response of the detector is represented in Table 1. For each neutron energy, the personal dose equivalent and the measured quality factor are listed[3]. To fulfil the calibration procedure for personal dosemeters, measurements have been made on a slab phantom made of polymethyl methacrylate (PMMA) of (30 cm x 30 cm x 15 cm). The personal dose equivalent response of the detector is comparable to the one obtained with usual TEPC's for area monitoring[4]. Hence, the results meet the ICRP recommendations.

Table 1 *Dose equivalent response and quality factor measured with the Multi-Cellular TEPC.*

Neutron Energies (monoenergetic, MeV)	0.144	0.250	0.565	1.2	2.5
Personal dose equivalent (Pulses.Sv^{-1})	$1.1x10^7$	$1.4x10^7$	$2.0x10^7$	$2.9x10^7$	$5.2x10^7$
Measured quality factor	13.7	13.5	16.0	13.0	11.6

In spite of these encouraging results, some technical problems remain to be solved. The detector is complicated to build (the many guard rings and welds of the wires) and microphonic effect, mainly due to the wires, is observed.

Most of those problems can be solved by replacing anode wires by rigid anodes. Among rigid anodes likely to be applied to proportional counters, the choice will be a compromise between different criteria: the gas gain achievable with reasonable potentials to keep proportional mode (without increasing the electrical consumption), simple handling (small size), simplicity of building, ruggedness, and cost. Three potential systems of detection have been selected from literature: 'microstrip'[5], 'gaps'[6,7] and 'needles'[8].

The features of each method have been optimised by means of computer simulations and experiments. The three systems are shown in Fig.1 and their different criteria are listed in Table 2. The drift plane distance is the distance between the converter and the multiplication system. The microstrip system is a set of alternating cathode and anode strips on a substrate. The gap system consists of a perforated substrate with copper electrodes: a cathode on the top of the substrate and an anode at the bottom of the gap.

Table 2 *Criteria of the systems actually used.*

Microstrips (mm)	Gaps (mm)	Needles (mm)
* Drift plane distance: 3	* Drift plane distance: 15	* dist (a,c) < 0.2 *[2]*
* Strips geometry:	*Gaps: i)circular hole: dist(a,c)=0.6*[2]*,Φ=0.5	* Needle shape:
(0.02,0.08,0.4) *[1]*	substrate=ceramic;	Φ = 0.6 ;
* Glass substrate:	ii)strips: dist(a,c)=1,width = 0.25,	height = 3.5
1E11 ohm.cm	substrate = plexiglass	

[1] stand for (anode width, cathode width, anode/anode pitch)
[2] dist(a,c) : anode to cathode distance

3 MODELISATIONS AND FIRST EXPERIMENTAL RESULTS

Two types of simulations are being used. The first one gives the electrostatic behaviour of the counter (potential V and electric field intensity E). The second one draws the shapes of microdosimetric spectra.

3.1 Electrostatic field inside the counters

The electrostatic treatment of the counter is based on solving Poisson's equation. Considering the cylindrical symmetry around the central axis of a channel of the MCC, the calculation's domain (Fig.1) is reduced to one half of the channel and one half of the converter.

The behaviour of the electric field intensities at two different points inside the counter is analysed. The first point, A, is located inside the multiplication region where the electric field lines converge. The second point, B, is the point where discharges are most likely to occur when high potentials are applied. int B represents the highest electric field intensity.

We plotted the electric field intensities at points A and B versus the anode/cathode potential difference for each system (Fig.2). Obviously, the microstrip is the most efficient method since low applied voltages create high electric field intensities. However, the discharge electric field intensities increase equally fast, which means that the discharge threshold is reached readily.

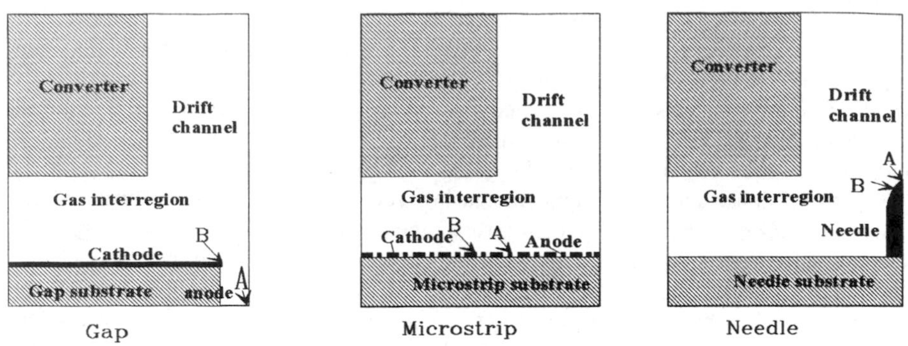

Figure 1 Calculation's domains of the three systems

The gap system has lower electric field intensities than the other two systems and intensities are essentially the same at points A and B. As a result, discharges inside the gap system occur at higher potentials which means that this system will perform over a larger potential range. The needles system, tested by Boutruche et al[2], has features in-between. The gain measurements will allow a justified choice to be made between the three systems.

3.2 Microdosimetric spectra shapes

From the previous Multi-Cellular TEPC (MCTEPC) experiments and the chord length distribution, it is shown that the shape of the spectra are different than those achieved with classical geometries (spherical or cylindrical). To improve the knowledge of the shape of the spectra, we computed the theoretical shape of the spectra for different neutron energies. In order to simplify the programs, only neutron elastic scattering with hydrogen is considered. Indeed, it is the most dominant contribution to kerma in the studied energy range (0.5 to 14.5 MeV)[9].

The results of the calculations are shown in Fig.3 for two geometries: spherical and multicellular. Differences in shape and especially large size events are due to long chord lengths existing in the MCTEPC[3]. It appears from the results that the proton edge method can be used to calibrate the spectra in terms of lineal energy[3].

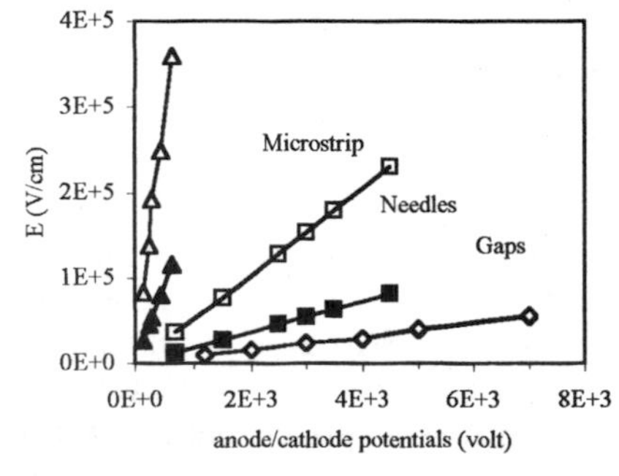

Figure 2 Electric Field intensities at multiplication (empty) and discharge points (plain) of the three systems

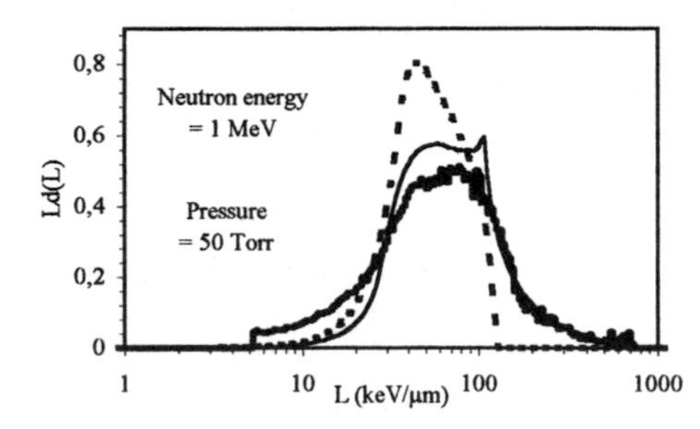

Figure 3 Shapes of micro-dosimetric spectra for MC (plain), spherical (large dots) and experimental spectrum (large line).

3.3 Experimental results

The gaps and the microstrip were tested using soft X-ray. The filling gas mixtures were Ar/CO2 (90/10) and Ar/isobutane (90/10), respectively.

The behaviour of the counter as the anode potential increases was studied for gaps (pressure set to 100 Torrs). A proportional mode is set for anode/cathode potential greater than 550 V. Moreover, the energy resolution is commonly 20 % at 5.4 keV. Typical values of the gain are 2000 to 4000. Table 3 shows the influence of pressure for a fixed gain. A larger amount of atoms inside the multiplication zone, by increasing the pressure, requires higher voltage to reach the fixed gain. This confirms the modelisation results for gap system: a large potential range is available before discharges occur.

Table 3 Potential needed to achieve a gain of 4000 as a function of pressure.

Pressure (Torr)	105	265	402	550	650
Anode/cathode potentials difference (Volt)	750	1080	1330	1580	1770

Microstrip has been tested under atmospheric pressure and the energy resolution reached is 20 % at 5.9 keV. The gain was about 600. Further experiments will be necessary to study its behaviour at lower pressure.

As a conclusion, microstrip may reach proportional regime with low electrical consumption but needles and gap are simpler to make and to handle. In any case, the long term stability under microdosimetric operating conditions must still be checked for the three systems before choosing one of them.

ACKNOWLEDGEMENTS
This work is supported by CEC contract number F13P-CT93-0072.

REFERENCES

1. International Commission on Radiological Protection, ICRP 1990 Recommendations of the ICRP Publication 60, Oxford Pergamon Press, 1991.
2. B. Boutruche, J.M. Bordy, J. Barthe, P. Ségur, G. Portal, New Concept of mini counter for individual neutron dosimetry, Rad.Prot.Dosim. 52, 335-338, 1994.
3. J.M. Bordy, J. Barthe, B. Boutruche, P. Ségur, A new proportional counter for individual neutron dosimetry, Rad. Prot. Dosim. 54, 369-372, 1994.
4. G. Dietze, H.G. Menzel, G. Bühler, Calibration of tissue-equivalent proportional counters used as radiation protection dosemeters, Rad.Prot.Dosim. 9, 245-249, 1984.
5. M.S. Dixit, F.G. Oakham, J.C. Armitage, J. Dubeau et al., Gas microstrip detectors on resistive plastic substrates, NIM in Physics research A 348, 365-371, 1994.
6. F. Bartol, M. Bordessoule, G. Chaplier, M. Lemonnier, S. Megtert, The CAT pixel proportional gas counter detector, J.Phys. III France 6, 337-347, 1996.
7. M. Lemonnier, Patent number : FR-9414158, 95941134.9-2208, Détecteurs de rayonnements ionisants à microcompteurs proportionnels.
8. G. Comby, J.F. Chalot, N. Anger, J.J. Beauval, J. Quidort, Contribution à l'étude des détecteurs multipointes à focalisation cathodique, Revue Phys. Appl. 16, 539-545, 1981.
9. R.S. Caswell, J.J. Coyne, M.L. Randolph, Kerma factors for neutron energy below 30 MeV, Int.J.Appl.Radiat.Isot. 33, 1227-1262, 1982.

Author Index

Allen J.E., 301
Allott R., 195
Almasi I., 353
Amols H.I., 262
Anachkova E., 353
Andreev S.G., 47, 133

Baek W.Y., 27
Bance D.A., 203
Bartczak W.M., 11
Bartels E., 176
Bartha T., 353
Becker R., 266
Beddoe A.H., 293, 305
Bednarek B., 399
Belli M., 191
Bergmann R., 240
Bhatt R.C., 357
Bilski P., 266, 407
Blomquist E., 129
Blomquist P., 125
Boei J.J.W.A., 160
Boguhn O., 176
Bolds K.D., 357
Bond V.P., 228
Bordy, J.M., 420
Bornfleth H., 143
Botchway S.W., 81
Boudaiffa B., 89
Bowey A.G., 323
Braby L.A., 315, 331
Brahme A., 125
Brandan M.E., 51
Brenner D.J., 3, 156, 327
Breskin A., 375
Bronic I.K., 365
Brooke S.L., 305
Budzanowski M., 266
Bullis J.E., 228

Carlsson J., 129
Cera F., 191
Chadwick K.H., 248
Chambaudet M., 282
Charbonnier, 4, 420
Charles M.W., 232
Charlier M., 77
Charlton D.E., 293
Chechik R., 375
Chen A.M., 156
Chepel V. Yu., 133
Cherubini R., 191
Close J.J., 301
Colautti P., 375
Cox R., 225
Crawford-Brown D., 180, 240
Cremer C., 143
Cremer T., 143
Cronkite E.P., 228
Cucinotta F.A., 35
Cunniffe S.M.T., 81

da Rosa L.A.R., 309
Dalla Vecchia M., 191
de Lara C., 93
DeLuca P.M. Jr., 278
Demonchy M., 15, 19
Detzler E., 176
Dietzel S., 143
Dingfelder M., 23
Distel B., 73
Distel L.V.R., 73
Dixit M.S., 383, 420
Dominguez I., 160

Dubeau J., 383, 420
Dugal P.C., 89
Durante M., 137

Eckl P., 180
Edwards A.A., 152, 172
Egger E., 274
Eils R., 143
Elkind M.M., 236
Emfietzoglou D., 379
English T., 347
Erdelyi K., 353
Evans T.M., 391
Eyrich W., 266

Fahey R.C., 57
Favaretto S., 191
Fazzalari N.L., 293
Filges D., 266
Folkard M., 323, 347
Ford J.R., 335
Forsberg M., 347
Franek B., 369
Friedland W., 43
Fritsch M., 266
Fromm M., 282
Fugger M., 274, 411
Furukawa K., 31

Gamboa-deBuen I., 51
Gantchev T., 85
Geard C.R., 327
Gerstenberg H.M., 357
Gialanella G., 137
Gong Y.F., 207
Goodhead D.T., 35, 195, 203, 335
Granzow M., 143
Greinert R., 176
Griffin C.S., 156, 164
Grigorova M., 160
Grimwood P.R., 232
Grindborg J.-E., 387
Groetz J.E., 282
Grossi G.F., 137
Grosswendt B., 27, 365
Gueraud B., 89

Hall E.J., 262, 327
Hantke D., 23
Harder D., 176
Harms-Ringdahl M., 199
Harvey A.N., 168
Hauffe J., 266
Hei T.K., 327
Heimgartner E., 339
Henshaw D.L., 301
Hill M.A., 164, 195, 203
Hofer K.G., 117
Hoflack, C., 420
Hofmann L., 339
Hofmann W., 180, 240
Höglund E., 129
Hong A., 297
Huels M.A., 85, 89
Hugot S., 77
Hummel A., 11
Hunting D., 89

Ianzini F., 191
Inokuti M., 23

Jacob P., 43
Jauch A., 143
Jelen K., 399
Jenkins G.J., 327
Jenner T.J., 93

Karlen D., 383
Katz R., 35, 282
Keitch P.A., 301
Kelemen A., 339
Kellerer A.M., 353
Khoury J., 89
Khvostunov I.K., 47, 133
Kiguchi K., 343
Klyachko D., 85
Kobayashi K., 65
Kobayashi Y., 343
Kobus H., 266
Kohler M., 339
Kowalski T.Z., 399
Kroc T.K., 278
Kronqvist U-S. E., 199
Kyllönen J.E., 361

Lahaye, T., 420
Langen K., 278
Langowski J., 143
Lappa A.V., 39
LaVerne J.A., 69
Leenhouts H.P., 248
Lemonnier, M., 420
Lennox A.J., 278
Leong A.S.-Y., 293
Levati L., 191
Lewensohn R., 199
Li J., 207
Li W.B., 207
Lin X., 117
Lindborg L., 361
Little M.P., 244
Löbrich M., 187
Lucas J.N., 156

Mancini E., 137
Marino, S.A., 228
Marsden S.J., 195, 335
Meijer A.E., 199
Merzagora M., 103, 137
Metting N.F., 331, 335
Meyer P., 282
Michael B.D., 323, 347
Michette A.G., 347
Milios J., 293
Mille R.C., 327
Miller R., 262
Milligan J.R., 57
Moiseenko V.V., 152
Monforti F., 137
Moosburger M., 266
Moschini G., 191
Moscovitch M., 379
Muirhead C.R., 244
Münkel Ch., 143

Namba H., 31
Natarajan A.T., 160
Nikjoo H., 3, 35, 81, 152
Noll M., 274
Nösterer M., 180

O'Donoghue J. A., 255
O'Neill P., 81, 93
Oakham F.G., 383
Oberhummer H., 240
Ohno S., 31
Olko P., 266, 383, 407
Oriel J., 403
Ottolenghi A., 103, 137

Paganetti H., 266
Palmans H., 270
Pansky A., 375
Paretzke H.G., 43
Perzl M., 43
Peterson H.P., 266
Pfauntsch J.S., 347
Pimblott S.M., 69
Pohn E., 411
Pomplun E., 15, 19
Prestwich W.V., 152
Prise K.M., 111, 323, 347
Pruchova H., 309, 369
Pszona S., 399
Pugliese M., 137
Pullar C., 323

Randers-Pehrson G., 327
Regulla D., 309
Reist H.W., 339
Richards S.R., 289
Richardson R.B., 297
Rodriguez-Villafuerte M., 51
Roos H., 353
Rossi H.H., 415
Ruiz S., 77
Rulikowska-Zarebska E., 399
Rydberg B., 125

Sabattier R., 77
Sachs R.K., 156
Samuelsson G., 361
Sanche L., 85, 89
Sapora O., 191
Savage J.R.K., 156, 164, 168
Savoye C., 77
Schettino G., 323, 347
Schmitz Th., 266
Schneiderman M.H., 117

Schöner W., 274, 411
Schulte K., 176
Schulte R.W.M., 211
Schüssler H., 73
Segur, P., 420
Seidlitz R., 309
Shchemelinin S., 375
Siebbeles L.D.A., 11
Simmons J.A., 289
Simone G., 191
Simpson P.J., 156
Smith F.A., 403
Somerville E.W., 383
Spitkovsky D.M., 133
Spotheim-Maurizot M., 77
Stenerlöw B., 129
Stepanek J., 339
Stevens D.L., 81, 164, 195, 203, 335
Stiller C.A., 244
Stinzing F., 266
Stork T., 43
Streffer C., 217
Surette R.A., 383
Swenberg C., 77
Sy D., 77

Tabocchini M.A., 191
Taguchi M., 31, 343
Takakura K., 65
Terrissol M., 11, 15, 19
Thieke C., 176
Tiveron P., 191
To L.B., 293
Torres B.A., 357
Townsend K.M.S., 335
Tucker C.E., 403
Turcu I.C.E., 195
Turner M.S., 293

Uehara S., 3
Uijt de Haag P.A.M., 248
Usami N., 65
Utteridge T.D., 293

Vana N., 274, 411
Verhaegen F., 270
Vermeulen S., 160
Virsik-Peuckert R.P., 176
Vojnovic B., 323, 347

Waker A.J., 89, 383
Wang C.K., 391
Ward J.F., 57
Watanabe H., 343
Watanabe R., 31, 65
Weinberger J., 262
Wells R.L., 236
Wheldon T.E., 255
Wilson J.W., 35
Wu D.C., 207
Wu L.J., 327
Wuu C.W., 228

Yamaguchi H., 97
Yamasaki S., 343

Zaider M., 228
Zhang X., 207
Zheng W.Z., 207
Zink D., 143

Subject Index

Alveolar sacs, 297
Apoptosis, 199
Auger electrons, 19

Brachytherapy, 262

Carcinogenesis, 217, 225
Cell cycle delay, 327
Cells
  haemopoietic stem cells, 293
  human fibroblasts, 164
  human lymphocytes, 152, 199
  lung, 289
  rat tracheal epithelial, 335
  V79 cells, 93, 137
Charged particle traversals in cells, 331, 335
Chromatid breaks, 168
Chromosomal aberrations, 143, 152, 160, 172, 176, 180, 327
Clustered DNA damage, 111, 133
Complex chromosome aberrations, 156, 164

DNA and chromatin models, 19, 77, 117, 143
DNA damage
  base damage, 85
  double strand break induction, 35, 69, 73, 81, 93, 103, 125, 187, 191
  protein crosslinks, 73
  repair of double strand breaks, 129
  short fragments of DNA, 43, 187
Diagnostic X-rays, 309

Electron energy and radiation damage, 69

Fluorescence *in situ* hybridisation, 156
Fricke dosimeter, 97

Gas counters, 369, 383, 399

Interaction cross sections
  in dimethyl ether, 27
  in liquid water, 23

Leukaemia, 228, 244
LiF:Mg,Cu,P solid state detectors, 407
Low doses and dose rates, 225
Low-energy electron (0-5000 eV) irradiation, 89
Lung cancer, 248

Mammalian cell irradiation rig, 203
Mean skeletal dose, 305
Microbeam, 315, 323, 327, 331, 343, 347
Microdosimetric distributions for target volumes, 47
Microstrip gas counters, 383, 399
Mixed-field radiation measurements, 357
Model of radiation carcinogenesis, 240, 244
Monochromatic X-rays, 65
Monte-Carlo simulations, 35, 39, 51, 369
Mutation, 191, 244, 248, 327

Nanodosimeter, 375, 391, 399

Oncogenic transformation, 180, 327

Paired TEPCs, 361
Pairwise interaction between DNA lesions, 176
Po-210 in human fetal tissues, 301
Proportional counter, 365, 420
Proton therapy, 266, 270, 274

Quality factors, 232

Radial dose, 31, 35
Radiation
  α-particles, 51, 207, 211, 289, 323, 327
  COSY proton beam, 191, 199
  heavy and light ions, 31, 137
  low energy electrons, 89
  neutrons, 93, 278, 420
  photons, 97, 387
  protons, 266, 270, 274, 323
  radon daughters, 248, 293
  ultrasoft X-rays, 81, 124, 164, 203, 347
Radiation-induced cell inactivation, 117, 125, 137, 191
Radiation quality, 129, 133, 152, 191, 195, 403
Radiation quality dosemeter, 403
Radiation risk assessment, 236
Radiation transport in DNA, 15
Radiobiological effectiveness
  of accelerated nitrogen-ions, 199
  of Re-188 beta particles, 262
Radionuclides, 301, 305
Radon, 248, 293
Radioprotection of DNA, 77
Radiosensitivity of cellular DNA, 57

Solid-state microdosimetry, 379
Spatial and temporal effects of high LET radiation, 195

Targeted radiotherapy, 255
Thermalized electrons, 11
Thermoluminescence, 51, 411
Track structure
  PARTRAC, 43
  simulations, 3, 31, 81, 103
Threshold dose, 217